Applied
Stochastic
Processes

Applied Stochastic Processes

A Biostatistical and Population Oriented Approach

Suddhendu Biswas

Department of Statistics
University of Delhi
Delhi

JOHN WILEY & SONS

NEW YORK • CHICHESTER • BRISBANE • TORONTO • SINGAPORE

First Published in 1995 by
WILEY EASTERN LIMITED
4835/24 Ansari Road, Daryaganj
New Delhi 110002, India.

Distributors:

Australia and New Zealand:
JACARANDA WILEY LIMITED
PO Box 1226, Milton Old 4064, Australia

Canada:
JOHN WILEY & SONS CANADA LIMITED
22 Worcester Road, Rexdale, Ontario, Canada

Europe and Africa:
JOHN WILEY & SONS LIMITED
Baffins Lane. Chichester, West Sussex, England

South East Asia:
JOHN WILEY & SONS (PTE) LIMITED
05-04, Block B, Union Industrial Building
37 Jalan Pemimpin, Singapore 2057

Africa and South Asia:
WILEY EASTERN LIMITED
4835/24 Ansari Road, Daryaganj
New Delhi 110002, India

North and South America and rest of the World:
JOHN WILEY & SONS INC.
605 Third Avenue, New York, NY 10158, USA

Library of Congress Cataloging-in-Publication Data

ISBN 0-470-22159-3 John Wiley & Sons, Inc.
ISBN 81-224-0691-2 Wiley Eastern Limited

Printed in India at Chaman Offset, New Delhi-110 002.

Dedicated
To
The Living Memory of my
Preceptor, Guide and Philosopher
Professor V.S. Huzurbazar.

Preface

The present book is a thoroughly extended, revised and generalised version of my earlier book entitled 'Stochastic Processes in Demography' and Applications' (Wiley Eastern Limited, 1988). The basic motivation of the book is to provide the background of the theoretical techniques of Applied Stochastic Processes as presented in the first six chapters of the Text; followed by the applications of the same in Stochastic theory of epidemics, Clinical drug trials, techniques of Demography (fertility and Mortality analysis) and models of Population growth, Survival and Competing Risk theory and Stochastic Processes in Genetics. The theoretical background to understand the applications of the techniques of Stochastic Processes in Biostatistics are Random Walk theory, and Markov Processes; Non Markov processes and Renewal theory, Martingales theory, Theory of Geiger Muller Counters and Palm Probability. Thus, the theory employed for the reading of the topics of Biostatistics can also be utilised as a by-product for the reader of Stochastic processes. While some of the chapters are repetitions of the corresponding topics occasionally with minor changes as in my earlier book; a majority of them are completely new topics and even sometimes the old topics are written in an altogetherly revised manner with larger number of applications as in Chapter four (Martingales theory), Chapter twelve (Stochastic Processes on Survival and Competing Risk Theory) etc. Chapters on Clinical trials, Classical and Carrier borne epidemic models (with Martingales applications) and a critical study of the bivariate exponential survival models (Marshall-Olkin, Freund, Downton, Gumbel, Block and Basu etc) as included in Chapter twelve, are topics, which to the best knowledge of the author, have not found their places in any text book of this kind so far.

The book is primarily meant for Post-graduate and higher level udergraduate students in Statistics, Biostatistics, Applied Probability with emphasis on Biosystems with a background of elementary probability, Advanced Calculus, Elements of Statistical Methodology and Statistical Inference. The entire text can be covered in a course of two semesters. Appendix covers some basic fundamentals as large sample-method for standard errors of Statistics, Cramer

Rao Inequality, Rao-Blackwell theorem, Central limit theorem and Laplace transforms.

It is a pertinent, although late occasion to express my indebtedness to my preceptor, guide, philosopher and a Senior colleague for sometime, Late Professor V.S. Huzurbazar, formerly the Professor and Head of the Department of Mathematics and Statistics, University of Poona for his noble inspiration, constant guidance and ever stretching helping hands even till a few days earlier to his last breathing in Denver, U.S.A. No words can express how I owe my existence to him. The book is, therefore, dedicated to the living memory of the great Stalwart with the profound grief that the same could not be done during his life time.

I acknowledge with thanks the great services rendered by my student Mr. Saroj Kumar Adhikari while editing and correcting the manuscript in his usual very meticuluous and systematic way. Similarly, the help rendered by my other students Ms. Sunita Chitkara, Ms. Rita Jain, Dr. Mariamma Thomas, Dr. Vijay Kumar Sehgal and Mr. Sanjoy Sengupta are thankfully acknowledged. A word of thanks to my daughter Swati who all along helped me in the write-up of the book and correction of galley proofs. Last, but not the least, I acknowledge my great appreciation and thanks to Mr. A Machwe of Wiley Eastern Limited for his innovative ideas of preparing a book of this form and his sincere and untiring efforts to make the book see the light of the day at the earliest in an elegant shape inspite of so much of hurdles.

SUDDHENDU BISWAS

15 August, 1993
University of Delhi
Delhi - 110007

Contents

Glossary of Notations and Symbols

V	:	For every
∃	:	There exists
⇒	:	Implies
⇐	:	Implied by
⇔	:	Implies and Implied by
∋	:	Such that
>	:	Greater than
<	:	Less than
≥	:	Greater or equal to
≤	:	Less or equal to
∧	:	Minimum
∈	:	Belongs to
∪	:	Union
∩	:	Intersection
∨	:	Maximum
≯	:	Not greater than
≮	:	Not less than
>>	:	Much greater
<<	:	Much smaller
r.v.	:	Random variable
p.g.f.	:	Probability generating function
m.g.f.	:	Moment generating function
c.f.	:	Characteristic function
c.d.f.	:	Cumulative distribution function
p.d.f.	:	Probability distribution (or density) function
$\sum$	:	Summation
$\prod$	:	Product
$L\left(f\left(t\right)\right)$	:	Laplace transform of $f(t)$.
iff	:	if and only if
n.s.	:	Necessary and sufficient
#	:	number of
$\int$	:	Integration
D	:	Differential Operator

Introduction to Stochastic Processes—Basic Concepts

1.0 Introduction

Of late, it is being increasingly felt that Stochastic modelling of any Physical, Biological, Social or Economic system (or Process) can give better representation, description and specification than a purely Deterministic model. This is because a Stochastic model takes into account of the uncertainity or the randomness which the process may experience (however perfect may be the deterministic law governing the general behaviour of the system); which is not considered in a purely mathematical modelling (without introducing the element of probability) of the system. Stochastic modelling takes into consideration of the basic element viz. random variables (r.v.s) and also the r.v.s which are dependent on time or other similar factors as age etc. Thus the collection of a time dependent or age dependent random variables over all conceivable points of time or age gives rise to what is known as "Stochastic process"; which is the methodology of Stochastic modelling of a system.

Since biological, socio-economic or psychological variables exhibit more departures from their expected values, the quantitative studies concerning different processes in these areas or disciplines, therefore, the provide more fertile fields for the application of Stochastic processes than the areas of more exact physical sciences. Thus, the movement of a population over time, affected by the factors like fertility, mortality, migration which changes randomly over time (apart from their systematic variations over time) can be taken up as a fruitful application of the tools of Stochastic processes. Similarly the growth of a malignant tumour over time by the charge and discharge of Carcinogenes in the body cells or the conversion of sensitive cells to drug ressistant cells during Chemotherapy, the spread of communicable disease like AIDS by immigration, the tendency that a population going to become homozygous in respect of a particular genetical character over time even under random mating, the generation of a sudden epidemic are live examples of application of Stochastic processes in Population or Biological Sciences.

The motivation of the present text is, therefore, to exhibit the techniques of Stochastic processes systematically, especially focus them highligting their potentials for applications with special emphasis on Bio-system or more precisely the demographic aspects of several Bio-systems.

1.1 Basic Prerequisites

1.1.1 Probability Generating Functions

Probability generating function (p.g.f) of a random variable X. If X is a discrete random variable, then $E(s^X)$ for $|s| < 1$, if existing, is defined to be the p.g.f. of the random variable X.

We have

$$E(s^X) = \phi(s); \tag{1.1}$$

and denoting $P(X = x_i) = p_i$, $i = 0, 1, 2, 3, ...,$

$$\phi(s) = p_0 + p_1 s + p_2 s^2 + ... + p_n s^n + ...$$

$$\phi'(s) = p_1 + 2p_2 s + 3p_3 s^2 + ...$$

$$\phi'(1) = p_1 + 2p_2 + 3p_3 + ... = E(X) = \mu_1'$$

$$\phi''(1) = 2p_2 + 3 \cdot 2p_3 + 4 \cdot 3p_4 + ...$$

$$= E[X(X-1)] = E(x^{(2)})$$

Proceeding in this way, we get

$$\phi^{(n)}(1) = E(X^{(n)}) = E[X(X-1)...(X-n+1)] \tag{1.2}$$

Again $\quad \phi(0) = p_0, \quad \phi'(0) = p_1, \quad \dfrac{\phi''(0)}{2!} = p_2$

$$\frac{\phi'''(0)}{3!} = p_3, ..., \frac{\phi^{(k)}(0)}{k} = p_k \tag{1.3}$$

holds.

If $g_1(s), g_2(s), ..., g_n(s)$ are probability generating functions of the independent discrete random variables $X_1, X_2, ..., X_n$, then $G_Z(s)$, the p.g.f. of $Z = X_1 + X_2 + ... + X_n$ is given by

$$G_Z(s) = E(s^Z) = E(s^{X_1 + X_2 + ... + X_n})$$

$$= E(s^{X_1}) E(s^{X_2}) ... E(s^{X_n}) = g_1(s) g_2(s) ... g_n(s) \tag{1.4}$$

This follows immediately.

Of particular interest in statistical analysis is the case in which independent r.v.s. $X_1, X_2, ... X_n$ have the same probability distribution $\{p_1\}$ and hence have same p.g.f. $G_Z(s)$. $Z = X_1 + X_2 + ... X_n$ is called the n-fold convolution of $X_1, X_2..., X_n$.

Now (1.4) $\Rightarrow$ $\qquad G_z(s) = [g(s)]^n \tag{1.5}$

In general the convolution of two functions $f_1(\cdot)$ and $f_2(\cdot)$ denoted as $f_1 * f_2$ is given by

$$f_1 * f_2 = \int_0^t f_1(u) \, dF_2(t - u) = \int_0^t f_2(u) \, dF_1(t - u) \qquad (1.6)$$

It represents the probability that the sum of the independent random variables U and V taking a value t.

Example 1.1   Show that the p.g.f. of a Bernoullian variate with parameter p is $(1 - p + ps)$ and that of a poisson variate with parameter λ is $e^{-\lambda(1-s)}$ and a binomial variate with parameters (n, p) is $(1 - p + ps)^n$.

Example 1.2
   (i)   If $[X \mid N = n] \sim B(n, p)$ and $N \sim B(\mu, \theta)$, then show that $X \sim B(\mu, \theta p)$.
   (ii)   If X_i's are i.i.d. random variables and $X_i \sim \Gamma(\alpha_i, \lambda)$, then $Z_n = \Sigma X_i \sim \Gamma(\Sigma\alpha_i, \lambda)$.

1.1.2   Multivariate Probability Generating Function

For several discrete random variables $X_1, X_2, ..., X_n$,

$$E(s_1^{X_1} s_2^{X_2} \dots s_n^{X_n}) = \phi(s_1, s_2, ..., s_n) \ \forall \ |S_i| < 1 \ (i = 1, 2, 3, ..., v) \qquad (1.7)$$

is called multivariate probability generating function.
   We have as usual

$$\left. \frac{\partial \phi(s_1 \, s_2 \, ... s_n)}{\partial s_i} \right|_{s_i = 1 (i = 1, 2, ..., n)} = E(X_i) \qquad (1.8)$$

$$\left. \frac{\partial^2 \phi(s_1 \, s_2 \, ... s_n)}{\partial s_i^2} \right|_{s_i = 1 (i = 1, 2, ..., n)} = E(X_i(X_i - 1)) \qquad (1.9)$$

$$\vdots \qquad\qquad \vdots$$

$$\left. \frac{\partial^k \phi}{\partial s_i^k} \right|_{s_i = 1 (i = 1, 2, ..., n)} = E[X_i (X_i - 1) \dots (X_i - k + 1)] \qquad (1.10)$$

$$\left. \frac{\partial^2 \phi}{\partial s_i \, \partial s_j} \right|_{s_i = 1, s_j = 1 (i \neq j)} = E[X_i X_j] \qquad (1.11)$$

$$\text{Cov}(X_i, X_j) = E(X_i X_j) - E(X_i) E(X_j)$$

$$= \left. \frac{\partial^2 \phi}{\partial s_i \, \partial s_j} \right|_{\substack{s_i = 1 \\ s_j = 1 \\ i \neq j}} - \left. \frac{\partial \phi}{\partial s_i} \right|_{s_i = 1} \left. \frac{\partial \phi}{\partial s_j} \right|_{s_j = 1} \qquad (1.12)$$

Also $P[X_1 = k_1, \ X_2 = k_2, \ ..., X_n = k_n]$

$$= \frac{1}{k_1!} \frac{\partial^{k_1} \phi}{\partial s_1^{k_1}} \bigg|_{\substack{s_i = 1 \\ s_1 = 0 \\ i \neq 1}} \frac{1}{k_2!} \frac{\partial^{k_2} \phi}{\partial s_2^{k_2}} \bigg|_{\substack{s_i = 1 \\ s_2 = 0 \\ i \neq 2}} \cdots \frac{1}{k_n!} \frac{\partial^{k_n} \phi}{\partial s_n^{k_n}} \bigg|_{\substack{s_i = 1 \\ s_n = 0 \\ i \neq n}} \tag{1.13}$$

In particular,

$$P[X_1 = k_1] = \frac{1}{k_1!} \frac{\partial^{k_1} \phi}{\partial s_1^{k_1}} \bigg|_{\substack{s_i = 1 \\ s_1 = 0 \\ i \neq 1}}$$

$$\vdots \qquad \vdots \qquad \vdots$$

$$P[X_n = k_n] = \frac{1}{k_n!} \frac{\partial^{k_n} \phi}{\partial s_2^{k_2}} \bigg|_{\substack{s_i = 1 \\ s_n = 0 \\ i \neq n}} \tag{1.14}$$

1.1.3 Hazard and Survival Functions

$h(z)$ is a hazard function if $h(z)\, dz$ is defined as the conditional probability of dying (in failing) between $(z, z + dz)$ given that a person has survived upto z.

$$h(z)\, dz = \frac{f(z)\, dz}{1 - F(z)} \tag{1.15}$$

where $f(\cdot)$ is the failure density function and $F(\cdot)$ is the c.d.f. of failure distribution. Denoting $1 - F(z), = R(z)$ the survival function (or reliability function), we have from (1.15)

$$- \log_e (1 - F(z)) \bigg|_0^t = \int_0^t h(z)\, dz$$

$$\Rightarrow \qquad - \log_e (1 - F(t)) = \int_0^t h(z)\, dz \qquad (\because F(0) = 0)$$

$$\Rightarrow \qquad \log_e R(t) = - \int_0^t h(z)\, dz$$

$$\Rightarrow \qquad R(t) = \exp\left[- \int_0^t h(z)\, dz \right] \tag{1.16}$$

where $R(t) = P[T > t]$.

1.1.4 Definition of Stochastic Process

An arbitrary infinite family of random variables $X(t)$ (or X_t) where $t \in T$ and T is a parametric space is called a *Stochastic Process*. t is called an indexing parameter.

A rigorous definition of stochastic process is subject to the satisfaction of

Kolmogorov's theorem (1930) known as the Fundamental Theorem of Stochastic Process which runs as follows:

If the joint distribution function of the r.v.'s $(X_{t_1}, X_{t_2}, ..., X_{t_n})$ is known for all finite n ($n = 1, 2, 3, ...$) and for all sets of values $(t_1, t_2, ..., t_n)$ belonging to T and if these distribution functions are compatible, then $\exists$ a probability field (Ω, B, P) where $\Omega \equiv$ sample space, $B \equiv$ Borel field of certain subsets of Ω i.e. set of random events and $P(A)$ is defined for the random events $A \in B$ and family of random variables X_t, $t \in T$ defined on Ω for which $P\{X_{t_1} \leq x_1, ..., X_{t_n} \leq x_n\}$ is equal to the prescribed distribution function $\forall$ n, $n = 1, 2, 3,...$ and $\forall$ $(t_1, t_2, ..., t_n) \in T$.

The values assumed by the process are called *states* and the set of possible values is known as '*state space*'. The set of possible values of the indexing parameter is called the '*parameter space*' which can be either discrete or continuous.

Even though in most of the physical problems, time is the natural index parameter, other kinds of parameters such as space etc. may also be used. A realization of a stochastic process $X(t)$, where $t \in T$ is the set of assigned values of $X(t)$ corresponding to t $\forall$ $t \in T$. The realization may be described by plotting $X(t)$ against t ($t = 0, 1, ...$).

Point Process

A point process is a stochastic process concerning random collection of point occurrences. If we impose certain emphasis on the points on which events occur rather than measuring the quantitative aspects of events (say, in the problem of intensity of shocks over points of time) then the domain of study becomes 'Point Processes' which falls within the purviews of 'Stochastic Processes.'

Three basic characteristics which are satisfied by some of the special kind of stochastic processes are as follows:

      (i) Stationarity   (ii) Orderliness

and   (iii) Absence of after effect

Stationarity implies that $\forall$ $t \geq 0$ and an integer $k \geq 0$ the probability of k number of events in an interval say $(\alpha, \alpha + t)$ denoted by $X_k(\alpha, \alpha + t)$ is independent of α, $\forall$ $\alpha \geq 0$. In other words, the probability of k occurrences will depend only on k and t and is independent of α. In other words, probability of a given number of events will be independent of the translation of the origin. More rigorously, a stochastic process $\{X_t; t \in T = (-\infty, \infty)\}$ is stationary if $\forall$ $\{t_1, t_2,..., t_n\}$, the joint p.d.f. of $X_{t_1} + \gamma, X_{t_2} + \gamma..., X_{t_n} + \gamma$ is same as $X_{t_1}, X_{t_2}, ..., X_{t_n}$ for $\gamma \in (-N, N)$ and stationary in the wide sense if Cov $(X_{t_1} + \gamma,..., X_{t_n} + \gamma)$ is a function of γ only.

Orderliness: We have

$$\sum_{k=0}^{\infty} X_k (\alpha, \alpha + t) = 1, \quad \forall \ \alpha \geq 0, t \geq 0$$

and k any non-negative integer.

Orderliness implies the probability of more than one event in small interval of time $(0, t)$ as $t \to 0$ is of the order zero.

$$\Rightarrow \quad \lim_{t \to 0} \frac{1 - X_0(\alpha, \alpha + t) - X_1(\alpha, \alpha + t)}{t} = 0$$

In case the process is stationary, the orderliness implies

$$\lim_{t \to 0} \frac{1 - X_0(\alpha, \alpha + t) - X_1(\alpha, \alpha + t)}{t} = \lim_{t \to 0} \frac{(1 - X_0(t) - X_1(t))}{t} = 0$$

and is independent of α.

Absence of after effect: If $X_k(\alpha, \alpha + t)$ is independent of the sequence of events in $(0, \alpha)$ then we call the point process to be independent from the effect of after effect. The absence of after effect thus implies the mutual independence of subsets of a point process taken over different points of time which do not overlap.

1.1.5 State Space and Parameter Space of a Stochastic Process

Let us take the example of a life table function l_x, the # survivors at each age $x(x = 0, 1, 2, ...) \cdot \{l_x\}$ may be regarded as a stochastic process, the index parameter x belonging to a discrete non-negative integer valued parameter space. The values of $l_x \; \forall \; x$ are also integer valued non-negative quantities which constitute the state space. Hence the state space is also discrete valued and non-negative. However, in general, a stochastic process may be classified on the basis of parametric space and state space as follows from the following table.

Some authors classified the stochastic process over discrete parameter space as 'Stochastic sequence' and 'Stochastic processes' over continuous parametric space as 'Stochastic process' in the proper sense.

Table 1.1

State space	Parametric space		Result
	Continuous example	Discrete example	
Continuous	1. The waiting time at age t given by $w(t)$ of a vital event over continuous parameter t 2. Diffusion Process	The waiting time at age t of a vital event say $w(t)$ when realizations are made over discrete points of time	Continuous valued Stochastic Processes
Discrete	The population size $p(t)$ when realizations are made over continuous point of time	The population size $p(t)$ when realizations are made in the middle of every year	Discrete valued Stochastic Processes

References

1. Bailey N.T.J. (1963): *Elements of Stochastic Processes with Applications to Natural Sciences*. John Wiley & Sons London, New york and Sydney.
2. Barlow R.E and Proschan F. (1965). *Mathematical Theory of Reliability*. Wiley. New York.
3. Bhat U.N. (1972): *Elements of Applied Stochastic Processes*. John Wiley & Sons, New York.
4. Biswas S. (1988): *Stochastic Processes in Demography and Applications*. Wiley Eastern, New Delhi.
5. Coleman Rodney (1974): *Stochastic Processes*. George Allen and Unwin Limited. London.
6. Chiang C.L. (1968): *Introduction to Stochastic Processes in Biostatistics*, John Wiley & Sons, New York.
7. Chiang C.L. (1980): *An introduction to Stochastic Processes and their Applications*, Kreiger, New York.
8. Cox D.R. (1960): *Renewal Theory*, Mathuen, London.
9. Cox D.R. And Miller H.D. (1965): *Theory of Stochastic Processes*.
10. Feller William (1968): *An Introduction to Probability Theory and Applications*, Vol I:, John Wiley & Sons, New York.
11. Hoel P.G., Port C.S. and Stone C.G. (1972): *Introduction to Stochastic Processes*, Houghton Miifin Company, Boston.
12. Karlin S. and Taylor H.M. (1975): *A First Course in Stochastic Processes*, Edition II, Academic Press, New York.
13. Ross M. and Sheldon H. (1983): *Stochastic Processes*, John Wiley & Sons New York, Chichester, Toronto and Singapore.
14. Takacs L. (1957): *Stochastic Process (Problems and Solutions)*—Translated by P. Zador, John Wiley & Sons, New York and Mathuen & Co.

Random Walk and Markov Process

2.0 Introduction

A discrete valued Stochastic process whose value may increase by 1 with probability p and decrease by 1 with probability q subject to the condition of maintaining the same value with probability $1 - (p + q)$ in the next unit of time (or in the next step) is a simple example of *Random walk in one dimension*. Further, the probability of going one step forward in unit time is independent of the probability of going one step backward with probability q. A simple example of such one dimensional random walk is available in general epidemic process; where the number of infected persons can increase by one with probability p during an infinitesimal time period (which may be taken as the unit of time) while one susceptible can be infected. The number of infected persons can decrease by one while one infected person is removed by death or by isolation through recovery. Finally, the size of the infected population may remain to be the same when there is no further infection or removal with probability $(1 - (p + q))$. We shall discuss several applications of one dimensional random walk in section 2.4. However, to provide necessary background for the same we introduce in the next section the concept of Simple Markov processes (and Markov chain).

2.1 Markov Process

A discrete valued stochastic process $X(t)$ is called a Markov process if for $t_0 < t_1 ... < t_i < ... < t_j \in T$ and for any integers $k_0, k_1, ..., k_j$,

$$P\{X(t_j) = k_j \,|\, X(t_0) = k_0, X(t_1) = k_1, ..., X(t_i) = k_i\}$$
$$= P\{X(t_j) = k_j \,|\, X(t_i) = k_i\} \tag{2.1}$$

Thus in a Markov process, given $X(t_i)$ (present or the current known value), the conditional probability distribution of $X(t_j)$ is independent of $X(t_0), X(t_1),$ ... $X(t_{i-1})$ (i.e. the past known values). However, it is not necessary that a Markov process should be discrete valued.

A continuous valued process will also be called a Markov process if

$$P\{k'_j \le X(t_j) \le k_j \mid X(t_0) = k_0, X(t_1) = k_1, ..., X(t_i) = k_i\}$$
$$= P\{k'_j \le X(t_j) \le k_j \mid X(t_i) = k_i\} \tag{2.2}$$

for any real number $k'_j \le k_j$ and $i < j$.

2.2 Markov Chain (M.C.)

If the parameter space of a Markov process is discrete then the Markov process is called a 'Markov Chain'.*

A definition of Markov chain may be given from life-table. If $\{l_x\}$ be the number of survivors at age $x = 0, 1, 2, ...$, then $\{l_x\}$ conforms to simple Markov chain, since

$$P\{l_x = k_x \mid l_0 = k_0, l_1 = k_1, ..., l_{x-1} = k_{x-1}\}$$
$$= P\{l_x = k_x \mid l_{x-1} = k_{x-1}\},$$
$$\forall \text{ positive integers } k_i, \ i = 0, 1, 2, ... x$$

A Markov process is said to be homogenous with respect to the parameter space if the transition probability $p_{ij}(t,\tau) = P[X(t) = j \mid X(\tau) = i]$ depends only on $(t - \tau)$ and not on t and τ separately.

2.3 Classification of the States of Markov Chain

The state 'j' of a Markov chain ($j = 0, 1, 2, ...$) is said to be *periodic* with period 'π' if

and
$$\left.\begin{array}{ll} p_{ij}^{(m\pi)} > 0 & \forall \ m = 1, 2, ... \\ p_{ij}^{(n)} = 0 & \forall \ n \ne m\pi \end{array}\right\} \tag{2.3}$$

where $p_{ij}^{(k)}$ represents the transition probability from i to j in k steps. If $\pi = 1$, then the state 'j' is called '*aperiodic*'.

Let the r.v. T_{jj} be the time at which the particle returns to state 'j' for the first time where $T_{jj} = 1$ if the particle stays in 'j' for a time unit. In this case, we say that the state 'j' is '*recurrent*' if $P[T_{jj} < \infty] = 1$ and '*transient*' if $P[T_{jj} < \infty] < 1$.

Again a state 'i' is called '*null recurrent*' if 'i' is recurrent and $E(T_{ii}) = \infty$ and '*positive recurrent*' if $E(T_{ii}) < \infty$.

Now a state 'i' leads to a state 'j' (which we write $i \to j$) if for some integer $k \ge 0$, $p_{ij}^{(k)} > 0$.

---

*Some authors define Markov process with finite or denumerable state space as Markov Chain. To reconcile the two definitions some authors have defined Markov chain as a discrete parameter Markov process with finite or countable number of state spaces.

We say the state 'i' and 'j' communicate if $i \to j$ and $j \to i$. Defining '$\sim$' as an equivalence relation which implies

(i) $i \to i$

(ii) $i \to j \Rightarrow j \to i$

(iii) $i \to j, j \to k \Rightarrow i \to k$

which are the conditions of *reflexivity, symmetry* and *transitivity* respectively. This defines an equivalence class among the states in the Markov chain. In other words, the entire set of the states in the Markov chain can be partitioned into equivalent classes '*irreducible classes*' of the Markov chain. Thus a Markov chain of just one class is called irreducible. It follows that while it is possible for a state belonging to an irreducible class to visit another state in another class, but it can never return. Otherwise two irreducible classes would communicate and form a single class. It can be seen that both periodicity and recurrence (or transience) are class properties.

For a Markov chain with finitely many states, it follows that at least one state must be 'recurrent' and every 'recurrent' state should necessarily be 'positive recurrent'.

2.4 Random Walk

Unrestricted Random Walk—A simple example of Stochastic Process.

Let us illustrate the idea of a simple unrestricted random walk by considering the number of carcinogene deposit (which can produce malignant tumour). A unit of carcinogene can be absorbed with probability p in a unit time and be discharged with probability q during a unit time. The problem is to obtain the distribution of the carcinogene upto a given time t which is clearly a problem of Stochastic process related to random walk (i.e., forward movement with an increase in the size of the population by one unit in each step with probability p and the backward movement with a decrease in the population size with probability q, where $p + q = 1$.

We require to obtain

$$P[Z_n = k \mid Z_0 = 0] = p_{0k}^{(n)} \tag{2.4}$$

where $Z_n = X_1 + ... + X_n$

subject to

$$P[X_i = 1] = p, \qquad P[X_i = -1] = q$$
$$\forall i = 1, 2, ..., n$$

$$\Rightarrow \qquad Z_n = Z_{n-1} + X_n \tag{2.5}$$

X_i's are all independent $\Rightarrow X_n$ is independent of Z_n.

Let us define r.v.^{s} $Y_i \ni Y_i = 1$ if $X_i = 0$ (2.6)

$$= 0, \text{ if } X_i = -1$$

Then $Y_i = \frac{1}{2}(X_i + 1)$ and Y_i can take only values 0 and 1, with probabilities q and p respectively. Therefore Y_i's are independent Bernoullian variates. Also,

$$\sum_{i=1}^{n} Y_i = \frac{1}{2}\sum_{i=1}^{n}(X_i + 1) = \frac{1}{2}Z_n + \frac{n}{2}$$

$$= \frac{1}{2}(Z_n + n) \tag{2.7}$$

and

$$\frac{1}{2}(Z_n + n) \sim B(n, p)$$

Now $P[Z_n = k] \Rightarrow P\left[\frac{1}{2}(Z_n + n) = \frac{1}{2}k + \frac{1}{2}n = \frac{1}{2}(k + n)\right]$ is given by

$$\binom{n}{\frac{1}{2}(k + n)} p^{\frac{1}{2}(k + n)} \cdot q^{n - \left(\frac{1}{2}(k + n)\right)} \tag{2.8}$$

or 0

according as $\frac{1}{2}(k + n) \in S$, the set of all positive integers,

or $\frac{1}{2}(k + n) \notin S$

Obviously,

$$E\left(\frac{1}{2}Z_n + \frac{n}{2}\right) = np$$

$$\Rightarrow \qquad E(Z_n) = 2\left(np - \frac{n}{2}\right) = 2np - n$$

$$= 2np - n(p + q) = n(p - q), \tag{2.9}$$

and

$$\text{Var}\left(\frac{Z_n}{2} + \frac{n}{2}\right) = npq = \frac{1}{4}\text{Var}(Z_n)$$

$$\Rightarrow \qquad \text{Var}(Z_n) = 4npq \tag{2.10}$$

The generating function of $p_{0k}^{(n)}$ is given by

$$P_{0k}(s) = \sum_{n=0}^{\infty} p_{0k}^{(n)} s^n \text{ which is not a probability generating function}$$

because $p_{0k}^{(n)}$ is not a probability distribution for a given n as

$$\sum_{n=0}^{\infty} p_{0k}^{(n)} \neq 1 \left(\because \sum_{n=0}^{\infty} p_{00}^{(n)} \geq p_{00}^{(1)} + p_{00}^{(2)} = 1 + \binom{2}{1}pq = 1 + 2pq > 1\right)$$

However, we can obtain $P_{0k}(s)$ for different $k = 0, 1, 2, ..$

$$P_{00}(s) = \sum_{n=0,2,4,\ldots} \binom{n}{\frac{1}{2}n} (p)^{\frac{n}{2}} (q)^{\frac{n}{2}} (s)^n$$

$$= \sum_{m=0}^{\infty} \binom{2m}{m} p^m q^m s^{2m}$$

$$= \sum_{m=0}^{\infty} \binom{2m}{m} (pqs^2)^m$$

$$= (1 - 4pqs^2)^{-\frac{1}{2}} \qquad (2.11)$$

$$p_{11}(s) = \sum_{n=1,3,5,\ldots} \binom{n}{\frac{1}{2}(n+1)} p^{\frac{1}{2}(n+1)} q^{\frac{1}{2}(n-1)} s^n$$

$$= \sum_{m=0}^{\infty} \binom{2m+1}{m+1} p^{m+1} q^m s^{2m+1} \qquad (2.12)$$

Now
$$\binom{2m+1}{m+1} = \binom{2m+1}{m} = \frac{(2m+1)!}{(m+1)!\, m!}$$

$$= \frac{[2(m+1)-1]}{(m+1)} \frac{(2m)!}{m!\, m!}$$

$$= \left(2 - \frac{1}{m+1}\right)\binom{2m}{m}$$

$$\therefore \sum_{m=0}^{\infty} \binom{2m+1}{m+1} p^{m+1} q^m s^{2m+1}$$

$$= \sum_{m=0}^{\infty} \left(2 - \frac{1}{m+1}\right)\binom{2m}{m} p^{m+1} q^m s^{2m+1}$$

$$= 2ps \sum_{m=0}^{\infty} \binom{2m}{m} (pqs^2)^m - \frac{1}{qs} \sum_{m=0}^{\infty} \binom{2m}{m} \frac{(pqs^2)^{m+1}}{m+1}$$

$$= 2ps(1 - 4pqs^2)^{-\frac{1}{2}} - \frac{1}{2qs}\left[1 - (1 - 4pqs^2)^{\frac{1}{2}}\right] \qquad (2.13)$$

$$\left[\because \quad \sum_{m=0}^{\infty} \binom{2m}{m} \frac{s^{m+1}}{m+1} = \sum_{m=0}^{\infty} \binom{2m}{m} \int_0^s x^n\, dx = \int_0^s \left\{ \sum_{m=0}^{\infty} \binom{2m}{m} x^n \right\} dx \right.$$

$$= \int_0^s \frac{dx}{\sqrt{1-4x}} \; = \; \frac{1}{2}(1 - \sqrt{1-4s}) \; \Bigg]$$

Alternately, $\qquad Z_0 = X_0 = i$

and $\qquad Z_n = X_0 + \ldots + X_n = j$

$\Rightarrow \qquad p_{ij}^{(n)} = P[X_1 + X_2 + \ldots + X_n = j - i \,|\, X_0 = i]$

The score of $(j - i)$ can be effected by $\dfrac{n+j-i}{2}$ positive jumps, each moving on

the right by '1' for each jump and $\dfrac{n-j+i}{2}$ negative jumps, each moving on the

left by '– 1' for each jump.

$$\therefore \quad p_{ij}^{(n)} = \binom{n}{\dfrac{n+j-i}{2}} p^{\frac{n+j-i}{2}} q^{\frac{n-j+i}{2}}, \text{ if } \frac{n+j-i}{2} \text{ is an integer;}$$

otherwise $p_{ij}^{(n)} = 0$. $\hfill (2.14)$

$$\Rightarrow \quad p_{00}^{(n)} = \binom{n}{\dfrac{n}{2}} p^{\frac{n}{2}}, \text{ if } \frac{n}{2} \text{ is an integer,} \; \Rightarrow p_{00}^{(2n)} = \binom{2n}{n} p^n q^n.$$

Again $\dbinom{2n}{n} = \dfrac{(2n)!}{n!\,n!} \cong \dfrac{4n}{(n\pi)^{1/2}}$ by Stirling's approximation to factorials.

$\therefore$ For $p_{00}^{(2n)} \cong \dfrac{(4pq)n}{(n\pi)^{1/2}} \to 0$ as $n \to \infty$ which implies '0' is a transient state,

where $p \neq q$

Otherwise, when $p = q = \dfrac{1}{2}$, $p_{00}^{(2n)} \cong \dfrac{1}{(n\pi)^{1/2}} \to \infty$ as $n \to \infty \Rightarrow$ '0' is a persistent

or recurrent state.

2.5 A Random Walk with Two Absorbing Barriers (Gambler's Ruin Problem)

Suppose two adversaries (A and B) start gambling with initial capitals Rs. 'a' and Rs. 'b' respectively. A will get Re.1 from B for each game of his win with probability p; or he has to give Re. 1 to B for each game for his loss with probability q. The outcome of each game is thus independent of any previous game played.

To obtain the probability that A will ultimately lose his entire capital and leave the game. Here $-a$ and b are the two absorbing states of the process in the sense that if A loses Rs. a he leaves the game or if A wins Rs. b then B leaves the game. The state space of the process is given by

$$\{-a, -a+1 \ldots 0, 1, 2, \ldots b\}.$$

If Z_n is the pay off of A at the nth game then the stopping time T of the game is given by

$$T = \min(T_1 T_2) \ni \left[\sum_{n=1}^{T_1} z_n = -a, \sum_{n=1}^{T_2} z_n = b \right] \tag{2.15}$$

Let Y be a random variable $\ni$

$$Y = 1, \text{ if } A \text{ is ruined i.e. } \sum_{n=1}^{\min(T_1, T_2) = T_1} z_n = -a$$

$$= 0, \text{ if otherwise, i.e. } \sum_{n=1}^{\min(T_1, T_2) = T_2} z_n = b$$

We require $P(Y = 1 \mid Z_0 = 0) = P_0$, say,
where we denote $P(Y = 1 \mid Z_0 = i) = p_i$ and $P(Y = 0 \mid Z_0 = i) = (1 - p_i)$.

Let

$$\sum_i s^Y P(Y \mid Z_0 = i) = G_i(s); i = 0, 1.$$

$$= s\, p_i + s^0 (1 - p_i) = [sp_i + (1 - p_i)]$$

$$= [1 - (1 - s) p_i] \tag{2.16}$$

Since Z_i's are independent r.v.s.,

$$\therefore \quad \sum s^Y P(Y \mid Z_0 = i, Z_1 = k)$$

$$= \sum s^Y P(Y \mid Z_1 = k) = G_k(s), \text{ say} \tag{2.17}$$

(This property is called the property of Markovity)

Now $G_i(0) = 1 - (1 - s) p_i \big|_{s=0} = 1 - p_i$

and $G_{z_i}(0) = 1 - p_{z_i}$

Again $p_i = E_{z_1 \mid z_0} = i(p_{z_i})$

$$= q\, p_{i-1} + p\, p_{i+1} \tag{2.18}$$

($\because$ if $Z_0 = i \Rightarrow Z_1 = i - 1$ with probability q

$$= i + 1 \text{ with probability } p)$$

$$\therefore \qquad p_i = q p_{i-1} + p\, p_{i+1} \qquad (\because p + q = 1)$$

$$\Rightarrow \qquad (p + q) p_i = q p_{i-1} + p\, p_{i+1}$$

$\Rightarrow \qquad p(p_i - p_{i+1}) = q(p_{i-1} - p_i)$ \hfill (2.19)

Putting $p_i - p_{i+1} = p_i'$ and $m = \dfrac{q}{p}$, we get

$$p_i' = \lambda \cdot p_{i-1}' = \lambda(\lambda p_{i-2}')$$

$$= \ldots = \lambda^i p_0' \qquad (2.20)$$

Again
$$p_i = P(Y = 1 \mid Z_c = i)$$
$$p_{-a} = P(Y = 1 \mid Z_0 = -a) = 1$$
$$p_b = P(Y = 1 \mid z_0 = b) = 0$$
$$p_{-a} - p_b = 1 - 0 = 1$$

Also

$$-1 = p_b - p_{-a} = \sum_{i=-a+1}^{b} (p_i - p_{i-1}) = \sum_{i=-a+1}^{b} p_i' \text{ identically.} \qquad (2.21)$$

Again $p_i' = \lambda^i p_0'$

$$-1 = \sum_{i=-a+1}^{b} \lambda^i p_0' = p_0' \sum_{i=-a+1}^{b} \lambda^i$$

$$\Rightarrow \qquad p_0' = -\frac{1}{\displaystyle\sum_{i=-a+1}^{b} \lambda^i} \qquad (2.22)$$

Again putting $a = 0$ in (2.21)

$$\Rightarrow \qquad p_b - p_0 = \sum_{i=1}^{b} p_i' = p_0' \sum_{i=1}^{b} \lambda^i$$

Noting $p_b = 0$, we have

$$-p_0 = p_0' \sum_{i=1}^{b} \lambda^i = -\frac{\displaystyle\sum_{i=1}^{b} \lambda^i}{\displaystyle\sum_{i=-a+1}^{b} \lambda^i}$$

$$\Rightarrow \qquad p_0 = \sum_{i=1}^{b} \lambda^i \Bigg/ \sum_{i=-a+1}^{1} \lambda^i = \frac{\lambda^b - 1}{\lambda^b - \lambda^{-a}}$$

$$\therefore \qquad p_0 = P(Y = 1 \mid Z_0 = 0) = P[\text{of ruin of } A]$$

$$= \frac{\lambda^b - 1}{\lambda^b - \lambda^{-a}}, \text{ where } \lambda = \frac{q}{p}$$

$$= \frac{\left(\frac{q}{p}\right)^b - 1}{\left(\frac{q}{p}\right)^b - \left(\frac{q}{p}\right)^{-a}} \tag{2.23}$$

In particular, if $q = p$,

$$p_0 = \frac{b}{a + b}$$

Therefore, the complementary probability that A wins the game

$$= 1 - \frac{\left(\frac{q}{p}\right)^b - 1}{\left(\frac{q}{p}\right)^b - \left(\frac{q}{p}\right)^{-a}} = \frac{1 - \left(\frac{q}{p}\right)^{-a}}{\left(\frac{q}{p}\right)^b - \left(\frac{q}{p}\right)^{-a}}$$

$$= \frac{\left(\frac{q}{p}\right)^a - 1}{\left(\frac{q}{p}\right)^{a+b} - \left(\frac{q}{p}\right)^0}$$

$$= \frac{1 - \left(\frac{q}{p}\right)^a}{1 - \left(\frac{q}{p}\right)^{a+b}} \tag{2.24}$$

In particular if $q = p = \frac{1}{2}$, the probability of A winning is $\left(1 - \frac{b}{a+1}\right)$ (vide 4.11, showing the arrival of the same result by using Martingales)

Remarks: It is certain that the game eventually ends.

Because $P(Y = 1 \mid Z_0 = 0) + P(Y = 0 \mid Z_0 = 0) = 1, \forall p, q$ subj. to $p + q = 1$.

Example 2.1 Obtain the probability distribution of the duration T of the game.

We obtain the p.g. f of the duration T given $Z_0 = i$. $(i = 0, 1, 2,...)$

Let us consider

$$E\left(s^T \mid Z_0 = i\right) = F_i(s); \qquad i = 0, 1, 2, ...$$

Then $\qquad F_i(s) = E_T(s^T \mid Z_0 = i)$

$$= E_{T, Z_1 \mid Z_0} = i(s^T)$$

$$= E_{T \mid Z_1} = j, Z_0 = i(s^T)$$

$$= E_{T'|Z_0} = j(W^{I+T'})$$

where $T = I + T'$ by the Markov property.

$$= s(E_{T|Z_0} = j(s^T)) = sE_j(F_j(s)) \qquad (2.25)$$

where $E_{T|Z_0} = j$ is denoted as $E_j \ \forall \ j = 0, 1, 2, \ldots$

Given $j = i + 1$ with probability p and $j = i - 1$ with probability q,

we have, $\qquad E_j(F_j(s)) = pF_{i+1}(s) + qF_{i-1}(s) \qquad (2.26)$

Putting (2.26) in (2.25)

$$F_i(s) = s(pF_{i+1}(s) + qF_{i-1}(s)) \qquad \ldots\ldots(2.27)$$

Let the trial solution of the difference equation (2.27) is,

$$F_i(s) = [v(s)]^i$$
$$(v(s))^i = sp(v(s))^{i+1} + sq(v(s))^{i-1}$$
$$\Rightarrow \qquad v = spv + sqv^2$$

$$\Rightarrow \qquad v = \left[1 \pm \sqrt{1 - 4pqs^2}\right]/2sp = v_1, v_2, \text{say}.$$

The general solution of (2.27) is given by

$$F_i(s) = A(v_1(s))^i + B(v_2(s))^i \qquad (2.28)$$

For $T = 0$, we have, by putting $i = -a$

$$F_{-a}(s) = E_T(s^T | Z_0 = -a) = s^0 = 1 \qquad (2.29)$$

Putting $i = b$, $F_b(s) = E_T(s^T | Z_0 = b) = s^0 = 1 \qquad (2.30)$

Putting $i = 0$, $F_0(s) = A + B = 1$

$$\therefore \qquad 1 = A(v_1(s))^{-a} + B(v_2(s))^{-a} \qquad (2.31)$$
$$1 = A(v_1(s))^b + B(v_2(s))^b \qquad (2.32)$$
and $\qquad 1 = A + B \qquad (2.33)$

whence A and B can be solved

as $\qquad A = \left[\dfrac{v_2^b - v_2^{-a}}{v_1^{-a} v_2^b - v_1^b v_2^{-a}}\right] \qquad (2.34)$

and $\qquad B = -\left[\dfrac{v_1^b - v_1^{-a}}{v_1^{-a} v_2^b - v_1^b v_2^{-a}}\right] \qquad (2.35)$

We require $F_0(s) = $ p.g. f of the process

$$= E(s^T | Z_0 = 0)$$

$$F_0(1) = p_0 + p_1 + \ldots + p_n = 1, \text{ where } p_i = p[T = i]$$

Let, $\qquad R(k) = P(T \geq k | Z_0 = 0) = \sum_{n=k}^{\infty} p_n \qquad (2.36)$

and $E(T) = \sum\limits_{k=0}^{\infty} R(k)$ is the expected duration of the game.

Example 2.2 Obtain the first passage time for the unrestricted one dimensional random walk.

Solution: Suppose $p > q$ (*i.e.* A's chance of loosing is more than his chance of gaining).

We require to obtain the distribution of the r.v. T_{ao} *i.e.* the first passage time from the state 'a' to *Zero* state. We have

$$T_{a,0} = T_{a,a-1} + T_{a-1,a-2} + \dots + T_{1,0} \qquad (2.37)$$

The r.v.s. $T_{j,j-1}$ are all independent; ($j = a, a-1, \dots, 1$)
The characteristic function of $T_{a,0}$ *viz* $\phi_{a0}(\theta)$

$$= [\phi_{10}(\theta)]^a$$

Let T be the time to the first event of a Poisson process of rate λ. Then $T \sim E(\lambda)$ (exponential with parameter λ).

Then
$$T_{10} = T + T_{00} = T \quad \text{with probability } q,$$
$$= T + T_{20} \quad \text{with probability } p.$$

$$\phi_{10}(\theta) = E_{T_{10}} [e^{i\theta} T_{10}]$$

$$= q E (e^{i\theta T}) + p E(e^{i\theta(T + T_{20})})$$

$$= q\psi(\theta) + p\psi(\theta) E(e^{i\theta T_{20}})$$

$$= \psi(\theta) [q + p(\phi_{20}(\theta))]$$

$$= \psi(\theta) [q + p (\phi_{10}(\theta))^2] \; (\because \phi_{a0}(a) = [\phi_{10}(0)]^a) \qquad (2.38)$$

Now
$$\psi(\theta) = \lambda \int\limits_{0}^{\infty} e^{i\theta x} e^{-\lambda x} \, dx = \frac{\lambda}{(\lambda - i\theta)} .$$

(2.38) $\Rightarrow$
$$\phi_{10}(\theta) = \psi(\theta)q + p\psi(\theta) [\phi_{10}(\theta)]^2 \qquad (2.39)$$

$\Rightarrow$
$$\phi_{10}(\theta) = \frac{\lambda}{\lambda - i\theta} q + p \frac{\lambda}{\lambda - i\theta} [\phi_{10}(\theta)]^2$$

$\Rightarrow$
$$p\psi [\phi_{10}(\theta)]^2 - \phi_{10}(\theta) + \psi q = 0$$

$\therefore$
$$\phi_{10}(\theta) = \frac{1 \pm \sqrt{1 - 4pq\psi^2}}{2p\psi} \qquad (2.40)$$

Again in consideration of $\phi_{10}(0) = 1 \; (p < q)$

$$\phi_{10}(\theta) = \frac{1 - \sqrt{4pq\psi^2}}{2p\psi}$$

$$\phi_{a0}(\theta) = [\phi_{10}(\theta)] \tag{2.41}$$

Hence we get the c.f. of the first passage time from a to 0.

Example 2.3   Show that unrestricted random walk with $p = q = 1/2$ is null recurrent.

Solution: We have for unrestricted random walk

$$P_{ii}(s) = P_{00}(s) = \left(1 - 4\left(\frac{1}{2}\right)\left(\frac{1}{2}\right)s^2\right)^{-1/2} \quad \text{(from (2.11))}$$

$$= (1 - s^2)^{-1/2} \tag{2.42}$$

Hence $\qquad\qquad P_{ii}(1) = \infty$

which shows that unrestricted random walk is recurrent.

Again $\qquad\qquad\qquad F_{ii}(s) = 1 - \{P_{ii}(s)\}^{-1}$

$$= 1 - (1 - s^2)^{1/2}$$

$$E(T_{ii}) = F_{ii}'(1) = \lim_{s \to 1} s\,(1 - s^2)^{-1/2} = \infty \tag{2.43}$$

which shows that a random walk is also null-recurrent.

Example 2.4   Consider an irreducible M.C. with transition probabilities

$$p_{0j} = r_j \qquad\qquad (j = 0, 1, 2, ...)$$

$$p_{ii} = p \qquad\qquad (i = 1, 2, 3, ...)$$

$$p_{i,\,i-1} = q = 1 - p \qquad (i = 1, 2, ...)$$

Show that M.C. is positive or null recurrent according as the mean of distribution of r_j is finite or infinite.

Solution: The transition probabilities show that given the position at the state i of $X(t)$ *i.e.* the next step is $(i-1)$ with probability q or it remains in the ith state is p $(p + q = 1)$.

Now, $\qquad\qquad T_{i0} = T_{i,\,i-1} + T_{i-1,\,i-2} + ... + T_{21} + T_{10}$

where each T_{ij}'s are distributed like T_{10}
Given $X = -1, T_{10} = 1$

$$E_{T_{10}\mid X = -1}(s^{T_{10}}) = s \text{ and}$$

given $X = 0$, $T_{10} = 1 + T_{10}'$ where T_{10}' is distributed like T_{10}, so that

$$E_{T_{10}\mid X = 0}(s^{T_{10}}) = E_{T_{10}'}(s^{1 + T_{10}'}) = sE_{T_{10}'}(s^{T_{10}'}) = sF_{10}(s) \tag{2.44}$$

where $\qquad\qquad F_{10}(s) = Es^{T_{10}} = E_X E_{T\mid X = x}(s^{T_{10}})$

$$= P[X = -1]\,s + P[X = 0]\,sF_{10}(s)$$

$$= qs + psF_{10}(s)$$

$$\therefore \qquad F_{10}(s) = \frac{qs}{1 - ps}$$

and
$$F_{i0}(s) = \{F_{10}(s)\}^i = \left(\frac{qs}{1 - ps}\right)^i \qquad (2.45)$$

Given the chain to be irreducible, the sequence $\{r_j\}$ must have an infinite number of non zero elements. Let us examine state '0'

Given $X = 0 \Rightarrow T_{00} = 1$, given $X = j \Rightarrow T_{00} = 1 + T_{j0}$

$$\therefore \qquad E_{T_{00} \mid X = j} \, s^{T_{00}} = E_{T_{j0}} \, s^{1 + T_{j0}} = s \{F_{10}(s)\}^j$$

$$\therefore \qquad F_{00}(s) = E_X \, E_{T_{00} \mid X} \, s^{T_{00}} = r_0 \, s + s \sum_{j=1}^{\infty} r_j \, \{F_{10}(s)\}^j$$

$$= j \sum_{j=0}^{\infty} \left(\frac{qs}{1 - ps}\right)^j r_j$$

$$= sR \{qs/(1 - ps)\} \text{ where } R(s) = \sum_{j=0}^{\infty} r_j s^j$$

$\therefore \; F_{00}(1) = R(1) = 1$, since $\{r_j\}$ is a distribution.

Further $F_{00}'(1) = 1 + \dfrac{1}{q} R'(1)$ where $R'(1)$ is the mean of the distribution of $\{r_j\}$.

$F_{00}'(1)$ is finite or infinite, according as $R'(1)$ is finite or infinite. This is precisely the condition for the State '0' as well that of M.C. to be null or positively recurrent.

State '0' is therefore recurrent and so the M.C. is recurrent.

2.6   Kolmogorov's Theorem

If a Markov Chain (M.C.) given by its transition matrix $P = [p_{ij}]$ is irreducible and aperiodic then $\exists$ one $[\pi] \ni$

$$\pi' = \pi' P \text{ holds} \Rightarrow \pi = p'\pi$$

then π is called the equilibrium distribution of the states. If the M.C. has an equilibrium distribution it is called '*Ergodic*'. In this case we have a single irreducible class and every state is recurrent and aperiodic. We deal with Chapman Kolmogorov equation in the next section which gives important solution among transition probabilities.

2.7   Chapman Kolmogorov Equation

Markov property implies important relations among the transition probabilities $p_{ij}(t, \tau)$.

Let ξ be a fixed point in (τ, t)

i.e.
$$\tau < \xi < t$$

and let $X(\tau), X(\xi)$ and $X(t)$ are the corresponding random variables. Because of the Markovity

$$P\{X(t) = k \,|\, X(\tau) = i, \; X(\xi) = j\} = P\{X(t) = k \,|\, X(\xi) = j\}$$
$$= p_{jk}(\xi, t)$$

Therefore, we have

$$P\{X(\xi) = j, X(t) = k \,|\, X(\tau) = i\}$$
$$= P\{X(t) = k \,|\, X(\xi) = j\} \, P\{X(\xi) = j \,|\, X(\tau) = i\}$$
$$= p_{jk}(\xi, t) \, p_{ij}(\tau, \xi) \tag{2.46}$$

The above represents a probability of transition from $X(\tau) = i$ to $X(t) = k$ through $X(\xi) = j$. Since passage from i to k can take place through all values of $\xi \in (\tau, t)$, therefore, we may write

$$p_{ik}(\tau, t) = \sum_{\xi \in (\tau, t)} p_{ij}(\tau, \xi) \, p_{jk}(\xi, t) \tag{2.47}$$

Equation 2.47 is known as a Chapman-Kolmogorov equation.
A Special Case: In case of homogeneous Markov Process

$$
\left.
\begin{aligned}
p_{ik}(\tau, t) &= p_{ik}(t - \tau) \\
p_{ij}(\tau, \xi) &= p_{ij}(\xi - \tau) \\[1em]
p_{ik}(\xi, t) &= p_{jk}(t - \xi)
\end{aligned}
\right\} \tag{2.48}
$$

and

Thus 2.47 reduces to

$$p_{ik}(t - \tau) = \sum_{\xi \in (\tau, t)} p_{ij}(\xi - \tau) \, p_{jk}(t - \xi) \tag{2.49}$$

which is the Chapman-Kolmogorov equation of a homogeneous Markov process.

2.8   Kolmogrorov Differential Equation

One may derive Kolmogorov differential equation from Chapman-Kolmogorov equation under the following regularity assumptions:

(i) $\forall$ integer i, $\exists$ a continuous function $v_{ii}(\tau)$ ∋

$$\lim_{\delta \to 0} \left(\frac{1 - p_{ii}(\tau, \tau + \delta)}{\delta} \right) = -v_{ii}(\tau) \quad \text{where } v_{ii}(\tau) \le 0 \tag{2.50}$$

(ii) $\forall$ pair of integers $i, j \; i \ne j \; \exists$ a continuous function $v_{ij}(\tau)$ ∋

$$\lim_{\delta \to 0} \frac{p_{ij}(\tau, \tau + \delta)}{\delta} = v_{ij}(\tau) \tag{2.51}$$

Further, for fixed j the passage from i to j in $(\tau; \tau + \delta)$ in (2.51) is uniform with respect to i. The function $v_{ij}(\tau)$ is called the intensity function of the process.

If we put $p_{ij}(\tau, \tau) = \delta_{ij}$ where δ_{ij} is kronekar delta.

$(2.50) \Rightarrow \qquad \lim_{\delta \to 0}\left[\frac{p_{ii}(\tau, \tau) - p_{ii}(\tau, \tau + \delta)}{\delta}\right] = -v_{ii}(\tau)$

$\Rightarrow \qquad \dfrac{d}{dt}(p_{ii}(t, t))\Big|_{t=\tau} = -v_{ii}(\tau)$

$\Rightarrow \qquad -\dfrac{d}{dt} p_{ii}(t, t)\Big|_{t=\tau} = v_{ii}(\tau) \tag{2.52}$

Similarly (2.51) can be written as

$$\lim_{\delta \to 0}\left(\frac{p_{ij}(\tau, \tau + \delta) - p_{ij}(\tau, \tau)}{\delta}\right) = v_{ij}(\tau)$$

$\Rightarrow \qquad \dfrac{d}{dt} p_{ij}(\tau, t)\Big|_{t=\tau} = v_{ij}(\tau) \tag{2.53}$

Further $(2.50) \Rightarrow \qquad 1 - p_{ii}(\tau, \tau + \delta) = -v_{ii}(\tau)\cdot\delta + o(\delta)$

$\Rightarrow \qquad 1 + v_{ii}(\tau)\,\delta + o(\delta) = p_{ii}(\tau, \tau + \delta) \tag{2.54}$

Further $(2.51) \Rightarrow \qquad p_{ij}(\tau, \tau + \delta) = v_{ij}(\tau)\cdot\delta + o(\delta) \tag{2.55}$

Noting $\qquad \displaystyle\sum_j p_{ij}(\tau, t) = 1 \; \forall \; \tau < t \tag{2.56}$

and putting $t = \tau + \delta$

$$\sum_j p_{ij}(\tau, t) = \sum_j p_{ij}(\tau, \tau + \delta) = p_{ii}(\tau, \tau + \delta) + \sum_{j \neq i} p_{ij}(\tau, \tau + \delta)$$

$$= 1 + v_{ii}(\tau)\cdot\delta + o(\delta) + \sum_{j \neq i} v_{ij}(\tau)\cdot\delta + o(\delta)$$

$$= 1 \tag{2.57}$$

$$\Rightarrow \delta v_{ii}(\tau) + o(\delta) = -\delta \sum_{j \neq i} v_{ij}(\tau)\cdot\delta t + o(\delta)$$

$$\Rightarrow v_{ii}(\tau) = -\sum_{j \neq i} v_{ij}(\tau) \tag{2.58}$$

Derivation of Kolmogorov Differential Equation

Let $\tau < t < t + \delta$. We have from Chapman-Kolmogorov equation

$$p_{ik}(\tau, t + \delta) = p_{ik}(\tau, t) p_{kk}(t, t + \delta) + \sum_{j \neq k} p_{ij}(\tau, t) p_{jk}(t, t + \delta)$$

where $\qquad\qquad p_{kk}(t, t + \delta) = 1 + v_{kk}(t)\cdot\delta + o(\delta)$

and $\qquad\qquad p_{jk}(t, t + \delta) = v_{jk}(t)\cdot\delta + o(\delta)$

$\Rightarrow \qquad\qquad p_{ik}(\tau, t + \delta) = p_{ik}(\tau, t)(1 + v_{kk}(t)\cdot\delta + o(\delta))$

$$+ \sum_{j \neq k} p_{ij}(\tau, t)(v_{jk}(t)\delta + o(\delta))$$

$\Rightarrow \qquad \dfrac{p_{ik}(\tau, t + \delta) - p_{ik}(\tau, t)}{\delta} = p_{ik}(\tau, t) v_{kk}(t)$

$$+ \frac{o(\delta)}{\delta} + \sum_{j \neq k} p_{ij}(\tau, t) \left(v_{jk}(\tau) + \frac{o(\delta)}{\delta}\right)$$

Taking limit as $\delta \to 0$ on both sides, under the regularity condition

$\Rightarrow \qquad \dfrac{\partial p_{ik}(\tau, t)}{\partial t} = p_{ik}(\tau, t) v_{kk}(t) + \sum_{j \neq k} p_{ij}(\tau, t) v_{jk}(t)$

$\Rightarrow \qquad \dfrac{\partial p_{ik}(\tau, t)}{\partial t} = \sum_{j} p_{ij}(\tau, t) v_{jk}(t) \qquad\qquad (2.59)$

Equation (2.59) represents Kolmogorov's forward equation.

Kolmogorov's Backward Equation

Again from Chapman Kolmogorov's equation, we have

$$p_{ik}(\tau - \delta, t) = \sum_{j} p_{ij}(\tau - \delta, \tau) p_{jk}(\tau, t); \delta > 0$$

$$= p_{ii}(\tau - \delta, \tau) p_{ik}(\tau, t) + \sum_{j \neq 1} p_{ij}(\tau - \delta, \tau) p_{jk}(\tau, t)$$

Also $\qquad\qquad p_{ij}(\tau - \delta, \tau) = 1 + v_{ii}(\tau - \delta)\delta + o(\delta)$

and $\qquad\qquad p_{ij}(\tau - \delta, \tau) = v_{ij}(\tau - \delta)\delta + o(\delta)$

$\Rightarrow \qquad p_{ik}(\tau - \delta, t) + [1 + v_{ii}(\tau - \delta)\delta + o(\delta)] p_{ik}(\tau, t)$

$$+ \sum_{j \neq i} (v_{ij}(\tau - \delta)\delta + o(\delta)) p_{jk}(\tau, t)$$

$$\frac{p_{ik}(\tau - \delta, t) - p_{ik}(\tau, t)}{-\delta} = \left[-v_{ii}(\tau - t) + \frac{o(\delta)}{-\delta}\right] p_{ik}(\tau, t)$$

$$-\sum_{j \neq i} v_{ij}(\tau - t) \, p_{jk}(\tau, t) - \frac{o(\delta)}{\delta} \sum_{j \neq i} p_{jk}(\tau, t)$$

Taking limit as $\delta \to 0$ on both sides

$$\Rightarrow \qquad \frac{\partial p_{ik}(\tau, t)}{\partial \tau} = -v_{ii}(\tau) \, p_{ik}(\tau, t) - \sum_{j \neq i} v_{ij}(\tau) \, p_{jk}(\tau, t)$$

$$\Rightarrow \qquad \frac{\partial p_{ik}(\tau, t)}{\partial \tau} = -\sum_{j=1} v_{ij}(\tau) \, p_{jk}(\tau, t) \qquad (2.60)$$

Therefore, Kolmogorov's forward equation is given by

$$\frac{\partial p_{ik}(\tau, t)}{\partial t} = \sum_{j} p_{ij}(\tau, t) \, v_{jk}(t)$$

with initial condition $p_{ij}(\tau, \tau) = \delta_{ij}$.
and Kolmogorov's backward equation is given by

$$\frac{\partial p_{ik}(\tau, t)}{\partial \tau} = -\sum p_{jk}(\tau, t) \, v_{ij}(\tau)$$

with initial condition $p_{ik}(t, t) = \delta_{ik}$.

Given that $\sum_{k} p_{ik}(\tau, t) = 1$, there exists a unique solution of $p_{ik}(\tau, t)$ satisfying both the forward and backward as well as Chapman Kolmogorov equations. In general $p_{ik}(\tau, t)$ is non-homogeneous with respect to time 't'; however, it may be shown that if the intensity functions $v_{ij}(t)$ are equal to constant ($v_{ij}(t)$ is independent of t) then v_{ij} is independent of time, and the process is independent of time. In this case

$$\left.\begin{array}{l} \dfrac{dp_{ik}(t)}{dt} = \sum p_{ij}(t) \, v_{jk} \\[2mm] \dfrac{dp_{ik}(t)}{dt} = -\sum v_{ij} \, p_{jk}(t) \end{array}\right\} \qquad (2.61)$$

where $p_{ik}(0) = \delta_{ik}$.

Example 2.5   Intensity matrix for a Poisson process

$$V = \begin{pmatrix} -\lambda & \lambda & 0 & \cdots & 0 \\ 0 & -\lambda & \lambda & \cdots & 0 \\ 0 & 0 & -\lambda & \cdots & 0 \\ \vdots & \vdots & \vdots & & \vdots \\ 0 & 0 & 0 & \cdots & -\lambda \end{pmatrix}$$

Kolmogorov forward equation is given as

$$\frac{\partial p_{ij}(\tau, t)}{\partial t} = p_{ij}(\tau, t)(-\lambda) + p_{i, j-1}(\tau, t)\lambda$$

$$\frac{\partial p_{ij}(\tau, t)}{\partial t} = \lambda p_{ij}(\tau, t) - \lambda p_{i+1, j}(\tau, t)$$

2.9 Application of Kolmogorov's Differential Equation in Population Models

(The general time independent birth and death processes with immigration)

Consider a Markov chain $X(t)\, t \in [0, \infty)$, where $X(t)$ represents the size of the population at time t.

If $X(t) = r\, (r = 0, 1, 2, ...)$ then during an infinitesimal time interval $(t, t + \delta t)$, $X(t)$ increases by unity (either by birth or by immigration with probability $\lambda_r\, \delta t + o(\delta t)$, decreases by one (on account of death or emigration) with probability $\mu_r\, \delta t + o(\delta t)$ or does not change (remains stationary with probability $1 - (\lambda_r + \mu_r)\, \delta t + o(\delta t)$.

Obtain the intensity matrix V of the process given that $\mu_0 = 0$. Also obtain the Kolmogorov forward equation and determine the equilibrium solution if it exists.

Solution: The state space of the Markov chain is $\{0, 1, 2, 3, ...\}$. We have

$$\left.\begin{array}{l} P\{X(t + \delta) = r + 1 \mid X(t) = r\} = \lambda_r \cdot \delta + o(\delta) \\[2mm] P\{X(t + \delta) = r \quad \mid X(t) = r\} = 1 - (\lambda_r + \mu_r) \cdot \delta + o(\delta) \\[2mm] P\{X(t + \delta) = r - 1 \mid X(t) = r\} = \mu_r \cdot \delta + o(\delta) \end{array}\right\} \quad (2.62)$$

$$P\{X(t + \delta) = r' \quad \mid X(t) = r\} = o(\delta) \text{ for } r' \neq r + 1, r, r - 1$$

The intensity matrix

$$V = \begin{array}{c} \\ 0 \\ 1 \\ 2 \\ 3 \\ \vdots \end{array}\begin{array}{c} \begin{array}{ccccc} 0 & 1 & 2 & 3 & 4... \end{array} \\ \left(\begin{array}{ccccc} -\lambda_0 & \lambda_0 & 0 & 0 & 0 \\ \mu_1 & -\mu_1 - \lambda_1 & \lambda_1 & 0 & 0 \\ 0 & \mu_2 & -\mu_2 - \lambda_2 & \lambda_2 & 0 \\ 0 & 0 & \mu_3 & -\mu_3 - \lambda_3 & \lambda_3 \\ \vdots & \vdots & \vdots & \vdots & \vdots \end{array}\right) \end{array} \quad (2.63)$$

Then the Kolmogorov's differential equations are

$$\frac{dp_{r0}}{dt} = -\lambda_0 p_{r0} + \mu_1 p_{r1}$$

$$\frac{dp_{r1}}{dt} = \lambda_{l-1} p_{r,l-1} - (\mu_l + \lambda_l) p_{rl} + \mu_{l+i} p_{r,l+1} \quad (l = 1, 2, ...) \Bigg\} \quad (2.64)$$

Let us write the equilibrium or steady state solution of $p_{rl}(t) \to \pi_l$ as $t \to \infty$. Then

$$0 = -\lambda_0 \pi_0 + \mu_1 \pi_1$$

$$0 = (\lambda_{l-1} \pi_{l-1} - \mu_l \pi_l) - (\lambda_l \pi_l - \mu_{l+1} \pi_{l+1}) \quad (l = 1, 2, ...)$$

Putting $I_l = \lambda_l \pi_l - \mu_{l+1} \pi_{l+1} \quad (l = 0, 1, 2, ...)$

$$0 = -I_0$$

$$0 = I_{l-1} - I_l$$

so that $I_l = 0 \; (l = 0, 1, 2, 3, ...)$

Therefore,
$$\frac{\pi_{l+1}}{\pi_l} = \frac{\lambda_l}{\mu_{l+1}}$$

$$\Rightarrow \qquad \pi_l = \left(\frac{\pi_l}{\pi_{l-1}} \cdot \frac{\pi_{l-1}}{\pi_{l-2}} \cdots \frac{\pi_1}{\pi_0}\right) \pi_0 = v_l \pi_0 \qquad (2.65)$$

where
$$v_l = \frac{\lambda_{l-1}}{\mu_l} \cdot \frac{\lambda_{l-2}}{\mu_{l-1}} \cdots \frac{\lambda_0}{\mu_1}, \quad v_0 = 1 \qquad (2.66)$$

Then
$$\sum_l \pi_l = 1 \Rightarrow 1 = \pi_0 \sum_{l=0}^{\infty} v_l \Rightarrow \pi_0 = \frac{1}{\sum\limits_{l=0}^{\infty} v_l}$$

$$\Rightarrow \qquad \pi_l = \frac{v_l}{\sum\limits_{l=0}^{\infty} v_l} \quad (l = 0, 1, 2, 3, ...) \qquad (2.67)$$

This will be a probability distribution if $\sum\limits_{l=0}^{\infty} v_l$ converges.

2.10 A Useful Result in Two State Markov-Chain Model in Continuous Time

If an event (say sickness) occurs with intensity λ and the complementary event (say recovery) with intensity μ then show that

$$(p_{ij}(t)) = \begin{pmatrix} \dfrac{\mu}{\lambda+\mu} + \dfrac{\lambda}{\lambda+\mu} e^{-(\lambda+\mu)t} & \dfrac{\lambda}{\lambda+\mu} - \dfrac{\lambda}{\lambda+\mu} e^{-(\lambda+\mu)t} \\[2mm] \dfrac{\mu}{\lambda+\mu} - \dfrac{\mu}{\lambda+\mu} e^{-(\lambda+\mu)t} & \dfrac{\lambda}{\lambda+\mu} + \dfrac{\mu}{\lambda+\mu} e^{-(\lambda+\mu)t} \end{pmatrix} \qquad (2.68)$$

where $p_{ij}(t) = P[X(t) = j \,|\, X(0) = i]$; $i, j = 0, 1$. Let $X(t)$ be the continuous time parameter Markov chain and

$$(\mu_{ij}(t)) = \begin{pmatrix} \dfrac{\mu t}{\lambda+\mu} + \dfrac{\lambda}{(\lambda+\mu)^2}(1 - e^{-(\lambda+\mu)t}) & \dfrac{\lambda t}{\lambda+\mu} - \dfrac{\lambda}{(\lambda+\mu)^2}(1 - e^{-(\lambda+\mu)t}) \\[2mm] \dfrac{\mu t}{\lambda+\mu} - \dfrac{\mu}{(\lambda+\mu)^2}(1 - e^{-(\lambda+\mu)t}) & \dfrac{\lambda t}{\lambda+\mu} + \dfrac{\mu}{(\lambda+\mu)^2}(1 - e^{-(\lambda+\mu)t}) \end{pmatrix}$$

$$(2.69)$$

where

$$\mu_{ij}(t) = E[X(t) = j \,|\, X(0) = i] \,;\, i, j = 0, 1, 2$$

i.e. expected amount of time in $(0; t]$ the system is in state 'j' given that it is initially in i.

Proof: We have the Kolmogorov's equation

$$p_{00}(t + \delta t) = p_{00}(t)(1 - (\lambda\delta t + o(\delta t)) + p_{01}(t)(\mu\delta t + o(\delta t))$$

$$\Rightarrow \quad p'_{00}(t) = -\lambda p_{00}(t) + \mu(1 - p_{00}(t)) \quad (\because p_{01}(t) = 1 - p_{00}(t))$$

$$\Rightarrow \quad -1 + SL(p_{00}(t)) = L(p'_{00}(t)) = \mu L(1) - (\lambda+\mu)L(p_{00}(t)) \qquad (2.70)$$

where 'L' stands for Laplace transform

$$\Rightarrow \qquad p_{00}(t) = \frac{\mu}{\lambda+\mu} + \frac{\lambda}{\lambda+\mu}(e^{-(\lambda+\mu)t})$$

$$\Rightarrow \qquad p_{01}(t) = 1 - p_{00}(t) = \frac{\lambda}{\lambda+\mu} - \frac{\lambda}{\lambda+\mu}(e^{-(\lambda+\mu)t})$$

By symmetry $p_{10}(t) = \dfrac{\mu}{\lambda+\mu} - \dfrac{\mu}{\lambda+\mu}(e^{-(\lambda+\mu)t})$, while interchanging λ with μ in $p_{01}(t)$.

Finally $\qquad p_{11}(t) = 1 - p_{10}(t)$

$$= \frac{\lambda}{\lambda+\mu} + \frac{\mu}{\lambda+\mu}(e^{-(\lambda+\mu)t}) \qquad (2.71)$$

This proves (2.68).
Further,

$$\mu_{11}(t) = \int_0^t p_{00}(\tau)\,d\tau = \frac{\mu t}{(\lambda+\mu)} + \frac{\lambda}{(\lambda+\mu)^2}(1 - e^{-(\lambda+\mu)t}) \quad (2.71')$$

$$\mu_{01}(t) = \int_0^t p_{01}(\tau)\,d\tau = \frac{\lambda t}{(\lambda+\mu)} - \frac{\lambda}{(\lambda+\mu)^2}(1 - e^{-(\lambda+\mu)t}) \quad (2.71'')$$

$$\mu_{10}(t) = \int_0^t p_{10}(\tau)\,d\tau = \frac{\mu t}{(\lambda+\mu)} - \frac{\mu}{(\lambda+\mu)^2}(1 - e^{-(\lambda+\mu)t}) \quad (2.71''')$$

and
$$\mu_{11}(t) = \int_0^t p_{11}(\tau)\,d\tau = \frac{\lambda t}{(\lambda+\mu)} + \frac{\mu}{(\lambda+\mu)^2}(1 - e^{-(\lambda+\mu)t}) \quad (2.71'''')$$

Since
$$p_{ij}(t) = P[X(t) = j \mid X(0) = i]$$

and
$$E[X(t) = j \mid X(0) = i] = \int_0^t p_{ij}(\tau)\,d\tau$$

Note, as $t \to \infty$, $p_{00}(\infty) = \dfrac{\mu}{\lambda+\mu}$, $p_{01}(\infty) = \dfrac{\lambda}{\lambda+\mu}$, $p_{10}(\infty) = \dfrac{\mu}{\lambda+\mu}$ and

$$p_{11}(\infty) = \frac{\lambda}{\lambda+\mu}$$

$$P[X(\infty) = 0] = \frac{\mu}{\lambda+\mu} \quad \text{and} \quad P[X(\infty) = 1] = \frac{\lambda}{\lambda+\mu}$$

are the stationary distributions of two states in the Markov Chain. These are also called 'Steady State Distribution' of the Markov Chain.

2.11   Certain Basic Stochastic Population Models Based on Markov Process

2.11.1   Poisson Process

Denoting by $\qquad p_k(t) = P[X(t) = k \mid (0, t)]$
Poisson process is given by Kolmogorov's equation

$$\left.\begin{aligned} p_k(t + \Delta) &= p_k(t)(1 - \lambda\Delta) + p_{k-1}(t)\,\lambda\Delta + o(\Delta) \\ k &= 1, 2, 3.. \\ p_0(t + \Delta) &= p_0(t)(1 - \lambda\Delta) + o(\Delta) \end{aligned}\right\} \quad (2.72)$$

where the probability of an event occurring in $(t, t + \Delta) = \lambda\Delta + o(\Delta)$
when λ is a constant, independent of t or $\Delta(0 \le \lambda < \infty)$ (condition of stationarity).

The probability of more than one event in $(t, t + \Delta)$ is of the order of zero (condition of orderliness).

$$\Rightarrow \qquad \frac{dp_k(t)}{dt} = -\lambda\, p_k(t) + \lambda\, p_{k-1}(t) \quad k = 1, 2, 3\ldots$$

and $$\frac{dp_0(t)}{dt} = -\lambda\, p_0(t) \tag{2.73}$$

The solution of (2.73) is given by

$$p_k(t) = e^{-\lambda t}\,\frac{(\lambda t)^k}{k!} \quad ; \quad k = 0, 1, 2, \ldots \tag{2.74}$$

2.11.2 Time Dependent Poisson Process

This is obtained by Kolmogorov's forward equation

$$p_n(t + \delta t) = p_{n-1}(t)\,(\lambda(t)\,\delta t + o(\delta t)) + p_n(t)\,(1 - [\lambda(t)\delta t + o(\delta t)])$$
$$n = 1, 2, 3\ldots$$

and $$p_0(t + \delta t) = p_0(t)\,(1 - \lambda(t)\,\delta t + o(\delta t)) \tag{2.75}$$

The solution is given by

$$p_n(t) = \exp\left(-\int_0^t \lambda(\tau)\,d\tau\right) \frac{\left[\displaystyle\int_0^t \lambda(\tau)\,d\tau\right]^n}{n!} \tag{2.75'}$$

Obviously we have

$$\lim_{\delta t \to 0} \frac{p_0(t + \delta t) - p_0(t)}{\delta t} = -\lambda(t)\,p_0(t) + \lim_{\delta t \to 0} \frac{o(\delta t)}{\delta t}$$

$$\Rightarrow \qquad \int_0^t \frac{p_0'(\tau)}{p_0(\tau)}\,d\tau = -\int_0^t \lambda(\tau)\,d\tau = -\Lambda(t) \quad \text{say.}$$

$$\Rightarrow \qquad p_0(t) = \exp\left(-\int_0^t \lambda(\tau)\,d\tau\right) = \exp(-\Lambda(t))$$

Similarly one can show

$$\lim_{\delta t \to 0} \frac{p_1(t + \delta t) - p_1(t)}{\delta t} = -\lambda(t)\,(p_1(t) - p_0(t)) + \lim_{\delta t \to 0} \frac{o(\delta t)}{\delta t}$$

$$\Rightarrow \qquad p_1'(t) + \lambda(t)\,p_1(t) = \lambda(t)\,p_0(t) \tag{2.76}$$

i.e. $\dfrac{dp_1(t)}{dt} + \lambda(t)\,p_1(t) = \lambda(t)\,p_0(t)$, which is a linear equation of order one.

$$\Rightarrow \qquad \exp\left(\int_0^t \lambda(\tau)\,d\tau\right)\frac{dp_1(t)}{dt} + \lambda(t)\,p_1(t)\exp\left(\int_0^t \lambda(\tau)\,d\tau\right)$$

$$= \lambda(t)\, p_0(t) \exp\left(\int_0^t \lambda(\tau)\, d\tau\right)$$

$$= \lambda(t) \quad \left(\because p_0(t) = \exp\left(-\int_0^t \lambda(\tau)\, d\tau\right)\right) \tag{2.77}$$

$$\Rightarrow \qquad \frac{d}{dt}\left(p_1(t) \exp\left(-\int_0^t \lambda(\tau)\, d\tau\right)\right) = \lambda(t)$$

$$\Rightarrow \qquad p_1(t) = \exp\left(-\int_0^t \lambda(\tau)\, d\tau\right)\left(\int_0^t \lambda(\tau)\, d\tau\right)$$

$$= \exp\left(-\int_0^t \lambda(\tau)\, d\tau\right) \Lambda(t) \tag{2.78}$$

Thus we see the structure of $p_n(t)$ may be given by

$$p_n(t) = \exp\left(-\int_0^t \lambda(\tau)\, d\tau\right) \frac{(f(t))^n}{n!} \tag{2.79}$$

where $f(t)$ is an unknown function whose structure is to be determined.
    Putting the trial solution

$$p_n(t) = \exp[-\Lambda(t)]\frac{(f(t))^n}{n!} \text{ in (2.75)}$$

we get $\qquad \dfrac{f'(t)}{f(t)} = -\lambda(t)$

$$\Rightarrow \qquad f(t) = \int_0^t \lambda(\tau)\, d\tau = \Lambda(t)$$

Thus $\qquad p_n(t) = \exp[-\Lambda(t)]\dfrac{(\Lambda(t))^n}{n!} \tag{2.80}$

where $\qquad \Lambda(t) = \int_0^t \lambda(\tau)\, d\tau$,   which proves the result.

It follows that

$$E[X(t)] = \int_0^\infty \lambda(\tau)\, d\tau = \Lambda(t) \tag{2.81}$$

2.11.3 Weighted Poisson Process

We define

$$p_n(t) = \int_0^\infty p_{n|\lambda}(t) f(\lambda) d\lambda$$

where $f(\lambda) = \dfrac{e^{-a\lambda} \lambda^{k-1} a^k}{\Gamma(k)}$; $\lambda > 0$ be the best prior distribution of λ;

$$a, k > 0$$

and
$$p_{n|\lambda}(t) = e^{-\lambda t} \frac{(\lambda t)^n}{n!}$$

$$\Rightarrow \qquad p_n(t) = \binom{n+k-1}{n}\left(\frac{t}{a+t}\right)^n \left(\frac{a}{a+t}\right)^k \qquad (2.82)$$

is a negative binomial distribution with parameters k and $p = \dfrac{a}{a+t}$

$$\therefore \qquad \left.\begin{array}{l} E(X(t)) = \dfrac{kt}{a} \\[2mm] \text{Var}\,(X(t)) = \dfrac{kt}{a}\left(1+\dfrac{t}{a}\right) \end{array}\right\} \qquad (2.83)$$

2.11.4 Pure Birth Process

It is defined as, given $X(t) = k$ and $X(0) = k_0$
  (i) the conditional probability that a new event will occur during $(t, t+\Delta)$ is $\lambda_k \Delta + o(\Delta)$.
  (ii) the conditional probability that more than one event will occur is $o(\Delta)$
The Kolmogorov equations are

$$\left.\begin{array}{l} \dfrac{dp_k(t)}{dt} = -\lambda_{k_0}\, p_{k_0}(t) \\[3mm] \dfrac{dp_k(t)}{dt} = -\lambda_k\, p_k(t) + \lambda_{k-1}\, p_{k-1}(t) \ \text{for } k > k_0 \end{array}\right\} \qquad (2.84)$$

and

Initial conditions are $p_{k_0}(0) = 1, p_k(0) = 0$ for $k \neq k_0$.
The solution is

$$p_k(t) = (-1)^{k-k_0} \lambda_{k_0} \cdots \lambda_{k-1}\left[\sum_{i=k_0}^{k} \frac{\exp(-\lambda_i t)}{\displaystyle\prod_{\substack{j=k_0 \\ j \neq i}}^{k} (\lambda_i - \lambda_j)}\right] \qquad (2.85)$$

2.11.5   Yule Process

It is a particular case of pure birth process by putting $\lambda_k = k\lambda$.

The Kolmogorov equations are

$$\frac{d}{dt}\, p_{k_0}(t) = -k_0 \lambda \, p_{k_0}(t)$$

and

$$\frac{d}{dt}\, p_k(t) = -k\lambda p_k(\lambda) + (k-1)\, \lambda \, p_{k-1}(t)$$

The solution is given by

$$p_k(t) = \binom{k-1}{k-k_0} \exp(-k_0 \lambda t)\,[(1-\exp(-\lambda t)^{k-k_0}]$$

$$= (-1)^{k-k_0}\, (k_0\lambda)\ldots((k-1)\lambda)\left[\sum_{i=k_0}^{k} \frac{e^{-i\lambda t}}{\displaystyle\prod_{\substack{i=k_0 \\ j\neq i}} (i\lambda - j\lambda)} \right] \qquad (2.86)$$

where

$$(k_0\lambda)\ldots((k-1)\lambda) = \lambda^{k-k_0}\binom{k-1}{k-k_0}(k-k_0)!$$

and

$$\prod_{\substack{j=k_0 \\ j\neq i}}^{k} (i\lambda - j\lambda) = \lambda^{k-k_0}\,(-1)^{k-i}\binom{k-k_0}{i-k_0}^{-1}(k-k_0)!$$

Here

$$\left.\begin{array}{l} E(X(t)) = k_0 e^{\lambda t} \\[2mm] \mathrm{Var}\,(X(t)) = k_0\, e^{\lambda t}\,(e^{\lambda t}-1) \end{array}\right\} \qquad (2.87)$$

2.11.6   Polya Processes

Polya process is a generalization of pure birth process in the sense that probability of birth in time $(t, t+\Delta)$ is

$$\lambda_k \cdot \Delta + o(\Delta)$$

and λ_k depends on k and t.

We set

$$\lambda_k(t) = \frac{\lambda + \lambda a k}{1 + \lambda a t}$$

where λ and a are non-negative constants.

We have the initial condition $X(0) = k_0$. Kolmogorov equations are

$$\frac{d}{dt} P_{k_0}(t) = -\frac{\lambda + \lambda a k_0}{1 + \lambda at} P_{k_0}(t)$$

and
$$\frac{d}{dt} P_k(t) = -\frac{\lambda + \lambda a k}{1 + \lambda at} P_k(t) + \frac{\lambda + \lambda a(k-1)}{1 + \lambda at} P_{k-1}(t) \qquad (2.88)$$
$$\text{for } k > k_0$$

The p.g.f. $\phi_x(s; t) = s^{k_0} \left[\dfrac{1/(1+\lambda at)}{1 - s(\lambda at/(1+\lambda at))} \right]^{k_0 + 1/a}$

and $X(t)$, excepting for the additive constant k_0, behaves like a negative binomial random variable for a given t, with parameters p and r where

$$p = \frac{1}{1 + \lambda at} \quad \text{and} \quad r = k_0 + \frac{1}{a}$$

2.12 The Discrete Branching Process

The process* described now may have several applications in population (art 3.6) problems as restricted population process. We consider both continuous and discrete forms. Let us assume that every individual (female) in a population gives rise to X number of births, where X is a discrete random variable having its p.g.f.

$$\phi(s) = E(s^X) \text{ for } |s| < 1 \qquad (2.89)$$

Further, we assume that individuals reproduce independently and finally die (or make themselves off from reproductive span). For the sake of simplicity, we assume that all the females give births (female) at a particular point of time (which is the mean age of child-bearing of the population) and cease to give birth after that period which is considered to be the end of the fertility span.

Let Z_n be the number of female children in the nth generation ($n = 1, 2, 3...$) and $\Pi_n(s)$ is the p.g.f. of Z_n.

Then it is easy to see

$$\Pi_{n+1}(s) = \Pi_n(\phi(s)) = \phi(\Pi_n(s)) \qquad (2.90)$$

To prove the same, we decompose

$$Z_{n+1} = X_1 + X_2 + ... + X_{Z_n} \qquad (2.91)$$

which is obtained directly by assuming that Z_n be the number of females of which first one gives rise to X_1 number of female births, second one gives rise to X_2 number of female births and Z_n gives to X_{Z_n} number of female births and hence (2.91) is an identity.

$$\therefore \quad \Pi_{n+1}(s) = E(s^{Z_{n+1}}) = E_{Z_n}(s^{X_1 + X_2 + ... + X_{Z_n}} | Z_n))$$

$$= E_{Z_n}[(\phi(s))^{Z_n}] \quad (\because X_1, X_2, ..., X_{Z_n} \text{ are all i.i.d.r.v.s.})$$

---

*Evolved by Francis Galton and H.W. Watson (1874).

$$, \quad = \Pi_n (\phi(s))$$

Also $\quad \Pi_{n+1}(s) = E_X [E(s^X)^{Z_n} | X]$

$$= E_X [E(s^{Z_n})^X | X]$$

$$= E_X [(\Pi_n(s))^X]$$

$$= \phi(\Pi_n(s)) \tag{2.92}$$

Hence combining (2.91) and (2.92)

$$\Rightarrow \quad \Pi_{n+1}(s) = \Pi_n(\phi(s)) = \phi(\Pi_n(s))$$

Next we are required to obtain $E(Z_n)$ and Var (Z_n) for population in the nth generation. Also $E(X) = \alpha = \phi'(1)$ and Var $(X) = \phi''(1) + \alpha - \alpha^2$

If $\quad Z_0 = 1, \quad E(Z_0) = 1$ and Var $(Z_0) = 0$

$$E(Z_{n+1} | Z_n = k) = E(X_1 + X_2 + ... + X_k)$$

$$= kE(X) = k\alpha, \text{ where } E(X) = \alpha = \phi'(1)$$

(The mean family size follows from the decomposition of (2.91), while X_i's being i.i.d.r.v.,)

$$\therefore \quad E(Z_{n+1}) = E_{Z_n}[E(Z_{n+1} | Z_n)] = E_{Z_n}(Z_n \alpha) = \alpha E(Z_n)$$

$$\Rightarrow \quad E(Z_{n+1}) = \alpha E(Z_n) = \alpha^2 E(Z_{n-1}) = ... = \alpha^{n+1} E(Z_0) = \alpha^{n+1}$$

Again $\quad$ Var $(Z_{n+1} | Z_n = k) = $ Var $(X_1 + X_2 + ... + X_k)$

$$= k \text{ Var } (X) \ (\because X_i \text{'are i.d.r.v's and distributed as } X)$$

$$= k\beta, \text{ where we denote Var } (X) = \beta.$$

Again $\quad$ Var $(X) = \phi''(1) + \alpha - \alpha^2$

$$\text{Var } (Z_{n+1}) = E[\text{Var}(Z_{n+1} | Z_n)] + \text{Var}[E(Z_{n+1} | Z_n)]$$

$$= E[Z_n \beta] + \text{Var } (\alpha Z_n)$$

$$\Rightarrow \quad \text{Var } (Z_{n+1}) = \beta E [Z_n] + \alpha^2 \text{ Var } (Z_n)$$

This is a recurrence relation.

$$\because \quad E(Z_n) = \alpha^n, \text{ therefore}$$

$$\text{Var } (Z_{n+1}) = \beta\alpha^n + \alpha^2 \text{ Var } (Z_n) \tag{2.93}$$

Putting $\quad n = 0$ in (2.93), we get

$$\text{Var } (Z_1) = \beta\alpha^0 + \alpha^2 \text{ Var } (Z_0) = \beta \ (\because \text{ Var } (Z_0) = 0)$$

Putting $\quad n = 1$ in (2.93), we get

$$\text{Var } (Z_2) = \beta\alpha + \alpha^2 \text{ Var}(Z_1) = \beta\alpha + \alpha^2\beta = \beta\alpha(1 + \alpha) \tag{2.94}$$

Proceeding in this way

$$\text{Var } (Z_3) = \beta\alpha^2 + \alpha^2 \text{ Var } (Z_2).$$

$$= \beta\alpha^2 + \alpha^2 \left(\beta\alpha + \beta\alpha^2\right)$$

$$= \beta\alpha^2 + \beta\alpha^3 + \beta\alpha^4 = \beta\alpha^2 \left(\frac{1-\alpha^3}{1-\alpha}\right) \tag{2.95}$$

$$\text{Var}\,(Z_4) = \beta\alpha^3 + \alpha^2 \left[\beta\alpha^2 + \beta\alpha^3 + \beta\alpha^4\right]$$

$$= \beta\alpha^3[1 + \alpha + \alpha^2 + \alpha^3] = \beta\alpha^3 \frac{(1-\alpha^4)}{1-\alpha} \tag{2.96}$$

It may be noted, if $\alpha = 1$ then $\text{Var}\,(Z_1) = \beta$

$$\text{Var}\,(Z_2) = 2\beta,\ \text{Var}\,(Z_3) = 3\beta \tag{2.97}$$

$\Rightarrow \qquad \text{Var}\,(Z_n) = n\beta$

Combining (2.93), (2.94), (2.95), (2.96) and (2.97), we get

$$\left. \begin{array}{l} \text{Var}\,(Z_n) = \dfrac{\beta\alpha^{n-1}(1-\alpha^n)}{1-\alpha},\ \text{provided } \alpha \neq 1 \\[2mm] \qquad\qquad = n\beta,\ \text{if } \alpha = 1 \end{array} \right\} \tag{2.98}$$

2.12.1 Some Results on Discrete Branching Process

We can easily see that the population process $\{Z_n; n = 0, 1, 2, ...\}$ conforms to a Markov Chain (M.C.).

Since $\qquad Z_{n+1} = X_1 + X_2 + ... + X_{Z_n}, \qquad \text{if } Z_n = 1, 2...$

$$= 0, \qquad\qquad\qquad \text{if } Z_n = 0$$

and X_i's are i.i.d.r.v.'s each having a common p.g.f. $\phi(s)$; the p.g.f. of

$$Z_{n+1} | Z_n = [\phi(s)]^{Z_n}$$

which entirely depends on Z_n. Hence $\{Z_n\}$ conforms to a simple Markov Chain.

Let $p_{ij} = P\{Z_{n+1} = j \,|\, Z_n = i\} = P\{Z_{n+1} = j \,|\, Z_n = i, Z_{n-1} = * ;\ Z_1 = *\}$

where $*$ indicates arbitrary values.

We have clearly $p_{0j} = \delta_{0j}$ (the Kronecker Delta) since the state 0 is absorbing.
Now let us denote $\quad P[X = j] = p_j, \quad \text{say} \quad j = 0, 1, 2...$
For p_{ij}, $i = 2, 3, 4, ...$, we have the p.g.f. of $[j \,|\, i\,]$ given by

$$\widetilde{\phi}_i(s) = \sum_{j=0}^{\infty} p_{ij}\, s^j \quad \text{for } |s| < 1 \tag{2.99}$$

Given $\qquad Z_n = i$ and $Z_{n+1} = X_1 + X_2 + ... + X_{Z_n},$ if $Z_n = 1, 2...$

$$= 0, \text{ if } Z_n = 0$$

and $\qquad\qquad X_i$'s are i.i.d.r.v.'s.

It follows that $\widetilde{\phi_i(s)} = E\,[s^{Z_{n+1}} | Z_n = i] = [\phi(s)]^i$. $\qquad\qquad$ (2.100)

Since $\qquad\qquad Z_{n+1} = X_1 + X_2 + \dots + X_{Z_n}$

$$E(s^{Z_{n+1}} | Z_n = i) = [E(s^{X_i})]^{Z_n} = [\phi(s)]^{Z_n} = [\phi(s)]^i$$

Hence comparing (2.99) and (2.100), it follows that p_{ij}, (i, j)th element of the transition matrix of the Markov chain is given by the coefficient of s^j on $[\phi(s)]^{Z_{n=i}}$. If $Z_0 = 1$, the probability that the population gets extinct on or before the nth generation is

$$p_{10}^{(n)} = P[Z_n = 0 | Z_0 = 1] = \pi_n \text{ (say)}$$

To observe the behaviour of π_n as n increases, we assume that,

$$0 < p_0 = p_{10} = \pi_1 \le 1$$

because if $p_0 = \pi_0 = 0$, no extinction is possible. Also the sequence $\{\pi_n\}$ is therefore bounded above and increasing. Hence $\{\pi_n\}$ must possess a limit. Let

$$\lim_{n \to \infty} \pi_n = \pi$$

$(\because \qquad \pi_1 = p_{10}^{(1)}, \ \pi_2 = p_{10}^{(2)} = p_{10}^{(1)} + P(Z_2 = 0 | Z_0 = 1, Z_1 \ne 0)$

$$\Rightarrow \pi_2 \ge \pi_1$$

Similarly $\qquad\qquad \pi_3 \ge \pi_2$ etc.)

$\Rightarrow \qquad\qquad \pi$ satisfies the equation $\pi = \phi(\pi)$ as $n \to \infty$

$(\because \quad \pi_n = \Pi_n(0) = \phi\{\Pi_{n-1}^{(0)}\} = \phi(\Pi_{n-1}))$ $\quad$ by (2.90)

Also $\qquad \phi(s) = E(s^X) \Rightarrow \phi\,(1) = 1$ satisfies $|s| = 1$

is always a solution of $\phi(\pi) = \pi$.

To search another solution, if exists say π^* of the same, let $\phi(\pi)$ be a power series in π with necessarily non-negative coefficients, (because the co-efficients represent probabilities and as such they should be non-negative). Therefore $\phi(\pi)$ is an increasing function of π $\forall \pi \in (0,1)$.

Again, we have for any non-zero solution π^* while denoting

$$\pi_1 = \phi(0), \ \pi_1 = \phi(0) < \phi(\pi^*) = \pi^* \text{ since } \pi^* > 0$$

Also

$$\pi_2 = \phi(\pi_1) < \phi(\pi^*) = \pi^*$$

since $\qquad\qquad \pi_1 < \pi^* \Rightarrow \phi(\pi_1) < \phi(\pi^*)$

ϕ being an increasing function of π in $0 < \pi < 1$.

Hence it follows that the chance of extinction given by π is the smallest positive root of the equation $\pi = \phi(\pi)$. Next, let us consider the case when the process starts with i number for females instead of 1. Obviously, it can be seen that chance of extinction is the probability that each of the i lines independently dies out; which is nothing but given by π^i where π represents the smallest root of the equation $\phi(\pi) = \pi$.

This leads to a simple iterative process of solution given by

$$\pi_{n+1} = \phi(\pi_n) \quad n = 1, 2, 3, \ldots$$

where π_n is the nth iterative solution and $\pi_1 = p_0$ satisfies the solution. This may be solved by successive iteration process using Newton Raphson method or the method of False Position.

2.13 Pure Death Process

It is analogous to birth process except that here $X(t)$ is decreased with increase in t.

Here

$$P[X(t) = k \mid X(0) = k_0] = p_k(t)$$

$$= \binom{k_0}{k} \exp\left[-k \int_0^t \mu(\tau)\, d\tau\right]\left[1 - \exp\left(\int_0^t \mu(\tau)\, d\tau\right)\right]^{k_0 - k} \tag{2.101}$$

where $\mu(t)$ is the intensity of a birth.

2.14 Birth and Death Processes

This is given by the Kolmogorov's equation

$$p_k(t + \Delta) = p_k(t)\{1 - (\lambda_k \Delta + \mu_k \Delta + o(\Delta))\}$$

$$+ p_{k-1}(t)\,\lambda_{k-1}(t)\cdot\Delta + p_{k+1}(t)\,\mu_{k+1}(t)\cdot\Delta + o(\Delta)$$

Here Kolmogorov's differential equations are

$$\left.\begin{array}{l}
\dfrac{d}{dt}\, p_0(t) = -[\lambda_0(t) + \mu_0(t)\, p_0(t) + \mu_1(t)\, p_1(t) \\[2mm]
\dfrac{d}{dt}\, p_k(t) = -[\lambda_k(t) + \mu_k(t)]\, p_k(t) + \lambda_{k-1}(t)\, p_{k-1}(t) \\[2mm]
\quad + \mu_{(k+1)}(t)\, p_{k+1}(t),\, k \geq 0
\end{array}\right\} \tag{2.102}$$

The initial conditions are $p_{k0}(0) = 1, p_k(0) = 0 \; \forall \; k \neq k_0$. The steady state solution is obtainable from Art. 2.9 when $r = k$.

2.14.1 Homogenous Birth and Death Process

Putting $\lambda_k(t) = k\lambda$ and $\mu_k(t) = k\mu$ in (2.102) we have,

$$\frac{\partial p_k}{\partial t} = -k(\lambda + \mu) p_k(t) + (k-1)\lambda p_{k-1}(t) + (k+1)\mu p_{k+1}(t)$$

and

$$\frac{\partial}{\partial t} p_0(t) = p_1(t)$$

p.g.f. $G_X(s; t)$ satisfies

$$\frac{\partial}{\partial t} G_X(s; t) + (\lambda s - \mu)(1 - s)) \frac{\partial}{\partial s} G_X(s; t) = 0$$

$\Rightarrow$ we have

$$\frac{dt}{1} = \frac{ds}{(\lambda s - \mu)(1 - s)} = \frac{dG_X(s; t)}{0}$$

as auxiliary equations of the partial differential equation.

$\Rightarrow$

$$G_X(s) = \alpha(t) \frac{[1 - (\alpha(t) + \beta(t)s]}{[1 - \beta(t)s]}$$

where

$$\alpha(t) = \frac{\mu(1 - e^{-(\lambda - \mu)t})}{\mu - \lambda e^{(\lambda - \mu)t}}$$

$$\beta(t) = \frac{\lambda(1 - e^{-(\lambda - \mu)t})}{\mu - \lambda e^{(\lambda - \mu)t}}$$

and

$$p_k(t) = \sum_{t=0}^{\min(\mu, k_0)} \binom{k_0}{1} \binom{k_0 + k - i - 1}{k - i} (\alpha(t))_{k-1}^{k_0 - i}$$

$$\cdot (\beta(t)) \cdot [1 - \alpha(t) + \beta(t)]^i \qquad (2.103)$$

2.15 Chiang's Illness Death Process

This process has two illness states S_1 and S_2;

S_1 = State of remaining ill, and

S_2 = State of remaining well

and r death states (absorbing) *viz*, $R_1, R_2, ..., R_\delta, ..., R_r$ ($\delta = 1, 2, ... r$).
   Two intensities have been defined $v_{\alpha\beta}$ and $\mu_{\alpha\delta}$, where

$$v_{\alpha\beta} \Delta + o(\Delta) = P \{\text{an individual in state } S_\alpha \text{ at time } \xi \text{ will be in state } S_\beta \text{ at time } \xi + \Delta\}$$

and

$$v_{\alpha\delta} \Delta + o(\Delta) = P \{\text{an individual in state } S_\alpha \text{ at time } \xi \text{ will be in state } R_\delta \text{ at time } \xi + \Delta\} \text{ where } \alpha \neq \beta$$

$$\alpha, \beta = 1, 2; \delta = 1, 2, ..., r$$

and
$$v_{\alpha\alpha} = -\left(v_{\alpha\beta} + \sum_{\delta=1}^{r} \mu_{\alpha\delta}\right)$$

Illness transition probability is defined as

$P_{\alpha\beta}(\tau, t) = P$ [an individual in state S_α at time τ will be in
state S_β at time t]; $\quad \alpha, \beta = 1, 2$

and the death transition probability is defined as

$Q_{\alpha\delta}(\tau, t) = P$ [an individual in state S_α at time τ will be in
state R_δ at time t]; $\alpha = 1, 2; \delta = 1, 2, ..., r$

Now initial conditions are

$$\left.\begin{array}{l} P_{\alpha\alpha}(\tau, \tau) = 1 \\[2mm] P_{\alpha\beta}(\tau, \tau) = 0 \end{array}\right\} \alpha, \beta = 1, 2 \\[4mm] Q_{\alpha\delta}(\tau, \tau) = 0; \; \alpha, \beta = 1, 2; \; \delta = 1, 2, .., r \right\} \tag{2.104}$$

The Kolmogorov equations are

$$P_{\alpha\alpha}(\tau, t+\Delta) = P_{\alpha\alpha}(\tau, t)[1 + v_{\alpha\alpha}\Delta + o(\Delta)]$$
$$+ P_{\alpha\beta}(\tau, t)[v_{\beta\alpha}\Delta + o(\Delta)] \tag{2.105}$$

and
$$P_{\alpha\beta}(\tau, t+\Delta) = P_{\alpha\beta}(\tau, t)[1 + v_{\beta\beta} + o(\Delta)]$$
$$+ P_{\alpha\alpha}(\tau, t)[v_{\alpha\beta}\Delta + o(\Delta)] \tag{2.106}$$

The process has been found to be homogeneous and the solution is of the form

$$P_{\alpha\alpha}(\tau, t) = P_{\alpha\alpha}(t-\tau) = P_{\alpha\alpha}(t) = \sum\left(\frac{\rho_i - v_{\beta\beta}}{\rho_i - \rho_j}\right)e^{\rho_i t} \tag{2.017}$$

$$P_{\alpha\beta}(\tau, t) = P_{\alpha\beta}(t-\tau) = P_{\alpha\beta}(t) = \sum\left(\frac{v_{\alpha\beta}}{\rho_i - \rho_j}\right)e^{\rho_i t} \tag{2.108}$$

where ρ_1 and ρ_2 satisfy the quadratic equation

$$\rho^2 - (v_{\alpha\alpha} + v_{\beta\beta})\rho + (v_{\alpha\alpha}v_{\beta\beta} - v_{\alpha\beta}v_{\beta\alpha}) = 0 \tag{2.109}$$

and $Q_{\alpha\delta}$ is given by

$$Q_{\alpha\delta} = \int_0^t P_{\alpha\alpha}(\tau)\mu_{\alpha\delta}\, d\tau + \int_0^t P_{\alpha\beta}(\tau)\mu_{\beta\delta}\, d\tau \tag{2.110}$$

Since an individual in illness state S_α may reach the state R_δ ($\delta = 1, 2, ...r$) directly from S_α or by way of S_β ($\beta \neq \alpha$), an individual in R_δ at time t may have reached that state at any time prior to t. Let us consider an infinitesimal time interval $(\tau, \tau + d\tau)$ for fixed τ, $0 < \tau \leq t$. The probability that an individual in state S_α at time '0' will reach the state R_δ in the interval $(\tau, \tau + d\tau)$ is

$$P_{\alpha\alpha}(\tau)\,\mu_{\alpha\delta}\,d\tau + P_{\alpha\beta}(\tau)\,\mu_{\beta\alpha}\,d\tau\;....(*)$$

As τ varies over the interval $(0, t)$ the corresponding events, whose probabilities given in (*) are mutually exclusive.

Example 2.6 Obtain the probability generating function (p.g.f.) of the simple Poisson process,

$$P[X = x\,|\,t] = \frac{e^{-\lambda t}(\lambda t)^x}{x!}$$

To obtain the p.g.f. of the process, we have (vide Art. 3.1)

$$\Pi_t(s) = s\int_0^t \Pi_{t-\tau}(s)\,f(\tau)\,d\tau + \int_t^\infty f(\tau)\,d\tau$$

Here $f(t) = \lambda e^{-\lambda t}$

$$\Rightarrow \qquad \Pi_t(s) = s\lambda \int_0^t \Pi_{t-u}(s)\,e^{-\lambda u}\,du + \lambda \int_t^\infty e^{-\lambda u}\,du$$

$$= s\lambda \int_0^t \Pi_{t-u}(s)\,e^{-\lambda u}\,du + \lambda\left.\frac{e^{-\lambda u}}{-\lambda}\right]_t^\infty$$

$$\Rightarrow \qquad \Pi_t(s)\,e^{\lambda t} = s\lambda \int_0^t \Pi_{t-u}(s)\,e^{-\lambda(u-t)}\,du + 1$$

Put $\qquad \Pi_t(s)\,e^{\lambda t} = A(t)$

$$\Rightarrow \qquad A(t) = s\lambda \int_0^t A(t-u)\,du + 1$$

Put $\qquad t - u = v$

$$\Rightarrow \qquad A(t) = s\lambda \int_0^t A(v)\,dv + 1$$

$$\therefore \qquad \frac{dA(t)}{dt} = s\lambda\left[\int_0^t \frac{d}{dt}(A(v))\,dv + \frac{dt}{dt}\,A(t)\right] = s\lambda\,A(t)$$

$$\Rightarrow \qquad \int_0^t \frac{dA(t)}{A(t)} = \int_0^t s\lambda\,dt$$

$$\Rightarrow \qquad \log A(t) = s\lambda t$$

$$\Rightarrow \qquad A(t) = e^{s\lambda t}$$

$\Rightarrow \qquad\qquad \Pi_t(s)\, e^{\lambda t} = e^{s\lambda t}$

$\Rightarrow \qquad\qquad \Pi_t(s) = e^{(s-1)\lambda t}$ is p.g.f. of the process

$$E(N_t) = U(t) = \left.\frac{d\Pi_t(s)}{ds}\right|_{s=1} = \left.\lambda t e^{\lambda t(s-1)}\right|_{s=1} = \lambda t$$

Again $\left.\dfrac{d^2\Pi_t(s)}{ds^2}\right|_{s=1} = E[N_t(N_t-1)] = E(N_t^{(2)})$, the second factorial moment

$$= \left.\lambda^2 t^2 e^{\lambda(s-1)}\right|_{s=1} = \lambda^2 t^2$$

$\therefore \qquad\quad \text{Var}(N_t) = E(N_t(N_t-1)) + E(N_t) - [E(N_t)]^2$

$$= \lambda^2 t^2 + \lambda t - (\lambda t)^2 = \lambda t$$

Therefore, $E(N_t) = \lambda t$ and $\text{Var}(N_t) = \lambda t$ for a Poisson process.

Example 2.7 *A linear growth model with immigration*

Such processes are important in the analysis of Population growth. Here we have $\lambda_n = \lambda n + \alpha$ and $\mu_n = \mu n$, λ_n represents the intensity of change in the positive direction as a result of instantaneous birth rate 'λ' applied over population size n and 'α' stands for immigration and μ_n represents the intensity of death in infinitesimal small period of time Δt μ_n given by

$$\mu_n\, \Delta t = n\mu\, \Delta t + 0(\Delta t)$$

The Kolmogorov equations become

$$p'_{i0}(t) = -\alpha p_{i0}(t) + \mu p_{i1}(t)$$

and $\qquad p'_{ij}(t) = [(\lambda(j-1)+\alpha)p_{i,j-1}(t) - ((\lambda+\mu)j+\alpha)p_{ij}(t)$

$$+ \mu(j+1)p_{i,j+1}(t)] \quad \text{for } j \geq 1$$

$\Rightarrow \displaystyle\sum_j s^j\, p'_{ij}(t)$

$$= \sum_j \lambda s^2(j-1)\, s^{j-2} p_{i,j-1}(t) + \sum_j \alpha s s^{j-1} p_{i,j-1}(t)$$

$$- \sum_j (\lambda+\mu)\, j\, s^j\, p_{ij}(t) - \alpha \sum_j s^j\, p_{ij}(t) + \mu(j+1)s^j p_{i,j+1}(t)$$

We set $p_j = 0\ \forall\, j < 0$ and carry on the summation from $j = -\infty$ to ∞ as follows.

Writing $\qquad\qquad \Pi_t(s) = \displaystyle\sum_{j=-\infty}^{\infty} p_{ij}(t)\, s^j$

$\Rightarrow \qquad \dfrac{\partial \Pi}{\partial t} = \lambda s^2 \dfrac{\partial \Pi}{\partial s} + \alpha s \Pi - (\lambda+\mu)s \dfrac{\partial \Pi}{\partial s} - \alpha \Pi + \mu \dfrac{\partial \Pi}{\partial s}$

i.e.,
$$\frac{\partial \Pi}{\partial t} + (\lambda s - \mu)(1 - s)\frac{\partial \Pi}{\partial s} = \alpha(s - 1)\Pi \qquad (2.111)$$

For this partial differential equation we have the auxiliary equations.

$$\frac{dt}{1} = \frac{ds}{(\lambda s - \mu)(1 - s)} = \frac{d\Pi}{-\alpha(1 - s)\Pi}$$

Assuming $\lambda \neq \mu$,

$$\Rightarrow \qquad (\lambda - \mu)\,dt = \frac{\lambda\,ds}{\lambda s - \mu} + \frac{ds}{1 - s} = d\log\left(\frac{\lambda s - \mu}{1 - s}\right)$$

$$\Rightarrow \qquad \frac{\lambda s - \mu}{1 - s}\, e^{-(\lambda - \mu)t} = \text{a constant, say, } c_1 \qquad (2.112)$$

and
$$\left(\frac{c}{\lambda}\right)\left(\frac{\lambda\,ds}{\lambda s - \mu}\right) = -\frac{d\Pi}{\Pi}$$

$$\Rightarrow \qquad (\lambda s - \mu)^{\alpha/\lambda}\,\Pi = \text{a constant, say, } c_2 \qquad (2.113)$$

Since partial differential equation of order one can have only one constant in its solution therefore c_1 and c_2 are functions of each other.

$\therefore$ We can write

$$(\lambda s - \mu)^{\alpha/\lambda}\,\Pi(s) = \phi\left\{\frac{\lambda s - \mu}{1 - s}\, e^{-(\lambda - \mu)t}\right\}$$

ϕ may be obtained from the initial conditions

i.e
$$\left.\begin{array}{c} Z_0 = i \text{ and } \Pi_0(s) = s^i, \\[2mm] r = \dfrac{\lambda s - \mu}{1 - s} \end{array}\right\} \qquad (2.114)$$

$$\therefore \qquad \phi(re^{-(\lambda - \mu)t}) = \left(\frac{\lambda(\mu + r)}{\lambda + r} - \mu\right)^{\alpha/\lambda}\Pi_t(s) \qquad (2.115)$$

For $t = 0$, $\Pi_0(s) = s^i$

$$\phi(r) = \left(\frac{\lambda(\mu + r)}{\lambda + r} - \mu\right)^{\frac{\alpha}{\lambda}}\left(\frac{\mu + r}{\lambda + r}\right)^i$$

$$\therefore \qquad \phi(re^{-(\lambda - \mu)t}) = \left(\frac{\lambda(\mu + re^{-(\lambda - \mu)t})}{\lambda + re^{-(\lambda - \mu)t}} - \mu\right)^{\frac{\alpha}{\lambda}}\left(\frac{\mu + re^{-(\lambda - \mu)t}}{\lambda + re^{-(\lambda - \mu)t}}\right)^i$$

$$= \left(\frac{\lambda(\mu + r)}{\lambda + r} - \mu\right)^{\frac{\alpha}{\lambda}}\Pi_t(s)$$

$\Rightarrow \qquad \Pi_t(s) = \left(\dfrac{\lambda(\mu + re^{-(\lambda - \mu)t})}{\lambda + re^{-(\lambda - \mu)t}} - \mu \right)^{\frac{\alpha}{\lambda}} \left(\dfrac{\mu + re^{-(\lambda - \mu)t}}{\lambda + re^{-(\lambda - \mu)t}} \right)^i .$

$$\left(\dfrac{\lambda(\mu + r)}{\lambda + r} - \mu \right)^{\frac{-\alpha}{\lambda}} \qquad (2.116)$$

Special cases:

$\qquad \lambda = \mu = 0 \Rightarrow$ Poisson process

$\qquad \lambda = 0 \Rightarrow$ Immigration-emigration process

$\qquad \alpha = \mu = 0 \Rightarrow$ Linear growth process or Yules' process

$\qquad \alpha = 0 \Rightarrow$ Linear birth and death process

One can find out $E(X(t))$ and Var $(X(t))$ from the p.g.f. $\Pi_t(s)$.

However, even without determining $\Pi_t(s)$ one can get from (2.111) by differentiating w.r. to 's' as follows

$$\Rightarrow \qquad \frac{\partial^2 \Pi}{\partial s\, \partial t} + (\lambda s - \mu)(1 - s)\frac{\partial^2 \Pi}{\partial s^2}$$

$$+ \{(\lambda s - \mu)(-1) + \lambda(1 - s) - \alpha(s - 1)\}\frac{\partial \Pi}{\partial s} = \alpha\Pi \qquad (2.117)$$

Now $\quad \Pi_t(1) = 1$

and $\qquad \dfrac{\partial \Pi_t(s)}{\partial s}\bigg|_{s=1} = E(X(t)) = M(t)$

Taking limit $s \to 1$

$$\frac{dM(t)}{dt} + (\mu - \lambda)M(t) = \alpha \qquad (2.118)$$

Using the initial condition

$\qquad M(t)\big|_{t=0} = E(X(0)) = k$

$$M(t) = \frac{\alpha}{\mu - \lambda} + \left(k - \frac{\alpha}{\mu - \lambda} \right) e^{-(\mu - \lambda)t}$$

$$= \frac{\alpha}{\lambda - \mu}[e^{(\lambda - \mu)t} - 1] + ke^{(\lambda - \mu)t} \qquad (2.119)$$

where $\lambda \neq \mu$.

Also

$$\frac{\partial^3 \Pi}{\partial^2 s\, \partial t}\bigg|_{s=1} = \frac{\partial}{\partial t}\left(\frac{\partial^2 \Pi}{\partial s^2} \right)\bigg|_{s=1}$$

$$= \frac{d}{dt}[E(X(t))(X(t) - 1)]$$

$$= \frac{d}{dt}[\text{Var}(X(t)) + E(X(t))^2 - E(X(t))]$$

$$= \frac{dV(t)}{dt} + (2M(t) - 1)\frac{dM(t)}{dt}$$

Differentiating (2.117) w.r. to s, we get

$$\frac{\partial^3 \Pi}{\partial^2 s\, \partial t} + (\lambda s - \mu)(1 - s)\frac{\partial^3 \Pi}{\partial s^3} + [\lambda(1 - s) + (\lambda s - \mu)(-1)]\frac{\partial^2 \Pi}{\partial s^2}$$

$$+ \frac{\partial^2 \Pi}{\partial s^2}\{(\lambda s - \mu)(-1) + \lambda(1 - s) - \alpha(s - 1)\}$$

$$+ \frac{\partial \Pi}{\partial s}\{(\lambda)(-1) - \lambda - \alpha\} = \alpha \frac{\partial \Pi}{\partial s} \qquad (2.120)$$

Proceeding to the limit $s \to 1$ on both sides of (2.120), we get

$$\frac{\partial^3 \Pi}{\partial s\, \partial t}\bigg|_{s=1} - (\lambda - \mu)\frac{\partial^2 \Pi}{\partial s^2}\bigg|_{s=1} + \frac{\partial^2 \Pi}{\partial s^2}\bigg|_{s=1}(\mu - \lambda) - \frac{\partial \Pi}{\partial s}\bigg|_{s=1}(\alpha + 2\lambda)$$

$$= \alpha \frac{\partial \Pi}{\partial s}\bigg|_{s=1}$$

$$\Rightarrow \qquad \frac{\partial^3 \Pi}{\partial s\, \partial t}\bigg|_{s=1} + (2\mu - 2\lambda)\frac{\partial^2 \Pi}{\partial s^2}\bigg|_{s=1} - \frac{\partial \Pi}{\partial s}\bigg|_{s=1}(2\alpha + 2\lambda) = 0.$$

$$\frac{dV(t)}{dt} + (2M(t) - 1)\frac{dM(t)}{dt} + 2(\mu - \lambda)[\text{Var}(X(t)) + E(X(t))^2 - E(X(t))]$$

$$- 2E(X(t))(\mu + \lambda) = 0$$

$$\Rightarrow \quad \frac{dV(t)}{dt} + (2M(t) - 1)\frac{dM(t)}{dt} + 2(\mu - \lambda)(V(t) + (M(t))^2 - M(t))$$

$$- 2M(t)(\alpha + \lambda) = 0.$$

Putting $\qquad M(t) = \frac{\alpha}{\lambda - \mu}[e^{(\lambda - \mu)t} - 1] + \mu e^{(\lambda - \mu)t}$

and $\qquad \frac{dM(t)}{dt} = \left(\frac{\alpha}{\lambda - \mu}\right)(\lambda - \mu)e^{(\lambda - \mu)t} + \mu(\lambda - \mu)e^{(\lambda - \mu)t}$

We get the first differential equation in $V(t)$ which when solved with initial condition $V(0) = 0$ gives Var $(X(t))$. This is left as an Exercise.

Example 2.8 Events occur at Poisson rate λ. Let N be the number of events in $(0, T)$ where T is a random variable. Obtain the p.g.f of N in terms of the characteristic function of T.

Solution: We have $E_{N\,|\,T\,=\,t}\,s^N = e^{-\lambda t(1-s)}$

$\therefore\qquad \phi_N(s) = E_N(s^N) \;=\; E_T\,\{E_{N\,|\,T\,=\,t}\,s^N\}$

$$= E_T\,\{\,e^{-\lambda\,t(1-s)}\}$$

$$= \int_0^\infty e^{-\lambda t(1-s)}\,f(t)\,dt$$

where $f(t)$ is the density function of T,

$$= \int_0^\infty e^{i\{i\lambda(1-s)\}}\,f(t)\,dt$$

$$= E_T\,\{\,e^{i\{i\lambda\,(1-s)T\}}\} \;=\; \phi_T\,\{i\lambda\,(1-s)\} \qquad (2.121)$$

where $\phi_T\,\{i\lambda(1-s)\}$ is the ch. function of the r.v. $[i\lambda\,(1-s)]$

Example 2.9   Two independent Poisson Processes have rates λ and μ respectively. Obtain the distribution of the r.v. $y(t) = Y$, the number of events in the first process which occur before the first event of the second process. The time T to the first event of the second process has c.f.

$$\phi_X(\theta) \;=\; \mu\int_0^\infty e^{i\theta t}e^{-\mu t}\,dt \;=\; \mu\int_0^\infty e^{-t(\mu-i\theta)}\,dt$$

$$= \mu\left[\frac{e^{-t(\mu-i\theta)}}{-(\mu-i\theta)}\right]_0^\infty \;=\; \frac{\mu}{\mu-i\theta}$$

Therefore, the probability generating function of the number of events in the first process upto $(0, T)$ is given by

$$G_T(s) \;=\; \phi_X\,\{i\lambda\,(1-s)\,\}$$

$$= \frac{\mu}{\mu-i(i\lambda(1-s))} \;=\; \frac{\mu}{\mu+\lambda(1-s)} \;=\; \frac{\mu}{\mu+\lambda-\lambda s}$$

$$= \frac{\lambda+\mu-\lambda}{\lambda+\mu-\lambda s} \;=\; \frac{1-\dfrac{\lambda}{\lambda+\mu}}{1-\dfrac{\lambda s}{\lambda+\mu}}$$

$$= \frac{\dfrac{\mu}{\mu+\lambda}}{1-\dfrac{\lambda s}{\mu+\lambda}} \;=\; \frac{\dfrac{\mu}{\mu+\lambda}}{1-s\left\{1-\dfrac{\mu}{\mu+\lambda}\right\}} \qquad (2.122)$$

References

1. Coleman Rodney (1974): *Stochastic Processes* George Allen and Unwin Ltd. London.
2. Chiang C.L. (1968): *Introduction to Stochastic Processes in Biostatistics,* John Wiley & Sons, New York.
3. Chiang C.L. (1980): *An Introduction to Stochastic Processes and their Applications,* Kreiger, New York.
4. Feller William (1951): *An Introduction to Probability Theory and its Applications,* Vol I, 3rd edition, John Wiley & Sons, New York.
5. Feller William (1968): *An Introduction to Probability Theory and its Applications,* Vol II, John Wiley & Sons
6. Karlin S. and Taylor H.M. (1975): *A First Course in Stochastic Processes,* Edition II, Academic Press, New York.
7. Ross M. and Sheldon H. (1983): *Stochastic Processes,* John Wiley & Sons New York, Chichester, Toronto and Singapore.
8. Takács L. (1957): *Stochastic Processes (Problems and Solutions)* - Translated by P. Zador, John Wiley & Sons, New York.

Non-Markov Process and Renewal Theory

3.0 Renewal Process

Let $S_n = X_1 + X_2 + ... + X_n$ represent the waiting time for the nth renewal, X_i represent the waiting time from the $(i-1)$th renewal to ith renewal. X_i's are independent random variables.

Now
$$P\,[S_n \le t] \; = \; P\,[N_t \ge n] \; = \; F_n\,(t) \tag{3.1}$$

where N_t represents the number of renewals in $(0, t)$ and $F_n\,(t)$ is the cumulative distribution function of S_n. Then

$$P\,[N_t = n] \; = \; P\,[N_t \ge n] - P\,[N_t \ge n + 1]$$
$$= \; F_n\,(t) - F_{n+1}\,(t)$$

$$E\,(N_t) \; = \; \sum_{n=0}^{\infty} n.P\,[N_t = n]$$

$$= \; \sum_{n=0}^{\infty} n\,[F_n\,(t) - F_{n+1}(t)]$$

$$= \; \sum_{n=1}^{\infty} F_n\,(t) \tag{3.2}$$

We define $E(N_t) = U(t)$ = renewal function or the average number of renewals in $(0, t)$

$$\Rightarrow \qquad \frac{dE\,(N_t)}{dt} \; = \; \frac{dU(t)}{dt} \; = \; \frac{d}{dt} \sum_{n=1}^{\infty} F_n\,(t) \; = \; \sum_{n=1}^{\infty} \frac{d}{dt} F_n\,(t)$$

$$= \; \sum_{n=1}^{\infty} f_n\,(t)$$

$$= \; u(t) \tag{3.3}$$

$u(t)$ is called the renewal density.

$u(t) dt = P[\text{a renewal will take place between } (t, t + dt)]$

and $\quad f_n(t)$ is the interval density for the nth renewal.

The Laplace transform of $u(t)$ is given by

$$L(u(t)) = \int_0^\infty e^{-st} u(t)\, dt = \int_0^\infty e^{-st} \sum_{n=1}^\infty f_n(t)\, dt$$

$$= \sum_{n=1}^\infty \int_0^\infty e^{-st} f_n(t)\, dt$$

$$= \sum_{n=1}^\infty L(f_n(t)) \tag{3.4}$$

Let the distribution of the waiting time X_1 be $f_1(t)$ and the common distribution X_i ($i = 2, 3, 4, ...$) be $f(t)$.

Then $\qquad\qquad f_n(t) = f_1(t) * [f(t)]^{*(n-1)} \tag{3.5}$

where $*$ stands for convolution.

$$L[f_n(t)] = L[f_1(t)]\, [L(f(t))]^{n-1} \tag{3.6}$$

$$L(u(t)) = \sum_{n=1}^\infty L(f_n(t)) = \sum_{n=1}^\infty L(f_1(t))\, [L(f(t)]^{n-1} \quad \text{from (3.4)}$$

$$= L(f_1(t)) \sum_{n=1}^\infty [L(f(t)]^{n-1}$$

$$= \frac{L(f_1(t))}{1 - L(f(t))} \tag{3.7}$$

Next $\qquad\qquad L(U(t)) = \int_0^\infty e^{-st} U(t)\, dt$

$$L(u(t)) = \int_0^\infty e^{-st} u(t)\, dt$$

$$= e^{-st} U(t)\, \Big|_0^\infty + s \int_0^\infty e^{-st} U(t)\, dt$$

$$= sL(U(t)) \quad (\because U(0) = 0) \tag{3.8}$$

From (3.7) and (3.8), we have

$$sL\,(U(t)) = \frac{L\,(f_1(t))}{1 - L\,(f(t))}$$

$$\Rightarrow \quad sL\,(U(t)) - sL\,(U(t))\,L\,(f(t)) = L\,(f_1(t))$$

$$\Rightarrow \quad L\,(U(t)) - L\,(U(t))\,L\,(f(t)) = \frac{1}{s}L\,(f_1(t)) \qquad (3.9)$$

Again $\quad L\,(f_1(t)) = \int\limits_0^\infty e^{-st} f_1(t)\,dt = e^{-st} F_1(t)\,\Big|_0^\infty + s\int\limits_0^\infty e^{-st} F_1(t)\,dt$

where $F_1(t)$ is the c.d.f. corresponding to the density function $f_1(t)$.

$$\Rightarrow \quad L\,(f_1(t)) = sL\,(F_1(t)) \qquad (3.10)$$

$$\frac{1}{s}L\,(f_1(t)) = L\,(F_1(t)) \qquad (3.11)$$

Putting (3.10) in (3.9)

$$\Rightarrow \quad L\,(U(t)) = L\,(U(t))\,L\,(f(t)) + L\,(F_1(t)) \qquad (3.11)$$

Taking Inverse Laplace transform on both sides $\Rightarrow$

$$U(t) = F_1(t) + \int\limits_0^t U\,(t - \tau)\,f(\tau)\,d\tau = F_1\,(t) + U(t) * f(\tau) \qquad (3.12)$$

which is the forward renewal equation.

Differentiating both sides with respect to $t \Rightarrow$

$$u(t) = f_1(t) + \int\limits_0^t u\,(t - \tau)\,f(\tau)\,d\tau \qquad (3.13)$$

(3.12) and (3.13) are the forward renewal equations.

3.1 Alternative Technique (Method of Probability Generating Function) for Forward and Backward Renewal Equations

Let τ be the time for the first renewal and N_t represent the number of renewals in $(0, t)$

If
> (i) $t < \tau$ then $N_t = 0$ clearly
> (ii) $t \geq \tau$ then $N_t = 1 + N_{t-\tau}$
$\qquad (3.14)$

Denote the probability generating function i.e. $E\,(s^{N_t})$ for any real variable N_t with $|s| < 1$ as $\Pi_t\,(s)$.

$$\begin{aligned} s^{N_t} &= 1 && \text{, if } t < \tau \\ s^{N_t} &= s^{N_{t-\tau}+1} && \text{, if } t \geq \tau \end{aligned} \Bigg\} \qquad (3.15)$$

$$\therefore \quad \Pi_t(s) = E(s^{N_t}) = E_\tau(E(s^{N_t}|\tau))$$

$$= \int_0^\infty E(s^{N_t}|\tau) f(\tau) \, d\tau$$

$$= \int_0^t E(s^{N_t}|\tau) f(\tau) \, d\tau + \int_t^\infty E(s^{N_t}|\tau) f(\tau) \, d\tau$$

$$= \int_0^t E(s^{N_{t-\tau}+1}) f(\tau) \, d\tau + \int_t^\infty f(\tau) \, d\tau$$

$$\Rightarrow \quad \Pi_t(s) = s \int_0^t \Pi_{t-\tau}(s) f(\tau) \, d\tau + \int_t^\infty f(\tau) \, d\tau \qquad (3.16)$$

which is a recurrence relation between $\Pi_t(s)$ and $\Pi_{t-\tau}(s)$.
Differentiating both sides w.r.t. s and putting $s = 1$

$$\Rightarrow \quad \frac{d\Pi_t(s)}{ds}\bigg|_{s=1} = \int_0^t \Pi_{t-\tau}(s)\bigg|_{s=1} f(\tau)\, d\tau + s\int_0^t \frac{d\Pi_{t-\tau}(s)}{ds}\bigg|_{s=1} f(\tau)\, d\tau$$

$$\Rightarrow \quad U(t) = F(t) + \int_0^t U(t-\tau) f(\tau)\, d\tau \qquad (3.17)$$

This equation is a special case of (3.12) when $f_1 = f$ and $F_1 = F$ and the process is a simple renewal whereas (3.12) corresponds to that of a general recurrent process.

Differentiating (3.17) with respect to $t \Rightarrow$

$$u(t) = f(t) + \int_0^t u(t-\tau) f(\tau)\, d\tau$$

which is again a special case of (3.13).

3.2 Distribution of Forward and Backward Recurrence Time

A. *Forward recurrence time:* Let S_n be the waiting time for the nth renewal and define

$$T_t^+ = S_n - t \quad \text{and} \quad T_t^- = t - S_{n-1}$$

then T_t^+ and T_t^- are called the forward and backward recurrence time respectively.

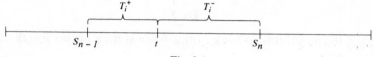

Fig. 3.1

The problem is to consider their joint distribution. Let $f(\cdot)$ and $u(\cdot)$ denote the interval and the renewal densities of the renewal process; and $g^+(\cdot), g^-(\cdot)$ denote the density of the forward and backward recurrence times respectively. Then

$$P[x \leq T_t^+ \leq x + dx] = g^+(x)\,dx$$

Further, let a renewal take place between v to $v + dv$ prior to t, (without any consideration of the backward recurrence time) with probability

$$u(v)\,dv, \quad 0 < v < t$$

$$P[T_t^+ \in (x, x + \Delta)] = g^+(x)\,\Delta + 0(\Delta)$$

$$= \{f(t+x)\,\Delta + 0(\Delta)\} + \int_0^t u(v)[\,f(t+x-v)\,\Delta + 0(\Delta)]\,dv$$

$$\Rightarrow \qquad g^+(x) = f(t+x) + \int_0^t u(v)\,f(t+x-v)\,dv \qquad (3.18)$$

Putting $\qquad u' = t - v$

$$\Rightarrow \qquad g^+(x) = f(t+x) + \int_0^t u(t-u')\,f(u'+x)\,du' \text{ holds} \qquad (3.19)$$

The first term on the right hand side gives the probability of a first renewal at $(t + x)$ and the second term gives the probability of a further renewal at $(t + x)$ given that the immediate past renewal took place at $v, 0 \leq v \leq t$. Note that (3.17) is true for fixed t.

As $\qquad t \to \infty, \quad u(t) \to \dfrac{1}{\mu} \,\forall\, t$

$$\Rightarrow \qquad g^+(x) \to g(x) = \int_0^\infty \frac{1}{\mu} f(u'+x)\,du' = \frac{1}{\mu} \int_x^\infty f(\omega)\,d\omega$$

$$= \frac{1 - F(x)}{\mu}$$

B. Backward recurrence time

We have the probability that there is one renewal between

$$(t - x) \text{ and } (t - x + dx) = u(t - x)\,dx$$

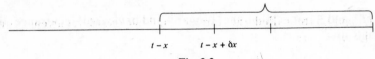

$$t - x \qquad t - x + \delta x$$

Fig. 3.2

Further, the probability of not having a renewal for a further length of time
$x = (1 - F(x))$

where
$$F(x) = \int_0^x f(t)\, dt,$$

the c.d.f. and $f(t)$ the interval density.

$\Rightarrow \qquad g^-(x)\, dx = u\,(t - x)\,(1 - F(x))\, dx,$

where $g^-(x)$ is the distribution of backward recurrence time.

$\Rightarrow \qquad\qquad g^-(x) = u(t - x)\,(1 - F(x))$ \hfill (3.20)

As $\quad t \to \infty, \quad u(t - x) \to \dfrac{1}{\mu} \Rightarrow \lim g^-(x) = \dfrac{1}{\mu}(1 - F(x))$ \hfill (3.20′)

which gives the distribution of the backward recurrence time as $t \to \infty$. One may
compare the results (3.20) with (3.19) as well as (3.20′).

Example 3.1

Let
$$f(t) = \lambda\, e^{-\lambda t}, \quad u(t) = \lambda$$

$$g^+(x) = \lambda e^{-\lambda(t + x)} + \lambda \int_0^t \lambda e^{-\lambda(t + x - v)}\, dv \quad \text{from (3.18)}$$

$$= \lambda e^{-\lambda(t + x)} + \lambda^2 \int_0^t e^{-\lambda(t + x - v)}\, dv$$

$$= \lambda e^{-\lambda(t + x)} + \frac{\lambda^2}{\lambda} \int_{t + x}^{x} e^{-\eta}\, d\eta, \text{ where } \eta = \lambda\,(t + x - v)$$

$$= \lambda\, e^{-\lambda\,(t + x)} - \lambda\, e^{-\lambda\,(t + x)} + \lambda\, e^{-\lambda x} = \lambda\, e^{-\lambda x}$$

$$\Rightarrow \quad E\,(T^+) = \int_0^\infty e^{-\lambda x}\, dx = \frac{1}{\lambda}$$

Note that $g^+(x) = \lambda e^{-\lambda x}$ does not depend on t; because of Loss of memory
property of the distribution. But in general $g^+(x)$ will depend not only on x but
also on value prior to t. This gives the renewal process a non Markovian
character.

Example 3.2 Let $u(t) = \lambda \ \forall \ t$ and is independent of t

$$f(x) = \lambda e^{-\lambda x}$$

$\Rightarrow \qquad 1 - F(x) = e^{-\lambda x}$

Then $\qquad g^-(x) = \lambda e^{-\lambda x}$

We have $\qquad E(T_t^-) = \int_0^t e^{-\lambda x} \, dx = \dfrac{1}{\lambda} - \dfrac{e^{-\lambda t}}{\lambda}$

and $\qquad E(T_t^+) + E(T_t^-) = \dfrac{2}{\lambda} - \dfrac{e^{-\lambda t}}{\lambda}$

whereas $E(T_t^+ + T_t^-)$ = Expected renewal time = $\dfrac{1}{\lambda}$ under Poisson input.

3.3   A Queueing Paradox

In Example 3.2 we have $E(T_t^+ + T_t^-) = \dfrac{2}{\lambda} - \dfrac{e^{-\lambda t}}{\lambda}$ for the waiting time distribution of a Poisson process. Where as the expectation of $(T_t^+ + T_t^-)$ being the mean inter-arrival time in a Poisson process should be equal to $\lambda \int_0^\infty t e^{-\lambda t} \, dt = \dfrac{1}{\lambda}$. How to explain this anomaly?

Note that, for $t \to \infty \quad E(T_t^+ + T_t^-) \to \dfrac{2}{\lambda}$, a value double of that of inter-renewal time in a Poisson process. In fact, as pointed out by Feller (1968), this arises due to sampling bias of choosing longer renewal intervals than the shorter one; this bias may even overestimate the renewal interval by double its actual length or measure.

By choosing an arbitrary point t and then measuring $(T_t^+ + T_t^-)$ to estimate the renewal interval leads to a higher likelihood of a lengthy renewal interval rather than a shorter interval. This phenomenon is known as 'length biased sampling' and occurs in a disguised form in a number of sampling situations.

3.4   Joint Distribution of the Forward and Backward Recurrence Time

Let T_t^+ and T_t^- be the forward and backward recurrence time respectively measured from an arbitrary point t.

The joint distribution of T_t^+ and T_t^- is obtained in the same way as the distribution of marginals derived earlier.

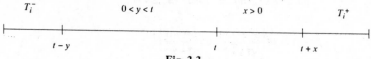

Fig. 3.3

In fact, for any $x > 0$ and $0 < y < t$ the event $E^* = [T_t^+ > x \cap T_t^- > y]$ occurs if there are no renewals in the intervals $(t-y, t+x)$ and this interval has probability that the time of a particular renewal, say, the nth renewal $Z_n > t + x$ given that the nth renewal has not taken place upto $(t - y)$.

$$P[T_t^+ > x \cap T_t^- > y] = \left\{ \frac{\overline{F}_n (t + x)}{\overline{F}_n (t - y)} \right\}, \quad n = 1, 2, 3 \qquad (3.21)$$

It is the probability that a renewal takes place at time $> t + x$ given that it has not taken place upto $(t - y)$.

For a Poisson process the same is given by

$$P[T_t^+ > x \cap T_t^- > y] = \frac{e^{-\lambda(t+x)}}{e^{-\lambda(t-y)}} = e^{-\lambda(x+y)} \qquad (3.22)$$

where

$$P[T_t^+ > x] = e^{-\lambda x} \qquad (3.23)$$

$$P[T_t^- > y] = e^{-\lambda y} \qquad (3.24)$$

$\therefore$

$$P[T_t^+ > x] P[T_t^- > y] = e^{-\lambda (x + y)} \qquad (3.25)$$

Thus for a Poisson process

$$P[T_t^+ > x \cap T_t^- > y] = P[T_t^+ > x] P[T_t^- > y] \qquad (3.26)$$

$\Rightarrow$ In Poisson process, distribution of forward and backward recurrence times are independent. In fact, it can be shown that this property is a characterization property of the process.

3.5 Asymptotic Renewal Theorem

We require to show

$$N_t \xrightarrow{P} \frac{t}{\mu} \Rightarrow \frac{N_t}{t} \xrightarrow{P} \frac{1}{\mu}$$

Proof: We have from (3.1)

$$P\{N_t \geq n\} = P\{S_n \leq t\}$$

where $\quad S_n = X_1 + X_2 + ... + X_n, \quad X_i$'s are i.i.d. r.v.'s

with $\quad E(X_i) = \mu$ and $\text{Var}(X_i) = \sigma^2$

$(3.1) \Rightarrow \quad P\{N_t < n\} = P\{S_n > t\}$

$\Rightarrow \quad 1 - P\{N_t \geq n\} = 1 - P\{S_n \leq t\}$

$$= 1 - P\left\{ \frac{S_n - n\mu}{\sigma \sqrt{n}} \leq \frac{t - n\mu}{\sigma \sqrt{n}} \right\}$$

$$= 1 - P\left\{ \frac{\left(\frac{S_n}{n} - \mu\right)}{\sigma/\sqrt{n}} \le \frac{\frac{t}{n} - \mu}{\sigma/\sqrt{n}} \right\}$$

Putting

$$\left(\frac{S_n - n\mu}{\sigma\sqrt{n}}\right) = Y_n$$

$$\Rightarrow \qquad 1 - P\{N_t \ge n\} = 1 - P\left\{Y_n \le \frac{t - n\mu}{\sigma\sqrt{n}}\right\} \qquad (3.27)$$

Since $\qquad S_n = X_1 + X_2 + ... + X_n, \; X_i\text{'s are i.i.d. r.v.'s}$

and $\qquad\qquad\qquad X_i \sim N(\mu, \sigma^2)$

$$\Rightarrow \qquad \left(\frac{S_n - n\mu}{\sigma\sqrt{n}}\right) = Y_n \overset{L}{\to} N(0, 1)$$

by Lindeberg Levy central limit theorem.

Also $\qquad\qquad n^{-1} S_n \overset{P}{\to} \mu \; \text{ as } n \to \infty$

By weak law of large numbers (WLLN)

$$\Rightarrow \qquad\qquad\qquad S_n \overset{P}{\to} n\mu$$

$$P\{N_t \ge n\} = P\{S_n \le t\}$$

$$\Rightarrow \qquad P\{N_t \ge n\} = P\{n\mu \le t\}$$

$$P\{N_t \ge n\} = P\left\{\frac{t}{\mu} \ge n\right\}$$

and $\qquad\qquad P\{N_t < n\} = P\left\{\frac{t}{\mu} < n\right\} \qquad \text{as } n \to \infty$

N_t behaves probabilistically as $\dfrac{t}{\mu}$

$$\Rightarrow \qquad \frac{N_t}{t} \overset{P}{\to} \frac{1}{\mu} \text{ with probability one. [Q. E. D.]}$$

We state another useful theorem known as "*Elementary Renewal Theorem*" (without proof).

$$E(N_t) \overset{P}{\to} \frac{t}{\mu} \text{ as } t \to \infty \qquad (3.28)$$

and $\qquad\qquad \dfrac{E(N_t)}{t} \overset{P}{\to} \dfrac{1}{\mu}$

where $\mu = E(X_n) < \infty$, the limit being interpreted as zero when $\mu = \infty$.

∴ Asymptiotic renewal density $u(t)$ tends to $\dfrac{1}{\mu}$ as $t \to \infty$.

3.6 Continuous Time Branching Process

Suppose that at time $t = 0$, the process starts with a single female birth, who may be assumed to be the founder of the generation and future female births are not considered over generations but over continuous time. Let Z_t be the population size at any time t. Let $\Pi_t(s)$ be the probability generating function (p.g.f) of Z_t. Let U be the time of founder's death (or the end of her fertility span). Let N be the number of female offsprings or children given by the founder. Both U and N are r.v.'s. Then we denote

$$E_N(s^N) = \phi(s)$$

Given that $U = u$, we have $\forall\ t < u$

$$Z_t = Z_0 = 1$$

while for $t > u$, we have,

$$Z_t = Z_{t-u}^{(1)} + Z_{t-u}^{(2)} + \ldots + Z_{t-u}^{(N)}, \quad \text{for } N = 1, 2, 3, \ldots$$

$$= 0, \quad \text{for } N = 0$$

where $Z_{t-u}^{(i)}$ represents the number of female births of the ith female during $(t - u)$ to t ($\forall\ t > u$) ($i = 1, 2, 3, \ldots, N$).

Each $Z_{t-u}^{(i)}$ are i.i.d.r.v's and distributed as Z_{t-u}. The p.g.f of Z_{t-u} is denoted as $\Pi_{t-u}(s)$.

Then,
$$E[s^{Z_t} \mid U = u < t] = E_N[E(s)^{N Z_{t-u}}] = E_N[E(s^{Z_{t-u}})^N]$$

$$= E_N[(\Pi_{t-u}(s))^N]$$

$$= \phi(\Pi_{t-u}(s))$$

Hence $\Pi_t(s) = E_{Z_t}(s^{Z_t})$

$$= E_U[E(s^{Z_t} \mid U = u)]$$

$$= \int_0^\infty f(u)\{E(s^{Z_t} \mid U = u)\,du, \quad \text{where } f(u) \text{ is the p.d.f. of } U.$$

$$= \int_0^t f(u)\,E(s^{Z_t} \mid U = u)\,du + \int_t^\infty f(u)\,E(s^{Z_t} \mid U = u)\,du$$

$$= \int_0^t f(u)\phi(\Pi_{t-u}(s))\,du + \int_t^\infty f(u)E(s)\,du \quad (\because Z_t = 1 \text{ if } t < u)$$

$$\Rightarrow \Pi_t(s) = \int_0^t f(u)\,\phi\,(\Pi_{t-u}(s))\,du + s\,(1 - F(t)) \qquad (3.29)$$

Example 3.3 An application of continuous time branching process (Biswas and Thomas (1992))

Let

(i) Z_t = population of AIDS patients at any time t ($t = 0$ represents the origin when a single AIDS patient immigrates in the population)

(ii) $\Pi_t(s) = E(s^{Z_t})$

(iii) U = a r.v. representing the time of detection/ isolation of an AIDS patient and $f(u)$ is the p.d.f. of U

$$= \lambda\,e^{-\lambda u}, \text{say} \quad 0 \le u < \infty$$
$$\text{and} \qquad \lambda > 0$$

(iv) N = No. of AIDS patients generated by infection from the initial immigrant before its isolation.

$$E_N(s^N) = \phi\,(s) \text{ say, and the p.d.f. of } N \text{ is } \Phi(N, t) = \frac{e^{-\mu t}(\mu t)^N}{N!}$$

To obtain $E(Z_t)$ and Var (Z_t).

We have from (3.29)

$$\Rightarrow \qquad \Pi_t(s) = \int_0^t f(u)\,\phi[\Pi_{t-u}(s)]\,du + s\,(1 - F(t))$$

Assuming N to be fixed initially,

$$\phi\,[\Pi_{t-u}(s)\,|\,N] = [\Pi_{t-u}(s)]^N$$
$$f(t) = \alpha e^{-\alpha t}$$

$$\Rightarrow \qquad \Pi_t(s) = \alpha \int_0^t e^{-\alpha u}\,[\Pi_{t-u}(s)]^N\,du + se^{-\alpha t}$$

$$\Rightarrow \qquad \Pi_t(s)\,e^{\alpha t} = \alpha \int_0^t e^{\alpha(t-u)}\,[\Pi_{t-u}(s)]^N\,du + s \qquad (3.30)$$

Differentiating (3.30) w.r.t 't' we have,

$$\frac{d\Pi_t(s)}{dt}\,e^{\alpha t} + \alpha\,e^{\alpha t}\,\Pi_t(s) = \alpha\,[e^{\alpha t}\,[\Pi_t(s)]^N]$$

$$\Rightarrow \qquad \frac{d\Pi_t(s)}{dt} + \alpha\,\Pi_t(s) = \alpha\,[\Pi_t(s)]^N \qquad (3.31)$$

$$\Rightarrow \quad [\Pi_t(s)]^{-N}\,\frac{d\Pi_t(s)}{dt} + \alpha\,[\Pi_t(s)]^{N-1} = \alpha \qquad (3.32)$$

Putting $\qquad\qquad [\Pi_t(s)]^{-N+1} = V$

$\Rightarrow \qquad\qquad \dfrac{dV}{dt} - \alpha\,(N-1)\,V = -\alpha\,(N-1)$ $\qquad\qquad$ (3.33)

$\therefore \qquad\qquad\qquad\qquad$ I.F. $= e^{-\alpha(N-1)t}$

Multiplying (3.33) by I.F. $\Rightarrow$

$$e^{-\alpha(N-1)t}\,\frac{dV}{dt} - V\alpha\,(N-1)\,e^{-\alpha(N-1)t} = -\alpha\,(N-1)\,e^{-\alpha(N-1)t}$$

$\Rightarrow \qquad\qquad \dfrac{d}{dt}\,[V\,e^{-\alpha(N-1)t}] = -\alpha\,(N-1)\,e^{-\alpha(N-1)t}$

$\Rightarrow \qquad\qquad V\,e^{-\alpha(N-1)t} = e^{-\alpha(N-1)t} + c$

$$\text{where } c \equiv \text{const. of integration.}$$

Putting $t = 0$ and noting

$$\Pi_t(s)\big|_{t=0} = \Pi_0(s) = s \qquad\qquad (3.34)$$

$\Rightarrow \qquad\qquad c = s^{1-N} - 1$

$\Rightarrow \qquad\qquad \Pi_t(s) = [1 + (s^{1-N} - 1)\,e^{\alpha(N-1)t}]^{\frac{1}{(1-N)}} \qquad (3.35)$

$$E\,(Z_t\,|\,N) = \frac{d\Pi_t(s)}{ds}\bigg|_{s=1} \qquad\qquad (3.36)$$

$$= e^{\alpha(N-1)t}$$

$$E\,(Z_t^2\,|\,N) = \frac{d^2\Pi_t(s)}{ds^2}\bigg|_{s=1} + E\,(Z_t\,|\,N)$$

$$= e^{\alpha(N-1)t}\,[N\,e^{\alpha(N-1)t} - N]\,N\,e^{\alpha(N-1)t}\,[e^{\alpha(N-1)t} - 1] + e^{\alpha(N-1)t}$$

$$\text{Var}\,(Z_t\,|\,N) = E\,(Z_t^2\,|\,N) - [E\,(Z_t\,|\,N)]^2$$

$$= (N-1)\,e^{\alpha(N-1)t}\,[e^{\alpha(N-1)t} - 1] \qquad (3.37)$$

$$E\,(Z_t) = E_N(E\,(Z_t\,|\,N))$$

$$= \sum_{N=1}^{\infty} e^{\alpha(N-1)t}\,e^{-\mu t}\,\frac{(\mu t)^N}{N!}$$

$$= e^{-t(\alpha+\mu)} \sum_{N=0}^{\infty} e^{\alpha N t}\,\frac{(\mu t)^N}{N!}$$

$$= e^{-t(\alpha+\mu)} \sum_{N=0}^{\infty} \frac{[e^{\alpha t}\,(\mu t)]^N}{N!}$$

$$= e^{-t(\alpha+\mu)} e^{e^{\alpha t}(\mu t)}$$

$$= e^{-[\alpha+\mu(1-e^{\alpha t})]t} \tag{3.38}$$

Again

$$\text{Var}(Z_t) = E_N[\text{Var}(Z_t \mid N)] + \text{Var}_N[E(Z_t \mid N)]$$

$$E_N[\text{Var}(Z_t \mid N)] = E[(N-1) e^{\alpha(N-1)t} [e^{\alpha(N-1)t} - 1]]$$

$$= E[(N-1) e^{2\alpha(N-1)t}] - E[(N-1) e^{\alpha(N-1)t}]$$

$$= E[N e^{2\alpha(N-1)t}] - E[e^{2\alpha(N-1)t}] - E[N e^{\alpha(N-1)t}]$$

$$+ E[e^{\alpha(N-1)t}]$$

We have $E[N e^{2\alpha(N-1)t}] = \sum_{N=0}^{\infty} N e^{2\alpha(N-1)t} e^{-\mu t} \cdot \dfrac{(\mu t)^t}{N!}$

$$= \mu t \, e^{-\mu t [1-e^{2\alpha t}]}$$

$$E[N e^{\alpha(N-1)t}] = \mu t \, e^{-\mu t [1-e^{2\alpha t}]}$$

$$E[e^{2\alpha(N-1)t}] = e^{-[2\alpha+\mu(1-e^{2\alpha t})]t}$$

$$E_N[\text{Var}(Z_t \mid N)] = \mu t e^{-\mu t[1-e^{2\alpha t}]} - e^{-[2\alpha+\mu[1-e^{2\alpha t}]]t}$$

$$- \mu t e^{-\mu t[1-e^{\alpha t}]} + e^{-[\alpha+\mu(1-e^{\alpha t})]t} \tag{3.39}$$

$$\text{Var}[E(Z_t \mid N)] = \text{Var}[e^{\alpha(N-1)t}]$$

$$= E[e^{2\alpha(N-1)t}] - [E(e^{\alpha(N-1)t})]^2$$

$$= e^{-[2\alpha+\mu[1-e^{2\alpha t}]]t} - e^{-2[\alpha+\mu(1-e^{\alpha t})]t} \tag{3.40}$$

$$\text{Var}(Z_t) = \mu t \, e^{-\mu t[1-e^{2\alpha t}]} - \mu t \, e^{-\mu t[1-e^{\alpha t}]} - e^{-[\alpha+\mu(1-e^{\alpha t})]t}$$

$$- e^{-2[\alpha+\mu(1-e^{\alpha t})]t} \tag{3.41}$$

Example 3.4 Obtain $\Pi_t(s)$ when $N = 2$ (the no. of offsprings) and the waiting time for detection is negative exponential with parameter λ.

Solution $\quad \Pi_t(s) = \int_0^t f(u) \, \phi[\Pi_{t-u}(s)] \, du + s(1 - F(t))$ (from 3.29)

$$= \int_0^t \lambda e^{-\lambda u} [\Pi_{t-u}(s)]^N \, du + s e^{-\lambda t}$$

$$= \int_0^t \lambda e^{-\lambda u} [\Pi_{t-u}(s)]^2 \, du + s e^{-\lambda t}, \quad \text{for } N = 2$$

$$\Rightarrow \qquad \Pi_t(s)\, e^{\lambda t} = \int_0^t \lambda\, e^{-\lambda(u-t)} \, [\Pi_{t-u}(s)]^2 \, du + s \tag{3.42}$$

Put $\qquad t - u = V \Rightarrow -du = dV$

$$\Rightarrow \qquad \Pi_t(s)\, e^{\lambda t} = \int_0^t \lambda\, e^{\lambda V} \, [\Pi_V(s)]^2 \, dV + s$$

$$\Rightarrow \qquad \frac{d\Pi_t(s)}{dt}\, e^{\lambda t} + \lambda\, e^{\lambda t}\, \Pi_t(s) = \lambda\, e^{\lambda t} \, [\Pi_t(s)]^2$$

$$\Rightarrow \qquad \frac{d\Pi_t(s)}{dt} + \lambda\, \Pi_t(s) = \lambda \, [\Pi_t(s)]^2$$

$$\Rightarrow \qquad \frac{d\Pi_t(s)}{dt} = \lambda\, \Pi_t(s)\, (\Pi_t(s) - 1)$$

$$\Rightarrow \qquad \lambda\, dt = \frac{d\Pi_t(s)}{\Pi_t(s)\,(\Pi_t(s)-1)} = \left(\frac{1}{\Pi_t(s)-1} - \frac{1}{\Pi_t(s)} \right) d\Pi_t(s)$$

$$\Rightarrow \qquad \lambda t + c = \log \left\{ \frac{\Pi_t(s)-1}{\Pi_t(s)} \right\}$$

$$\Pi_0(s) = s \Rightarrow c = \log \frac{s-1}{s} \Rightarrow \Pi_t(s) = \frac{s\, e^{-\lambda t}}{1-(1-e^{-\lambda t})\, s} \tag{3.43}$$

$$\Rightarrow X(t) - 1 \sim G(e^{-\lambda t}) \tag{3.44}$$

3.7   Diffusion Process

Diffusion processes belong to the class of Markov Process $\{X_t : t \in T\}$ with both parameter space as well as state space continuous while a small change in t leading to a small change in X_t (or $X(t)$). A realization of the process can be thought of a system, moving quite erratically in a continuous medium while its progress depends only on its current position.

An important class of diffusion processes is known as "*Gaussian process*". If $\{X(t) : t \in T\}$ is a process whose state space is real line $R^{(1)}$ (one dimensional) and the parameter space is a subset of $R^{(1)}$ then it is known as a Gaussian process if $\forall$ finite n the joint distribution of $(X(t_1), ..., X(t_n))$ is multivariate normal. Again a Gaussian process is *stationary* if its covariance function viz. Cov $(T_i, T_j) = g(T_i, T_j) = g(T_i - T_j)$. Finally, the Brownian Motion (Wienner Process) with parameter λ $(\lambda > 0)$ $\{X(t); t \in R\}$ is the Gaussian process with independent increments with

(i) $E(X(t)) = 0$

(ii) $g(T_i, T_j) = \lambda \min (T_i, T_j)$.

The Brownian motion is useful in several biological situations concerning stochastic models especially in diffusion theory of Populations models.

Example 3.5   An example in Brownian Motion:

If $\{X_t\}$ is a Brownian motion with parameter θ then prove that $\{Z_t\} = \{e^{-\lambda t} \times e^{c\lambda t}\}$ is a stationary Gaussian Process and determine its covariance function where c is a constant.

Since X_t for a given t is normally distributed random variable, so is $\alpha(t) \times \beta(t)$ where $\alpha(t)$ and $\beta(t)$ are real constants, the process Z_t is, therefore, Gaussian.

Now
$$E(Z_t) = E(e^{-\lambda t} X_{e^{c\lambda t}}) = e^{-\lambda t} E(X_{e^{c\lambda t}})$$

$$= 0$$

as
$$E(X_t) = 0 \quad \forall\, t.$$

Therefore,

$$\text{Cov}(Z_t, Z_{t+\tau})$$

$$= E(Z_t, Z_{t+\tau}) - \underbrace{E(Z_t)}_{=0}\ \underbrace{E(Z_{t+\tau})}_{=0}$$

$$= E\left[e^{-\lambda t} X_{e^{c\lambda t}}\, e^{-\lambda(t+\tau)} X_{e^{c\lambda(t+\tau)}}\right]$$

$$= e^{-\lambda(ct+\tau)}\, \text{Cov}\left[X_{e^{c\lambda t}}, X_{e^{c\lambda(t+\tau)}}\right]$$

$$= \theta\, e^{-\lambda(ct+\tau)} \min\left(e^{c\lambda t}, e^{c\lambda(t+\tau)}\right)$$

$$= \theta\, e^{-\lambda|\tau|}$$

Clearly, this does not depend on t. Therefore, it's a stationary Gaussian Process.

Remarks:

These processes can arise as the limiting form of a Random walk.

Let us consider the unrestricted random walk; and assume that the particle takes independent infinitesimal steps of length Δx to the right with probability p and to the left with probability $q = 1 - p$ after short time intervals of length Δt. We may assume that during a time interval of length t, the displacement X_t consists of approximately $\dfrac{t}{\Delta t}$ independent steps.

Therefore
$$E(X_t) \cong \frac{1}{\Delta t}(p-q)\,\Delta x = (p-q)t\,\frac{\Delta x}{\Delta t}$$

$$\text{Var}(X_t) \cong \frac{1}{\Delta t}4pq\,(\Delta x)^2 = 4pq\,t\,\frac{(\Delta x)^2}{\Delta t}$$

If we allow Δx and $\Delta t \to 0$ so that

$$\frac{(\Delta x)^2}{\Delta t} = \theta \quad \text{and}$$

$$p = \frac{1}{2} + \frac{\Pi}{2\theta}\,\Delta x \quad \text{and} \quad q = \frac{1}{2} - \frac{\Pi}{2\theta}\,\Delta x$$

where Π and θ are constants,

then $\qquad E(X_t) \rightarrow \Pi\, t$ and $\mathrm{Var}(Z_t) \rightarrow \theta t$

$\therefore$ By Central Limit Theorem

$$(X_t - \Pi t)/\sqrt{\theta t} \;\sim\; N(0, 1).$$

References

1. Bailey N.T.J. (1963): *Elements of Stochastic Processes and their Applications in Natural Sciences*. John Wiley & Sons. London, New York and Sydney.
2. Biswas S. (1988): *Stochastic Processes in Demography and Applications,* Wiley Eastern & Co.
3. Biswas S. and Thomas M. (1992): *On a Continuous Branching Process Model for Testing the Efficacy of Some Control Programmes for AIDS* under publication in *Biometric* Journal, Vol 37, No:1.
4. Coleman Rodney (1974): *Stochastic Processes*. George Allen and Unwin Limited, London.
5. Chiang C.L. (1968): *Introduction to Stochastic Processes in Biostatistics*. John Wiley & Sons, New York.
6. Chiang C.L. (1980): *An Introduction to Stochastic Processes and their Applications*. Kreiger, New York.
7. Doob D.L. (1953): *Stochastic Processes*, Wiley.
8. Feller William (1951): *An Introduction to Probability Theory and its Applications*. Vol I, 3rd edition, John Wiley & Sons.
9. Karlin S. and Taylor H.M. (1957): *A First Course in Stochastic Processes,* Edition II, Academic Press, New York.

Martingales

4.0 Introduction

Martingales Theory is very useful in some of the problems on population models (like epidemic models) and competing risk theory; especially in finding out stopping rules for which optional sampling theorem and optimal stopping theorems, (an outline of which are given in this section) may be employed.

Below, we outline the basic concepts of martingales and catalogue some of those results which are useful in applications in Stochastic Population models.

Definition: A stochastic process $\{X_n : n = 0, 1, 2, 3, ...\}$ defined on discrete parameter space is a Martingale if $\forall\ n = 0, 1, 2, 3, ...\}$

(i) $E(|X_n|) < \infty$

(ii) $E\{X_{n+1} | X_0, X_1, X_2, ..., X_n\} = X_n$ (4.1)

hold.

A more general definition: Let $\{X_n : n = 0, 1, 2, 3, ...\}$ and $\{Y_n : n = 0, 1, 2, ...\}$ are two related stochastic processes. We say $\{X_n\}$ is Martingale with respect to $\{Y_n\}$ if $\hspace{3cm} \forall\ n = 0, 1, 2, 3, ...$

(i) $E(|X_n|) < \infty$

(ii) $E\{X_{n+1} | Y_0, Y_1, ..., Y_n\} = X_n$ (4.2)

One may well imagine $\{Y_0, Y_1, .., Y_n\}$ as the information or history upto the stage n.

A result for a Martingale:

$$E(X_n) = E(X_0) \text{ holds for } \forall\ n$$

This follows from $E(X_{n+1}) = E(E(X_{n+1} | Y_0, Y_1, ... Y_n))$

$$= E(X_n) \text{ since } \{X_n\} \text{ is a Martingale.}$$

4.1   Examples on Martingales

Example 4.1.1   Let $Y_0 = 0$ and $E(Y_i) = 0$ subject to $E(|Y_i|) < \infty$ $(i = 1, 2 ... n)$ and Y_i's are i. i.d.r.v's, then $X_n = Y_0 + .. + Y_n \Rightarrow \{X_n\}$ is a Martingale.

Proof. Y_i's are i.i.d. r.v's . $(i = 1, 2, ... n)$

$$E(Y_i) = 0, \quad Y_0 = 0$$

$$\therefore \quad E(X_{n+1} \mid Y_0 \dots Y_n)$$

$$= E(\underbrace{Y_0 + \dots + Y_n}_{= X_n} + Y_{n+1} \mid Y_0 \, Y_1 \dots Y_n)$$

$$= E(Y_{n+1} \mid Y_0 \, Y_1 \dots Y_n) + X_n$$

$$= E(Y_{n+1}) + X_n \quad (\because Y_{n+1} \text{ is independent of } Y_0 \, Y_1 \dots Y_n)$$

$$= X_n$$

$\Rightarrow \{X_n\}$ is a Martingale w.r. to $\{Y_0 \, Y_1 \dots Y_n\}$

Example 4.1.2 *(Martingales from Markov Chain)*

If $Y_0 \, Y_1 \dots Y_n \dots$ be a Markov Chain (M.C.) whose transition matrix P has the characteristic root λ and the corresponding right eigen vector f then if we denote

$$X_n = \lambda^{-n} f(Y_n), \quad n = 0, 1, 2, \dots$$

where $f(Y_n)$ represents the element in the Y_n^{th} row of the column vector f, then X_n is a Martingale.

Proof: $\{Y_n\}$ is a Markov Chain

$$E[f(Y_{n+1} \mid Y_n)] = \sum_{j \in s} P_{Y_{n,\, j}} f(j) \dots \quad ; s \equiv \text{state space} \tag{4.3}$$

$P = [p_{ij}]$ is the transition matrix of the M.C.

Again $\qquad\qquad \lambda f = Pf$

$$\Rightarrow \qquad\qquad \lambda f(i) = \sum_{j \in s} p_{ij} \, f(j)$$

$$\Rightarrow \qquad\qquad \lambda f(Y_n) = \sum_{j \in s} p_{Y_{n,\, j}} \, f(j) \tag{4.4}$$

Comparing (4.3) and (4.4)

$$\Rightarrow \qquad\qquad E[f(Y_{n+1} \mid Y_n)] = \lambda f(Y_n) \tag{4.5}$$

Again $\qquad E[X_{n+1} \mid Y_0, Y_1 \dots, Y_n]$

$$= E[X_{n+1} \mid Y_n] \text{, by the property of Markovity}$$

$$= E[\lambda^{-n-1} f(Y_{n+1}) \mid Y_n]$$

$$= \lambda^{-n-1} E[f(Y_{n+1} \mid Y_n)] = \lambda^{-n-1} \lambda f(Y_n) \quad \text{from (4.5)}$$

$$= \lambda^{-n} f(Y_n) = X_n$$

$\Rightarrow \{X_n\}$ is a Martingale w.r. to $\{Y_0, Y_1 \dots, Y_n \dots\}$

Example 4.1.3 *(Branching process)*

Let $\{Y_n\}$ be a branching process given by

$$Y_{n+1} = Z^{(n)}(1) + Z^{(n)}(2) + \dots + Z^{(n)}(Y_0)$$

representing the number of offsprings in the $(n + 1)^{th}$ generation; where $Z_n(i)$ represents the number of offsprings of the ith individual in the nth generation and $Z_n(i)$'s are i.i.d r.v.s with $E(Z_n(i)) = m \ \forall \ i$ in the nth generation. Then $X_n = m^{-n} Y_n$ is a Martingale.

Proof: $E[X_{n+1} | Y_0, Y_1 ..., Y_n]$

$$= E[m^{-n-1} Y_{n+1} | Y_0, Y_1 ..., Y_n]$$

$$= m^{-n-1} E[Y_{n+1} | Y_n]$$

$$= m^{-n-1} Y_n \cdot m = m^{-n} Y_n = X_n$$

Hence $\{X_n\}$ is a Martingale w.r. to $\{Y_0, Y_1 ..., Y_n\}$.

Example 4.1.4 (*Uniform Distribution*)

If Y_n be uniformly distributed in $[0, 1]$ and let $[Y_{n+1} | Y_n]$ be distributed uniformly in $(Y_n, 1)$, then

$$X_n = 2^n (1 - Y_n) \text{ is a Martingale.}$$

Proof:
$$Y_n \xrightarrow{U} (0,1)$$

$$Y_{n+1} | Y_n \xrightarrow{U} (Y_n, 1)$$

then to show $X_n = 2^n (1 - Y_n)$ is a Martingale.

$$E[X_{n+1} | Y_0, Y_1 ..., Y_n] = E[X_{n+1} | Y_n]$$

$$= E[2^{n+1}(1 - Y_{n+1}) | Y_n]$$

$$= 2^{n+1} - 2^{n+1} E(Y_{n+1} | Y_n)$$

$$= 2^{n+1} - 2^{n+1} \frac{(1 + Y_n)}{2}$$

$$\left(\because E(Y_{n+1} | Y_n) = \frac{1}{1 - Y_n} \int_{Y_n}^{1} Y_{n+1} \, dY_{n+1} = \frac{1}{1 - Y_n} \frac{1 - Y_n^2}{2} = \frac{1 + Y_n}{2} \right)$$

$$= 2^{n+1} \left[1 - \frac{1 + Y_n}{2}\right] = 2^{n+1} \frac{(1 - Y_n)}{2} = 2^n (1 - Y_n) = X_n$$

Hence $\{X_n\}$ is a Martingale.

Example 4.1.5 (*Likelihood Ratio as a Martingale*)

Let $Y_0, Y_1,... Y_n$ .. be i.i.d.r.v.s
and f_0 and f_1 be the p.d.f under H_0 (null hypothesis) and H_1 (alternate hypothesis) and define

$$X_n = \frac{f_1(Y_0) f_1(Y_1) ... f_1(Y_n)}{f_0(Y_0) f_0(Y_1) ... f_0(Y_n)} \quad n = 0, 1, 2, ..$$

where $f_0(Y_i) > 0$ and $f_1(Y_i) > 0 \ (i = 0, 1, 2...n)$

$$\Rightarrow \{X_n\} \text{ is a Martingale.}$$

Proof: We have

$$E(X_{n+1} \mid Y_0 \ldots Y_n) = E\left\{ \frac{f_1(Y_{n+1})}{f_0(Y_{n+1})} X_n \middle| Y_0 \ldots Y_n \right\}$$

$$= E\left\{ \frac{f_1(Y_{n+1})}{f_0(Y_{n+1})} X_n \middle| X_n \right\}$$

($\because X_n$ is given since $Y_0, Y_1 \ldots, Y_n$ are given and Y_{n+1} is independent of Y_0, Y_1, $\ldots, Y_n$)

$$= X_n \int \frac{f_1(Y_{n+1})}{f_0(Y_{n+1})} f_0(Y_{n+1}) \, dY_{n+1}$$

$$= X_n, \text{ identically}$$

$\Rightarrow \{X_n\}$ is a Martingale.

Example 4.1.6   (*Wald's Martingale*)

Let $\phi(\lambda) = E[e^{\lambda Y_k}]$ exists for some $\lambda \neq 0$
where $\{Y_k\}$'s are i.i.d.r.v.'s

$$(k = 1, 2, 3 \ldots) \text{ and } Y_0 = 0$$

then, $X_n = (\phi(\lambda))^{-n} e^{\lambda(Y_1 + Y_2 + \ldots + Y_n)}$

is a Martingale w.r. to $S_n = Y_1 + Y_2 + \ldots + Y_n$

Proof: We note that $S_n = Y_1 + Y_2 + \ldots + Y_n$ conforms to a Markov processes.

i.e. $P(S_n \mid S_1, S_2 \ldots, S_{n-1}) = P(S_n \mid S_{n-1})$

since Y_i's are i.i.d.r.v.'s.

We maintain that $\phi(\lambda)$ is an eigen value of Markov process $S_n = Y_1 + Y_2 + \ldots + Y_n$ and $f(y) = e^{\lambda y}$ being the corresponding eigen function.

To prove the same, it is sufficient to show that

$$\phi(\lambda) f(x) = \int f(y) \, dF(y - x)$$

is satisfied by $f(y) = e^{\lambda y}$.
On substitution of $f(y) = e^{\lambda y}$, L.H.S. of the above becomes

$$= \phi(\lambda) e^{\lambda x} \text{ and the R.H.S.} = \int e^{\lambda y} \, dF(y - x)$$

Putting $y - x = \xi$ and noting for given X, distribution of Y is same as that of ξ, we have

$$\text{R.H.S.} = e^{\lambda x} \int e^{\lambda \xi} \, dF(\xi)$$

$$= e^{\lambda x} E[e^{\lambda y}] = e^{\lambda x} \phi(\lambda) = \text{L.H.S.}$$

Hence $\phi(\lambda)$ is an eigen value corresponding to the eigen function $f(y) = e^{\lambda y}$

Therefore, $\qquad X_n = [\phi(\lambda)]^{-n} e^{\lambda(Y_1 + Y_2 + \dots + Y_n)}$

is a Martingale w.r. to the Markov process formed by the partial sum $S_n = Y_1 + Y_2 + \dots + Y_n$ (vide Example (4.1.2)).

This establishes that $X_n = [\phi(\lambda)]^{-n} e^{\lambda(Y_1 + Y_2 + \dots + Y_n)}$ is a Martingale w.r. to $Y_1 + Y_2 + \dots + Y_n = S_n$, which is known as Wald's Martingale.

Example 4.1.7 (*Radon Nikodym derivative as a Martingale*)

Let $\qquad\qquad Z \xrightarrow{U} [0, 1] \quad (U \equiv$ uniformly distributed)

and Y_n be a sequence of r.v.'s,

$$Y_n = \frac{K}{2^n}, \qquad n = 1, 2, \dots$$

and choose k (uniquely) so that

$$Y_n = \frac{k}{2^n} \le Z < \frac{k+1}{2^n} = Y_n + 2^{-n}$$

As n increases Y_n provides more and more information about Z.

i.e. $\qquad\qquad \left[\lim_{n \to \infty} Y_n = Z \right].$

Next let f be a bounded function in $[0, 1]$ and let us define

$$X_n = \frac{f(Y_n + 2^{-n}) - f(Y_n)}{2^{-n}}$$

(Note, in the limit X_n as $n \to \infty$ it becomes the derivative of f at Y_n, provided the limit exists; X_∞ is called Radon Nikodym Derivative of f evaluated at Z).

It can be proved that $\{X_n\}$ is a Martingale w.r. to $\{Y_n\}$.
We may note that the conditional distribution of Z given $Y_0, Y_1 \dots, Y_n$ has a uniform distribution in $(Y_n, Y_n + 2^{-n})$.

$\therefore \quad E(Y_{n+1} \mid Y_n)$

$$= 2^{(n+1)} \int_{Y_n}^{Y_n + 2^{-(n+1)}} Y_{n+1}\, dY_{n+1} = 2^{(n+1)} \left[\frac{(Y_n + 2^{-(n+1)})^2}{2} - \frac{Y_n^2}{2} \right]$$

$$= \frac{2^{(n+1)}}{2} [(Y_n + 2^{-(n+1)} - Y_n)(2Y_n + 2^{-(n+1)})]$$

$$= \frac{2^{(n+1)}}{2} [2^{-(n+1)} (2) (Y_n + 2^{-n})]$$

$$= \frac{1}{2} [Y_n + (Y_n + 2^{-n})] = \frac{1}{2} Y_n + \frac{1}{2}(Y_n + 2^{-n})$$

This shows that Y_{n+1} can take the value of Y_n or $Y_n + 2^{-n}$ with equal probability $1/2$.

Also, $E(X_{n+1} | Y_0 Y_1 \dots Y_n)$

$$= 2^{n+1} E[f(Y_{n+1} + 2^{-(n+1)}) - f(Y_{n+1}) | Y_0 \dots Y_n]$$

$$= 2^{n+1} \left[\frac{1}{2} (f(Y_n + 2^{-(n+1)}) - f(Y_n)) \right.$$

$$\left. + \frac{1}{2} (f(Y_n + 2^{-(n+1)} + 2^{-(n+1)}) - f(Y_n + 2^{-(n+1)})) \right]$$

$$= 2^{n+1} \left[\frac{1}{2} (f(Y_n + 2^{-(n+1)}) - f(Y_n)) \right.$$

$$\left. + \frac{1}{2} [(f(Y_n + 2^{-n}) - f(Y_n + 2^{-(n+1)})] \right]$$

$$= \frac{2^{n+1}}{2} [f(Y_n + 2^{-n}) - f(Y_n)]$$

$$= 2^n [f(Y_n + 2^{-n}) - f(Y_n)]$$

$$= X_n$$

Again $$X_n = \frac{f(Y_n + 2^{-n}) - f(Y_n)}{2^{-n}}$$

is approximately the derivative of f at Z. In fact, it can be proved that under a fairly general conditions $\lim_{n \to \infty} X_n = X_\infty(Z)$ is the derivative of function of a random variable Z.

Example 4.1.8   *(Doob's Martingale)*

Let $Y_0, Y_1 \dots$ be a sequence of r.v.'s and suppose X be another, so that $E[|X|] < \infty$.

Then $X_n = E[X | Y_0, \dots Y_n]$ is a Martingale w.r. to $\{Y_n\}$.
This is called Doob's Martingale.

Proof: $E[|X_n|] = E[E(X | Y_0, \dots, Y_n)]$

$$\leq E[E(|X| | Y_0, \dots, Y_n)]$$

$$= E(|X1|) < \infty.$$

Again

$$E[X_{n+1} | Y_0, Y_1 \dots, Y_n]$$

$$= E[E[X | Y_0, \dots, Y_{n+1} | Y_0, Y_1 \dots, Y_n]]$$

$$= E[X | Y_0, \dots, Y_n] = X_n$$

which shows that $\{X_n\}$ is a Martingale.

Example 4.1.9 (*Generation of Martingales by inverse of linear functions*)
Consider a linear function $Y_{n+1} = l_{n+1}(z) = a_n + b_n z$; $b_n \neq 0$,

$$\Rightarrow \qquad l_{n+1}^{-1}(Y_{n+1}) = z = \frac{Y_{n+1} - a_n}{b_n}$$

Similarly $\quad Y_n = l_n(z) = a_{n-1} + b_{n-1} z$; $\quad b_{n-1} \neq 0$

$$\Rightarrow \qquad l_n^{-1}(Y_n) = \frac{Y_n - a_{n-1}}{b_{n-1}}$$

$$\Rightarrow \qquad l_n^{-1}(l_{n+1}^{-1}(Y_{n+1})) = l_n^{-1}\left(\frac{Y_{n+1} - a_n}{b_n}\right)$$

$$= \frac{\dfrac{Y_{n+1} - a_n}{b_n} - a_{n-1}}{b_{n-1}}$$

is a linear function of Y_{n+1} involving (a_n, b_n) and (a_{n-1}, b_{n-1}).
Proceeding in this way

$$l_1^{-1}(l_2^{-1} \dots l_n^{-1}(l_{n+1}^{-1}(Y_{n+1}))) \text{ is a linear function of}$$

$$Y_{n+1}, (a_n, b_n), (a_{n-1}, b_{n-1}), \dots (a_1, b_1)$$

Let us define $\qquad l_1^{-1}(l_2^{-1} \dots l_n^{-1}(l_{n+1}^{-1}(Y)))$

$$= L_{n+1}(Y)$$

$$\Rightarrow \qquad l_1^{-1}(l_2^{-1} \dots l_{n-1}^{-1} l_n^{-1}(l_{n+1}^{-1}(Y_n)))$$

$$= l_1^{-1} \dots l_n^{-1}(Y_n) = L_n(Y_n)$$

$$\Rightarrow \qquad L_{n+1}(l_{n+1}(Y_n)) = L_n(Y_n)$$

Then it can be shown immediately that $X_n = KL_n(Y_n)$ is a Martingale w.r. to $[Y_0, Y_1 \dots, Y_n]$ for arbitrary constant K.

Proof: $\quad E[X_{n+1} | Y_0, \dots, Y_n]$

$$= E[KL_{n+1}(Y_{n+1}) | Y_0, \dots Y_n]$$

$$= KL_{n+1}[E(Y_{n+1} | Y_0, Y_1 \dots, Y_n)]$$

$$= KL_{n+1}[E(a_n + b_n Y_n) | Y_n] ;$$

$$L_{n+1} \text{ being a linear function}$$

$$= KL_{n+1}(l_{n+1}(Y_n)) \quad \text{where } l_{n+1}(Y) = a_n + b_n Y$$

$$= KL_n(Y_n)$$

$$= X_n$$

$\Rightarrow \{X_n\}$ is a Martingale.

Illustration:

Let $Y_0 \sim U(0, 1)$ (uniformly distributed in $(0, 1)$)

$$Y_1 | Y_0 \sim U(Y_0, 1)$$

$$Y_2 | Y_1 \sim U(Y_1, 1)$$

$$\dots\dots\dots\dots\dots\dots\dots$$

$$Y_n | Y_{n-1} \sim U(Y_{n-1}, 1)$$

then $X_n = 2^n (1 - Y_n)$ is a Martingale w.r. to $Y_0, Y_1, \dots, Y_n$.

Proof:

We have $f(y_0) = 1$ $; 0 \le y_0 \le 1$

$$f(y_1 | Y_0) = \frac{1}{1 - Y_0} \qquad ; Y_0 \le y_1 \le 1$$

$$f(y_2 | Y_1) = \frac{1}{1 - Y_1} \qquad ; Y_1 \le y_2 \le 1$$

$$\vdots \qquad\qquad \vdots \qquad\qquad \vdots$$

$$f(y_n | Y_{n-1}) = \frac{1}{1 - Y_{n-1}} \qquad ; Y_{n-1} \le y_n \le 1$$

Therefore, $E(Y_1 | Y_0) = \int_{Y_0}^{1} y_1 f(y_1 | Y_0) \, dy_1 = \dfrac{(1 - Y_0^2)/2}{(1 - Y_0)} = \dfrac{1 + Y_0}{2}$

$$= \frac{1}{2} + \frac{Y_0}{2}$$

Thus if we write

$$E(Y_1 | Y_0) = a_0 + b_0 Y_0 \qquad\qquad \Rightarrow a_0 = 1/2, b_0 = 1/2$$

$$E(Y_2 | Y_1) = a_1 + b_1 Y_1 \qquad\qquad \Rightarrow a_1 = 1/2, b_1 = 1/2$$

$$E(Y_n | Y_{n-1}) = a_{n-1} - b_{n-1} Y_{n-1} \qquad \Rightarrow a_{n-1} = 1/2, b_{n-1} = 1/2$$

Let $y_1 = l_1(Z) = a_0 + b_0 Z$

$$= \frac{1}{2} + \frac{1}{2} Z$$

$$\Rightarrow \qquad Z = l_1^{-1}(y_1) = \frac{y_1 - a_0}{b_0} = \frac{y_1 - \dfrac{1}{2}}{\dfrac{1}{2}}$$

$$= (2y_1 - 1)$$

$$y_2 = l_2(Z) = a_1 + b_1 Z$$

$\Rightarrow$
$$I_2^{-1}(y_2) = \frac{y_2 - a_1}{b_1} = \frac{y_2 - \frac{1}{2}}{\frac{1}{2}}$$

$$= (2y_2 - 1)$$

$\therefore$
$$I_1^{-1}(I_2^{-1}(y_2)) = I_1^{-1}(2y_2 - 1)$$

$$= 2(2y_2 - 1) - 1$$

$$= 2^2(y_2 - 1) + 1$$

Also $\quad I_1^{-1}(I_2^{-1}(I_3^{-1}(y_3)))$

$$= 2^3 y_3 - (2^3 - 1) = 2^3(y_3 - 1) + 1$$

$\Rightarrow \quad I_1^{-1}(I_2^{-1}(I_3^{-1} \ldots (I_n^{-1}(y_n))))$

$$= 2^n y_n - (2^n - 1)$$

$$= 2^n(y_n - 1) + 1$$

$\Rightarrow \qquad X_n = [2^n(y_n - 1) + 1] \quad$ is a Martingale.

Again $\quad X_n' = (1 - X_n) = 2^n(1 - Y_n)$ can be proved to be a Martingale.

$$E(X_{n+1}' \mid Y_0 \ldots Y_n) = E(1 - X_n \mid Y_0 \ldots Y_n)$$

$$= 1 - X_n = X_n'$$

Hence $\qquad X_n' = 2^n(1 - Y_n)$ is a Martingale.

4.2 Sub Martingales and Super Martingales

Let $\{X_n \, ; n = 0, 1, 2, \ldots\}$ and $\{Y_n \, ; n = 0, 1, 2, \ldots\}$ be two Stochastic Processes. Then $\{X_n\}$ is called *Super Martingales* w.r.t. $\{Y_n\}$ if $\forall \, n$,

   (i) $\quad E[X_n^-] > -\infty$ where $x^- = \text{Inf} \{x, 0\}$

   (ii) $\quad E[X_{n+1} \mid Y_0, Y_1, \ldots, Y_n] \leq X_n \qquad\qquad (4.6)$

   (iii) $\quad X_n$ is a function of $Y_0, Y_1 \ldots, Y_n$

Similarly $\{X_n\}$ is called a *Sub Martingale* w.r.t. $\{Y_n\}$ if

   (i) $\quad E[X_n^+] \leq \infty$ where $x^+ = \text{Sup} \{x, 0\}$

   (ii) $\quad E[X_{n+1} \mid Y_0, Y_1, \ldots, Y_n] \geq X_n$

   (iii) $\quad X_n$ is a function of $Y_0, Y_1 \ldots, Y_n \qquad\qquad (4.7)$

If $\{X_n\}$ is a Super Martingale w.r. to $\{Y_n\}$ then $\{-X_n\}$ is a Sub Martingale w.r. to $\{Y_n\}$. Similarly, if $\{X_n\}$ is a Martingale w.r.t. $\{Y_n\}$, then it is a Sub as well as a Super Martingale.

4.3 Convex Functions and Martingales

If ϕ is a Convex function* and if $\{X_n\}$ is a Martingale w.r.t. Y_n
$\Rightarrow \{\phi(X_n)\}$ is a Sub Martingale w.r. to Y_n provided $[\phi(X_n^+)] < \infty \ \forall \ n$

The corresponding results by passing $\{X_n\}$ to $- \{X_n\}$ holds for Super martingales.

4.4 Markov Time (or Stopping Time)

Let $Y_0, Y_1 ......, Y_i ....$ be the pay-off of a gambler at the first, second..... ith .... trials of the game respectively. Assuming the game is fair $E(Y_i) = 0 \ \forall \ i = 1, 2.....$ Then

$$X_n = Y_0 + Y_1 + ... + Y_n \quad (Y_0 = 0)$$

is the total amount in the hand of gambler after the nth trial of the game. Now if $\sum_{i=0}^{n} Y_i$ goes below a certain level, he may decide to leave the game. On the other hand he may still like to leave the game even if $\sum_{i=0}^{n} Y_i$ is sufficiently high. X_n is a Martingale under the above assumption.

In case he wants to leave the game at time $T = n$, T is called the stopping time or Markov time of the Martingale.

4.4.1 Examples of Markov Time

Example (i)[*] Constant time $T = k$ is a Markov time.
Because for $\{Y_n\}$ if we define

$$I(Y_0, Y_1 ..., Y_n) = 1 \qquad ; \quad \text{if } n = k$$
$$\{T = n\} = 0 \qquad ; \quad \text{if } n \ne k$$

the condition is fulfilled.

Example (ii)   The first time the process $Y_0, Y_1...$ reaches a subset A of the state space is a Markov time.

i.e. $$T(A) = \text{Min } \{n : Y_n \in A\}$$

Here $$T_{(A) = n} \{ Y_0 \ Y_1 \ ... \ Y_n \} = 1 \text{ if } Y_j \notin A, j = 0, 1, 2 ... n - 1$$
$$= 0, \text{ otherwise}$$

---

*i.e. for two points $x_1, x_2 \in I$, an interval
$\alpha\phi(x_1) + (1 - \alpha) \phi (x_2) \ge \phi(\alpha x_1 + (1 - \alpha)x_2)$ holds for $0 < \alpha < 1$
and $\ \forall \ x_1, x_2 \in I$, an interval.
or *for n points* in an interval I viz $x_1 \ x_2 \ ... \ x_n$

and $$\alpha_0 \ge 0, \sum_{i=1}^{n} \alpha_i = 1$$

$$\sum_{i=1}^{n} \alpha_i \ \phi(x_i) \ge \phi\left(\sum_{i=1}^{n} \alpha_i \ x_i \right) \text{ holds}$$

4.5 Optional Sampling Theorem

Suppose $\{X_n\}$ is a Martingale and T is a Markov time w.r.t. $\{Y_n\}$

Then
$$E[X_0] = E[X_{T \wedge n}]$$

$$= \lim_{n \to \infty} E[X_{T \wedge n}] \qquad (4.8)$$

If $\quad T < \infty \quad$ then $\lim_{n \to \infty} X_{T \wedge n} = X_T$

Actually $\qquad X_{T \wedge n} = X_T \quad$ wherever $\quad n > T.$

Thus whenever we can justify the interchange of limit $n \to \infty$ and expectations exist, the following result holds.

$$E[X_0] = \lim_{n \to \infty} E[X_{T \wedge n}]$$

$$= E[\lim_{n \to \infty} X_{T \wedge n}] = E[X_T] \qquad (4.9)$$

4.6 Optional Stopping Theorem

Let $\{X_n\}$ be a Martingale and T be a Markov time.
If (i) $P\{T < \infty\} = 1$
  (ii) $E[|X_T|] < \infty$
  (iii) $\lim_{n \to \infty} E[X_n I_{(T > n)}] = 0$

then
$$E[X_T] = E[X_0], \qquad (4.10)$$

where $I_{(T > n)}$ represents the indicator function for $T > n.$

Proof of Optional Stopping Theorem
We require the proofs of following Lemmas.

Lemma I Let $\{X_n\}$ be a Martingale (super) and T be denoted as Markov time with respect to a related process $\{Y_n\}$

Then $\qquad\qquad \forall \ n \geq k$

$$E[X_n I_{(T = k)}] \leq E[X_k I_{(T = k)}] \qquad (4.10.1)$$

The inequality holds if $\{X_n\}$ is a Super Martingale.
Proof: We have

$$E[X_n I_{(T = k)}] = E[E[X_n I_{(T = k)} (Y_0 Y_1 \dots Y_n) | Y_0 Y_1 \dots Y_k]$$

$$= E[I_{(t = k)} E(X_n | Y_0 Y_1, \dots Y_k)]$$

$$\leq E[I_{(T = k)} X_k] \quad (\because E(X_n | Y_0, \dots, Y_k)$$

if $\{X_n\}$ is a Super Martingale,
otherwise $E(X_n | Y_0. Y_i .. Y_k) = X_k$ for a Martingale

Lemma II If $\{X_n\}$ is a (super) Martingale and T is a Markov time, then $\forall \; n = 1, 2, 3, ...$

$$E[X_0] \geq E[X_{T \wedge n}] \geq E[X_n]$$

the inequality holds in the case of Super Martingale and the equality holds for a Martingale.

Proof: We have,

$$E[X_{T \wedge n}] = \sum_{k=0}^{n-1} E[X_T I_{(T=k)}] + E[X_n I_{(T \geq n)}]$$

$$E[X_{T \wedge n}] = E[X_1] + E[X_2] + ... + E[X_{n-1}]$$
$$\text{when } T < n \text{ i.e. } T \leq n - 1$$

$$E[X_{T \wedge n}] = E[X_n I_{(T \geq n)}] \text{ when } T \geq n).$$

$$= \sum_{k=0}^{n-1} E[X_k I_{(T=k)}] + E[X_n I_{(T \geq n)}]$$

$$(\because X_T = X_k \text{ when } T = k),$$

Again under the conditions of Lemma I (4.10.1)

$$E[X_n I_{(T=k)}] \leq E[X_k I_{(T=k)}]$$

$$\Rightarrow \quad E[X_k I_{(T=k)}] \geq E[X_n I_{(T=k)}]$$

$$E[X_{T \wedge n}] \geq \sum_{k=0}^{n-1} \underbrace{E[X_n I_{(T=k)}]}_{=0} + E[X_n I_{(T \geq n)}]$$

(Since, $[X_n I_{(T \geq n)}] = X_n$ when $T = n$)

$$= E[X_n]$$

Again for a Martingale $E[X_n] = E[X_0]$
which completes the proof in the case of Martingale. For a Super Martingale, it has already been shown that

$$E[X_{T \wedge n}] \geq E[X_n]$$

and now we require to establish

$$E[X_0] \geq E[X_{T \wedge n}]$$

Now let us define the sequence

$$\tilde{X}_n = \sum_{k=1}^{n} \{X_k - E[X_k \mid Y_0, Y_1, ..., Y_{k-1}]\}$$

$$(\because \; E[E(X_k \mid Y_0 \, Y_1 \, ... \, Y_{k-1})] = X_k)$$

$\tilde{X}_n$ is a Martingale. Also $\tilde{X}_0 = 0$,

$$E(\tilde{X}_n) = 0$$

$$\therefore \quad E(\tilde{X}_{T\wedge n}) = 0 = E\left[\sum_{k=1}^{T\wedge n} \{X_k - E[X_k \mid Y_0, ..., Y_{k-1}]\}\right]$$

Again $\qquad E[X_k \mid Y_0 ... Y_{k-1}] \leq X_{k-1}$

the inequality holds for a Super Martingale.

$$E(\tilde{X}_{T\wedge N}) = 0 \geq E\left[\sum_{k=1}^{T\wedge n} \{X_k - X_{k-1}\}\right]$$

$$= E[X_1 - X_0 + X_2 - X_1 + ... + X_{T\wedge n} - X_{T\wedge n-1}]$$

$$= E[X_{T\wedge n} - X_0]$$

$$\therefore \qquad E[X_{T\wedge n} - X_0] \leq 0$$

$$\therefore \qquad E[X_0] \geq E[X_{T\wedge n}]$$

which completes the proof.

Lemma III   Let W be an arbitrary random variable satisfying $E[\mid W \mid] < \infty$ and T is stopping time for which

$$P\{T < \infty\} = 1$$

$$\Rightarrow \qquad \lim_{n \to \infty} E[W \, I_{(T > n)}] = 0$$

$$\lim_{n \to \infty} E[W \, I_{(T \leq n)}] = E(W)$$

Proof:

We have $\qquad E[\mid W \mid] \geq E[\mid W \mid I_{(T \leq n)}]$

$$= \sum_{r=0}^{n} E[\mid W \mid \mid T = r] P[T = r]$$

$$= E[\mid W \mid]$$

Hence $\qquad \lim_{n \to \infty} E[\mid W \mid I_{(T \leq n)}] = E[\mid W \mid]$

Also we have $\qquad |E[W] - E[W \, I_{(T \leq n)}]| \geq 0$

$$\Rightarrow \qquad |E[W \, I_{(T > n)}]| \geq 0$$

i.e. $\qquad 0 \le |E[WI_{(T>n)}]| \le |E(W|I_{(T>n)})|$

But $\qquad\qquad E(W|I_{(T>n)}) \to 0$ as $n \to \infty$

$\Rightarrow \qquad E(W|I_{(T>n)}) \le |E(WI_{(T>n)})| \le 0$ as $n \to \infty$

$\Rightarrow \qquad \lim_{n \to \infty} E(W I_{(T>n)}) = 0 \qquad$ (Proved)

Proof of the Main Theorem. It may be noted that the condition $E\{|X_T|\} < \infty$ must hold independently and not merely the condition viz. $E[|X_n|] < \infty \ \forall \ n$. Now $\forall \ n$, we have

$$E[X_T] = E[X_T I(T \le n)] + E[X_T I_{(T>n)}]$$
$$= E[X_{T \wedge n}] - E[X_n I_{(T>n)}] + E[X_T I_{(T>n)}]$$

By Lemma I, $E[X_{T \wedge n}] = E[X_0]$

Also by the condition (iii) of the theorem

$$E[X_n I_{(T>n)}] = 0.$$

Also by putting $W = X_T$ the third term on the R.H.S. above $E[X_T I_{(T>n)}] = E[W I_{(T>N)}] = 0$ as $n \to \infty$ by Lemma III.

$\therefore \qquad\qquad E[X_T] = \lim_{n \to \infty} E[X_{T \wedge n}] = E(X_0)$

This proves optional stopping theorem

Example (iii) *A problem of random walk on drug trials :* Let Y_i's are i.i.d. r.v.'s representing the response of the ith patient w.r.t. a certain drug. We have

$$Y_i = 1 \text{ for success, } Y_i = -1 \text{ for a failure and } Y_0 = 0.$$

$\therefore \qquad\qquad P\{Y_i = 1\} = p, \quad P\{Y_i = -1\} = q$

$\qquad\qquad\qquad S_n = Y_1 + \dots + Y_n, \quad S_0 = 0, \ n \ge 1$

A stopping rule is given by stopping time $T \ni$

$$T = \min_n \left[\sum_{i=1}^{n} Y_i = -b, \sum_{i=1}^{n} Y_i = a \right]; \text{ where } a \text{ and } b \text{ are positive integers,}$$

so that a drug is discontinued if $\sum_{i=1}^{n} Y_i = -b$ is obtained first and continued in the next sample if $\sum_{i=1}^{n} Y_i = a$, then obtain the probability that the drug is discontinued and the expected duration of the same. Here we have

$$S'_n = (Y_1 - E(Y_1)) + (Y_2 - E(Y_2)) + \dots + (Y_n - E(Y_n)) \text{ is a Martingale over } \{Y_n\}$$

and $S''_n = \left(\frac{q}{p}\right)^{S_n}$ is also a Martingale.

We have $\qquad E(Y_1) = E(Y_2) = \ldots\ldots = E(Y_n)$

$$= p \times 1 + q \times (-1) = p - q$$

By optional stopping theorem

$$E(S'_n) = E\left(\sum_{i=1}^{n} Y_i - n(p-q)\right) = E(S'_0) = 0$$

$$E(S'_T) = \pi(-b) + (1-\pi)\,a = E(T)(p-q)$$

$$-\pi(a+b) + a = E(T)(p-q)$$

$$\pi = \frac{a - E(T)(p-q)}{(a+b)} \qquad\qquad (4.11)$$

If $\qquad q = p \Rightarrow \pi = \frac{a}{a+b}.$

Again $\qquad E(S''_T) = \pi\left(\frac{q}{p}\right)^{-b} + (1-\pi)\left(\frac{q}{p}\right)^{a}$

$$= E(S''_0) = \left(\frac{q}{p}\right)^{S_0} = 1$$

$\therefore \qquad\qquad \pi\left(\frac{q}{p}\right)^{-b} + (1-\pi)\left(\frac{q}{p}\right)^{a} = 1$

$\Rightarrow \qquad\qquad \pi\left[\left(\frac{q}{p}\right)^{-b} - \left(\frac{q}{p}\right)^{a}\right] = 1 - \left(\frac{q}{p}\right)^{a}$

$\Rightarrow \qquad\qquad \pi = \dfrac{1 - \left(\frac{q}{p}\right)^{a}}{\left(\frac{q}{p}\right)^{b} - \left(\frac{q}{p}\right)^{a}} \qquad\qquad (4.12)$

Equating (4.11) and (4.12) we get $E(T)$ as,

$$E(T) = \frac{a - E(T)(p-q)}{(a+b)} = \frac{1 - \left(\frac{q}{p}\right)^{a}}{\left(\frac{q}{p}\right)^{b} - \left(\frac{q}{p}\right)^{a}}$$

$\Rightarrow \qquad [a - E(T)(P-q)] = \dfrac{(a+b)\left[1 - \left(\frac{q}{p}\right)^{a}\right]}{\left(\frac{q}{p}\right)^{b} - \left(\frac{q}{p}\right)^{a}}$

$$\Rightarrow \quad \left\{ \frac{a - (a+b)\left[1 - \left(\frac{q}{p}\right)^a\right]}{\left(\frac{q}{p}\right)^b - \left(\frac{q}{p}\right)^a} \right\} \frac{1}{p-q}$$

$$= E(T), \quad (\because q \neq p).$$

Again

$$Z_n = S_n'^2 - n \quad \text{is a Martingale}$$

i.e.

$$Z_n = \left[\sum_{i=1}^{n}(Y_i - (p-q))\right]^2 - n \text{ is a Martingale}$$

$$E(Z_T) = \sum_{i=1}^{T}[Y_i - (p-q)]^2 - E(T) = E(Z_0) = 0$$

$$E\left(\sum_{i=1}^{T} Y_i^2\right) = E(T) - E(T)(p-q)^2 + 2(p-q)E\left[\sum_{i=1}^{T} Y_i\right]$$

$$\Rightarrow \quad E\left(\sum_{i=1}^{T} Y_i^2\right) = E(T) - E(T)(p-q)^2 + (p-q)\{\pi(-b) + (1-\pi)a\}$$

$$E(T)\left[\text{Var}(Y) + [E(Y)]^2\right]$$
$$= E(T) - E(T)(p-q)^2 + 2(p-q)\{\pi(-b) + (1-\pi)a\}$$

$$\Rightarrow \quad E(T)\left[\text{Var}(Y) + [(p-q)]^2\right]$$
$$= E(T) - E(T)(p-q)^2 + 2(p-q)[\pi(-b) + (1-\pi)a]$$

On substitution of $E(T)$, and π one can obtain Var (Y).

4.7 An Important Identity: Wald's Identity

$$E[\phi(\theta)^{-T}\{e^{\theta S_T}\}] = 1$$

where T is the stopping time.

$$\phi(\theta) = E[e^{\theta Y_i}] \text{ exists when } Y_i\text{'s are i.i.d. r.v.s.(non-degenerate)}$$

and $Y_0 = 0,$

$$T = \min[n \,;\, S_n \leq -a \text{ or } S_n \geq a]$$

where $S_n = \sum_{i=0}^{n} Y_i$ \hfill (4.13)

4.7.1 Expected Stopping Time using Wald's Identity:

We have

$$E[(\phi(\theta))^{-n} e^{\theta S_n}] = 1$$

where $\qquad \phi(\theta) = E[e^{\theta Y}]; \quad S_n = Y_1 + Y_2 + ... + Y_n.$

Hence by optional sampling rule for $T = n$ the stopping time

$$E[(\phi(\theta))^{-T} e^{\theta S_T}] = 1. \qquad (4.14)$$

$$\phi(0) = 1; \quad \phi'(0) = E(Y).$$

Differentiating (4.14) w.r.t. θ and putting $\theta = 0$

$$\Rightarrow \quad E[(-T)(\phi(\theta))^{-T-1} \phi'(\theta) e^{\theta S_T} + (\phi(\theta))^{-T} S_T e^{\theta S_T}]_{\theta=0} = 0$$

$$\Rightarrow \qquad\qquad -E(T) E(Y) + E(S_T) = 0$$

$$\Rightarrow \qquad\qquad E(T) = \frac{E(S_T)}{E(Y)} \qquad\qquad (4.15)$$

4.8 Continuous Parameter Martingale

Martingales w.r.t. σ fields:

Hitherto, we have considered $E(X \mid Y_0, Y_1, ..., Y_n)$ where $X, Y_0, Y_1, ... , Y_n$ possesses joint continuous density; or are jointly distributed discrete r.v.s.

However, the analysis extended to more complex expression like,

$$E(X \mid Y_0, ..., Y_n, ...) \text{ or } E(X \mid Y(u); \ 0 \le u \le t)$$

becomes more complicated. In the former case we have the conditional expectation under the conditioning of denumerably infinite number of r.v.s, while in the latter the conditioning is restricted to a function of absolutely continuous r.v. in an interval which has non-denumerably infinite number of points.

Hence the traditional way of defining a conditional expectation under conditioning of a finite number of r.v.s. cannot be extended in either of the above two cases. A valid approach to define the above conditional expectations is with respect to certain collections called σ-fields of events.

Let $\{X_n\}_0^\infty$ be a sequence of real r.v.s. on the probability space $(\Omega, \mathcal{B}, \mathcal{P})$ and $\{\mathcal{F}_n\}$ be a sequence of sub σ-fields $\ni \mathcal{F}_0 \subset \mathcal{F}_1 \subset \mathcal{F}_2 ... \subset \mathcal{F}_n \subset \mathcal{F}_{n+1} .... \subset \mathcal{F}$. We say $\{X_n\}$ is adopted to $\mathcal{F}_n$ if $\forall n, X_n$ is $\mathcal{F}_n$ measurable. What do we understand by the same? Let us illustrate the concept by giving an example.

Suppose $Y_0, Y_2, ..., Y_n ....$ are also defined on $(\Omega, \mathcal{F}, \mathcal{P})$ and $\mathcal{F}_n$ is a σ-field generated by $\{Y_0, Y_1, ..., Y_n\}$.

(Thus Y_0 generates $\mathcal{F}_0$, $\{Y_0, Y_1\}$ generates $\mathcal{F}_1 \supset \mathcal{F}_0$ etc.)

Then $\qquad\qquad \mathcal{F}_n \subset \mathcal{F}_{n+1}... \subset \mathcal{F}_n. \forall n$

If we define $X_n = g_n (Y_0, Y_1, ..., Y_n)$ for a sequence of Borel measurable function $g_n(\cdot)$ then we say $\{X_n\}$ is adopted to $\mathcal{F}_n$.

We may again think of $\mathcal{F}_n$ as containing information available at stage n ; just as we did earlier with $(Y_0, Y_1, ..., Y_n)$.

Then X_n is measurable w.r.t. $\mathcal{F}_n = \mathcal{F}(Y_0, Y_1,..., Y_n)$ iff as in our earlier stage, X_n is determined by $Y_0, Y_1,..., Y_n$. The relation $\mathcal{F}_n \subset \mathcal{F}_{n+1}$ expresses the increase in information as n progresses.

4.8.1 Definition of a Sub Martingale, Super Martingale and Martingales over a σ-field

Let $\{X_n\}$ be a sequence of r.v.s. defined in $(\Omega, \mathcal{F}, P)$ and $\mathcal{F}_n$ be a sequence of sub σ-fields with $\mathcal{F}_n \subset \mathcal{F}_{n+1},.... \subset \mathcal{F}$ $\forall$ n then $\{X_n\}$ is a *Sub Martingale* w.r.t. $\{\mathcal{F}_n\}$ if

   (i) X_n is adopted to $\{\mathcal{F}_n\}$ (i.e. $\exists$ X_n which is $\mathcal{F}_n$ measurable),

   (ii) $E[X_n^+] < \infty, E[X_{n+1} | \mathcal{F}_n] \geq X_n$ $\forall$ n;

where $x^+ = \text{Max}(0, x)$

If $\{-X_n\}$ is a Sub Martingale w.r.t. to $\mathcal{F}_n$ then $\{X_n\}$ is a Super Martingale.

If both $\{-X_n\}$ and $\{X_n\}$ are Sub Martingales, then $\{X_n\}$ is a Martingale w.r.t. $\{\mathcal{F}_n\}$.

Martingales optional sampling. theorem (stated earlier) and convergence theorems are both valid in continuous time as follows.

4.8.2 Optional Sampling Theorem

If $\{X(t) ; t \geq 0\}$ is a Sub-Martingale w.r.t. $\mathcal{F}_t$, then

$$E[X(0) \leq E[X(T \wedge t)] \leq E(X(t)), \text{ for all Markov time } T.$$

For Super Martingale, we have

$$E[X(0)] \geq E[X(T \wedge t)] \geq E[X(t)]$$

and for a Martingale,

$$E[X(0)] = E[X(T \wedge t)] = E[X(t)]$$

4.8.3 Martingale Convergence Theorem

Let $\{X_n\}$ be a Martingale satisfying

$$\underset{n \geq 0}{\text{Sup}} \, E[|X_n|] < \infty$$

then $\exists$ a r.v. X_∞ to which X_n converges as $n \to \infty$, with probability one,

i.e,
$$P\left\{ \lim_{n \to \infty} X_n = X_\infty \right\} = 1.$$

Further, if $\{X_n\}$ is Martingale and is uniformly integrable, then $\{X_n\}$ converges to the mean,

i.e.,
$$\lim_{n \to \infty} E\left[|X_n - X_\infty|\right] = 0$$

Example 4.1.10 To show that,

$$Z(t) = f(X(t)) = 1 + \frac{\mu_1}{\lambda_1} + \frac{\mu_1 \mu_2}{\lambda_1 \lambda_2} + \dots + \frac{\mu_1 \mu_2 \cdot \cdot \mu_{X(t)-1}}{\lambda_1 \lambda_2 \cdot \cdot \lambda_{X(t)-1}}$$

is a Martingale over $\mathcal{F}_t = \mathcal{F}(X(u)\,;\, 0 \le u \le t)$ provided $E(Z_t)$ is finite.

Proof: Let $s < t$ and $i \ge 1$, where i is the size of the population. Let us define

$$g_i(t) = E[Z(t)\,|\,X(u);\, 0 \le u \le s;\, X(s) = i]$$
$$= E[Z(t)\,|\,X(s) = i],\ \text{because of the Markovity of the process } X(t).$$

Then $g_i(t+h) = E[Z(t+h)\,|\,X(s) = i]$

$$= \sum_{k=0}^{\infty} E\left[Z(t+h)\,|\,X(t)=k\right] P\left[X(t)=k\,|\,X(s)=i\right]$$

$$= \sum_{k=0}^{\infty} \{\, E(Z(t)\,|\,X(t)=k)\, P(X(t)=k\,|\,X(s)=i)$$

$$+ h\, E\,(Z\cdot(t)\,|\,X(t) = k)\, P\,(X(t) = k\,|\,X(s) = i) + o\,(h)\}$$

$$\Rightarrow g_i(t+h) = \sum_{k=0}^{\infty} E[Z(t)\,|\,X(t)=k]\, P[X(t)=k\,|\,X(s)=i]$$

$$+ h\, \sum_{k=0}^{\infty} E[Z'(t)\,|\,X(t)=k]\, P[X(t)=k\,|\,X(s)=i] + o(h)$$

$$= g_i(t) + h\, \sum_{k=0}^{\infty} \Big\{\, \lambda(k)\,[f(k+1)-f(k)] + \mu(k)[f(k-1)-f(k)]$$

$$+\ [1-(\lambda(k)+\mu(k))]\,[f(k)-f(k)]\, \Big\} \times P[X(t) = k\,|\,X(s) = i] + o(h)$$

$\{\because \quad Z(t)\,|\,X(t) = k = f[X(t)]_{X(t)=k} = f(k).$

and $Z'(t) = \dfrac{f(k+1)-f(k)}{1} + o(h)$ with prob. $\lambda(k)$

$$= \dfrac{f(k-1)-f(k)}{1} + o(h)\ \text{with prob. } \mu(k)$$

$$= f(k) - f(k) + o(h)\ \text{with prob. } 1-(\lambda(k)+\mu(k)) + o(h)\}$$

$$\Rightarrow \qquad \lim_{h \to 0} \frac{g_i(t+h) - g_i(t)}{h}$$

$$= \sum_{k=0}^{\infty} [\lambda(k)(f(k+1) - f(k)) - \mu(k)(f(k) - f(k-1))]$$

$$P[X(t) = k \,|\, X(s) = i]$$

$$\Rightarrow \qquad g_i'(t) = \sum_{k=0}^{\infty} \left\{ \lambda_k \left[\frac{\mu_1 \mu_2 \cdots \mu_k}{\lambda_1 \lambda_2 \cdots \lambda_k} \right] - \mu_k \left[\frac{\mu_1 \cdots \mu_{k-1}}{\lambda_1 \cdots \lambda_{k-1}} \right] \right\}$$

$$\times P[X(t) = k \,|\, X(s) = i]$$

$$= 0, \text{ identically.}$$

Again $g_i'(t) = 0 \Rightarrow g_i(t) = E[Z(t) \,|\, X(s) = i]$ is a constant for $t > 0$.

Letting $t \downarrow s$, we have

$$g_i(s) = E[Z(s) \,|\, X(s) = i]$$

$$= g_i(t) = E[Z(t) \,|\, X(t) = i]$$

$\therefore$ $Z_i(t)$ is a Martingale. $\qquad\qquad$ [Proved].

4.9 Stopping Rule for Continuous Parameter Martingale

4.9.1 A problem on stopping rule for Poisson process

We have

$$Y(t) = X(t) - \lambda t$$

is a Martingale over $\mathcal{F}(t) = \mathcal{F}(X(t))$ where $X(t)$ is a Poisson Process. We fix up a positive integer a and let T_a be the first time $X(t)$ reaches a.

Set $\qquad\qquad X(0) = 0 \Rightarrow X(T_a) = a$

$\Rightarrow E[Y(T_a)] = E[X(T_a) - \lambda(T_a)] = Y(0)$ by the optional sampling theorem.

Also, $\qquad\qquad Y(0) = X(0) - \lambda \times 0 = 0$

$$\Rightarrow \qquad\qquad E[X(T_a)] = \lambda E[T_a]$$

$$\Rightarrow \qquad\qquad a = \lambda E[T_a]$$

$$\Rightarrow \qquad\qquad \frac{a}{\lambda} = E[T_a] \qquad\qquad\qquad (4.16)$$

Note that $\frac{a}{\lambda}$ is expected time of a Gamma distribution which is the 'a'-fold convolution of a Negative Exponential distribution with parameter λ.
Also,

$$U(t) = Y^2(t) - \lambda t$$

is a Martingale, if $X(t) - \lambda t = Y(t)$, where $X(t)$ is a Poisson process. If T_a is the stopping time then

$$E[U(T_a)] = E[Y^2(T_a)] - \lambda E[T_a]$$
$$= E[U(0)] = 0$$
$\Rightarrow \qquad E[Y^2(T_a)] = \lambda E[T_a] = a$

Also $\qquad Y(T_a) = X(T_a) - \lambda T_a$

$$E[Y^2(T_a)] = E(X(T_a) - \lambda T_a)^2$$

$$= E[\lambda E(T_a) - \lambda T_a]^2 \quad (\because X(T_a) = a, \; E(T_a) = \frac{a}{\lambda}$$

$$\therefore X(T_a) = \lambda E(T_a))$$

$$E[Y^3(T_a)] = \lambda^2 E[T_a - E(T_a)]^2$$
$\Rightarrow \qquad a = \lambda^2 \, \text{Var} \, (T_a)$

$\Rightarrow \qquad \dfrac{a}{\lambda^2} = \text{Var} \, (T_a) \qquad\qquad\qquad (4.17)$

Finally

$$E(V(T_a)) = E\{\exp[-\theta X(T_a) + \lambda T_a(1 - e^{-\theta})]\} = E(V(0))] = 1$$

Thus $\quad V(t) = \exp[-\theta X(t) + \lambda t(1 - e^{-\theta})]$ is a Martingale

$\Rightarrow \qquad 1 = E[\exp[-\theta X(T_a)] \exp\{-\alpha T_a\}], \text{ where } \alpha = -\lambda(1 - e^{-\theta})$

$\Rightarrow \qquad 1 = e^{-\theta a} E[\exp(-\alpha T_a)]$

$\Rightarrow \qquad E[\exp(-\alpha T_a)] = e^{\theta a}$

Again $\qquad\qquad \alpha = -\lambda(1 - e^{-\theta})$

$\Rightarrow \qquad \dfrac{\alpha + \lambda}{\lambda} = e^{-\theta}$

$\Rightarrow \qquad \dfrac{\lambda}{\alpha + \lambda} = e^{\theta}$

$\Rightarrow \qquad \left(\dfrac{\lambda}{\alpha + \lambda}\right)^a = e^{a\theta}$

$\therefore \qquad E[\exp(-\alpha T_a)] = \left(\dfrac{\lambda}{\alpha + \lambda}\right)^a \qquad\qquad (4.18)$

The results (4.16), (4.17) and (4.18) show that T_a has a Gamma distribution with parameters a and λ.

4.9.2 Stopping Rule for Birth and Death Process

If $\lambda_i = \lambda(i) = \lambda$ and $\mu_i = \mu(i) = \mu$ for a birth and death process $X(t)$

and define
$$f(j) = 1 + \frac{\mu_1}{\lambda_1} + \frac{\mu_1 \mu_2}{\lambda_1 \lambda_2} + \dots + \frac{\mu_1 \mu_2 \dots \mu_{j-1}}{\lambda_1 \lambda_2 \dots \lambda_{j-1}}$$

$$= 1 + \frac{\mu}{\lambda} + \left(\frac{\mu}{\lambda}\right)^2 + \dots + \left(\frac{\mu}{\lambda}\right)^{j-1}$$

Then $Z(t) = f[X(t)]$ is a Martingale over $\mathcal{F}(t)$ (vide Example 4.1.10)
Putting

$$\left.\begin{array}{l} \lambda_i = \lambda \\ \\ \mu_i = \mu \end{array}\right\}; \quad i = 1, 2 \dots n$$

$$Z(t) = f[X(t)] = 1 + \frac{\mu}{\lambda} + \frac{\mu^2}{\lambda^2} + \dots + \frac{\mu^{X(t)-1}}{\lambda^{X(t)-1}} \qquad (4.19)$$

$$= \frac{\left(\frac{\mu}{\lambda}\right)^{X(t)} - 1}{\left[\left(\frac{\mu}{\lambda}\right) - 1\right]} = \frac{\left[\left(\frac{\mu}{\lambda}\right)^{X(t)} - 1\right]}{\mu - \lambda} \cdot \lambda \qquad (4.20)$$

is a Martingale over $\mathcal{F}(t) = \mathcal{F}(X(t))$

Suppose we define a stopping time,

$$T_{0,m} = \min \{t \geq 0, X(t) = 0 \text{ or } X(t) = m\} \text{ given } X(0) = i$$

and T_a is the probability that the process is absorbed at '0' before reaching m, given that the initial state $X(0) = i$.

Then applying optional sampling theorem, we have

$$E[Z(T_{0,m})] = E[f(X(T_{0,m})] = E[f(X(0))] = E[f(i)] = f(i)$$

Also

$$E[Z(T_{0,m})] \equiv \text{mean size of } f[X(T_{0,m})]$$

when $X(T_{0,m})$ attains '0' and 'm' respectively i.e. $f[X(T_{0,m})]$ attaining $f(0)$ and $f(m)$ respectively with steady state probabilities $V(i)$, $1 - V(i)$ given that $f(0) = i$. In other words,

$$f(i) = (1 - V(i)) f(m) + V(i) \times 0$$

$$\Rightarrow \qquad V(i) = \frac{f(m) - f(i)}{f(m)}$$

and
$$1 - V(i) = \frac{f(i)}{f(m)}$$

It appears that $V(i)$ proportions of the population is reaching state '0' (before reaching 'm') and $(1 - V(i))$ proportion is reaching 'm' before reaching '0'. To obtain the expected stopping time, let us consider another Martingale, say,

$$Y(t) = g[X(t)] - \int_0^t \{ [\lambda [X(u)][g(X(u)) + 1] - g(X(u))]$$

$$- \mu [X(u) [g(X(u)) - g(X(u) - 1)] \} \, du \qquad (4.21)$$

where $X(t)$ is a birth and death process.

Let $g(i)$ be so chosen, so that the second term under the sign of integration in (4.21) becomes unity. This greatly simplifies the Martingale structure. Then we have much simplified expression. Under the same, we have

$$Y(t) = g[X(t)] - t \quad \text{is a Martingale on } \mathcal{F}(X(t))$$

Again, by the optional sampling theorem

$$E[Y(S_T)] = E[g(X(S_T))] - E(S_T) = E[Y(0)]$$

where S_T is the stopping time. Given $X(S_T)$ and $X(0)$, for the birth and death processes, we get $E[Y(S_T)]$ and $E[Y(0)]$.

It can be seen that

$$g(i+1) = \frac{1}{\lambda_1} + \frac{1}{\lambda_2}\left(1 + \frac{\mu_2}{\lambda_1}\right) + \frac{1}{\lambda_3}\left(1 + \frac{\mu_3}{\lambda_2}\left(1 + \frac{\mu_2}{\lambda_1}\right)\right) + \dots$$

$$+ \frac{1}{\lambda_i}\left[1 + \frac{\mu_i}{\lambda_i - 1}\left(1 + \frac{\mu_i - 1}{\lambda_i - 3}\right)\left(1 + \frac{\mu_i - 2}{\lambda_i - 3}\right)\left(1 + \frac{\mu_i - 4}{\lambda_i - 5}\right)\dots\right.$$

$$\dots \left.\left(1 + \frac{\mu_4}{\lambda_3}\left(1 + \frac{\mu_2}{\lambda_1}\right)\right)\right] \qquad (4.22)$$

subject to $g(1) = 0$ and $g(0) = 0$, $i = 1, 2, 3, \dots$ is a solution of the difference equation corresponding to the integrand on the right hand side of equation (4.21) equated to unity, (vide. Biswas and Pachal (1987)). These results enable us to obtain $E(S_T)$ i.e. expected stopping time. This kind of stopping rule is useful in several situations; inclusive of making an application of evolving a stopping rule of sterilizing mothers for family welfare programme based on the number of surviving children.

4.10 Cox's Regression Model and Martingales Theory

Cox's regression model (1975) based on the method of '*Partial Likelihood*' (a new method of estimation) plays a very important role in analyzing the data in a more realistic way on survival or fertility or any other kind involving population characteristics using stochastic models. The speciality of this model may be illustrated by stating a problem in morbidity as follows.

Suppose we are interested in the survival of patients suffering from hypertension. The hazard rate at any point of time need not necessarily depend on time only but also on a host of explanatory variables or covariates, some of which may not be expressed in quantitative form. For example, other conditions affecting the hazard rate of an individual may be factors as age, blood pressure, occupation, general health condition and the presence of other related complications etc. Cox's regression model not only takes into account of all the explanatory variables but also is based on estimational technique which estimates the parameters concerning the individual coveriates, independent of the parameters concerning the hazard rate at time t.

To put the idea in concrete form, let us take the simplest form of the Cox's model,

$$\lambda_j(t) = \lambda_0(t) \exp [\beta' \, z_j(t)] \tag{4.23}$$

where

$\lambda_j(t)$ = Hazard rate of the jth individual at any time t.

$\lambda_0(t)$ = Hazard function with respect to time only ignoring the other covariates

i.e. keeping $z_{ji} = 0 \; \forall \; i = 1, 2, ... \, p; \; j = 1, 2, ... \, n$

where

$$z'_j = (z_{j1}, z_{j2} ... z_{jp})$$

is the p-component covariate vector of the jth individual.

Corresponding to every individual $j = 1, 2, ... \, n$ the random variable T_j is the time of death which in a randomly consored sample is observed upto a period c_j for the jth individual. Taking logarithm on both sides of (4.23), we have

$$\log \lambda_j(t) = \log \lambda_0(t) + [\beta_1 z_{j1}(t) + \beta_2 z_{j2}(t) + ... + \beta_p z_{jp}(t)]$$

The equation could have been treated as a single equation *log-linear model* for the estimation of β_i's ($i = 1, 2... \, p$) as well as $\lambda_0(t)$. However, the innovativness of Cox's approach is that Cox proposed the estimates of the regression parameters $\beta_1, \beta_2, ...\beta_p$ independent of $\lambda_0(t)$. As such, he introduced the method of '*partial likelihood*' which may be explained in the following lines.

Let $R(t) = \{j : T_j \geq t, c_j\}$ be the risk set i.e. the set of individuals exposed to the risk under observations (the jth individual observed between $(0, c_j)$).

Given that we have a set of n persons in the sample and that a person dies in the set; the probability that the jth person dies (assuming that the death occurs independently) is given by

$$\frac{\lambda_0(t) \exp [\beta' \, z_j(t)]}{\sum\limits_{j \in R(t)} \lambda_0(t) \exp [\beta' \, z_j(t)]} = \frac{\exp [\beta' \, z_j(t)]}{\sum\limits_{j \in R(t)} \exp [\beta' \, z_j(t)]} = P_L \tag{4.24}$$

Cox defined,

$$L(\beta) = \prod_{T_j \leq c_j} \prod_{j} \frac{\exp [\beta' \, z_j(t)]}{\sum\limits_{j \in R(t)} \exp [\beta' \, z_j(t)]} \tag{4.25}$$

$$j = 1, 2, ... \, n$$

the product being extended over all $j \forall T_j \leq C_j$; $j = 1, 2...n$ as partial likelihood estimating the parameters by the method of maximum likelihood and conjectured that the method would give estimates of $\beta_1, \beta_2,...,\beta_p$ which would have otherwise the asymptotic properties of the maximum likelihood estimators. However, there had been a lot of controversies among statisticians with regard to the validity of Cox's conjecture especially during 1975-79. The controversies arise because, certainly $L(\beta)$ is not a conditional likelihood for β based on the conditional distribution of the data given some statistic. Nor is it a marginal likelihood based on the marginal distribution and on the reduction of data. Cox (1975), therefore, introduced the notion of partial likelihood to remedy the defect and showed that $L(\beta)$ is an example of partial likelihood. The estimates obtained by the partial likelihood do possess the asymptotic properties of maximum likelihood estimators which can be shown by using the properties of Martingales; and especially by using central limit theorem of Martingales viz., derivative of the log partial likelihood conforms to a Martingale.

We present below certain other concepts of Martingales theory and the relevant results of Martingales theory enabling the readers to get an insight into this matter justifying the *'partial likelihood'* method of estimation. The approach followed is due to Gill (1984).

4.11 Martingales: Some More Results

As defined in 4.8 we have a continuous parameter Martingale M given by

$$M = \{ M(t); t \geq 0 \}$$

which is a stochastic process whose increment in an interval (u,v) given the past upto u has expectation zero;

$$E[M(v) - M(u) \mid \mathcal{F}_u] = 0, \quad \forall \, 0 \leq u < v < \infty \qquad (4.26)$$

Also given $\mathcal{F}_u$, $M(u)$ is fixed.

Let u is just before t and v is just after it

i.e. $$u = t - \frac{\Delta t}{2}; \; v = t + \frac{\Delta t}{2}$$

and $$\Delta t \rightarrow 0$$

Then (4.26) reduces to

$$E\left[\lim_{\Delta t \rightarrow 0} M\left(t + \frac{\Delta t}{2}\right) - M\left(t - \frac{\Delta t}{2}\right) \Big|_{\mathcal{F}_{t - \frac{\Delta t}{2}}} \right] = 0$$

$$= E\left[dM\left(t\right) \mid \mathcal{F}_{t-} \right] = 0 \qquad (4.27)$$

Hence an alternative definition of Martingale is also given by (4.27).

Next let us consider a multivariate counting process

$$N(t) = [N_1(t), N_2(t), \dots N_n(t)]'$$

and the corresponding intensity vector

$$\Lambda(t) = [\Lambda_1(t), \dots \Lambda_2(t) \dots \Lambda_n(t)]'$$

and given $N_t(t)$, a counting process such that

$$dN_i(t) \,|\, \mathcal{F}_t = 1 \quad \text{with probability } \Lambda_i(t)\, dt$$

$$= 0 \quad \text{with probability } (1 - \Lambda_i(t)dt)$$

$$\forall \; i = 1, 2, \dots n$$

$$\Rightarrow \qquad E(dN_i(t) \,|\, \mathcal{F}_t] = \Lambda_t(t)dt.$$

Next let as define another process $M_i(t)$ э

$$dM_i(t) = dN_i(t) - \Lambda_i(t)dt \tag{4.28}$$

$$\Rightarrow \qquad E(dM_i(t)) = E(dN_i(t) \,|\, \mathcal{F}_i) - \Lambda_i(t)dt$$

$$= \Lambda_i(t)dt - \Lambda_i(t)dt$$

$$= 0$$

$$\Rightarrow \qquad E(dM_i(t)) = 0$$

$$\Rightarrow \qquad M_i(t) \text{ is a Martingale.}$$

Integrating (4.28) we have

$$M_i(t) = N_i(t) - \int_0^t \Lambda_i(\tau)\, d\tau. \tag{4.29}$$

4.12   Predictable Variation Process of Martingale

We denote the predictable variation of a process $M_i(t)$ as $< M_i > (t)$ which is defined as

$$d < M_i > (t) = E[(dM_i(t))^2 \,|\, \mathcal{F}_{t-}] = \text{Var } \{dM_i(t)\, \mathcal{F}_{t-}\} \tag{4.30}$$

It follows that Var $\{dM_i(t) \,|\, \mathcal{F}_{t-}\}$ is predictable and non decreasing over t. We can conceive it to be the sum of the conditional variance of the increments of $M_i(t)$ taken over infinitesimal intervals which partition $(0, t)$ where each conditional variance of the increment taken over such small intervals is obtained while taking the history, happened upto the beginning of the previous interval.

Now $\qquad E(dN_i(t) \,|\, \mathcal{F}_{t-}) = \Lambda_i(t)\, dt.$

$$\text{Var } (dN_i(t) \,|\, \mathcal{F}_{t-}) = \Lambda_i(t)\, dt\, (1 - \Lambda_i(t)\, dt)$$

$$= \Lambda_i(t)\, dt \tag{4.31}$$

due to orderliness of the process (viz. not more than one event in the infinitesimal interval).

Hence we write from (4.28) and (4.31)

$$< M_i > (t) = \int_0^t \Lambda_i(\gamma) \, d\gamma = < N_i > t \tag{4.32}$$

Combining (4.29) and (4.32) $\Rightarrow$

$$M_i(t) = N_i(t) - < M_i > (t)$$

$$\Rightarrow \qquad M_i(t) + < M_i > (t) = N_i(t) \tag{4.33}$$

for a counting process $N_i(t)$.

This gives a technique of constructing Martingales from counting process.

4.13 Martingales from Different Processes

Using (4.29) and (4.33) in art. 4.11 and 4.12 respectively we can construct Martingales on continuous parameter space.

A. Poisson process

If $\{X(t); t \geq 0\}$ is a Pure Birth process having birth parameter λ then using art. 4.11 and 4.12, we can show that,

$$
\left.
\begin{aligned}
Y(t) &= X(t) - \lambda t \\[2mm]
U(t) &= Y^2(t) - \lambda t \\[2mm]
V(t) &= \exp[-\theta X(t)] + \lambda t (1 - e^{-\theta})
\end{aligned}
\right\} \tag{4.34}
$$

and

are all Martingales (with respect to $\mathcal{F}_u = \mathcal{F}(X(u) : 0 \leq u < t)$.

B. Birth process

Suppose $\{X(t) : t \geq 0\}$ is a Pure Birth process having birth parameters

$$\lambda(t) \geq 0, \text{ for } i \geq 0.$$

Assume, for convenience only $X(0) = 0$.
Then

$$
\left.
\begin{aligned}
(i)\ \ Y(t) &= X(t) - \int_0^t \lambda[X(u)] \, du \\[3mm]
(ii)\ \ V(t) &= \exp[\theta X(t) + (1 - e^\theta)] \int_0^t \lambda[X(u)] \, du
\end{aligned}
\right\} \tag{4.35}
$$

are Martingales (with respect to $\mathcal{F}_u = \mathcal{F}(X(u) \ 0 \leq u < t))$.

C. Birth and Death Processes

Let $X(t)$ be a density dependent birth and death process, with parameters

$$\lambda\,[X(t) = i] = \lambda_i, \quad \mu\,[X(t) = i] = \mu_i$$

Let $g\,(X(i) = i) = g(i)$, $i = 0, 1, 2, \ldots$ be arbitrary, provided the expectation of $Y(t)$ as given below

(i) $$Y(t) = g\,[X(t) - \int_0^t \{\lambda\,[X(u)]\,[g(X(u) + 1) - g(X(u))] - \mu\,[X(u)] \cdot$$

$$[g(X(u) - 1) - g(X(u))]\}du$$

is finite; then $\{Y(t)\}$ is a Martingale, over $\mathcal{F}(t) = \mathcal{F}(X(t))$.

(ii) If $$f(j) = 1 + \frac{\mu_1}{\lambda_1} + \frac{\mu_1\,\mu_2}{\lambda_1\,\lambda_2} + \ldots + \frac{\mu_1\,\mu_2 \ldots \mu_{j-1}}{\lambda_1\,\lambda_2 \ldots \lambda_{j-1}}$$

and $$Z(t) = f\,[X(t)]$$

then $Z(t)$ is a Martingale over $\mathcal{F}(X(t))$,

where $\lambda_t = \lambda(t)$ and $\mu_t = \mu(t)$ are the intensities of birth and death under the population size $X(t)$. (Proved earlier in Ex. 4.1.10).

4.14 A Theorem of Martingale Transform Process

Let us define a process

$$\overline{M}(t) = \{\overline{M}(t); t \geq 0\}$$

where we define

$$\overline{M}(t) = \int_0^t H(s)\,dM(s) \tag{4.36}$$

where M is a Martingale and H is a predictable process (either a constant, non-random or deterministic process).

$\Rightarrow$ $$d\overline{M}(t) = H(t)\,dM(t)$$

$$E\,\{d\overline{M}(t)\,|\,\mathcal{F}_{t-}\} = E\,\{\,H(t)\,dM(t)\,|\,\mathcal{F}_{t-}\}$$

$$= H(t)\,E\{dM(t)\,|\,\mathcal{F}_{t-}\}$$

because H is a predictable process.

$$= H(t) \times 0$$

$$(\because\ M \text{ is a Martingale})$$

Hence $\overline{M}$ is a Martingale.

Thus we prove that any predictable function integrated with respect to a

Martingale gives rise to another Martingale. This is the theorem of Martingale transform process.

Further,

$$\text{Var}\{d\overline{M}(t)|\mathcal{F}_{t-}\} = \text{Var}\{H(t)\,dM(t)|\mathcal{F}_{t-}\}$$
$$= (H(t))^2\,\text{Var}\,(dM(t)|\mathcal{F}_{t-})$$
$$= (H(t))^2\,d < M > (t) \tag{4.37}$$

4.15 Martingale Central Limit Theorem

Let us define a time transformed Brownian motion (vide 3.7) $W(t)$; $t \geq 0$ as a stochastic process with the following properties.

(a) The realizations of $W(t)$ are continuous functions subject to

$$W(0) = 0.$$

(b) $\forall$ $t_1, t_2, ..., t_n$, $W(t_1)$, $W(t_2)$, ... $W(t_n)$ are multivariate normally distributed with zero means and independent increments. We have, therefore, for $s < t$, $W(t) - W(s)$ is independent of $W(s)$; (In fact $W(t)$ is independent of say, $U \,\forall\, u < s$).

By independence of increments, the conditional variance $dM(t)$ given the path of W on $(0, t)$ does not depend on the past.

(c) Also $E[dW(t)|\mathcal{F}_{t-}] = 0$

$\Rightarrow W$ is a continuous Martingale with predictable variation process $< W > (t)$ equal to some deterministic function say, A. These properties (a), (b) and (c) characterize the distribution of W as Gaussian.

Further for a sequence of Martingales $M^{(n)}$ such that $n = 1, 2, ..., \exists$

(i) jumps of $M^{(n)} \to 0$ as $n \to \infty$ ($\Rightarrow M^{(n)}$ is nearly continuous).

(ii) the predictable variation process of $M^{(n)}$ converges in distribution to W as $n \to \infty$. In particular $M^{(n)}(t)$ is asymptotically distributed with mean zero and variance $\Lambda(t)$ and the increments of $M^{(n)}$ are conventionally independent.

4.16 Validity of Cox's Partial Likelihood

Using Martingale transform theorem, Gill (1984) has shown that

$$\int_{t=0}^{u} n^{-\frac{1}{2}} [Z_i(t) - E_0(t)]\,dN_i(t) \tag{4.38}$$

where

$$E_0(t) = \frac{\sum_j Z_j(t)\exp(\beta' Z_j(t))}{\sum_j \exp(\beta' Z_j(t))}$$

is a Martingale,

and, therefore
$$\sum_{i=1}^{n} \int_{i=0}^{u} n^{-1/2} [Z_i(t) - E_0(u)] \, dN_i(t)$$

can be taken as a sum of n vector Martingales and hence also is a Martingale M.

Further, denoting by $c(\hat{\beta}, t)$ the estimated log of Cox's likelihood evaluated at time t ($\hat{\beta}$, being the partial likelihood estimator) and

$$X(\hat{\beta}, t) = n^{-1} [c(\hat{\beta}, t) - c(\beta, t)]$$

it is proved that $X(\hat{\beta}, 1)$ converges in probability to a function of β which is concave with a unique maximum at β (the true parameter vector β).

Hence
$$\hat{\beta} \xrightarrow{p} \beta.$$

Thus by Martingale central limit theorem we may conclude that large sample maximum likelihood status is applicable to $\hat{\beta}$ when n is so large, so that

$$\frac{1}{n} \sum_{i=1}^{n} Z_i^r(t) \exp[\beta' Z_j(t)], \quad r = 0, 1, 2 \ldots$$

are almost non-random for all t and $\forall \, \hat{\beta}$ close to β.

This justifies the asymptotic maximum likelihood properties of partial likelihood estimators of β.

We have thus given only an outline of the proof. For details of the proof the reader is referred to Gill (1984).

4.17   Applications of Martingales in Epidemic Models

Here we propose to discuss some applications of Martingales theory in stochastic epidemic models. While we show here some elementary applications of Martingales in classical, simple and general epidemic model, an illustrative application of Martingales in Carrier borne epidemic model as developed by Picard (1984) will be taken up in 7.2.4.

Two basic results of Martingales on continuous parameter space will be used here. The results pertaining to simple and general epidemic models are due to Niels Becker (1989).

First result :   If $N(t)$ is a counting process
$$E\{dN(t) \mid \mathcal{F}(t)\} = \Lambda(t) \, dt$$

and
$$M(t) = N(t) - \int_{0}^{t} \Lambda(\tau) \, d\tau$$

is a zero mean Martingale [i.e. $E(M(t)) = E(M(0)) = 0$]

$$\Rightarrow \qquad \text{Var}(M(t)) = E\left[\int_{0}^{t} \Lambda(\tau) \, d\tau \right] = E(N(t)) \qquad (4.39)$$

Proof: We denote

$$d < M > (t) = E\,(dM^2(t)\,|\,\mathcal{F}_{t-})$$

Now we have,

$$(M(t) + dM(t))^2 = M^2(t) - 2M(t)dM(t) + dM^2(t),\text{ identically.}$$

Also $$E\,(dM^2(t)\,|\,\mathcal{F}_{t-}) = \text{Var}\,(dM(t))$$

since $$E\,(dM(t)\,|\,\mathcal{F}_{t-}) = 0$$

$$\text{Var}\,(dM(t)\,|\,\mathcal{F}_{t-})$$

$$= \text{Var}\,(dN(t) - \Lambda(t)dt\,|\,\mathcal{F}_{t-})$$

$$= \text{Var}\,(dN(t)\,|\,\mathcal{F}_{t-})$$

Also $$d < M > t = E\,(dM^2(t)\,|\,\mathcal{F}_{t-})$$

$$= \text{Var}\,(dM(t)\,|\,\mathcal{F}_{t-})$$

$$= \text{Var}\,(dN(t)\,|\,\mathcal{F}_{t-})$$

$$= \Lambda(t)dt\,(1 - \lambda(t)\,dt)$$

$$\cong \Lambda(t)\,dt$$

by the principle of orderliness of the process.

$$\therefore \qquad < M >(t) = \int_0^t \Lambda(\tau)\,d\tau$$

Also $$E\,(N(t)) = \int_0^t \Lambda(\tau)\,d\tau$$

Hence $$\text{Var}\,(M(t)) = E\,(N(t)) \qquad\qquad (4.40)$$

Second result :   Let $H(x)$ be a predictable variation process.
Then as shown in (4.36)

$$\overline{M}(t) = \int_0^t H(\tau)\,dM(\tau)$$

is a Martingale with respect to $\mathcal{F}_{t-}$.
From (4.39)

$$\text{Var}\,(d\overline{M}(t)\,|\,\mathcal{F}_{t-}) = \text{Var}\,(H(t)\,dM(t)\,|\,\mathcal{F}_{t-})$$

$$= (H(t))^2 \cdot \text{Var}\,(dM(t)\,|\,\mathcal{F}_{t-})$$

$$= (H(t))^2 \cdot d < M > (t)$$

$$= (H(t))^2\,E\,(dN(t)\,|\,\mathcal{F}_{t-})$$

We can put the result as

$$\text{Var}\left(\overline{H}(t)\right) = \text{Var}\left[\int_0^t H(\tau)\, dM(\tau)\right]$$

$$= E\left[\int_0^t H^2(\tau)\, d < M > (\tau)\right]$$

$$= E\left[\int_0^t H^2(\tau)\, dN(\tau)\right] \qquad (4.41)$$

Application on Classical Epidemic Models:

4.17.1 Simple Epidemic Model

Case I: When the progress of the epidemic is known at any time *t*:
Suppose our problem lies in estimating the infection rate γ of simple epidemic model (a model which considers only susceptibles and inefectives which are the time dependent variables in the epidemic process. The sub population which is removed by death, immunity or cure is not taken into consideration in simple epidemic model but only in general epidemic model which we will consider in the next subsection).

Assume at time $t = 0$ we have i # infectives and s # susceptibles. If at a time t, $N(t)$ # individuals are infected,
then at time t we have

$$\text{# Susceptibles: } S(t) = s - N(t)$$

$$\text{# infected: } I(t) = i + N(t)$$

i.e. we assume that the size of the population remains $s + i$ till the end of the epidemic. This kind of situation will arise when the period of immunity following recovery is zero and there is no mortality in the infection so caused.

Case II :
In the next place, let us consider again a simple epidemic where the progress of the epidemic is known only at a time T (not at intermediate points). Let us define a predictable function

$$H(\tau) = \frac{K(\tau-)}{I(\tau-)\, S(\tau-)}$$

Note that with the introduction of $\tau-$, $H(\tau)$ becomes predictable. This implies in $\int H(\tau)\, dN(\tau)$, we put the values of $I(\tau-)$ and $S(\tau-)$ just prior to each of the jump occurs,

and $\qquad K(\tau-) = 1, \quad$ if $S(\tau) > 0$

$$= 0, \quad \text{if otherwise.}$$

With this set up we have,

$$\int_0^T H(\tau)\,dN(\tau) = \frac{1}{i\,s}\cdot 1 + \frac{1}{(i+1)(s-1)}\cdot 1 + \frac{1}{(i+2)(s-2)}\cdot 1 +$$

$$\ldots + \frac{1}{(S(T)-1)(I(T)+1)}\cdot 1 \qquad \begin{cases} I(0-0) = i \\ \\ S(0-0) = s \end{cases}$$

Now $\int_0^T H(\tau)\,dM(\tau) = \overline{M}(t)$ will also be a Martingale over $\mathcal{F}_{t\,-}$;

where

$$M(t) = N(t) - \int_0^t \Lambda(\tau)\,d\tau$$

$$\overline{M}(t) = \int_0^t H(\tau)\,d\left[N(\tau) - \int_0^t \Lambda(\tau)\,d\tau \right]$$

We assume orderliness in the process i.e. during an infinitesimal time one or none can be infected i.e. if γ is the the infection rate, then

$$P\{dN(t) = 1 \mid \mathcal{F}_{t\,-}\} = \gamma\, I(t)\, S(t) + o(st)$$
$$P\{dN(t) = 0 \mid \mathcal{F}_{t\,-}\} = 1 - \{\gamma\, I(t)\, S(t) + o(st)\}$$

Then we have

$$M(t) = N(t) - \int_0^t \gamma\, I(\tau)\, S(\tau)\,d\tau$$

is a Martingale over $\mathcal{F}_{t\,-}$ with zero mean.
If T is the stopping time of the epidemic, then

$$E\,(M(t)) = M(0) \text{ by optional sampling theorem}$$

$$\Rightarrow \qquad E\,(M(t)) = E\left[N(T) - \gamma \int_0^T I(\tau)\, S(\tau)\,d\tau \right]$$

$$= M(0) \qquad (\because N(0) = 0)$$

$$= 0$$

$$\therefore \qquad \hat{\gamma} = \frac{N(T)}{\displaystyle\int_0^T I(t)\, S(\tau)\,d\tau} \tag{4.42}$$

Also sampling variance of $\hat{\gamma}$ can be obtained by using the first result as.

$$\text{Var}\,(M(t)\,|\,\mathcal{F}_{t\,-}) \;=\; E\,(N(t)\,|\,\mathcal{F}_{t\,-}).$$

We have

$$\widehat{M(t)} \;=\; \widehat{N(t)} - \int_0^t \hat{\gamma}\,I(\tau)\,S(\tau)\,d\tau$$

$$E\,(\hat{N}(T)) \;=\; E\!\left[\hat{\gamma}\int_0^T I(\tau)\,S(\tau)\,d\tau\right]$$

Also $\quad \text{Var}\,(\hat{M}(T)) \;=\; E\,(\hat{N}(T))^2 - \left[E\!\left(\hat{\gamma}\int_0^T I(\tau)\,S(\tau)\,d\tau\right)\right]^2$

$$= E\,(\hat{\gamma}^2)\,E\left[\int_0^T I(\tau)\,S(\tau)\,d\tau\right]^2$$

$$- [E\,(\hat{\gamma})]^2\,E\left[\int_0^T I(\tau)\,S(\tau)\,d\tau\right]^2$$

$$\text{Var}\,(N(\hat{T})) \;=\; \text{Var}\,(\hat{\gamma})\,E\left[\int_0^T I(\tau)\,S(\tau)\,d\tau\right]^2$$

$$\Rightarrow \qquad \text{Var}\,(\hat{\gamma}) \;=\; \frac{\text{Var}\,(M(\hat{T})}{E\left[\int_0^T I(\tau)\,S(\tau)\,d\tau\right]^2}$$

$$= \frac{E\,(\hat{N}(T))}{E\left[\int_0^T I(\tau)\,S(\tau)\,d\tau\right]^2}$$

Hence $\qquad \text{Var}\,(\hat{\gamma}) \;=\; \dfrac{\hat{N}(T)}{\left[\int_0^T I(\tau)\,S(\tau)\,d\tau\right]^2}$

$$(\text{S.E.}\,(\hat{\gamma})) \;=\; \sqrt{\frac{\hat{N}(T)}{\left[\int_0^T I(\tau)\,S(\tau)\,d\tau\right]^2}} \;=\; \frac{(\hat{N}(T))^{1/2}}{\int_0^T I(\tau)\,S(\tau)\,d\tau} \qquad (4.42')$$

Note (4.42) by Martingales is the same as the maximum likelihood estimator (m.l.e.) of γ.

$$= \int_0^t H(\tau)\, dN(\tau) - \int_0^t H(\tau)\, \Lambda(\tau)\, d\tau$$

$$= \int_0^t H(\tau)\, dN(\tau) - \gamma \int_0^t H(\tau)\, I(\tau)\, S(\tau)\, d\tau$$

where, $\Lambda(\tau)\, d\tau = \gamma I(\tau)\, S(\tau)\, d\tau$.

$$= \int_0^t H(\tau)\, dN(\tau) - \gamma \int_0^t \frac{k(\tau-)}{I(\tau-)\, S(\tau-)}\, I(\tau)\, S(\tau)\, d\tau$$

$$= \int_0^t H(\tau)\, dN(\tau) - \gamma \int_0^t k(\tau-)\, d\tau \tag{4.43}$$

$$\therefore\ E\,(\widehat{M(T)}) = E\left[\int_0^T H(\tau)\, dN(\tau) - \gamma \int_0^T k(\tau-)\, d\tau \right]$$

$$= \overline{M}(0) = 0 \tag{4.44}$$

$$\Rightarrow\quad E\left[\int_0^T H(\tau)\, dN(\tau) \right] = \gamma \int_0^T k(\tau-)\, d\tau$$

$$= E\left[(\hat\gamma) \int_0^T k(\tau-)\, d\tau \right]$$

$$\hat\gamma = \int_0^T H(\tau)\, dN(\tau) \bigg/ \int_0^T k(\tau-)\, d\tau$$

If $H(\tau) = 1$,

then
$$\hat\gamma = \frac{N(T)}{\int_0^T I(\tau)\, S(\tau)\, d\tau}; \quad (N(0) = 0) \tag{4.45}$$

which is same as (4.42)

Next
$$\mathrm{Var}\,(\overline{M}(t)) = \mathrm{Var}\left[\int_0^t H(\tau)\, dN(\tau) \right]$$

$$= E\left[\int_0^t H^2(\tau)\, dN(\tau) \right] \quad \text{by (4.41)} \tag{4.46}$$

Now Var $(\overline{M}(t))$ = Var $\left[\int\limits_0^T H(\tau)\, dN(\tau) - \gamma \int\limits_0^T k(\tau -)\, d\tau\right]$ by (4.44)

$$= \text{Var}\left[\int\limits_0^T \frac{k(\tau -)}{I(\tau -)\, S(\tau -)}\, I(\tau)\, S(\tau)\, d\tau - \gamma \int\limits_0^T k(\tau -)\, d\tau\right]$$

$$= \text{Var}\left[\hat{\gamma} \int\limits_0^T k(\tau -)\, d\tau - \gamma \int\limits_0^T k(\tau -)\, d\tau\right]$$

$$= E\,(\hat{\gamma} - \gamma)^2 \left[\int\limits_0^T k(\tau -)\, d\tau\right]^2 \tag{4.47}$$

In view of (4.46) and (4.47)

$$E\left[\int\limits_0^T H^2(\tau)\, dN(\tau)\right] = \text{Var}\,(\hat{\gamma})\left[\int\limits_0^T k(\tau -)\, d\tau\right]$$

$$\therefore \qquad\qquad \text{Var}\,(\hat{\gamma}) = \int\limits_0^T \frac{H^2(\tau)\, dN(\tau)}{S^T\, k(\tau)\, d\tau} \tag{4.48}$$

Special Case :
In case γ is time dependent i.e., $\gamma = \beta(t)$, then we can write

$$\overline{M}(t) = \int\limits_0^t H(\tau)\, dN(\tau)$$

$$= \int\limits_0^t H(\tau)\, d\left[N(\tau) - \int\limits_0^t \gamma(\tau)\, S(\tau)\, d\tau\right]$$

$$= \int\limits_0^t H(\tau)\, dN(\tau) - \int\limits_0^t H(\tau)\, \gamma(\tau)\, I(\tau)\, S(\tau)\, d\tau$$

$$= \int\limits_0^t H(\tau)\, dN(\tau) - \int\limits_0^t \frac{k(\tau -)}{I(\tau -)\, S(\tau -)}\, \gamma(\tau)\, I(\tau)\, S(\tau)\, d\tau$$

$$\overline{M}(t) = \int\limits_0^t H(\tau)\, dN(\tau) - \int\limits_0^t k(\tau)\, \gamma(\tau)\, d\tau$$

Again $\int\limits_0^t H(\tau)\, dN(\tau)$ can be regarded as an estimate of $\int\limits_0^t \gamma(\tau)\, d\tau$ with a standard error of

$$\sqrt{\int_0^t \gamma^2(\tau)\, dN(\tau)}$$

$$\left\{ \because \mathrm{Var}\left[\int_0^t \gamma(\tau)\, d\tau\right] = E\left[\int_0^t \gamma^2(\tau)\, dN(\tau)\right] \right\}$$

$$\left\{ \because \qquad \int_0^t M(\tau)\, dN(\tau) = \int_0^t H(\tau)\, \gamma(\tau)\, I(\tau)\, S(\tau)\, d\tau \right.$$

$$= \int_0^t H(\tau)\, \gamma(\tau)\, d\tau = \int_0^t \gamma(\tau)\, d\tau$$

where
$$k(\tau) = 1 \quad \text{if} \quad S(\tau) > 0$$
$$k(\tau) = 0 \quad \text{if} \quad S(\tau) = 0$$

4.17.2 General Epidemic Model

Consider the three types of sub populations viz, susceptible, infective and isolated or removed. The model is made realistic by further assuming that the infected individuals pass through a latent period of arbitrary duration.

Let us consider a population of i # infectives and s # susceptibles at time $t = 0$ (i.e. at start). Let $N(t)$ be the # individuals who are further infected in $(0, t]$; obviously $N(0) = 0$. Denote the number of susceptibles at time t as $S(t)$ Among the persons who have been infected upto time t $I(t)$ are in their infections period and $R(t)$ are removed by time t (by death or cure).

Now
$$P\{\delta N(t) = 1, \delta R(t) = 0\} = \beta I(t)\, S(t)\, \delta t + o\,(\delta t),$$

where $\beta \equiv$ Infection rate

$$P\{\delta N(t) = 0, \delta R(t) = 1\} = \gamma I(t)\, \delta t + o(\delta t)$$
$$P\{\delta N(t) = 0, \delta R(t) = 0\} = 1 - [\beta(t)\, S(t)\, \delta t + o(\delta t)] \qquad (4.49)$$

We denote
$$\theta = \frac{\beta}{\gamma} = \frac{\text{Infection rate}}{\text{Removal rate}} \qquad (4.50)$$

$$\gamma = \text{Removal rate} \cong \frac{1}{\text{Mean duration of the infection period}}$$

$\hat{\theta}$ = infection rate X Mean duration of infection period, the potential that an infective has a letter reasons of potential is while infecting a Susceptible

$$(4.51)$$

A better measure of potential is $S\theta$, where S is the number of susceptibles which is known as a threshold parameter.

Now $N(t)$ (the number of infected persons in $(0, t]$) and $R(t)$ (the number of isolated persons in $(0, t]$) are counting processes with parameters $\beta\, I(t)\, S(t)$ and $\gamma\, I(t)$ respectively.

Then
$$M_1(t) \;=\; N(t) - \int_0^t \beta\, I(\tau)\, S(\tau)\, d\tau \qquad\qquad (4.52)$$

and
$$M_2(t) \;=\; R(t) - \int_0^t \gamma\, I(\tau)\, d\tau \qquad\qquad (4.53)$$

are Martingales w r.t. $\mathcal{F}_{t-}$. The means of both the Martingales are zero.

$$E\,(N(t)) \;=\; \int_0^t \beta\, I(\tau)\, S(\tau)\, d\tau \qquad\qquad (4.54)$$

and
$$E\,(R(t)) \;=\; \int_0^t \gamma\, I(\tau)\, d\tau \qquad\qquad (4.55)$$

While our object is to estimate the parameter θ we note that we may not get the transition or changes in the values of $S(t)$ and $I(t)$ over the period of epidemic. Hence the estimating equations using Martingales must express θ as a function of observables.

Following Becker (1989), let us construct a predictable function as

$$H(\tau) \;=\; \frac{k(\tau -)}{S(\tau -)} \qquad\qquad (4.56)$$

where $k(\tau)$ is unity when $S(\tau) < 0$, otherwise $k(\tau) = 0$.

$H(\tau)$ is a predictable function which when integrated with respect to $M_1(t)$ gives a Martingale

$$\overline{M}(t) \;=\; \int_0^t H(\tau)\, dM_1(\tau) \quad \text{is a Martingale on } \mathcal{F}_{t-}$$

i.e.
$$\overline{M}(t) \;=\; \int_0^t \frac{k(\tau -)}{S(\tau -)}\, d\!\left(N(\tau) - \beta \int_0^t I(\tau)\, s(\tau)\, d\tau \right)$$

$$=\; \int_0^t \frac{k(\tau -)}{S(\tau -)}\, d\!\left(N(\tau) - \beta \int_0^t I(\tau)\, s(\tau)\, d\tau \right) \qquad \text{from (4.52)}$$

$$= \int_0^t H(\tau)\, dN(\tau) - \beta \int_0^t k(\tau -)I(\tau)\, d\tau \qquad (4.57)$$

Again when $S(\tau) > 0 \Rightarrow k(\tau -) = 1$

$$\therefore \qquad \int_0^t k(\tau -)\, I(\tau)\, d\tau = \int_0^t I(\tau)\, d\tau \quad \text{whenever } S(\tau) > 0$$

Also
$$\int_0^t H(\tau)\, dN(\tau) = \int_0^t \frac{k(\tau -)}{S(\tau -)}\, dN(\tau)$$

$$= \frac{1}{S} \cdot 1 + \frac{1}{(S-1)} \cdot 1 + \dots + \frac{1}{S(t -)} \qquad (4.58)$$

is completed by $S(t -)$.

Now consider the Martingale

$$M = \bar{M} - \theta\, M_2$$

which has zero mean, since $\bar{M}$ has zero mean and M_2 has also zero mean.

$$M = \bar{M} - \theta\, M_2$$

$$= \int_0^t H(\tau)\, dN(\tau) - \beta \int_0^t k(\tau -)\, I(\tau -)\, d\tau$$

$$- \theta \left[R(t) - \gamma \int_0^t I(\tau)\, d\tau \right]$$

$$= \int_0^t H(\tau)\, dN(\tau) - \theta\, R(t) - \beta \int_0^t k(\tau -)\, I(\tau -)\, d\tau$$

$$+ \beta \int_0^t I(\tau)\, d\tau \quad (\because \theta\gamma = \beta)$$

$$= \int_0^t H(\tau)\, dN(\tau) - \theta\, R(t) + \beta \int_0^t I(\tau)\, (1 - k(\tau))\, d\tau \qquad (4.59)$$

If we assume that we know

$$\int_0^T H(\tau)\, dN(\tau) = \frac{1}{S} + \frac{1}{(S-1)} + \dots + \frac{1}{(S_T - 1)} \qquad (4.60)$$

and $R(T)$ (i.e. No of persons isolated),

but we do not know $\quad\int\limits_0^T I(\tau)\,(1-k(\tau))\,d\tau$

where T is the stopping time of the epidemic when the infection process of the epidemic ends. T precisely represents time when either every infected individual has been removed or every susceptible has been infected. Following Becker (1989) we take,

$$T = \inf\{\,t \ge 0\,;\, S(t)\,[i + N(t) - R(t)]\} = 0. \qquad (4.61)$$

T is minimum too when either $S(T) = 0$ or $[i + N(T) - R(T)] = 0$
i.e. when the size of the susceptibles is zero or the total number of infected persons are removed.
Now

$$M(T) = \int\limits_0^T H(\tau)\,dN(\tau) - \theta\,R(T) + \beta\int\limits_0^T I(\tau)\,(1-k(\tau))\,d\tau \qquad (4.62)$$

where $\qquad M(t) = \overline{M} - \theta\,M_2$ is a Martingale

Now consider $\qquad\int\limits_0^T I(\tau)\,(1-k(\tau))\,d\tau$

We have, $k(\tau) = 1$ when $S(\tau) > 0$

$$\therefore \qquad M(T) = \int\limits_0^t H(\tau)\,dN(\tau) - \theta\,R(T) \qquad (4.63)$$

$$\Rightarrow \qquad \int\limits_0^T I(\tau)\,(1-k(\tau))\,d\tau = 0$$

or if $\qquad S(\tau) = 0\ $ then $\ k(\tau) = 0$

$$\Rightarrow \qquad \int\limits_0^T I(\tau)\,d\tau = 0$$

By Martingale optional sampling theorem

$$E\,(M(T)) = 0 \Rightarrow E\left[\int\limits_0^T H(\tau)\,dN(\tau)\right] - E\,[\hat\theta\,R(T)] = 0$$

$$\Rightarrow \hat\theta = \frac{\displaystyle\int\limits_0^T H(\tau)\,dN(\tau)}{R(T)} \qquad (4.64)$$

wher $\qquad \int\limits_0^T H(\tau)\,dN(\tau) = \dfrac{1}{S} + \dfrac{1}{(S-1)} + \ldots + \dfrac{1}{(S_T - 1)}$ from (4.60)

$\Rightarrow \qquad \hat{\theta} = \left\{ \dfrac{1}{S} + \dfrac{1}{(S-1)} + \ldots + \dfrac{1}{(S_T - 1)} \right\} R(T)$

The estimate of θ as given in (4.64) is valid when the date pertains to a single epidemic in a uniformly mixing community. To obtain the sampling variance of $\hat{\theta}$ we proceed as follows:

We have

$$\text{Var}\,(M(t)) = \text{Var}\,(\overline{M}(t) - \theta\,M_2(t))$$

$$= \text{Var}\,(\overline{M}(t)) + \theta^2\,\text{Var}\,(M_2(t)) - 2\theta\,\text{Cov}\,[\overline{M}(t),\,M_2(t)]$$

Since $\qquad \text{Cov}\,[\overline{M}(t),\,M_2(t)] = 0$

It follows that

$$\text{Var}\,(M(t)) = \text{Var}\,(\overline{M}(t)) + \theta^2\,\text{Var}\,(M_2(t))$$

$\Rightarrow \qquad \text{Var}\,(M(t)) = E\left[\int\limits_0^t H^2(\tau)\,dN(\tau) \right] + \theta^2\,E\,[R(t)] \qquad (4.65)$

($\because R(t)$ corresponds to the counting process in the formation of the Martingale $M_2(t)$)

Again,

$$\text{Var}\,(M(T)) = \text{Var}\left[\int\limits_0^T H(\tau)\,dN(\tau) - \hat{\theta}\,(R(T)) \right] \qquad \text{by (4.63)}$$

$$= \text{Var}\,(\hat{\theta}\,(R(T)))$$

$$= \text{Var}\,(\hat{\theta})\,(R(T))^2 \qquad (4.66)$$

Equating (4.65) and (4.66) we have

$$\left[\int\limits_0^T H^2(\tau)\,dN(\tau) \right] + \hat{\theta}^2\,[R(T)] = \text{Var}\,(\hat{\theta})\,(R(T))^2$$

$\Rightarrow \qquad \text{Var}\,(\hat{\theta}) = \dfrac{\int\limits_0^T H^2(\tau)\,dN(\tau) + \hat{\theta}^2\,R(T)}{(R(T))^2} \qquad (4.67)$

where $\qquad \int\limits_0^T H^2(\tau)\,dN(\tau) = \dfrac{1}{S^2} + \dfrac{1}{(S-1)^2} + \ldots + \dfrac{1}{\{S_T + 1\}^2}$

$$\Rightarrow \quad \text{Var}\,(\hat{\theta}) \;=\; \frac{\dfrac{1}{S^2} + \dfrac{1}{(S-1)^2} + \ldots + \dfrac{1}{(S_T+1)^2} + \hat{\theta}^2\,R(T)}{(R(T))^2} \tag{4.67'}$$

Now by using Martingales central limit theorem

$$\frac{\hat{\theta} - \theta}{\text{S.E.}\,(\hat{\theta})}$$

is approximately normally distributed where S is large. The result can be used to obtain the confidence interval of $\hat{\theta}$.

4.18 Some More Applications of Martingales in Estimating the Duration and the Size of the Mild Epidemic

For mild epidemic like conjunctivitis, for which the period of sickness is about a week and the period following recovery is considerably large, the average duration of mild epidemic is obtainable again by Martingales stopping rule assuming the case fatality is of the order of zero. The result is due to Biswas and Thomas (1989).

Let us consider a branching process

$$Y_{n+1} \;=\; Z_1^{(n)} + Z_2^{(n)} + \ldots + Z_{Y_n}^{(n)} \tag{4.68}$$

where Y_n represents the # newly infected persons in the nth unit of time (say, a week), ($n = 0, 1, 2\ldots$). $Z_i^{(n)}$ represents the number of newly infected persons from the ith infective ($i = 1, 2 \ldots Y_n$) in the $(n+1)^{\text{th}}$ unit of time.

Assuming that the epidemic starts with $Y_0 = i_0$ # infectives introduced in the population by immigration or by other similar sources, we have,

$$\left.\begin{aligned} E\,(Z_i^{(n)}) &= m & i = 1, 2, ..Y_n \\[4pt] \text{Var}\,(Z_i^{(n)}) &= \beta & \text{say} \\[4pt] \text{Cov}\,(Z_i^{(n)}, Z_j^{(n)}) &= 0 & \forall\, i \neq j \end{aligned}\right\} \tag{4.69}$$

Under the above assumptions, we have, by simple branching process

$$E\,(Y_n \mid Y_0 = i_0) \;=\; i_0 \cdot m^n \tag{4.70}$$

and

$$\left.\begin{aligned} \text{Var}\,(Y_n \mid Y_0 = i_0) &= \beta n \;; & \text{if } m = 1 \\[4pt] &= \beta\, m^{n-1}\!\left(\frac{1-m^n}{1-m}\right) ; & \text{if } m \neq 1 \end{aligned}\right\} \tag{4.71}$$

(vide "Stochastic Processes in Demography and applications"—Biswas (1988) John Wiley/Wiley Eastern, Page 21)

It is shown that $X_n = m^{-n}Y_n$ is a Martingale over $\mathcal{F}_n = \mathcal{F}(X_n)$ where $X_n \in \sigma$ field.

The Martingale is valid when the process is supercritical i.e. $m > 1$. For $m < 1$ as $n \to \infty$ $E(X_n)$ does not exist and when $m = 1$, $X_n = Y_n \to 0$ a.s. and $\text{Var}(X_n) \to \infty$ indicating a substantial instability. Hence we consider X_n when, $m > 1$ only.

Let T be the stopping time of the Martingale which will depend on Y_n. If at $n = T$ we have

$$Y_{T=0} | Y_i = i_0$$

then it may imply the stopping time of the epidemic; on the other hand, the epidemic may stop if

$$Y_0 + Y_1 + \dots + Y_T = M.$$

where M is the size of the population over which the epidemic is spread.

Let π_0 be the probability that the epidemic stops before affecting every individual in the population.

Then $\qquad\qquad Y_T = 0 \qquad$ with probability π_0

or, $\qquad\qquad = M - (Y_0 + Y_1 + \dots + Y_{T-1})$

$$= M - (i_0 + i_0 \cdot m + i_0 \cdot m^2 + \dots + i_0 \cdot m^{T-1})$$

$$= M - \frac{i_0}{m-1}(m^T - 1) \quad \text{with probability } (1 - \pi_0)$$

The stopping time

$$T = \min\left[t: Y_t = 0, \ Y_t = M - \frac{i_0}{m-1}(m^T - 1)\right]$$

$\Rightarrow \qquad E(Y_T | T) = E(Y_T = 0 | T) + E(Y_T \neq 0 | T)$

$$= \pi_0 \times 0 + (1 - \pi_0)\left[M - \frac{i_0}{m-1}(m^T - 1)\right] \qquad (4.72)$$

By optional sampling theorem

$$E(X_T) = E(m^{-T}Y_T) = E(X_0) = E(Y_0) = i_0$$

$\Rightarrow \qquad i_0 = E(m^{-T}Y_T)$

$$= E(m^{-T}(1 - \pi_0))\left[M - \frac{i_0}{m-1}(m^T - 1)\right]$$

$\Rightarrow \qquad \dfrac{i_0}{1 - \pi_0} = M E(m^{-T}) - \dfrac{i_0}{m-1} + \dfrac{i_0}{m-1} E(m^{-T})$

$\Rightarrow \qquad \dfrac{i_0(m - \pi_0)}{(1 - \pi_0)(m-1)} = \left(M + \dfrac{i_0}{m-1}\right) E(m^{-T})$

$$\Rightarrow \qquad E(m^{-T}) = \frac{i_0 (m - \pi_0)}{(1 - \pi_0)(m - 1)} \times \frac{1}{M + \dfrac{i_0}{m - 1}}$$

$$\Rightarrow \qquad E(1 - T \log_e m) \cong \frac{i_0 (m - \pi_0)}{(1 - \pi_0)(m - 1)} \times \frac{1}{M + \dfrac{i_0}{m - 1}}$$

(by neglecting higher power of $\log_e m$ in the expansion of m^{-T})

$$\Rightarrow \qquad \log_e m \, E(T) = 1 - \frac{i_0 (m - \pi_0)}{(1 - \pi_0)(m - 1)} \times \frac{1}{M + \dfrac{i_0}{m - 1}}$$

$$= \frac{M - \dfrac{i_0}{1 - \pi_0}}{M + \dfrac{i_0}{m - 1}}$$

$$\Rightarrow \qquad E(T) \cong \frac{M - \dfrac{i_0}{i - \pi_0}}{M + \dfrac{i_0}{m - 1}} \times \frac{1}{\log_e m} \qquad (4.73)$$

which gives an approximation of the duration of epidemic for a supercritical process. Note that for critical and subcritical processes i.e. for $m = 1$ and $0 < m < 1$, $\pi_0 = 1$ and hence $E(T)$ is indeterminate.

Evaluation of π_0

Duration of epidemic can be calculated if π_0 is known for different values of m. Now by the property of simple branching process, if the epidemic starts with i_0 number of infected persons and $\Phi(s)$ represents the p.g.f. of the distribution of the newly infected persons per unit time then,

$$\pi_0 = S^{i_0}$$

where S is the smallest root lying between 0 and 1 of the equation:

$$\Phi(S) = S$$

Let us assume that each infected person again infects k new persons (k is a random variable) following Binomial distribution with parameters n and p. Then s is the smallest root of the equation

$$(q + ps)^n - s = 0.$$

and $m = np$

π_0 for different values of i_0, n and p have been worked out by Biswas and Thomas (1989) which show the duration of epidemic for a supercritical as well as subcritical process.

References

1. Becker Neils G (1989): *Analysis of Infectious Disease Data.* Chapman and Hall, London, New york.
2. Biswas S and Pachal T.K. (1987): *On a method of estimating the stopping time of the basis of number of surviving children for a sterilization programme—A Martingale approach.* Proceedings of the Seminar of Stochastic Modelling and Decision making, University of Delhi, Khama Publishers, New Delhi.
3. Biswas S (1988): *Stochastic Processes in Demography and applications.* Wiley Eastern/John Wiley & Sons.
4. Biswas Suddhendu and Noor Hamed Saad (1989): *A Note on the application of Martingales in a problem of Non communicable epidemics*— Biometric journal, Vol 31, No 4, page 487 - 494.
5. Biswas Suddhendu and Ebraheem Nather Abas (1989): *A Martingale approach to the problem of extinction of some migration free closed population* - *SCIMA* Vol 18, no. 3, page 113 - 121.
6. Biswas S and Thomas M (1989):*A Martingale approach to the investigation of the duration of mild epidemic.* Biometric Journal, vol 31, No. 4, page 477-486.
7. Biswas S and Thomas M (1992): *On a Martingale approach for the estimation of the parameters of a Bivariate carrier borne epidemic model in presence of a linear relationship between carriers and susceptibles.* Biometric Journal, Vol 34, No. 2, page 219-230.
8. Biswas S and Thomas M (1993): *On a Martingale based approach for the estimation of the parameters affecting the growth of a tumour,* Sankhya, Series B, Vol 55, Part 2, page 219-228.
9. Gill R.D. (1984): *Understanding Cox's regression model—A Martingale approach Journal of American Statistical Association,* Vol 79, Page 441 - 447.
10. Doob J.L. (1953): Stochastic Processes, Wiley, New York. Hoel C.G. Port C.S. and Stone C.J. (1972): *Introduction to Stochastic Processes.* Houghton Miffin Company, Boston.
11. Karlin S and Taylor H.M. (1975): *A first course in Stochastic Processes.* Academic Press, New York.
12. Ross , Sheldon H (1983): *Stochastic Processes.* John Wiley & Sons. New York. Chichester, Toronto and Singapore.
13. Selke T and Siegmund D(1983): *Sequential Analysis of proportional hazard model* - Biometrika, Vol 70, No 2, page 315 - 326.
14. Takács L (1957): *Stochastic Processes (Problems and Solutions).* Translated by P. Zador, John Wiley & Sons, New York, Mathuen & Co.

Counter Theory and Age-Replacement Policy

5.0 Introduction

A Geiger-Müller or G.M. Counter model is a registering mechanism that detects the presence of radioactive material. Impulses emitted by the radioactive material arrive at the counter, but because of inertia, the counter does not register some of the impulses. Suppose that an impulse arrives at a fixed time t the counter registers the impulse. The registration results in a dead time of duration say π, during which impulses will not be registered by the counter. In general, the first impulse to arrive after the termination of the dead time π, again will be registered by the counter resulting in a dead time of length, say, π_2 and so on. However, in simpler cases, impulses arriving during dead time do not prolong it so that each period of dead time is caused by a registered impulse only. This is called a G.M. Counter of type I (Pyke 1958).

In a counter of type II, (Smith 1958) each arriving impulse causes a dead time so that arrivals during a period of dead time prolong it further. Theory of counter model arose from certain physical processes. It can be applied to many biological, social, demographic and industrial processes.

5.1 Demographic Problem of Counter Model with Fixed Dead Time π (Counter Model of Type I with Fixed Dead Time)

Problem

A sickness (say lead infection) takes place with intensity λ subject to the condition that every sickness is followed by fixed exposure π (dead time) during which no further sickness takes place, then obtain the probability distribution of the number of sickness in $(0, t]$.

Solution: Let $\tau_1 < \tau_2 < ... < \tau_n < ...$ be the renewal times (waiting time of sickness) and with sickness rate λ (Poisson intensity) with negative exponential density function,

$$f(t) = \lambda e^{-\lambda t}; \quad 0 \le t < \infty, \lambda > 0$$

$$\Rightarrow \qquad P[\tau_1 < x] = 1 - e^{-\lambda x}, \ \lambda \ge 0, 0 \le x < \infty \qquad (5.1)$$

$\Rightarrow$ $\qquad$ $P\left[\tau_n - \tau_{n-1} \leq x\right] = 1 - e^{-\lambda(x-\pi)}$ for $x \geq \pi$ $\qquad$ (5.2)

Let $W(t, n)$ = probability of not more than n events or sicknesses upto time t.

$$W(t, n) = P\left[\tau_{n+1} > t\right] = 1 - P\left[\tau_{n+1} \leq t\right]$$
$$= 1 - F_{n+1}(t) = R_{n+1}(t) \qquad (5.3)$$

where $F_{n+1}(t)$ is the cumulative distribution function (c.d.f.) of the random variable τ_{n+1} and $R_{n+1}(t) = 1 - F_{n+1}(t)$ is the corresponding survival function. Denoting $L(\cdot)$ as the Laplace transform

$$L\left(W(t, n)\right) = \int_0^\infty e^{-st} \left(1 - F_{n+1}(t)\right) dt$$

$$= \frac{1}{s} - \frac{1}{s} \cdot L(f_{n+1}(t)) \left(\text{Since } L\left(F_{n+1}(t)\right) = \frac{1}{s} L\left(f_{n+1}(t)\right)\right) \qquad (5.4)$$

where $f_{n+1}(t)$ is the interval density function of $(n+1)$th renewal (order of sicknesses).

We have $\qquad$ $f_{n+1}(t) = f_1 * \{f^{(n)}\}*$

where f_1 is the density function of τ_1 and f the density function of $(\tau_r - \tau_{r-1})$; $r = 2, 3, \ldots$ and $*$ stands for convolution; $\{f^{(n)}\}*$ stands for n-fold convolution of f.

$$L\left[f_{n+1}(t)\right] = L\left[f_1(t)\right] \{L[f(t)]\}^n$$

Now $\qquad$ $L\left[f_1(t)\right] = \dfrac{\lambda}{\lambda + s}$

and $\qquad$ $L\left[f(t)\right] = \int_\pi^\infty e^{-st} \lambda e^{-\lambda(t-\pi)} dt = \dfrac{\lambda e^{-\pi s}}{(\lambda + s)}$ $\qquad$ (5.5)

$\therefore$ $\qquad$ $L\left[f_{n+1}(t)\right] = \dfrac{\lambda^{n+1}}{(\lambda + s)^{n+1}} \cdot e^{-n\pi s}$ $\qquad$ (5.6)

Putting (5.5) and (5.6) in (5.4), we get

$$L\left[W(t, n)\right] = \frac{1}{s} - \frac{1}{s} \cdot \frac{\lambda^{n+1}}{(\lambda + s)^{n+1}} \cdot e^{-n\pi s} \qquad (5.7)$$

By taking inverse Laplace transform

$$W(t, n) = 1 - \lambda \int_{n\pi}^t \frac{e^{-\lambda(u - n\pi)} \lambda^n (u - n\pi)^n du}{\Gamma(n+1)} \qquad (5.8)$$

Also using the result $\displaystyle\sum_{j=0}^n e^{-\mu} \frac{\mu^j}{j!} = 1 - \int_0^\mu \frac{e^{-z} z^n dz}{\Gamma(n+1)}$ $\qquad$ (5.9) and putting $\mu = \lambda(t - n\pi)$ on both sides of (5.9), we get

$$\sum_{j=0}^{n} e^{-\lambda(t-n\pi)} \frac{[\lambda(t-n\pi)]^{j}}{j!} = 1 - \frac{1}{\Gamma(n+1)} \int_{0}^{\lambda(t-n\pi)} e^{-z} z^{n} dz \qquad (5.10)$$

Further, on substitution of

$$z = \lambda(u - n\pi)$$

$$z = 0 \Rightarrow u = n\pi$$

$$z = \lambda(t - n\pi) \Rightarrow u = t$$

$$\Rightarrow \sum_{j=0}^{n} e^{-\lambda(t-n\pi)} \frac{[\lambda(t-n\pi)]^{j}}{j!} = 1 - \frac{\lambda}{\Gamma(n+1)} \int_{n\pi}^{t} e^{-\lambda(t-n\pi)} [\lambda(t-n\pi)]^{n} du$$

$$(5.11)$$

Comparing (5.8) with (5.11), we have

$$W(t, n) = \sum_{j=0}^{n} e^{-\lambda(t-n\pi)} \frac{[\lambda(t-n\pi)]^{j}}{j!} \qquad (5.12)$$

Therefore $\qquad P[X = n] = W(t, n) - W(t, n - 1)$

$$= \sum_{j=0}^{n} e^{-\lambda(t-n\pi)} \frac{[\lambda(t-n\pi)]^{j}}{j!} - \sum_{j=0}^{n-1} e^{-\lambda(t-(n-1)\pi)} \frac{[\lambda(t-(n-1)\pi)]^{j}}{j!} \qquad (5.13)$$

and $n \le \left[\frac{t}{\pi}\right]$ where $\left[\frac{t}{\pi}\right]$ refers the greatest integer contained in $\frac{t}{\pi}$.

5.1.1 A Problem on Counter Model (Biswas, Nair and Nautiyal, 1983)

Show that the conditional probability of having n sicknesses at $x_1 < x_2 < ... < x_n$ subject to the condition that no sickness can occur between $(x_i, x_i + \pi)$ $(i = 1, 2, ..., n)$ is given by

$$\prod_{i=1}^{n-1} \frac{\phi(x_i) f(x_n)}{\prod_{i=1}^{n} \{1 - (F(x_i + \pi) - F(x_i))\}} \qquad (5.14)$$

where $\qquad \phi(x_i)$ is the hazard rate at $x = x_i$
$f(x_i)$ is the waiting time density function.
$F(\cdot)$ is the cumulative distribution function, at $X = x_i$, $i = 1, 2, .. n$
We have $\qquad P[X_1 < x_1, X_2 < x_2] = F(x_1, x_2)$

$$= \int_{0}^{x_1} \frac{F(x_2) - F(x)}{1 - F(x)} dF(x)$$

$$= \int_0^{x_1} \frac{F(x_2)}{\overline{F}(x)} \, dF(x) - \int_0^{x_1} \frac{F(x)}{\overline{F}(x)} \, dF(x)$$

where $$\overline{F}(x) = 1 - F(x)$$

$$= F(x_2) \int_0^{x_1} \frac{dF(x)}{\overline{F}(x)} + [F(x_1) + \log_e (1 - F(x_1))]$$

$$= F(x_2) \int_0^{x_1} \phi(x) \, dx + [F(x_1) + \log_e (1 - F(x_1))]$$

where $$\phi(x) = \frac{f(x)}{\overline{F}(x)}$$

and $$f(x) = \frac{dF(x)}{dx}$$

$\Rightarrow$ $$F(x_1, x_2) = F(X_2) \, \Phi(x_1) + F(x_1) - \int_0^{x_1} \phi(x) \, dx$$

$$\left(\because \log_e (1 - F(x_1)) = \log_e \overline{F}(x_1) = \log_e \left(\exp\left[-\int_0^{x_1} \phi(x) \, dx \right] \right) \right.$$

$$\left. = -\int_0^{x_1} \phi(x) \, dx \right)$$

and $$\int_0^{x_1} \phi(x) \, dx = \Phi(x_1)$$

$\Rightarrow$ $$F(x_1, x_2) = F(x_2) \, \Phi(x_1) + F(x_1) - \Phi(x_1)$$

$\therefore$ $$f(x_1, x_2) = \frac{\partial^2 F(x_1, x_2)}{\partial x_1 \, \partial x_2} = f(x_2) \phi(x_1)$$

Similarly, one can show that

$$f(x_1, x_2, x_3) = \phi(x_1) \, \phi(x_2) f(x_3)$$

and $$f(x_1, x_2, \ldots x_n) = \phi(x_1) \, \phi(x_2) \ldots \phi(x_{n-1}) f(x_n)$$

Now the conditional probability of n sicknesses at $x_1 < x_2 < \ldots < x_n$ subject to the condition that no sicknesses or event can occur between $(x_i, x_i + \pi)$ $(i = 1, 2, 3, \ldots n)$ is given by

$$= \prod_{i=1}^{n-1} \frac{\phi(x_i) f(x_n)}{\prod\limits_{i=1}^{n} \{1-(F(x_i+\pi)-F(x_i))\}}$$

5.2   Counter Model Type I with Random Variable Dead Time

A particle arrives at $t = 0$ and locks the counter for a dead time of duration Y_1. With the registration of the particle the counter is blocked for a time of length say Y_2. The next particle to be registerd is that of the first arrival once the counter is freed. The process is repeated where the successive locking times denoted by $Y_1, Y_2, Y_3 \ldots$ are assumed independent with a common distribution $P\{Y_k < y\} = G(y)$ and independent of the arrival process.

   To obtain the waiting time distribution of the inter-arrival of the process.

Let
$$Z = Y + \gamma_y$$

where Y is the dead time following the registration and γ_y is the residual life time before the counter is locked.

   $g(\cdot)$ and $f(\cdot)$ denote the density function of the dead time and the residual free time (before the system is locked again)

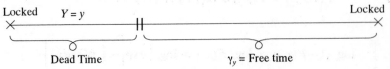

Locked $Y = y$ Locked

Dead Time $\gamma_y =$ Free time

Fig. 5.1

We have,

$$P\{Z \leq Y + \gamma_y \leq z + dz\} = \psi(z)\,dz = \int_0^z g(y) f(z-y)\,dy \qquad (5.15)$$

A special case: If $g(y) = \mu e^{-\mu y}$
$$f(z-y) = \lambda \exp[-\lambda(z-y)]$$

then, $$\psi(z) = \frac{\lambda\mu\, e^{-\mu z}}{(\mu-\lambda)} \{1 - \exp[-(\mu-\lambda)z)]\}; \quad \lambda \neq \mu$$

represents the waiting time distributions between two consecutive registrations.

   Biswas, Nautiyal and Tyagi (1983) obtain the p.g.f. of the process as

$$\Pi_t(s) = \frac{1}{2}\left[1 + \frac{(\lambda+\mu)}{(\lambda^2+\mu^2-2\lambda\mu+4\lambda\mu s)}\right]$$

$$\times \exp\left\{-\left(\frac{\lambda+\mu}{2}\right)t - \left(\frac{\mu-\lambda}{2}\right)t\left[1 + \frac{4\lambda\mu s}{(\mu-\lambda)^2}\right]^{\frac{1}{2}}\right\}$$

$$+ \frac{1}{2}\left[1 - \frac{(\lambda + \mu)}{(\lambda^2 + \mu^2 - 2\lambda\mu + 4\lambda\mu s)^{1/2}}\right]$$

$$\times \exp\left\{-\left(\frac{\lambda + \mu}{2}\right)t - t\left(\frac{\mu - \lambda}{2}\right)\left[1 + \frac{4\lambda\mu s}{(\mu - \lambda)^2}\right]^{\frac{1}{2}}\right\} \quad (5.16)$$

5.3   Counter Model Type II with Fixed Dead Time

Here the locking mechanism is more complicated. As before an incoming signal is registered if it arrives when the counter is free. In type II counter every arriving signal can prolong the dead time in the counter, increasing the associated locking times.

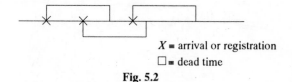

$X \equiv$ arrival or registration
$\square \equiv$ dead time

Fig. 5.2

Let $\tau_1 < \tau_2 < ... < \tau_n$ are the registration times of the first, second and nth renewal and arrivals take place with Poisson rate λ. The length of the dead time following every registration is π. We assume $\{\tau_n - \tau_{n-1}\}$ are the i.i.d.r.v.'s.

$$P\{\tau_1 \leq x\} = 1 + e^{-\lambda x} \quad (5.17)$$

$$P\{(\tau_n - \tau_{n-1}) \leq x\} = F(x) \quad (5.18)$$

where the structure of $F(x)$ is to be obtained.

Let
$$P\{x \leq \tau_n - \tau_{n-1} \leq x + dx\} = f(x)\, dx$$

Let
$$\phi(s) = \int_0^\infty e^{-st} f(t)\, dt = L(f(t))$$

Let
$$u(t) = \frac{dU(t)}{dt}$$

be the renewal density of the process and $U(t)$ be the renewal function.

Then
$$L[u(t)] = \frac{L(f_1(t))}{1 - L(f(t))}$$

$$L[f_1(t)] = \lambda \int_0^\infty e^{-st} e^{-\lambda t}\, dt = \frac{\lambda}{\lambda + s} \quad (5.19)$$

Also
$$L[u(t)] = L\left[\frac{dU(t)}{dt}\right] = \frac{\left(\dfrac{\lambda}{\lambda + s}\right)}{1 - \phi(s)}$$

$$\Rightarrow \qquad 1 - \phi(s) = \frac{\lambda}{\lambda + s} \left[\int_0^\infty e^{-st} \frac{dU(t)}{dt} \cdot dt \right]^{-1}$$

$$\Rightarrow \qquad 1 - \frac{\lambda}{\lambda + s} \left[\int_0^\infty e^{-st} \frac{dU(t)}{dt} \cdot dt \right]^{-1} = \phi(s) \qquad (5.20)$$

Next to obtain $\dfrac{dU(t)}{dt}$, we proceed as follows:
We have

$$U(t + \delta t) - U(t) = e^{-\lambda t} \lambda \cdot \delta t + o(\delta t) \quad \text{if } t < \pi \qquad (5.21)$$

given that initially the counter is free and

$$U(t + \delta t) - U(t) = e^{-\lambda \pi} \lambda \cdot \delta t + o(\delta t) \quad \text{if } t \geq \pi \qquad (5.22)$$

$$\Rightarrow \qquad \begin{aligned} \frac{dU(t)}{dt} &= \lambda e^{-\lambda t} \quad ; \text{ if } t < \pi \\[2mm] &= \lambda e^{-\lambda \pi} \quad ; \text{ if } t \geq \pi \end{aligned} \left. \right\} \qquad (5.23)$$

$$\therefore \qquad L\left[\frac{dU(t)}{dt}\right] = \int_0^\infty e^{-st} \frac{dU(t)}{dt} \cdot dt$$

$$= \int_0^\pi e^{-st} \frac{dU(t)}{dt} \cdot dt + \int_\pi^\infty e^{-st} \frac{dU(t)}{dt} \cdot dt$$

$$= \lambda \int_0^\pi e^{-st} e^{-\lambda t} \, dt + \int_\pi^\infty e^{-st} \lambda e^{-\lambda \pi} \, dt$$

$$L[u(t)] = \frac{\lambda \exp(-(\lambda + s)\pi)}{s} - \frac{\lambda \exp(-(\lambda + s)\pi)}{(\lambda + s)} + \frac{\lambda}{\lambda + s} \qquad (5.24)$$

$$\Rightarrow u(t) = \frac{dU(t)}{dt} = L^{-1}\left[\frac{\lambda}{\lambda + s} + \lambda e^{-\lambda \pi}\left(\frac{1}{s} - \frac{1}{\lambda + s}\right)e^{-s\pi}\right] \qquad (5.25)$$

Now
$$L^{-1}\left(\frac{\lambda}{\lambda + s}\right) = \lambda e^{-\lambda t}$$

$$\therefore \quad L^{-1}\left(\lambda e^{-\lambda \pi}\left(\frac{1}{s} - \frac{1}{\lambda + s}\right)\right) = \lambda e^{-\lambda \pi}(1 - e^{-\lambda t})$$

$$L^{-1}\left[\lambda e^{-\lambda \pi}\left(\frac{1}{s} - \frac{1}{\lambda + s}\right)e^{-s\pi}\right] = \lambda e^{-\lambda \pi}\{1 - \exp[-\lambda(t - \pi)]\} \quad ; \text{ if } t > \pi$$

$$= \lambda e^{-\lambda \pi} \quad ; \text{ if } t \leq \pi$$

by the second shifting property of Laplace transform.

$$\therefore \quad u(t) = \lambda e^{-\lambda t} + \lambda e^{-\lambda \pi} (1 - \exp[-\lambda (t - \pi)]) \quad ; \text{if } t \geq \pi$$
$$= \lambda e^{-\lambda t} + \lambda e^{-\lambda \pi} \qquad\qquad\qquad ; \text{if } t < \pi \quad\quad (5.26)$$

$$L (U (t)) = \frac{1}{s} L \left(\frac{dU (t)}{dt} \right) = \frac{1}{s} L (u(t))$$

$$L (U (t)) = \frac{1}{s} \left[\frac{\lambda}{\lambda + s} + \lambda \left(\frac{1}{s} - \frac{1}{\lambda + s} \right) \exp(-(\lambda + s) \pi) \right]$$

$$\Rightarrow U(t) = (1 - e^{-\lambda t}) + \lambda e^{-\lambda \pi} \left[(t - \pi) - \frac{1}{\lambda} (1 - \exp\{-\lambda (t - \pi)\}) \right] \text{ if } t \geq \pi$$

$$= (1 - e^{-\lambda t}) \qquad \text{if } t < \pi \qquad\qquad (5.27)$$

For the waiting time distribution $f(\cdot)$, the reader is referred to Takacs (1960).

Biswas (1980) has shown an application of the model to describe the morbid condition in the human system, in which one morbid condition in presence of another only prolongs the morbidity period (dead time). The average number of fresh morbid spells in $(0, t)$ under a given morbidity rate λ and sickness period π (which may got extended when further complication occurs with the same rate λ) is given by (5.27).

5.4 Counter Model Type II with Random Variable Dead Time

The counter process is quite complicated and difficult to analyze in general form. However, the results are known only for process with Poisson inputs with arrival rate λ.

Let $p(t)$ be the probability that the counter is free at a time t (i.e. registration is possible).

We require to show that

$$p(t) = \exp\left(-\lambda \int_0^t [1 - G(y)] \, dy \right) \qquad\qquad (5.28)$$

where $G(y)$ represents the c.d.f. of the dead time distribution. We assume a result (without proof) to prove (5.28) as follows:

[Given n occurrences of a Poisson process in the interval $(0, t)$ the distribution of occurrence times is the same as that of n independent random variables taken from a uniform distribution in $(0, t)$].

Proof: The counter is free at time $T = t$ if all dead periods engendered by these signals have been terminated before t.

Let $G(t - y) = P$ [dead time commencing in y will end before time t]

$$P \text{ [induced period culminated prior to } t] = \frac{\int_0^t G(t - y) \, dy}{\int_0^t dy} \qquad\qquad (5.29)$$

Since the locking times are assumed to be independent of the arrival process, we have

$$
P\{\text{counter is free at time } t \mid n \text{ signals in } (0, t)\} = \left[\frac{\displaystyle\int_0^t G(t - y)\, dy}{\displaystyle\int_0^t dy} \right]^n
\tag{5.30}
$$

But the number of signals arriving during the interval $(0, t)$ has a Poisson distribution with mean λt.

From the law of total probability

$$
p(t) = \sum_{j=0}^{\infty} \left\{ \frac{1}{t} \int_0^t G(t - y)\, dy \right\}^j \frac{(\lambda t)^j\, e^{-\lambda t}}{j!}
$$

$$
= \sum_{j=0}^{\infty} \left\{ \frac{\lambda t}{t} \int_0^t G(t - y)\, dy \right\}^j \frac{e^{-\lambda t}}{j!}
\tag{5.31}
$$

$$
= e^{-\lambda t} \sum_{j=0}^{\infty} \frac{\left\{ \lambda \int_0^t G(t - y)\, dy \right\}^j}{j!}
$$

$$
= e^{-\lambda t} \exp\left(\lambda \int_0^t G(t - y)\, dy \right)
$$

$$
= \exp\left(-\lambda \int_0^t dy \right) \exp\left(\lambda \int_0^t G(t - y)\, dy \right)
$$

$$
= \exp\left(-\lambda \int_0^t dy \right) \exp\left(\lambda \int_0^t G(\tau)\, d\tau \right)
$$

$$
\therefore \qquad p(t) = \exp\left(-\lambda \int_0^t (1 - G(y))\, dy \right)
\tag{5.32}
$$

Continuing with our assumption $p(t)$ can be related to $U(t)$

and we have,
$$
\frac{dU(t)}{dt} = \lambda\, p(t)
$$

where $U(t)$ represents the renewal function in $(0, t)$.

Now P [a signal appearing in $(t, t + \delta t)$] $= p(t) \lambda \, \delta t + o \, (\delta t)$

$\therefore$ $U(t + \delta t) = U(t) + \lambda \, \delta t \, p(t) + 0 \, (1 - p \, \lambda \, \delta t) + o(\delta t)$

and $\displaystyle \lim_{\delta t \to 0} \frac{U(t + \delta t) - U(t)}{\delta t} = \lambda \, p(t)$

$\Rightarrow$ $\displaystyle \frac{dU(t)}{dt} = \lambda \, p(t)$

$\Rightarrow$ $\displaystyle U(t) = \lambda \int_0^t p(t) \, d\tau \quad (\because \; U(0) = 0)$ \hfill (5.33)

Putting (5.32) in (5.33), we get

$$U(t) = \lambda \int_0^t \exp\left(-\lambda \int_0^\tau [1 - G(y)] \, dy \right) d\tau \tag{5.34}$$

This proves the result.

5.4.1   A Problem of Counter Model Type II with Random Variable Dead Time

Obtain the expectation and variance of inter-registration time in a counter model type II with random variable dead time. (Takac's 1960; Biswas, Nautiyal and Tyagi 1983).

We have $\displaystyle p(\tau) = \exp\left(-\lambda \int_0^\tau [1 - G(y)] \, dy \right) = P$ [the counter is open]

$$u(\tau) = \lambda \exp\left(-\lambda \int_0^\tau [1 - G(y)] \, dy \right) = \text{Renewal density at } \tau$$

$$u(\infty) = \lambda \exp\left(-\lambda \int_0^\infty [1 - G(y)] \, dy \right) = \lambda e^{-2\alpha} \tag{5.35}$$

where $\displaystyle \int_0^\infty (1 - G(y)) \, dy = \alpha = \text{mean of the dead time.}$

By asymptotic renewal theorem, denoting μ as the mean of the inter-arrival time, we have

$$\mu \to \frac{1}{u(\tau)} \quad \text{as } \tau \to \infty$$

$$\Rightarrow \qquad \mu \;\rightarrow\; \frac{1}{u(\infty)} \;=\; \frac{1}{\lambda\,e^{-\lambda\alpha}} \;=\; \frac{e^{\lambda\alpha}}{\lambda} \tag{5.36}$$

Thus
$$\mu \;=\; \frac{e^{\lambda\alpha}}{\lambda}$$

For the variance σ^2 of the inter-arrival time, we have

$$U(t) \;=\; \left[\frac{t}{\mu} + \frac{\sigma^2 - \mu^2}{2\mu^2} + 0(1)\right] \tag{5.37}$$

holds for large values of t.

(Vide Cox D.R. Renewal Theory, Mathuen monograph)

$$\Rightarrow \qquad 2\mu^2\,U(t) \;=\; 2\mu t + (\sigma^2 - \mu^2) + o(2\mu^2)$$

$$\Rightarrow \qquad 2\mu^2\lambda \int_0^t \exp\left[-\lambda \int_0^\tau (1 - G(y))\,dy\right] d\tau$$

$$=\; 2\mu \int_0^t d\tau + (\sigma^2 - \mu^2) + o(2\mu^2)$$

$$\Rightarrow \qquad 2\mu^2\lambda \left[\int_0^t \left(\exp\left(-\lambda \int_0^\tau (1 - G(y))\,dy\right) - \frac{1}{\lambda\mu}\right) d\tau\right]$$

$$+\; \mu^2 + o(2\mu^2) \;=\; \sigma^2$$

Denoting
$$\mu^2 + o(2\mu^2) \;=\; \sigma_0^2, \text{ we have}$$

$$\sigma^2 \;=\; \sigma_0^2 + 2\mu^2\lambda \left[\int_0^t \left(\exp\left(-\lambda \int_0^\tau (1 - G(y))\,dy\right) - \frac{1}{\lambda\mu}\right) d\tau\right]$$

$$\sigma^2 \;=\; \sigma_0^2 + 2\mu^2\lambda \left[\int_0^t \left(p_0(\tau) - \frac{1}{\lambda\mu}\right) d\tau\right] \tag{5.38}$$

where
$$p_0(\tau) \;=\; \exp\left(-\lambda \int_0^\tau (1 - G(y))\,dy\right)$$

$$=\; P\,[\text{the counter is open}]$$

Again as $\tau \rightarrow \infty$

$$p_0(\tau) \;\rightarrow\; \exp\left(-\lambda \int_0^\infty (1 - G(y))\,dy \;=\; e^{-\lambda\alpha} \;=\; \frac{1}{e^{\lambda\alpha}}\right)$$

Hence $\sigma^2 = \sigma_0^2$ as $\tau \to \infty$

To evaluate σ_0^2 we proceed as follows :

$$L(u(t)) = \frac{L(f_1(t))}{1 - L(f(t))}$$

$$L(u(t)) - \phi(s) L(u(t)) = \frac{\lambda}{\lambda + s} \quad (\text{since } L(f(t)) = \phi(s)))$$

and $$L(f_1(t)) = \frac{\lambda}{\lambda + s}$$

$\Rightarrow$ $$L(u(t))(1 - \phi(s)) = \frac{\lambda}{\lambda + s}$$

$\Rightarrow$ $$(1 - \phi(s)) = \frac{\lambda}{\lambda + s}(L(u(t)))^{-1}$$

$$1 - \frac{\lambda}{\lambda + s}\{L(u(t))\}^{-1} = \phi(s)$$

$\Rightarrow$ $$1 - \frac{\lambda}{\lambda + s}\left\{\int_0^\infty e^{-st}\lambda\left[\exp\left(-\lambda\int_0^\tau (1 - G(y))\,dy\right)\right]dt\right\}^{-1} = \phi(s)$$

$\Rightarrow$ $$\phi(s) = 1 - \frac{\lambda}{\lambda + s}\left\{\int_0^\infty \exp\left(-\lambda\int_0^\infty (1 - G(y))\,dy\right)\lambda e^{-st}dt\right\}^{-1}$$

as $\tau \to \infty$

$$= 1 - \frac{\lambda}{\lambda + s}\left\{\int_0^\infty \lambda e^{-st}e^{-\lambda\alpha}dt\right\}^{-1}$$

$$= 1 - \frac{\lambda}{\lambda + s}\left(\lambda e^{-\lambda\alpha}\frac{1}{s}\right)^{-1}$$

$$= 1 - \frac{\lambda}{\lambda + s}\cdot\frac{1}{\lambda}e^{\lambda\alpha}\cdot s$$

$$= 1 - \frac{s}{\lambda + s}e^{\lambda\alpha} \tag{5.39}$$

Also

$$\phi'(s) = \left(\frac{(\lambda + s) - s}{(\lambda + s)^2}\right)e^{\lambda\alpha} = \frac{\lambda}{(\lambda + s)^2}e^{\lambda\alpha}$$

$$\Rightarrow \qquad \phi'(0) = \frac{e^{\lambda\alpha}}{\lambda} = \mu \tag{5.40}$$

$$\phi''(0) = \left.\frac{2(\lambda+s)\lambda}{(\lambda+s)^4}e^{\lambda\alpha}\right|_{s=0} = \left.\frac{2\lambda^2}{(\lambda+s)^4}e^{\lambda\alpha}\right|_{s=0} = \frac{2}{\lambda^2}e^{\lambda\alpha} \tag{5.41}$$

$$\therefore \qquad \sigma_0^2 = \phi''(0) - [\phi'(0)]^2$$

$$= \frac{2}{\lambda^2}e^{\lambda\alpha} - \mu^2 = \frac{2}{\lambda}\mu - \mu^2 = \frac{2\mu - \lambda\mu^2}{\lambda} \tag{5.42}$$

Putting (5.42) in (5.38), we have

$$\sigma^2 = \left(\frac{2\mu - \lambda\mu^2}{\lambda}\right) + 2\mu^2\lambda\left[\int_0^t \left(p_0(\tau) - \frac{1}{\lambda\mu}\right)d\tau\right] \tag{5.43}$$

The present technique of derivation of (5.43) is due to Biswas *et al* (1983).

5.5   Some Problems on Counter Model Type I

Problem : 1   (Cox and Isham (1980))

We assume that inter-registration time begins with a dead time. If X is a random variable representing the inter-registration time interval in a G.M. Counter type I and f_X stands for the density function of X then,

$$f_X(x) = \int_0^x dy\, g(y) \left\{ f_Z(x) + \int_0^y dz\, h_Z(z)\, f_Z(x-z) \right\} \tag{5.44}$$

where

$g(y)$ = density function of dead time
$h_Z(z)$ = renewal density corresponding to f_Z
$Z \equiv$ a r.v. measuring the probability of arrival from the point of last arrived.

Proof:
The first term corresponds to the case when the first arrival comes at x with probability $f_Z(x)$ when the dead time is over. The probability of the same is

$f_Z(x)\int_0^x g(y)\,dy$. The second case corresponds to the case when the last but one

arrival occurs during the dead time Z where $0 \le Z \le y$. Thereafter, the last arrival is after a time interval $x - z$ with probability:

$$\int_0^y f_Z(x-z)\,h_Z(z)\,dz \int_0^x g(y)\,dy$$

We have from Renewal Theory,

$$h_Z(x) = f_Z(x) + \int_0^x f_Z(x - z) h_Z(z) dz \qquad (5.45)$$

(forward integral equation).

Now (5.44) $\Rightarrow$

$$f_X(x) = f_Z(x) \int_0^x g(y) dy + \int_0^y h_Z(z) f_Z(x - z) \int_0^x g(y) dy$$

$$= \int_0^x g(y) dy \left[f_Z(x) + \int_0^y h_Z(z) f_Z(x - z) dz \right]$$

Using (5.45) in the above $\Rightarrow$

$$= \int_0^x g(y) dy \left[h_Z(x) - \int_0^x h_Z(z) f_Z(x - z) dz + \int_0^y h_Z(z) f_Z(x - z) dz \right]$$

$$= \int_0^x g(y) dy \left[h_Z(x) - \int_y^x h_Z(z) f_Z(x - z) dz \right] \quad (\because x \geq y)$$

$$\therefore \qquad F_X(x) = \int_0^x g(y) dy \left\{ \int_0^x h_Z(t) dt - \int_y^x h_Z(z) \left[\int_0^x f_Z(t - z) dt \right] dz \right\}$$

For $x \geq y$

$$= \int_0^x g(y) dy \left[\int_0^x h_Z(t) dt - \int_y^x F_Z(x - z) h_Z(z) dz \right]$$

$$= \int_0^x g(y) dy \left[\int_y^x h_Z(t) dz - \int_y^x h_Z(z) F_Z(x - z) dt \right]$$

$$= \int_0^x g(y) dy \left[\int_y^x h_Z(z) [1 - F_Z(x - z)] dz \right]$$

$$= \int_0^x g(y) dy \left[\int_y^x h_Z(z) R_Z(x - z) dz \right] \qquad (5.46)$$

where $R_Z(x - z) = 1 - F_Z(x - z)$

Integrating by parts, we have

$$
F_X(x) = \int_y^x h_Z(z) \, R_Z(x-z) \, dz \int g(y) \, dy \Bigg|_{y=0}^{y=x}
$$

$$
- \int_0^x \left\{ \frac{d}{dy} \int_y^x h_Z(z) \, R_Z(x-z) \, dz \int g(y) \, dy \right\} dy
$$

$$
= I_1 - I_2 , \text{ say}
$$

Now $I_1 = 0$ as $G(0) = 0$

$$
= - \int_0^x (-) h_Z(y) \, R_Z(x-y) \, G(y) \, dy
$$

$$
\therefore \qquad F_X(x) = \int_0^x G(y) \, F_Z(x-y) \, h_Z(y) \, dy
$$

Problem : 2   (Cox and Isham (1980))

Show that

$$
\mu_X = E(X) = \mu_Z \left\{ 1 + \int_0^\infty H_Z(y) \, g(y) \, dy \right\} \tag{5.47}
$$

where μ_X is the expected time between two consecutive registrations and μ_Z, that between consecutive arrivals.

$H_Z(y)$ is the renewal function (i.e., the average number of arrivals in $(0, y]$)
$g(\cdot) \equiv$ density function of dead time,
Consider

$$
E(X \mid Y = y) = \mu_Z H_Z(y) + \mu_Z \cdot 1
$$

$$
= \mu_Z (1 + H_Z(y))
$$

$$
E(X) = \int_0^\infty \mu_Z (1 + H_Z(y)) \cdot g(y) \, dy
$$

$$
= \mu_Z \int_0^\infty g(y) \, dy + \mu_Z \int_0^\infty H_Z(y) \, g(y) \, dy
$$

Since
$$
\int_0^\infty g(y) \, dy = 1
$$

$$E(X) = \mu_X = \mu_Z \left[1 + \int_0^\infty H_Z(y)\, g(y)\, dy \right]$$

Problem : 3 Show that if $p(t)$ is the probability that the counter is open given that it begins with a dead time at $t = 0$, then

$$1 - p(t) = \overline{G(t)} + \int_0^t \overline{G}(t - \tau)\, h_X(\tau)\, d\tau$$

where $$\overline{G(t)} = 1 - G(t), \quad G(t) = \int_0^t g(u)\, du$$

As $\tau \to \infty$, $H_X(\tau) \cong \dfrac{\tau}{\mu_X} \Rightarrow$ as $t \to \infty$, $p(t) = 1 - \dfrac{\mu_\gamma}{\mu_X}$

Problem : 4   (Biswas and Nauharia (1981))

Obtain the waiting time distribution for the nth arrival when the arrival rate (Poisson) is weighted by gamma distribution

$$\phi(\lambda) = \frac{a^k e^{-a\lambda} \lambda^{k-1}}{\Gamma(k)}, \quad 0 \le \lambda \le \infty;\ a, k > 0$$

and between two arrivals, there is a fixed dead time with length π.
Hints: We have

$$f(t \mid \lambda) = \lambda e^{-\lambda t}$$

$$L\left(f_n(t \mid \lambda)\right) = \left(\frac{\lambda}{\lambda + s}\right)^n$$

$\Rightarrow$ $$f_n(t \mid \lambda) = \frac{\lambda^n e^{-\lambda t} t^{n-1}}{\Gamma(n)}$$

$$f_n(t) = \frac{a^k t^{n-1}}{\Gamma(n)\,\Gamma(k)} \int_0^\infty e^{-\lambda(a+t)} \lambda^{n+k-1}\, d\lambda$$

$$= \frac{a^k}{\beta(n,k)} \cdot \frac{t^{n-1}}{(a+t)^{k+n}} = \frac{a^k}{B(n,k)} \cdot \frac{t^{n-1}}{(a+t)^{k+n-1}}$$

Incorporating a dead time of length π following each interval, we have, the distribution of the waiting time t_n,

$$f(t_n) = \frac{a^k}{B(n,k)} \cdot \frac{(t_n + (n-1)\pi)^{n-1}}{(a + t_n + (n-1)\pi)^{k+n-1}}$$

which is beta-distribution of type II.

Problem : 5   (Bivariate G.M. Counter)

When one organ in the parallel bi-component system fails and is under repair (treatment) for a fixed period π, the hazard rate λ (under Poisson input) is increased to λ' ($\lambda' > \lambda$). The system failure occurs when the surviving organ fails during the time of repair. Show that the probability of the first system failure at the first failure of either of the two organs is given by

$$p^{(1)} = 2\left(\frac{\lambda}{\lambda + \lambda'}\right)(1 - e^{-\lambda\pi}) + 2\lambda'\pi - \frac{\lambda}{\lambda'}(1 - e^{\lambda\pi})$$

(Biswas and Nair (1984))

Solution:

Let $f(\cdot) \equiv$ interval density of the failure distribution
$u(\cdot) \equiv$ renewal density of the failure distribution
$F(\cdot) \equiv$ c.d.f. of the failure distribution
$\overline{F}(.) \equiv$ survival function of the failure distribution

Then the probability that the first organ fails between x_1 to $x_1 + dx_1$, first time, is $f(x_1)\, dx_1$. Again let P [that the second organ fails during the time of repairment π [or treatment) of the first organ] i.e. between x_1 to $x_1 + \pi$

$$= \phi(x_1, \pi)$$

If we assume the two separate Geiger Müller counters for the first and the second organ than we can find

$$\phi(x_1, \pi) = \int\limits_{x_1}^{x_1 + \pi} \left[\overline{F}(t) + \int\limits_{\pi}^{t} u(t - y)\, \overline{F}(y - \pi)\, dy\right] u(t)\, dt$$

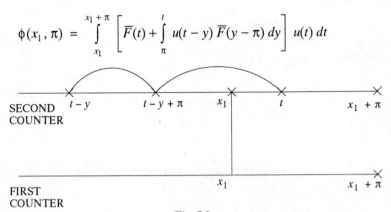

Fig. 5.3

In order that a failure of the second component at some time $t \in (x_1, x_1 + \pi)$ occurs, we may have

   (i)   either no failure of the second organ takes place upto t (see Fig. 5.3) and then a failure occurs at $t + 0$ where $t \in (x_1, x_1 + \pi)$

   (ii) or a failure of the second organ occurred first time at $(t - y) < x_1$ and it was sent for repair for a period π (i.e. counter remains locked till a

period $t - y + \pi$) and then no failure occurred between $(t - y + \pi)$ to t.
The probability of this event in the second counter is given by

$$\int_{x_1}^{x_1 + \pi} \overline{F}(t)\, u(t)\, dt + \int_{x_1}^{x_1 + \pi} \left[\int_{\pi}^{t} u(t - y)\, \overline{F}(t - (t - y + \pi))\, dy \right] u(t)\, dt$$

$$= \int_{x_1}^{x_1 + \pi} \left[\overline{F}(t) + \int_{\pi}^{t} u(t - y)\, \overline{F}(y - \pi)\, dy \right] u(t)\, dt$$

Hence the probability that either of the two organs fails first and the surviving organ also fails during treatment of the first failing component is given by

$$p^{(1)} = 2 \int_{0}^{\infty} f(x_1) \left\{ \int_{x_1}^{x_1 + \pi} \left[\overline{F}(t) + \int_{\pi}^{t} u(t - y)\, \overline{F}(y - \pi)\, dy \right] u(t)\, dt \right\} dx_1$$

Finally, putting $\quad \overline{F}(t) = e^{-\lambda t}, \quad u(t - y) = \lambda,$

$$f(x) = \lambda\, e^{-\lambda x} \text{ etc., the result follows.}$$

5.6 Models on Age Replacement Policy (A.R.P)

Age replacement policy in reliability is a policy which replaces an item upon reaching a fixed age T or failure whichever is earlier. An analogous example exists in Biostatistics. Suppose a heart patient visits a clinic for check up periodically, at a regular time interval of T days or whenever he faces some complications, whichever is earlier. Each clinic check up on each complication is considered to be analogous to one failure deliberately in the reliability of the system. The motivation of A.R.P is to increase system longevity.

Let us now define a failure distribution $F_T(x)$ under the age replacement policy such that

$$\left. \begin{aligned} F_T(x) &= F(x) && \text{for } x < T \\ &= 1 && \text{for } x \geq T \end{aligned} \right\} \tag{5.48}$$

when $F(x)$ is the c.d.f. of a natural failure distribution. The mean renewal time

$$\mu_T = \int_{0}^{\infty} R_T(x)\, dx = \int_{0}^{\infty} (1 - F_T(x))\, dx$$

where $R_T(x) = 1 - F_T(x)$ being the reliability function

$$\mu_T = \int_{0}^{T} (1 - F_T(x))\, dx + \int_{T}^{\infty} (1 - F_T(x))\, dx$$

$$= \int_0^T (1 - F_T(x))\,dx \quad (\because F_T(x) = 1 \text{ everywhere in } (T, \infty))$$

$$= \int_0^T (1 - F(x))\,dx \tag{5.49}$$

$$\Rightarrow \qquad \mu = \int_0^\infty (1 - F(x))\,dx \Rightarrow \mu_T \le \mu \Rightarrow \frac{1}{\mu_T} \ge \frac{1}{\mu} \tag{5.50}$$

Let $Y_0, Y_1, Y_2 \ldots$ are the true failure points where we take $Y_0 = 0$

Then
$$Y_1 = NT + Z \tag{5.51}$$

$N \equiv$ number of times A.R.P. is practised (a.r.v.) and Z the failure time following the last A.R.P.

$$P[N > k] = (1 - F(T))^k \tag{5.52}$$

$$P[Z \le x] = \frac{F(z)}{F(T)}; \ 0 \le z \le t \tag{5.53}$$

$$\Rightarrow \qquad E(N) = \sum_{n=1}^\infty P[N = n] \cdot n \tag{5.54}$$

Now $P[N = n] = P[N \ge n] - P[N \ge n + 1] = (1 - F(T))^n - (1 - F(T))^{n+1}$

$$\therefore \qquad E(N) = \sum_{n=1}^\infty [(1 - F(T))^n - (1 - F(T))^{n+1}] \cdot n$$

$$= (1 - F(T)) + (1 - F(T))^2 + (1 - F(T))^3 + \ldots$$

$$= \frac{1 - F(T)}{F(T)} \tag{5.55}$$

$$\therefore \qquad E(Y_1) = E(N)\,T + E(Z) \quad \text{(from (5.51))}$$

$$= \frac{1 - F(T)}{F(T)}\,T + \int_0^T \left[\frac{1 - F(z)}{F(T)}\right] dz$$

$$= \frac{1}{F(T)} \left\{ \{T\,(1 - F(T))\} + \int_0^T [F(T) - F(z)\,dz \right\}$$

$$= \frac{1}{F(T)} \left\{ T - \int_0^T F(z)\,dz \right\}$$

$$\Rightarrow \qquad E(Y_1) = \frac{1}{F(T)}\left[\int_0^T (1 - F(z))\, dz\right] \qquad\qquad (5.56)$$

Example 5.1 Show that the expected time for true failure in an age replacement policy (A.R.P.) is the same as without A.R.P. in case of exponential failure rate.

Example 5.2 Preventive Maintenance Scheme in a Weibull type model (Biswas and Monthir (1991))

Show that the expected true time to failure under Preventive Maintenance Scheme (TTF) in a Weibull model with hazard rate

$$h(t) = \lambda\, t^{\alpha - 1}$$

is approximately given by

$$E\,(TTF) = \frac{\lambda^{\frac{1}{\alpha}-1}\,\lambda^{-\frac{1}{\alpha}}}{1 - e^{-\lambda T^\alpha/\alpha}}\left\{\Gamma\!\left(\frac{1}{\alpha}\right) - \left(\frac{e^{-\frac{\lambda T^\alpha}{\alpha}}\left(\frac{\lambda T^\alpha}{\alpha}\right)^{1/2}}{\frac{\lambda T^\alpha}{\alpha} - \frac{1}{\alpha} + 1}\right)\right.$$

$$\left.\left(1 - \frac{\frac{1}{\alpha}-1}{\left(\frac{\lambda T^\alpha}{\alpha} - \frac{1}{\alpha} + 1\right)^2 + \frac{2\lambda T^\alpha}{\alpha}}\right)\right\}$$

where T is the time for periodic check up.

Solution : We have

$$h(z) = \lambda\, z^{\alpha - 1}$$

$$R(z) = 1 - F(z) = e^{-\frac{\lambda z^\alpha}{\alpha}}\; ; \; F(T) = 1 - e^{-\frac{\lambda T^\alpha}{\alpha}}$$

$$E(TTF) = \frac{\displaystyle\int_0^T e^{-\lambda z^\alpha/\alpha}}{1 - e^{-\lambda T^\alpha/\alpha}}$$

$$\equiv \frac{(\alpha)^{\frac{1}{\alpha}-1}\displaystyle\int_0^{\frac{\lambda T^\alpha}{\alpha}} e^{-\theta}\,\theta^{\frac{1}{\alpha}-1}\, d\theta}{(\lambda)^{\frac{1}{\alpha}}\,(1 - e^{-\lambda T^\alpha/\alpha})}$$

Again

$$\int_0^{\frac{\lambda T^\alpha}{\alpha}} e^{-\theta}\,\theta^{\frac{1}{\alpha}-1}\, d\theta = \Gamma\!\left(\frac{1}{\alpha}\right) - \int_{\frac{\lambda T^\alpha}{\alpha}}^{\infty} e^{-\theta}\,\theta^{\frac{1}{\alpha}-1}\, d\theta$$

Again by approximatation used by Gray et.al (1969).

$$\frac{\displaystyle\int_{\frac{\lambda T^\alpha}{\alpha}}^{\infty} e^{-\theta}\,\theta^{\frac{1}{\alpha}-1}\,d\theta}{-\dfrac{\lambda T^\alpha}{\alpha}\left(\dfrac{\lambda T^\alpha}{\alpha}\right)^{\frac{1}{\alpha}-1}}$$

References

1. Biswas S and Nauhari Indu (1980): *A note on the development of some interrupted waiting time distributions.* Pure and Applied Mathematica Sciences. Vol XI, No. I, March 1980, page 83-90.
2. Biswas S (1973): A *note on the generalization of William Brass model.* Demography, August 1973, Vol. 10, No. 3 Page 459-467.
3. Biswas S (1980): *On the extension of some results of Counter models with Poisson inputs and their applications.* Journal of Indian Statistical Association, Vol 18, page 45.53.
4. Biswas S, Nautiyal B.L. and Tyagi R.N.S. (1983): *Some results of a Stochastic Process associated with Gaiger Muller Counter model.* Sankhya, Series B, Vol 45, Part I, page 271-283.
5. Biswas S, Nautiyal B.L. and Tyagi R.N.S. (1981): *A stochastic epidemic model based on G.M. Counter type II for the evaluation of epidemic indices.* Gujrat Statistical review, Vol IX, No. I (1981), page 17-28.
6. Biswas S, Nair G, and Nautiyal B.L. (1983): *On a probability model of the number of conceptions classified by the nature of terminations based on a bivariate Geiger Muller Counter.* Demography India, Vol.. 12, No. 2, July - Dec 1983, page 289-300.
7. Biswas S and Nair G (1984): *A generalization of Freund's model for a repairable paired component based on a bivariate Geiger Müller Counter.* Micro-electronics and Reliability, Vol 24, No. 4, page 671-675.
8. Biswas Suddhendu and Noor Hamed Saad (1988): *On a probability model for classifying the unknown causes of death due to two Competing risks into assignable causes.* Biometric Journal, Vol. 30, No. 7, page 827-833.
9. Biswas S and Monthir A (1992): *On optimal time of periodic check-up Preventive maintenance scheme (PMS) under bath tub and Weibull type of failure rates.* Int. Journal of Systems Science, Vol. 22 No. 12, page 2651-2661.
10. Grey, H.L., Thompson, R.W. and McWilliams, G.V.: *A new approximation for x^2 - integral-Mathematics of Computation,* Vol 23, 1961, p. 85-89.
11. Karlin S and Taylor H.M. (1975): *A first course in Stochastic Processes,* Edition II, Academic Press, New York.
12. Takács Lajos (1951): *On the occurrence of coincidence phenomenal in case of happenings with arbitrary distribution law of duration.* Acta Mathematica, Hungary, Vol. 2, page 276-297.
13. Takács L (1957): *On some probability problems concerning theory of counters.* Acta Mathematica, Hungary, Vol. 8, page 127-138.
14. Takács L. (1960): *Stochastic Processes (Problems and Solutions):* Translated by P. Zador, John Wiley & Sons inc, New York; Mathuen and Co.

Palm Probability and Its Applications

6.0 Introduction

The stochastic model can be used
  (i) in predicting the number of events (like births, accidental shocks etc.) during a fixed period of time say $(0, t)$.
  (ii) in predicting the waiting time distribution between two consecutive events or between any two events, say ith and jth events.

The same type of problems can be solved by an entirely new technique known as "*palm probability*" which was hitherto used in queuing processes" (Khintchine (1960)). Palm probability is defined as the conditional probability of a specified number of events in a time interval given that an event has been happened at the begining of the interval. Cox and Isham's (1980) treatment of palm-distributions is described below.

For $u < v$, let $N (u, v)$ be a random variable giving the number of events occurring in (u, v) and $\{X_i\}$ be the sequence of the intervals between successive events in a process starting from an arbitrary point.

Consider the survivor function

$$K_X(x) = P[X < x]$$

$$= \lim_{\delta \to 0^+} P[N(0, x) = 0 \mid N(-\delta, 0) < 0] \qquad (6.1)$$

which is the limiting probability that, given an event occurs immediately before the origin, the next event of the process occurs after the instant (i.e. the system will survive further for more than a period x).

Now, by stationarity condition

$$P[N(0, x) = 0 \cap N(-\delta, 0) > 0]$$

$$= P[N(0, x) = 0] - P[N(-\delta, x) = 0]$$

$$= P(N(x) = 0) - P[N(x + \delta) = 0]. \qquad (6.2)$$

$$\Rightarrow \quad P[N(0, x) = 0 \mid N(-\delta, 0) < 0]$$

$$= \frac{P[N(0, x) = 0 \cap N(-\delta, 0) > 0]}{P[N(-\delta, 0) > 0]}$$

$$= \frac{P[N(x) = 0] - P[N(x + \delta) = 0]}{P[N(\delta) > 0]}$$

$$\Rightarrow \quad P[N(0, x) = 0 \mid N(-\delta, 0) > 0] \, \delta^{-1} \, P[N(\delta) < 0]$$

$$= \delta^{-1} [P(N(x) = 0) - P(N(x + \delta) = 0)]$$

$$= -\delta^{-1} [P[N(x + \delta) = 0] - P[N(x) = 0]] \qquad (6.3)$$

We define

$$\lim_{\delta \to 0^+} \delta^{-1} P[N(\delta) > 0] = \lambda \qquad (6.4)$$

as the occurrence parameter λ of the process, which we assume to be finite. Further, we denote,

$$P[N(x) = k] = p_k(x); \quad k = 0, 1, 2, \dots \qquad (6.5)$$

as the distribution of $N(x)$, then in the limit as $\delta \to 0^+$

$$K_X(x) = P[X > x] = \lim_{\delta \to 0^+} P[N(0, x) = 0 \mid N(-\delta, 0) < 0]$$

$$\Rightarrow P[N(0, x) = 0 \mid N(-\delta, 0) > 0] \, \delta^{-1} \, P[N(\delta) > 0]$$

$$= K_X(x) \, \delta^{-1} \, P[N(\delta) < 0]$$

$$= -\delta^{-1} \{P(N(x + \delta) = 0) - P(N(x) = 0)\} \quad \text{from (6.3)}$$

$$\Rightarrow \lambda K_X(x) = -D_x \{p_0(x)\} \quad \text{as } \delta \to 0^+ \qquad (6.6)$$

where D_X denotes the derivative with respect to x.

The equation (6.6) links the distribution of the interval between successive events with survivor function $K_X(x)$, to that of forward recurrence time survivor function $p_0(x)$ (where $p_0(x)$ is the probability that, starting from an arbitrary time instant, there are no events in the following interval of length x).

Example 6.1   For a Poisson process of rate λ, $p_0(x) = e^{-\lambda x} =$ survivor function w.r.t. counting process.

$$\therefore \qquad D_X[p_0(x)] = -\lambda e^{-\lambda x}$$

$$\Rightarrow \qquad -D_X[p_0(x)] = \lambda e^{-\lambda x}$$

$$K_X(x) \lambda = \lambda e^{-\lambda x}$$

$$\Rightarrow \qquad K_X(x) = e^{-\lambda x} = \text{survivor function w.r.t. waiting time}$$

$$F_X(x) = 1 - K_X(x) = 1 - e^{-\lambda x}$$

$$\frac{d}{dx}[F_X(x)] = \lambda e^{-\lambda x}$$

is the distribution of inter-arrival time.

Note that $p_0(x) = K_X(x)$ does not necessarily imply that the process is Poisson.

6.1   More General Results on Palm Probability

More general results connecting distributions of events 'conditional on a point at the origin' with those where the origin is an arbitrary instant may be obtained. We shall now assume that the process to be considered is completely stationary and has a finite occurrence parameter v and is orderly, so that v is equal to the rate λ of the process. Then for each $x > 0$, the Palm distribution is a discrete distribution defined by

$$\Pi_k(x) = \lim_{\delta \to 0^+} P[N(0, x) = k \mid N(-\delta, 0) > 0], \quad k = 0, 1, 2,... \tag{6.7}$$

In a careful mathematical development the existence of v and $\Pi_k(x)$, and more generally of other limiting probabilities of the number of events B given that

$$N(-\delta, 0) > 0$$

is of the form
$$\lim_{\delta \to 0^+} P[B \mid N(-\delta, 0) > 0] \tag{6.8}$$

which has to be proved.

Let the probability measure Π be defined for the number of events B on those processes which have a point at the origin. The measure Π can then be shown to satisfy

$$\Pi(B) = \lim_{\delta \to 0^+} P[B \mid N(-\delta, 0) > 0] \tag{6.9}$$

for a wide class of events B. The measure Π is called the *Palm measure of process*. In equation (6.6) the distribution of the interval measured from an arbitrary time instant to the next point of the process, is linked to that of the interval between successive points. Similarly, the functions $\Pi_k(x)$ defining the Palm distribution given in (6.7), which specify the distribution of the number of events in an interval of length x given by a point at the origin, can be connected with the functions $p_k(x)$ which give the distribution of the number of events in an interval of length x starting at an arbitrary origin. These connecting equations are known as *Palm-Khintchine equations* and may be derived as follows.

Since the process is orderly, if $k > 0$, as $\delta \to 0^+$

$$p_k(x + \delta) = P[N(-\delta, x) = k]$$
$$= P[N(-\delta, 0) = 0, N(0, x) = k]$$
$$+ P[N(-\delta, 0) = 1, N(0, x) = k - 1] + o(\delta)$$
$$= p_k(x) - P[N(-\delta, 0) > 0, N(0, x) = k]$$
$$+ P[N(-\delta, 0) > 0, N(0, x) = k - 1] + o(\delta)$$

so that

$$p_k(x + \delta) - p_k(x) = - P[N(-\delta, 0) > 0, N(0, x) = k]$$
$$+ P[N(-\delta, 0) > 0, N(0, x) = k - 1] + o(\delta)$$
$$= - P[N(-\delta, 0) > 0] P[N(0, x) = k$$
$$\mid [N(-\delta, 0) > 0] + P[N(-\delta, 0) > 0]$$

$$P\left[N(0, x) = k - 1 \mid N(-\delta, 0) >\right] + o(\delta)$$

$$\therefore \quad \delta^{-1}\left[p_k(x + \delta) - p_k(x)\right] = -\delta^{-1} P\left[N(-\delta, 0) > 0\right] P\left[N(0, x) = k \right.$$
$$\left. \mid N(-\delta, 0) > 0\right] + \delta^{-1} P\left[N(-\delta, 0) > 0\right]$$
$$- P\left[N(0, x) = k - 1 \mid N(-\delta, 0]\right] + o(1)$$
$$= -\delta^{-1}\{P(N(-\delta, 0) > 0) P[N(0, x) = k$$
$$\mid N(-\delta, 0) > 0] - P[N(-\delta, 0) > 0]$$
$$P\left[N(0, x) = k - 1 \mid N(-\delta, 0) > 0\right] + o(1)$$

Hence

$$\delta^{-1}[p_k(x + \delta) - p_k(x)] = -\delta^{-1}\{P(N(-\delta, 0) > 0)\}\{P(N(0, x) = k$$
$$\mid N(-\delta, 0) > 0)\} - P[N(0, x) = k - 1 \mid N(-\delta, 0) > 0] + o(1)$$

Hence taking limit as $\delta \to 0^+$

$$D_X\{p_X(x)\} = -\delta^{-1} P\left[N(-\delta, 0) > 0\right] \left[\Pi_k(x) - \Pi_{k-1}(x)\right] \qquad (6.10)$$

$$= -\lambda\left[\Pi_k(x) - \Pi_{k-1}(x)\right] \quad \text{from (6.4) and (6.7)}$$

where D_X denotes the right hand derivative. The corresponding equation for $k = 0$ has already been derived in and is given by

$$D_X\{p_0(x)\} = -\lambda\,\Pi_0(x) \qquad (6.11)$$

The integral forms of (6.10) and (6.11) are

$$p_k(x) = -\lambda \int_0^x \left[\Pi_k(u) - \Pi_{k-1}(u)\right] du \quad (k = 1, 2, \ldots) \qquad (6.12)$$

$$p_0(x) = 1 - \lambda \int_0^x \Pi_0(u)\, du \qquad (6.13)$$

It follows from (6.10) and (6.11) that

$$-\frac{1}{\lambda} D_X\left[p_0(x) + \ldots + p_k(x)\right] = -\frac{1}{\lambda} D_X\{P[N(x) \le k]\}$$

$$= \Pi_k(x)$$

Therefore, the probability of having exactly k events in $(0, x)$ starting from an event at 0, can be obtained by differentiating the probability of getting no more than k events in $(0, x)$ where 0 is an arbitrary time instant. Alternatively, from (6.12) and (6.13) we have

$$p_0(x) + p_1(x) + \ldots + p_k(x) = P[N(x) \le k]$$

$$= 1 - \lambda \int_0^x \Pi_k(u)\, du \qquad (6.14)$$

So that the probability of getting not more than k events in $(0, x)$, where 0 is an arbitrary instant can be obtained by integrating the probability of exactly k events in the interval when there is an event at 0. In addition the right hand side of (6.14) is equal to

$$1 - \lambda \int_0^x \Pi_k(u)\, du = \lambda \int_x^\infty \Pi_k(u)\, du$$

$$\left\{ \because \left(p_0(x) + p_1(x) + \dots + p_k(x) = 1 - \lambda \int_0^x \Pi_k(u)\, du \right. \right.$$

$$\Rightarrow \qquad p_0(\infty) + \dots + p_k(\infty) = 1 - \lambda \int_0^\infty \Pi_k(u)\, du$$

$$\Rightarrow \qquad \lambda \int_0^\infty \Pi_k(u)\, du = 1$$

$$\therefore \quad 1 - \lambda \int_0^x \Pi_k(u)\, du = \lambda \int_0^\infty \Pi_k(u)\, du - \lambda \int_0^x \Pi_k(u)\, du$$

$$\left. \left. = \lambda \int_x^\infty \Pi_k(u)\, du \right) \right\}$$

Again $\lambda \int_x^\infty \Pi_k(u)\, du = \lambda \int_0^\infty \Pi_k(y + x)\, dy$ by putting $u = y + x$

Hence

$$p_0(x) + \dots + p_k(x) = P[N(x) \le k]$$

$$= \lambda \int_0^\infty \Pi_k(y + k)\, dy \qquad (6.15)$$

We may justify (6.15) by the argument that if '0' is an arbitrary time instant and there are not more than k events in $(0, x)$, then there must exist an event with co-ordinate y, for same $y > 0$, such that there are exactly k events in $(-y, x)$. Since the process is orderly, the probability of an event in $(-y, -y + \delta)$ is $\lambda\delta + o(\delta)$ and therefore

$$P[N(x) \le k] = \lambda \int_0^\infty \Pi_k(y + x)\, dy$$

The equations (6.10), (6.11), (6.12) and (6.14) can be summarized by using probability generating functions. For, if we define

$$G(z, x) = \sum_{k=0}^{\infty} z^k p_k(x)$$

and

$$G_0(z, x) = \sum_{k=0}^{\infty} z^k \Pi_k(x)$$

so that G refers to an arbitrary origin while G_0 refers to the situation given by a point at the origin.

In fact

$$\sum_{i=k}^{\infty} \Pi_i(x) = G^{(k)}(x)$$

where the right hand side denotes the cumulative distribution of $X_1 + X_2 + \ldots + X_k$ obtained by k-fold convolution.

$$\Rightarrow \qquad \Pi_k(x) = G^{(k)}(x) - G^{(k+1)}(x)$$

and the $p_k(x)$ are given by (6.12) & (6.13).

We have

$$D_X p_k(x) = -\lambda [\Pi_k(x) - \Pi_{k-1}(x)], \quad D_X p_0(x) = -\lambda \Pi_0(x)$$

$$\Rightarrow \qquad D_X \sum_{k=0}^{\infty} p_k(x) z^k = -\lambda \sum_{k=0}^{\infty} z^k [\Pi_k(x) - \Pi_{k-1}(x)]$$

$$= -\lambda \sum_{k=0}^{\infty} z^k \Pi_k(x) + z\lambda \sum_{k=1}^{\infty} z^{k-1} \Pi_{k-1}(x)$$

$$= -\lambda G_0(z, x) + \lambda z G_0(z, x)$$

$$= -\lambda (1-z) G_0(z, x)$$

or equivalently

$$G(z, x) = 1 - \lambda (1-z) \int_0^x G_0(z, u) \, du \tag{6.16}$$

A simple example is provided by a renewal process in which the intervals between successive events are independently and identically distributed with density g. Then the Palm probabilities refer to a process starting with a point at, or just before the origin i.e. to an ordinary renewal process, whereas the probabilities $\{p_k(x)\}$ refer to a process starting from an arbitrary time origin i.e. to an equilibrium renewal process. In this case, the Palm probabilities are readily computed via

$$P[N(t) > n] = P[T_{n+1} \le t]$$

6.2 Development of the Waiting Time Distribution Model Based on Palm Probability

Notations used are as follows:

   (i) $\phi_k(t)$ = conditional probability of k number of shocks in $(0, t]$ given that the first shock occurred at $T = 0$; $\phi_k(t)$ ($k = 0, 1, 2, ..., n$) is a Palm probability measure

  (ii) $V_k(t)$ = unconditional probability of k events in $(0, t)$; $k = 0, 1, 2, ..$ n

 (iii) Let the shocks occur with Poisson intensity λ; but λ varies from individual to individual following a gamma distribution whose density function is given by

$$\psi(\lambda) = \frac{a^k e^{-at} \lambda^{k-1}}{\Gamma(k)} \; ; \quad a, k > 0$$

 (iv) $G(z, t)$ = probability generating function (p.g.f.) of $V_k(t)$; $k = 0, 1, 2, ...$

  (v) $G_0(z, t)$ = p.g.f. of $\phi_k(t)$, $k = 0, 1, 2, ...$

Then

$$V_n(t \mid \lambda) = E\left[\frac{e^{-\lambda t}(\lambda t)^n}{n!} \bigg| \lambda \right]$$

$$\Rightarrow \quad V_n(t) = \int_0^\infty V_n(t \mid \lambda) \psi(\lambda) \, d\lambda$$

$$= \int_0^\infty \frac{e^{-\lambda t}(\lambda t)^n}{n!} \frac{a^k e^{-a\lambda} \lambda^{k-1}}{\Gamma(k)} \, d\lambda$$

$$= \frac{a^k t^n}{n! \Gamma(k)} \int_0^\infty e^{-(a+t)\lambda} \lambda^{n+k-1} \, d\lambda$$

$$= \frac{a^k t^n}{n! \Gamma(k)} \frac{\Gamma(n+k)}{(a+t)^{n+k}} \tag{6.17}$$

Further,

$$G(z, t) = \text{p.g.f. of } V_n(t)$$

$$= \sum_{n=0}^\infty z^n V_n(t)$$

$$= \sum_{n=0}^\infty z^n \frac{a^k t^n}{n! \Gamma(k)} \frac{\Gamma(n+k)}{(a+t)^{n+k}} \quad \text{from (6.17)}$$

$$= \frac{a^k}{(a+t)^k \, \Gamma(k)} \left[(k-1)! + \frac{k!zt}{(a+t)} + \frac{(k+1)!}{2!} \frac{(zt)^2}{(a+t)^2} + .. \right]$$

$$= \frac{a^k}{(a+t)^k} \left\{ \left[1 + k \cdot \frac{zt}{(a+t)} + \frac{k(k+1)}{2!} \frac{(zt)^2}{(a+t)^2} + .. \right] \right\}$$

$$= \frac{a^k}{(a+t)^k} \left(1 - \frac{zt}{a+t} \right)^{-k} = \left(\frac{a}{a+t} \right)^k \left(\frac{a+t}{a+t-zt} \right)^k$$

$$= \left\{ \frac{a+t}{a+t(1-z)} \right\}^k \qquad (6.18)$$

Also

$$G_0(z, t) = \text{p.g.f. of } \phi_k(t)$$

$$= \sum_{k=0}^{\infty} z^k \, \phi_k(t)$$

$$= \phi_0(t) + z \, \phi_1(t) + z^2 \, \phi_2(t) + \dots \qquad (6.19)$$

$$\frac{\partial G(z, t)}{\partial t} = -\frac{k}{a}(1-z) \, G_0(z, t) \qquad (6.20)$$

(6.18) and (6.19)

$$\Rightarrow \qquad a^k \frac{\partial}{\partial t} \{a + t(1-z)\}^{-k} = -\frac{k}{a}(1-z) \, G_0(z, t)$$

$$\Rightarrow \qquad a^k \{ -k(a + t(1-z))^{-k-1}(1-z) \}$$

$$= -\frac{k}{a}(1-z) \, G_0(z, t)$$

$$\Rightarrow \qquad G_0(z, t) = \frac{a^{k+1}}{\{a + t(1-z)\}^{k+1}} \qquad (6.21)$$

Let T_n be the waiting time for the nth event given that the first event has occurred at $T = 0$. Then

$$F_n(t) = P[T_n \le t] = \sum_{N=n}^{\infty} \phi_N(t) \qquad (6.22)$$

Also defining

$$H_0(z, t) = \text{p.g.f. of } F_n(t) = \sum_{n=0}^{\infty} z^n F_n(t)$$

$$= F_0(t) + z \, F_1(t) + z^2 \, F_2(t) + \dots$$

$$= \sum_{N=0}^{\infty} \phi_N(t) + z \sum_{N=1}^{\infty} \phi_N(t) + z^2 \sum_{N=2}^{\infty} \phi_N(t) + \dots \qquad \text{from (6.22)}$$

$$= \sum_{N=0}^{\infty} \phi_N(t) + z \sum_{N=0}^{\infty} [\phi_N(t) - \phi_0(t)] + z^2 \sum_{N=0}^{\infty} [\phi_N(t) - \phi_0(t) - \phi_1(t)]$$

$$+ z^3 \left[\sum_{N=0}^{\infty} \phi_N(t) - \phi_0(t) - \phi_1(t) - \phi_2(t) \right] + \dots$$

$$= (1 + z + z^2 + z^3 + \dots) \sum_{N=0}^{\infty} \phi_N(t) - z[\phi_0(t) + z\phi_1(t) + z^2\phi_2(t) + \dots]$$

$$- z^2 \phi_0(t) [1 + z + z^2 + \dots] - z^3 \phi_1(t) [1 + z + z^2 + \dots] + \dots$$

$$= (1-z)^{-1} \sum_{N=0}^{\infty} \phi_N(t) - z\, G_0(z,t) - \frac{z^2}{1-z} \phi_0(t) - \frac{z^3}{1-z} \phi_1(t) + \dots$$

$$\text{using (6.19)}$$

$$= (1-z)^{-1} - z\, G_0(z,t) - \frac{z^2}{1-z} G_0(z,t)$$

$$= \frac{1 - z(1-z) G_0(z,t) - z^2 G_0(z,t)}{1-z}$$

$$= \frac{1 - z\, G_0(z,t)}{1-z} \qquad (6.23)$$

$$\therefore \qquad H_0(z,t) = \sum_{n=0}^{\infty} z^n F_n(t) = \frac{1 - z\, G_0(z,t)}{1-z}$$

$$\Rightarrow \qquad \frac{\partial}{\partial t} H_0(z,t) = \sum_{n=0}^{\infty} z^n \frac{\partial}{\partial t} F_n(t)$$

$$\Rightarrow \qquad \frac{\partial}{\partial t} \left[\frac{1 - z\, G_0(z,t)}{1-z} \right] = \sum_{n=0}^{\infty} z^n f_n(t)$$

$$\Rightarrow \qquad -\frac{z}{1-z} \frac{\partial}{\partial t} G_0(z,t) = \sum_{n=0}^{\infty} z^n f_n(t) \qquad (6.24)$$

Also $$\qquad G_0(z,t) = \frac{a^{k+1}}{[a + t(1-z)]^{k+1}}, \qquad \text{from (6.21)}$$

$$\Rightarrow \qquad \frac{\partial}{\partial t} G_0(z, t) = a^{k+1} \left[-(k+1) \{ a + t(1-z) \}^{-k-2} (1-z) \right]$$

$$= -\frac{(k+1) a^{k+1} (1-z)}{[a + t(1-z)]^{k+2}} \qquad (6.25)$$

(6.24) and (6.25)

$$\Rightarrow \sum_{n=0}^{\infty} z^n f_n(t) = -\frac{z}{1-z} \frac{\{-(k+1) a^{k+1} (1-z)\}}{[a+t(1-z)]^{k+2}} = \frac{z(k+1) a^{k+1}}{[a+t(1-z)]^{k+2}}$$

$$= \frac{z(k+1)}{a} \left[\frac{a}{a+t(1-z)} \right]^{k+2} = \frac{z(k+1)}{a} \left[\frac{a}{a+t-tz} \right]^{k+2}$$

$$= \frac{z(k+1)}{a} \left(\frac{a}{a'-tz} \right)^{k+2} \qquad \text{where } a' = a+t$$

$$= \frac{z(k+1)}{a} \left(\frac{1}{\dfrac{a'}{a} - \dfrac{tz}{a}} \right)^{k+2} = \frac{z(k+1)}{a} \left(\frac{a'}{a} - \frac{tz}{a} \right)^{-(k+2)}$$

$$= \frac{z(k+1)}{a} \left(\frac{a'}{a} \right)^{-(k+2)} \left[1 - \frac{tz}{a'} \right]^{-(k+2)}$$

$$= \frac{z(k+1)}{a} \left(\frac{a'}{a} \right)^{-(k+2)} \left\{ 1 + (k+2)\frac{tz}{a'} + \frac{(k+1)(k+3)}{1.2} \left(\frac{tz}{a'} \right) + \right.$$

$$\left. \dots + \frac{(k+2)(k+3)\dots(k+n)}{1.2\dots(n-1)} \left(\frac{tz}{a'} \right)^{n-1} + \dots \right. \qquad (6.26)$$

Equating coefficients of z^n on the both sides of (6.26), we have,

$$f_n(t) = \frac{(k+1)}{a} \left(\frac{a'}{a} \right)^{-(k+2)} \frac{(k+2)(k+3)\dots(k+n) \cdot t^{n-1}}{1.2\dots(n-1)(a')^{n-1}}$$

$$= \frac{(k+1)(k+2)(k+3)\dots(k+n)}{(n-1)!} \frac{a^{k+1}}{(a+t)^{n+k+1}}, \qquad (6.27)$$

$$(\because a' = a+t)$$

a result which has been derived alternatively by Biswas and Pachal (1983), which represents the waiting time of the *n*th shock given that the first shock occurs at $T = 0$. The present proof is however, based on more rigorous foundation.

Note that

$$f_1(t) = \frac{(k+1)\,a^{k+1}}{(a+t)^{k+2}} \tag{6.28}$$

as compared to the unconditional density function of the waiting time distribution of the first shock as

$$f(t) = \frac{a^k}{\Gamma(k)} \int_0^\infty \lambda e^{-a\lambda}\, \lambda^{k-1}\, e^{-\lambda t}\, d\lambda$$

$$= \frac{a^k}{\Gamma(k)}\, \frac{\Gamma(k+1)}{(a+t)^{k+1}} = \frac{k a^k}{(a+t)^{k+1}} \tag{6.29}$$

A comparison of (6.28) and (6.29) shows that the hazard rate is $\dfrac{k+1}{a}$ in case

of (6.28) in comparison to $\dfrac{k}{a}$ in (6.29) $(k, a > 0)$. This shows that following each

shock the hazard rate of the system increases by $\dfrac{1}{a}(a > 0)$.

A generalization of the Palm probability distribution when intensity is time dependent is obtained by Biswas and Nair (1985).

References

1. Biswas S and Nair G (1985): *On the development of successive damage model based on Palm probability*. Micro-electronics and Reliability, Vol. 24, No. 4, page 671-675.
2. Biswas S and Nair G (1986): *Palm probabilistic technique on the prediction of the arrival time of last fatal shock based on the data of earlier shocks*. Journal of Agricultural Research Statistics, Vol. XXXVIII, Nos 2, August 1986, page 240-248.
3. Biswas S and Pachal T.K. (1983): *On the application of Palm probability for obtaining Inter-arrival time distribution a weighted Poison process*. Calcutta Statistical Association Bulletin, Vol. 12, March & June 1983, Nos 125-126, page III-115.
4. Biswas Suddhendu and Sehgal Vijay Kumar (1988): *On the correlation between inter-arrival delays of shocks*. Micro -electronics and Reliability, Vol. 28, No. 2 page 189-192.
5. Cox D.R. and Isham V (1980): *Point Processes*. Chapman and Hall, Monographs on Statistics and Applied Probability.
6. Cox D.R. and Oakeds D (1983): *Analysis of Survival Data*. Chapman and Hall. Monographs on Statistics and Applied Probability.
7. Khintchine A.Y. (1960): *Mathematical models in the theory of Queueing*. (Translated by D.M. Andrews and H.M. Quenouille)-Charles Griffin Monograph, London.

Stochastic Epidemic Processes

7.0   Random Variable Technique

Below we present a technique of handling Stochastic differential equations; especially in Population, Ecological and Epidemiological models due to Bailey (1963). This technique is often applied in the formation of stochastic Differential Equations; i.e. to write down the differential equation concerning p.g.f. (or the m.g.f) with respect to time, thus enabling us to study the process. The most useful application of this kind of technique lies in providing a description of the population process (Demographic, Ecological, epidemiological vis-a-vis the solution of the differential equation involving p.g.f. or m.g.f. of the process.

Let r.v.s. $X(t)$ and $X(t + \Delta t)$ represent the size of the population of individuals under observation at time t and $(t + \Delta t)$ respectively.

We write

$$\Delta X(t) = [X(t + \Delta t) - X(t)] \tag{7.1}$$

Further, we assume $X(t)$ and $\Delta X(t) \; \forall \; t$ are independently distributed. Since the change of a new element joining the population (or leaving the same) is not only independent of the previous states of the system, but also of present state therefore, with this assumption, the p.g.f. $\phi (s, t)$ of the stochastic process $X(t)$ viz.

$$E (s^{X(t)}) = \phi (s, t)$$

can be written as

$$\phi (s \; ; t + \Delta t) = \phi (s \; ; t) \, \Delta \, \phi (s, t) \tag{7.2}$$

$$\Delta \, \phi (s \; ; t) = E [s^{\Delta X(t)}]$$

We have $\Delta(X(t)) = 1$ with probability $\lambda \, \Delta t + o \, (\Delta t)$

$$= 0 \text{ with probability } (1 - (\lambda \Delta t + o \, (\Delta t))$$

as per the set up of a Poisson Process with parameter λt.

$$\therefore \qquad \Delta \phi (s \; ; t) = E [s^{\Delta X(t)}]$$

$$= (1 - \lambda \, \Delta t) \, s^0 + \lambda \, \Delta t s + o(\Delta t)$$

$$= (1 - \lambda\, \Delta t) + \lambda\, \Delta ts + o(\Delta t)$$

$$= (1 + \lambda\, (s - 1)\, \Delta t) + o(\Delta t) \tag{7.3}$$

Putting (7.3) in (7.2) $\Rightarrow$

$$\phi\,(s\,;\,t + \Delta t) = \phi\,(s\,;\,t)\,[1 + \lambda\,(s - 1)\,\Delta t] + o(\Delta t)$$

$$\Rightarrow \lim_{\Delta t \to 0} \frac{\phi(s\,;\,t + \Delta t) - \phi(s\,;\,t)}{\Delta t} = \lambda(s - 1)\,\phi\,(s\,;\,t)$$

$$\Rightarrow \frac{\partial \phi\,(s\,;\,t)}{\partial t} = \lambda(s - 1)\,\phi\,(s\,;\,t) \tag{7.4}$$

By similar reasoning, we can get similar result for moment generating and cumulant generating function denoted by $M\,(\theta\,;\,t)$ and $K\,(\theta\,;\,t)$ respectively valid for a Poisson Process.

These are given by

$$\left.\begin{array}{l} \dfrac{\partial M\,(\theta\,;\,t)}{\partial t} = \lambda\,(e^{\theta} - 1)\,M\,(\theta\,;\,t) \\[4mm] \text{and} \qquad \dfrac{\partial K\,(\theta\,;\,t)}{\partial t} = \lambda\,(e^{\theta} - 1), \text{obviously} \end{array}\right\} \tag{7.5}$$

However (7.4) and (7.5) have been derived on the basis of assumption of Poisson Process that $X(t)$, and $\Delta X(t)$, are independent. In actual situation, $\Delta X(t)$ need not be independent of $X(t)$. To generalize the model we assume a finite # transitions (other than '0' and '1') in the interval

$$(t, t + \Delta t) \quad \forall \quad t.$$

However assumption of $\Delta X(t)$ taking any finite # values is assumed subject to the condition $\Delta X(t)$ is small (condition of orderliness).

$$\left\{\begin{array}{l} \text{Let} \quad P\,\{\Delta X\,(t) = j \mid X(t)\} = X \\[2mm] \text{and} \quad P\,\{\Delta X\,(t) = 0 \mid X(t)\} = X \end{array}\right\} \begin{array}{l} = f_j\,(X)\,\Delta t + o(\Delta t), \quad j = 1, 2, 3,\ldots \\[2mm] = (1 - \displaystyle\sum_{j \neq 0} f_j\,(X)\,\Delta t + o(\Delta t)) \end{array} \tag{7.6}$$

Then the m.g.f.

$$M\,(\theta\,;\,t + \Delta t) = E_{t + \Delta t}\,\{\exp\,(\theta\,X(t + \Delta t))\}$$

$$= E_{t + \Delta t}\{\exp\,(\theta\,X(t) + \theta\,\Delta X(t))\},$$

by neglecting higher powers of $\Delta(t)$

$$= E_t\,[\exp\,(\theta\,X(t))]\,E_{\Delta t \mid t}\,(\exp\,(\theta\,\Delta X(t))) \tag{7.7}$$

by the theorem of expectation of two dependent events.

From (7.7), it follows that

$$\frac{\partial M(\theta;t)}{\partial t} = \lim_{\Delta t \to 0} \frac{M(\theta, t + \Delta t) - M(\theta, t)}{\Delta t}$$

$$= \lim_{\Delta t \to 0} \frac{1}{\Delta t} [E_t \{\exp(\theta X(t))\} E_{\Delta t \mid t} (\exp(\theta \Delta X(t))\}$$

$$- E_t \{\exp(\theta X(t))\}]$$

$$= E_t \left[\exp(\theta X(t)) \lim_{\Delta t \to 0} E_{\Delta t \mid t} \left(\frac{\exp(\theta \Delta X(t)) - 1}{\Delta t} \right) \right]$$

Assuming $\qquad \lim_{\Delta t \to 0} E_{\Delta t \mid t} \left(\dfrac{\exp(\theta \Delta X(t)) - 1}{\Delta t} \right)$

has a limit say, $\psi(\theta, t, X(t))$.

We can write

$$\frac{\partial M(\theta, t)}{\partial t} = E_t \{\exp(\theta X(t)) \psi(\theta, t, X)\}, \quad \text{where } X(t) = X$$

$$= \psi\left(\theta, t, \frac{\partial}{\partial \theta}\right) E_t (\exp(\theta X(t))) = \psi\left(\theta, t, \frac{\partial}{\partial \theta}\right) M(\theta, t) \quad (7.8)$$

where $\dfrac{\partial}{\partial \theta}$ is operative only on $M(\theta, t)$ and the differential and expectation on operators are commutable, i.e. $\dfrac{\partial}{\partial \theta} E_t \equiv E_t \dfrac{\partial}{\partial \theta}$ holds, symbolically,

$$\left(\because \psi\left(\theta, t, \frac{\partial}{\partial \theta}\right) M(\theta, t) = \psi\left(\theta, t, \frac{\partial}{\partial \theta}\right) E_t (\exp(\theta X(t))) \right.$$

$$\left. = \psi\left(\theta, t, \frac{\partial}{\partial \theta} E_t (\exp(\theta X(t))) \right) \right)$$

assuming that the operator $\dfrac{\partial}{\partial \theta}$ is applicable only on $M(\theta, t)$.

$$= \psi\left(\theta, t, E_t \frac{\partial}{\partial \theta} (\exp(\theta X(t))) \right)$$

$$= \psi(\theta, t, E_t (X(t) \exp(\theta X(t))))$$

$$= E_t [\exp(\theta X(t)) \psi(\theta, t, X(t))] = \text{L.H.S. of (7.8)}$$

Again

$$\psi(\theta, t, X) = \lim_{\Delta t \to 0} E_{\Delta t \mid t} \frac{[\exp(\theta \Delta X(t)) - 1]}{\Delta t} \left\{ 1 - \left(\sum_{j \neq 0} f_j(X) \Delta t + o(\Delta t) \right) \right\}$$

$$+ \lim_{\Delta t \to 0} \frac{\left(\sum_{j \neq 0} e^{\theta j} f_j(X) \Delta t + o(\Delta t) \right) - 1}{\Delta t} \quad \text{(from (7.6))}$$

$$= \lim_{\Delta t \to 0} \sum_{j \neq 0} \frac{[(e^{\theta j} - 1) f_j(X)] \Delta t}{\Delta t}$$

$$= \sum_{j \neq 0} (e^{\theta j} - 1) f_j(X) \tag{7.9}$$

Thus

$$\psi(\theta, t, X) = \sum_{j \neq 0} (e^{\theta j} - 1) f_j(X)$$

$$\Rightarrow \psi\left(\theta, t, \frac{\partial}{\partial \theta}\right) = \sum_{j \neq 0} (e^{\theta j} - 1) f_j\left(\frac{\partial}{\partial \theta}\right)$$

Therefore using (7.8) $\Rightarrow$

$$\frac{\partial M(\theta, t)}{\partial t} = \sum_{j \neq 0} (e^{\theta j} - 1) f_j\left(\frac{\partial}{\partial \theta}\right) M(\theta, t) \tag{7.10}$$

Putting $\quad e^{\theta} = s \Rightarrow \dfrac{\partial}{\partial \theta}(e^{\theta}) = \dfrac{ds}{\partial \theta} = \dfrac{\partial}{\partial \theta}(s) = \left[s \dfrac{\partial}{\partial \theta}\right] e^{\theta}$

$$\therefore \qquad \left(\because e^{\theta} \frac{\partial \theta}{\partial s} = 1 \right) \Rightarrow \frac{\partial}{\partial \theta} = s \frac{\partial}{\partial \theta}, \text{ symbolically} \tag{7.11}$$

We can write (7.10) as

$$\frac{\partial \phi(s, t)}{\partial t} = \sum_{j \neq 0} (s^j - 1) f_j\left(s \frac{\partial}{\partial s}\right) \phi(s, t) \tag{7.12}$$

Extending the results for the bivariate cases and denoting

$$P\{\Delta X(t) = j, \ \Delta Y(t) = k \,|\, X(t), \ Y(t)\} = f_{jk}(X(t), Y(t)) \Delta t + 0(\Delta t)$$

by excluding the case of both $(j, k) = (0, 0)$.

We can get similarly for the m.g.f. and p.g.f.

$$\frac{\partial M(\theta, \psi, t)}{\partial t} = \sum_{j, k} \{\exp(j\theta + k\psi) - 1\} f_{jk}\left(\frac{\partial}{\partial \theta}, \frac{\partial}{\partial \psi}\right) M(\theta, \psi, t) \tag{7.13}$$

and $\quad \dfrac{\partial \phi(x, y, t)}{\partial t} = \displaystyle\sum_{j, k} (x^j y^k - 1) f_{jk}\left(x \dfrac{\partial}{\partial x}, y \dfrac{\partial}{\partial y}\right) \phi(x, y, t) \tag{7.14}$

respectively. The results are of considerable applications in the development of two dimensional epidemic models.

7.1   An Illustration in Epidemic Model

Let $X(t)$ = number of susceptibles and $Y(t)$ = number of infectives. We denote $P\{X(t) = u, Y(t) = v\} = p_{uv}(t)$ = joint probability of u number of susceptibles and v number of infectives in time t. Denote p.g.f.

$$\phi(s_1, s_2, t) = \sum_{u, v} p_{u, v}(t) s_1^u s_2^v$$

In this kind of epidemic model, we have two types of transitions in infinitesimal small period denoted by $(j, k) = (-1, 1)$ or $(0, -1)$; the former one denotes number of susceptibles decreased by one and consequently the number of infectives increased by one and in the latter, there is no change in the number of susceptibles while the number of infectives decreases by one because the infectives being removed because of death or cure.

In the former case the infection rate is β and in the latter case the removal rate is γ.

Thus
$$f_{-1, 1}(x, y) = \beta xy \text{ and in the latter case}$$

$$f_{0, -1}(x, y) = \gamma y$$

$$\frac{\partial \phi(s_1, s_2, t)}{\partial t} = \left[(s_1^{-1} s_2 - 1)\beta \left(s_1 \frac{\partial}{\partial s_1}\right)\left(s_2 \frac{\partial}{\partial s_2}\right)\right.$$

$$\left. + (s_2^{-1} - 1)\gamma \left(s_2 \frac{\partial}{\partial s_2}\right)\right]$$

by putting $j = -1$ and $k = 1$ and $j = 0$, $k = -1$ respectively while replacing x, y by s_1 and s_2 respectively in (7.14) we have

$$\Rightarrow \frac{\partial \phi(s_1, s_2, t)}{\partial t} = \beta(s_2^2 - s_1 s_2)\frac{\partial^2 \phi}{\partial s_1 \partial s_2} + (1 - s_2)\gamma \frac{\partial \phi}{\partial s_2} \qquad (7.14a)$$

This gives the differential equation of the Bivariate p.g.f.

7.2   Stochastic Epidemic Models

The classical epidemic model takes into consideration of three mutually exclusive segments of the population viz.
   (i)   Population of Susceptibles,
   (ii)   Population of Infected persons, and
   (iii)   Population of Isolated or removed individuals by death or recovery (for which immunity against the same disease lasts for sometime further).
However, a *simple stochastic model* considers only first two segments of the population while ignoring the third segment. The premises for that is justified in case of diseases which are not fatal and the immunity period following

recovery can be considered to be negligible (like Influenza, Cough, Conjunctivitis etc.....)

Whereas in a general epidemic model we consider all the three segments.

Again an epidemic model may be deterministic or Stochastic according as we do not or do introduce probability element in the Mathematical formulation of the model. In stochastic models we consider the random fluctuations in the epidemic patterns while systematic factors are held at fixed levels. Thus stochastic models are also Deterministic models, on the average in many cases although exceptions remain.

7.2.1 Simple Stochastic Model

Development of the model.

Let the number of susceptibles and infectives be $X(t)$ and $Y(t)$ respectively at any time t $(0 \le t < \infty)$ $t \in T$ (continuous parameter space) while state spaces given $X(t)$ or $Y(t)$ are discrete and positive integer valued.

Let β be the infection rate, independent of time which implies that the number of infections during an infinitesimal interval of time $(t, t + \delta t)$ is $\beta X(t) Y(t) \delta t + o(\delta t)$ subject to $X(0) + Y(0) = n + 1$.

Let us assume $X(0) = n$ and $Y(0) = 1$, the initial number of infective with which an epidemic starts.

We denote by

$$p_r(t) = P[X(t) = r \mid X(0) = n ; Y(0) = 1] \tag{7.15}$$

Then clearly

$$p_r(t + \delta t) = p_r(t)[1 - \beta r(n + 1 - r)\,\delta t + o(\delta t)] + p_{r+1}(t)$$

$$[\beta(r + 1)(n + 1 - (r + 1))\,\delta t + o(\delta t)] + o(\delta t)$$

$$\Rightarrow \quad [p_r(t + \delta t) - p_r(t)] = -\beta r(n - r + 1)\,\delta t\, p_r(t) + \beta(r + 1)$$

$$(n - r)\, p_{r+1}(t) + o(\delta t)$$

$$\Rightarrow \quad \frac{p_r(t + \delta t) - p_r(t)}{\delta t} = -\beta r(n - r + 1)\,\delta t + \beta(r + 1)(n - r)\, p_{r+1}(t) + o(\delta t)$$

$$\Rightarrow \quad \lim_{\delta t \to 0} \frac{p_r(t + \delta t) - p_r(t)}{\delta t} = -\beta r(n + 1 - r)\, p_r(t) + \beta(r + 1)(n - r)\, p_{r+1}(t)$$

$$\Rightarrow \quad \frac{dp_r(t)}{dt} = -\beta r(n - 1 + r)\, p_r(t) + \beta(r + 1)(n - r)\, p_{r+1}(t) \tag{7.16}$$

Putting $\qquad \tau = \beta t \tag{7.17}$

$$\Rightarrow \qquad p_r(\tau) = \beta\, p_r(t)$$

$$\Rightarrow \qquad \frac{\beta \, dp_r(t)}{dt} = \frac{dp_r(\tau)}{dt}$$

$$= \frac{dp_r(\tau)}{d\tau} \frac{d\tau}{dt} = \beta \frac{dp_r(\tau)}{d\tau}$$

$$\frac{dp_r(t)}{dt} = \frac{dp_r(\tau)}{d\tau} \qquad\qquad (7.18)$$

Using (7.17), (7.18) in (7.16) we get

$$\frac{dp_r(\tau)}{d\tau} = -r(n - r + 1) \, p_r(\tau) + (r + 1)(n - r) \, p_{r+1}(\tau)$$

$$0 \le r \le n + 1 \qquad\qquad (7.19)$$

Let us take $r = n$.
Then $(7.19) \Rightarrow$

$$\frac{dp_n(\tau)}{dT} = -n \, p_n(\tau)$$

Taking Laplace transform on both sides, we get,

$$-1 + SL(p_n(\tau)) = -nL(p_n(T))$$

$$\Rightarrow \qquad (S + n) L(p_r(\tau)) = 1$$

$$\Rightarrow \qquad L(p_n(\tau)) = \frac{1}{S + n}$$

$$\Rightarrow \qquad p_n(\tau) = L^{-1}\left(\frac{1}{s + n}\right) = e^{-\tau n}$$

$$\Rightarrow \qquad p_n(t) = e^{-n\beta t} \qquad (\because \tau = \beta t)$$

$$= P[\text{no infection upto } \tau]. \qquad\qquad (7.20)$$

Again considering

$$\frac{d \, p_r(\tau)}{d\tau} = -r(n - r + 1) \, p_r(\tau) + (r + 1)(n - r) \, p_{r+1}(\tau)$$

Taking L-transform on both sides $\Rightarrow$

$$SL(p_r(\tau)) = -r(n - r + 1) L(p_r(\tau)) + (r + 1)(n - r) L(p_{r+1}(\tau))$$

$$\Rightarrow \quad (S + r(n - r + 1)) L(p_r(\tau)) = (r + 1)(n - r) L(p_{r+1}(\tau))$$

$$\Rightarrow \qquad L(p_r(\tau)) = \frac{(r + 1)(n - r)}{(s + r(n - r + 1))} L(p_{r+1}(\tau)) \qquad\qquad (7.21)$$

Putting $r = n - 1$ in (7.21),

$$L(p_{n-1}(\tau)) = \frac{(n - 1 + 1)(n - (n - 1))}{(S + (n - 1)(n + 1 - (n - 1)))} L(p_n(\tau))$$

$$= \frac{n}{(S + 2(n-1))} \cdot \frac{1}{(S+n)}$$

Putting $r = n - 2$

$$L\left(p_{n-2}(\tau)\right) = \frac{(n-2+1)(n-(n-2))}{S+(n-2)(n+1-(n-2))} \, L\left(p_{n-2}(\tau)\right)$$

$$= \frac{(n-1)2}{S+3(n-2)} \cdot \frac{n}{S+2(n-1)} \cdot \frac{1}{S+n}$$

Again putting $r = n - 3$

$$\Rightarrow L\left(p_{n-3}(\tau)\right) = \frac{(n-3+1)(n-(n-3))}{S+(n-3)[n+1-(n-3)]} \cdot L\left(p_{n-2}(\tau)\right)$$

$$= \frac{3(n-2)}{S+4(n-3)} \frac{2(n-1)}{S+3(n-2)} \frac{n}{S+2(n-1)} \frac{1}{S+n}$$

$$= \frac{(1\cdot 2\cdot 3)\, n\,(n-1)(n-2)}{(S+n)(S+2(n-1))(S+3(n-2))(S+4(n-3))}.$$

By similar way

$$L(p_{n-(n-r)}(\tau)) = L\left(p_r(\tau)\right)$$

$$=: \frac{1\cdot 2\cdot 3\cdots (n-r)\, n\,(n-1)\ldots(n-(n-1-r))}{(S+n)(S+2(n-1))\ldots[S+(n-r+1)(n-(n-r)]}$$

or $L\left(p_r(\tau)\right) = \dfrac{(n-r)!\,n!}{r!}\left\{ \displaystyle\prod_{j=1}^{n-r+1}(S+j(n-(j-1)))\right\}^{-1}$ (7.22)

$$= \frac{C_{r,1}}{(s+r)} + \frac{C_{r,2}}{(s+2(n-1))} + \ldots + \frac{C_{r,n-r+1}}{(s+(n-r+1)r)}$$

Then, $\displaystyle\lim_{s\to -k(n-k+1)}(S+k(n-k+1)\,L\left(p_r(\tau)\right) = C_{rk}$, say,

where $\dfrac{n!\,(n-r)!}{r!}\left\{\displaystyle\lim_{s\to -k(n-k+1)}\left[\prod_{j=1}^{n-r+1}(S+j(n-k+1))\right]^{-1}\right\} = C_{rk}$

$= \dfrac{n!\,(n-r)!}{r!}\, [(-k(n-k+1)+r)(-k(n-k+1)+2(n-1))(-k(n-k+1)$

$\qquad + 3(n-2))\,(-k(n-k+1)+(k-1)\,(n-k+2))]$

$\qquad (-k(n-k+1)+(k+1)(n-k))(-k(n-k+1)+(k+2))$

$\qquad (n-k-1)(-k(n-k+1)+(n-r-1)\,(n-(n-r-1)+1))$

$$(-k(n-k+1)+(n-r)(n-(n-r)+1)$$

$$(-k(n-k+1)+(n-r+1)(n-(n-r+1)+1)]-1$$

$$=\frac{(n-r)!\,n!}{r!}\left\{\left[\begin{array}{c}-k\,(n-k+1)\\+k\end{array}\right]+\frac{(n-k)}{}\right\}\begin{array}{c}[(-k(n-k+1)+2k)\\+2(n-k-1)\end{array}$$

$$\{-k(n-k+1)+3k+3(n-k-1)\}$$

$$\ldots\,(-k(n-k+1)+k(k-1)+(k-1)(n-(k-1)+1-k))$$

$$\ldots\ldots\ldots\ldots\ldots\ldots\ldots\ldots\ldots\ldots\ldots\ldots\ldots\ldots$$

$$(-k(n-k+1)+k(n-r-1))+(n-r-1)(n-(n-r-1)-$$

$$k+1)(-k(n-k+1)+k(n-r))+(n-r)(n-(n-r)+1-k))$$

$$((-k(n-k+1)+k(n-r+1))+(n-r+1)(n-(n-r+1)$$

$$-k+1)]^{-1}$$

$$=\frac{(n-r)!\,n!}{r!}[((-1)(n-k)(k-1))((-)(n-k-1)(k-2))((-)(n-k+1)(k-3))$$

$$\ldots\ldots(-1)\,(n-2k+2)\,.\,1]^{-1}\left[\frac{(n-2k+1)}{(n-2k+1)}\right]^{-1}\,[(n-2k)\,(1)\,\ldots$$

$$(r-k+2)(n-r-k-1)]^{-1}\,[(r-k+1)(n-r-k)]^{-1}$$

$$[(r-k)(n-r-k+1)]^{-1}\,[(r-k-1)\ldots1]^{-1}[(r-k-1)\ldots1)]$$

$$=\frac{(n-r)!\,n!}{r!}\,[(-1)^{k-1}\,((k-1)(k-2)\ldots\,1)^{-1}(n-2k+1)\,\{(n-k)$$

$$(n-k-1)(n-k-2)\ldots(n-2k+2)\,(n-2k+1)(n-2k)\ldots$$

$$(r-k+1)(r-k)(r-k-1)\ldots\frac{1}{[(r-k+1)\ldots}\,\}^{-1}$$

$$\frac{1}{[(n-r-k+1)\,(n-r-k)\ldots1)]}$$

$$\therefore\qquad C_{r,k}=\frac{(n-r)!\,n!}{r!}\frac{(-1)^{k-1}}{(k-1)!}\frac{(n-2k+1)\cdot(r-k+1)}{(n-k)!\,(n-r-k+1)!}\qquad(7.23)$$

$$p_r(\tau)=C_{r,1}\,e^{-n\tau}+C_{r,2}\,e^{-2(n-1)\tau}+\ldots+C_{r,n-r+1}\,e^{-r(n-r+1)\tau},\ r>\frac{1}{2}n$$

$$=p[r\text{ susceptibles in time }\tau=\beta t]$$

$$=\sum_{r=1}^{n-r+1}C_{r,k}\,e^{-k(n-k+r)},\qquad\text{for }r>\frac{1}{2}n\qquad(7.24)$$

where $C_{r,k}$ is given as in (7.23).

Duration of an Epidemic: One aspect of the simple stochastic epidemic that turns out to be fairly tractable is the duration of time i.e. the time that elapses before all susceptibles become infected.

Given that there are j infectives and $(n-j+1)$ susceptibles, the chance of a new infection during an infinitesimal interval $\delta\tau$ is given by

$$j(n-j+1)\,\delta\tau$$

Therefore, the interval τ_j between the occurence of the jth infection and $(j+1)^{th}$ infection has a negative exponential distribution given by

$$f(\tau_j) = j(n-j+1)e^{-j(n-j+1)\tau_j} \qquad (7.25)$$

It is clear that the r.v. τ_j's are all independently distributed. Moreover, the duration of time T of the epidemic is given by

$$T = \sum_{j=1}^{n} \tau_j$$

The rth cumulant of the distribution of T is the sum of the rth cumulants of the distribution of τ_j.
Now

$$E(e^{tX}) = \lambda \int_0^{\infty} e^{tX} e^{-\lambda x}\, dx = \lambda \int_0^{\infty} e^{-x(\lambda-t)}\, dx$$

$$= \lambda \left[-\frac{e^{-\lambda(\lambda-t)x}}{(\lambda-t)} \right]_0^{\infty} = \frac{\lambda}{\lambda-t} = 1 - \frac{1}{\frac{t}{\lambda}} \qquad (7.26)$$

$$\therefore \quad \log E(e^{tX}) = \log \frac{1}{1-\frac{t}{\lambda}}$$

$$= -\log\left(1-\frac{t}{\lambda}\right) = \frac{t}{\lambda} + \frac{1}{2}\frac{t^2}{\lambda^2} + \dots + \frac{1}{r}\frac{t^r}{r!} + \dots$$

$$= \text{coeff of } \frac{t^r}{r!} \text{ in } \log E(e^{tX})$$

$$= \frac{(r-1)!}{\lambda^r} = r^{th} \text{ cumulant of a negative exponential}$$

variate with parameter λ. $\qquad (7.27)$

$\therefore$ r^{th} cumulant of a negative exponential variate T_j with parameter $j(n-j+1)$

$$= \frac{(r-1)!}{j^r(n-j+1)^r}.$$

$\therefore$ K_r = Sum of the rth cumulant of T_j

$$K_r = \sum_{j=1}^{n} \frac{(r-1)!}{j^r (n-j+1)^r} \tag{7.28}$$

For small n one can compute K_r directly by using the above form.

However, for large n, it is advantageous to use asymptotic expressions. It can be shown that neglecting terms of higher order of n^{-1} (Bailey (1963).

$$K_1 = \frac{2 (\log n)}{n+1};$$

where

$$\zeta = \lim_{n \to \infty} \left(1 + \frac{1}{2} + \frac{1}{3} + \dots + \frac{1}{n} - \log n \right)$$

$$= .5772.57 \text{ is Euler's constant}$$

and

$$K_r = \frac{2(r-1)!}{nr} \zeta(r)$$

where

$$\zeta(r) = \sum_{j=1}^{\infty} j^{-r} \quad \text{for } |j| > 1, \text{ where } \zeta(r) \text{ is Riemann's Zeta}$$

function.

It follows that the coefficient of variation of the duration of Epidemic is

$$\frac{K_2^{\frac{1}{2}}}{K_1} \xrightarrow{\text{asympt.}} \frac{\pi}{2\sqrt{3}} \log n$$

while skewness and Kurtosis take the limiting form

$$\lim_{n \to \infty} \gamma_1 = \lim_{n \to \infty} \frac{K_3}{K_2^2} = \frac{2^{1/2} \zeta(s)}{(\zeta(2))^{1/2}} = 0.806.$$

$$\lim_{n \to \infty} \gamma_2 = \lim_{n \to \infty} \frac{K_4}{K_2^2} = \frac{3 \zeta(4)}{(\zeta(2))^2} = 1.200.$$

Therefore, even for infinitely large number of observations the departure from normality is noticeable. For sample of moderate size, Bailey states that the coefficient of variation is of the order of 27% say for $n = 20$. The implication is that there exists quite significant variation in epidemiological behaviour among individuals.

7.2.2 General Epidemic Model

We have by the random variable technique applied on $X(t)$ = no. of susceptibles at any time t and $Y(t)$ = no. of infectives at any time t subject to $X(t) + Y(t) + Z(t)$

= a fixed number where $Z(t)$ = no. of individuals isolated or removed at a time t, the p.g.f. of the Bivariate stochastic process given by (vide Art. 7.1)

$$\frac{\partial \phi(s_1 \, s_2 \, t)}{\partial t} = \beta (s_1^2 - s_1 s_2) \frac{\partial^2 \phi}{\partial s_1 \, \partial s_2} + (1 - s_2) \, \gamma \, \frac{\partial \phi}{\partial s_2} \quad \text{(see 7.14a)}$$

where β and γ are the infection and the removal rates respectively. Putting $\frac{\beta}{\gamma} = \rho$ and $\tau = \beta t$, we get

$$\frac{\partial \phi(s_1 \, s_2 \, t)}{\partial \tau} = (s_2^2 - s_1 s_2) \frac{\partial^2 \phi}{\partial s_1 \, \partial s_2} + \rho \, (1 - s_2) \, \frac{\partial \phi}{\partial s_2} \quad (7.29)$$

Putting
$$\phi \, (s_1 \, s_2 \, t) = \sum_{u, \, v} s_1^u \, s_2^v \, p(u, v) \quad (7.30)$$

where
$$p \, (u, v) = P \, [X(t) = u, \; Y(t) = v].$$

$$\Rightarrow \quad \frac{\partial \phi}{\partial s_1} = \sum_{u, \, v} u \, S_1^{u-1} \, S_2^v \, p(u, v)$$

and
$$\frac{\partial \phi}{\partial s_2} = \sum_{u, \, v} S_1^u \, v \, S_2^{v-1} \, p(u, v) \quad \left.\right\} \quad (7.31)$$

$$\frac{\partial^2 \phi}{\partial s_1 \, \partial s_2} = \sum_{u, \, v} uv \, S_1^{u-1} \, S_2^{v-1} \, p(u, v)$$

Then (7.29) $\Rightarrow \sum S_1^u \, S_2^v \, \dfrac{\partial p(u, v)}{\partial \tau} = (S_2^2 - S_1 S_2) \sum uv \, S_1^{u-1} \, S_2^{v-1} \, p(u, v)$

$$+ \rho \, (1 - S_2) \sum S_1^{uv} \, S_2^{v-1} \, p(u, v).$$

Equating coefficients of $S_1^u \, S_2^v$ on both sides $\Rightarrow$

$$\frac{\partial p(u, v)}{\partial \tau} = (u + 1)(v - 1) \, p \, (u + 1, v - 1) - uvp \, (u, v)$$

$$+ \, e(v + 1) \, p(u, v + 1) - \rho vp \, (u, v)$$

$$= (u + 1)(v - 1) \, p_{u+1, v-1} - v \, (p + u) \, P_{u, \, v}$$

$$+ \rho \, (v + 1) \, p_{u, \, v+1}. \quad (7.32)$$

The general solution of the above differential equation pertaining to classical general epidemic model remained unknown for a century. However, Gani (1965) and Siskind (1965) independently obtained the general solution of the differential equation subject to the boundary conditions:

$$u = n, v = a, \ u(t) \le n, \ v(t) \le n + a \text{ and } u + v \le n + a. \tag{7.33}$$

In this section we attempt to present how special solution to the general epidemic model may be obtainable by assuming initial or boundary condition. The technique is due to Bailey:

(7.32) & (7.33) $\Rightarrow$

$$\frac{\partial p(n, a)}{\partial \tau} = (n+1)(a-1) \underbrace{p_{n+1, a-1}}_{= 0} \quad (\because n+1 \ge n = \# \text{ susceptibles})$$

$$- a(\rho + n) p_{n, a} + \rho(a+1) \underbrace{p_{n, a+1}}_{= 0} \quad (\because n + a + 1 \ge n + a)$$

where $n + a = $ total size of the population

$$\Rightarrow \qquad \frac{\partial p(n, a)}{\partial \tau} = -a(\rho + n) p_{n, a}. \tag{7.34}$$

Taking Laplace transform on both sides of (7.34)

$$-1 + SLp(n, a) = -a(\rho + n) L(p_{n, a})$$

$$(S + a(\rho + n)) L(p(n, a)) = 1$$

$$L(p_{n, a}) = \frac{1}{S + a(\rho + n)}$$

$$\therefore \qquad p_{n, a} = e^{-a(\rho + n)} = e^{-a(\rho + n)\beta t}. \tag{7.35}$$

It is not difficult to see that the general solution of (7.32) is of the form:

$$p(u, v) = C + \sum_{r=1}^{\infty} \alpha_r e^{-\alpha_r t} = C + \phi(t), \text{ say} \tag{7.36}$$

If we put $u = n - w$, $v = 0$, as $t \to \infty$ in (7.36),

we get
$$\lim_{t \to \infty} p(n - w, 0) = c. \tag{7.37}$$

Again putting $u = n - w$, $v = 0$ in (7.36) and taking L-transforms

$$L(p_{n-w}, 0) = \frac{c}{S} + \int_0^{\infty} e^{-St} \phi(t) \, dt$$

$$\lim_{S \to 0} SL(p_{n-w}, 0) = c. \tag{7.38}$$

Thus comparing (7.37) and (7.38), we have

$$\lim_{t \to \infty} p(n - w, 0) = \lim_{S \to 0} SLp(n - w, 0) \tag{7.38a}$$

Let
$$p_w = \lim_{t \to \infty} p(n - w, 0)$$

is the asymptotic probability of W persons being affected.
To obtain p_W:

2nd Part:

We consider Laplace transforms on both sides of (7.32)

$$SLp\,(u, v) = (u + 1)(v - 1)\,Lp\,(u + 1, v - 1) - v\,(\rho + u)\,Lp\,(u, v)$$
$$+ \rho\,(v + 1)\,Lp\,(u, v + 1)$$
$$\Rightarrow (u + 1)\,(v - 1)\,Lp\,(u + 1, v - 1) - (S + v(\rho + u))\,Lp\,(u, v)$$
$$+ \rho\,(v + 1)\,p(u, v + 1) = 0. \tag{7.39}$$

To obtain P_w, we put $u = n - w$, $v = 0$ in (7.39), we get

$$- SL\,(p_{n-w},\,0) + \rho\,(L\,(p_{n-w},\,1)) = 0 \quad (\because p_{n-w+1,-1} = 0)$$

$$\Rightarrow \qquad \lim_{S \to 0} SL(p_{n-w,0}) = \lim_{S \to 0} \rho L(p_{n-w,1})$$

$$\therefore \qquad\qquad P_W = \rho \lim_{S \to 0} L(p_{n-w,1}) \quad \text{(from 7.38a)}$$

$$P_W = \rho\,[f_{n-w,\,1}] \tag{7.40}$$

is the probability that an epidemic stops with w number of individuals being affected. In particular. $P_n = \rho \lim_{S \to 0} L(p_{0,1}) = \rho f_{0,1}$

In the next place, from (7.39),

$$(u+1)(v-1) \lim_{S \to 0} q_{u+1,v-1} - v(u+\rho) \lim_{S \to 0} q_{u,v} + \rho(v+1) \lim_{S \to 0} q_{u,v+1} = 0.$$

where $q_{u,v} = L\,(p\,(u, v))$.

$$\Rightarrow (u + 1)(v + 1)\,f_{u+1,v-1} - v\,(u + \rho)\,f_{u,v} + \rho\,(v + 1)\,f_{u,v+1} = 0. \tag{7.41}$$

Bailey employed the substitution

$$f_{u,v} = \frac{n!\,(n+\rho-1)!\,\rho^{n+a-u-v}}{v(u)!\,(n+\rho)!}\,I_{u,v}. \tag{7.42}$$

Then the equation (7.41) reduces to,

$$\frac{(u+\rho)!}{u!}\,I_{u+1,v-1} - \frac{(u+\rho)!}{u!}\,I_{u,v} + \frac{(u+\rho-1)}{u!}\,I_{u,v+1} = 0.$$

$$\Rightarrow \qquad I_{u+1,v-1} - I_{u,v} + (u+\rho)^{-1}\,I_{u,v+1} = 0 \tag{7.43}$$

We can put

$$I_{u,v} = I_{u+1,v-1} + (u+\rho)^{-1}\,I_{u,v+1}$$
$$I_{u,v+1} = I_{u+1,v} + (u+\rho)^{-1}\,I_{u,v+2}$$
$$I_{u,v+2} = I_{u+1,v+1} + (u+\rho)^{-1}\,I_{u,v+3}$$
$$I_{u,v+3} = I_{u+1,v+2} + (u+\rho)^{-1}\,I_{u,v+4}$$

Thus, $I_{u,v} = I_{u+1,v-1} + (u+\rho)^{-1} I_{u,v+1}$

$\qquad = I_{u+1,v-1} + (u+\rho)^{-1} [I_{u+1,v} + (u+\rho)^{-1} I_{u,v+2}]$

$\qquad = I_{u+1,v-1} + (u+\rho)^{-1} I_{u+1,v} + (u+\rho)^{-2} I_{u,v+2}$

$\qquad = I_{u+1,v-1} + (u+\rho)^{-1} I_{u+1,v} + (u+\rho)^{-2} [I_{u+1,v+1}$
$\qquad\qquad + (u+\rho)^{-1} I_{u,v+3}]$

$\qquad = I_{u+1,v-1} + (u+\rho)^{-1} I_{u+1,v} + (u+\rho)^{-2} I_{u+1,v+1}$
$\qquad\qquad + (u+\rho)^{3} I_{u,v+3}$

$\qquad = I_{u+1,v-1} + (u+\rho)^{-1} I_{u+1,v} + (u+\rho)^{-2} I_{u+1,v+1}$
$\qquad\qquad + (u+\rho)^{-2} [I_{u+1,v+2} + (u+\rho)^{-1} I_{u,v+u}]$

$\qquad = I_{u+1,v-1} + (u+\rho)^{-1} I_{u+1,v} + (u+\rho)^{-2} I_{u+1,v+1}$
$\qquad\qquad + (u+\rho)^{-2} I_{u+1,v+2} + (u+\rho)^{-3} I_{u,v+4}$

Proceeding in this way, we may write $I_{u,v}$ as a linear function of $I_{u+1,k}$ where $k = v-1 \dots; n+a-u-1$ so that $u+v \le n+a$

i.e. $\qquad\qquad u+1+k = n+a$

$\Rightarrow \qquad\qquad k = n+a-u-1$

$\therefore$ Otherwise $\qquad u+1+k = u+v$

$\Rightarrow \qquad\qquad k = v-1$

$I_{u,v} = I_{u+1,v-1} + (u+\rho)^{-1} I_{u+1,v} + (u+\rho)^{-2} I_{u+1,v+1}$

$\qquad + (u+\rho)^{-3} I_{u+1,v+2} \dots. + (u+\rho)^{-[(k-v)+1]^{v+2}} I_{u+1,v+(k-v)}$

$\qquad + (u+\rho)^{-(n+a-u-v)} I_{u+1,(n+a-u-1)}$

$\therefore \qquad I_{u,v} = \sum_{k=v-1}^{n+a-u-1} (u+\rho)^{-k+v-1} I_{u+1,k}$ \hfill (7.44)

Also $\quad I_{u,1} = (u+\rho)^{-1} I_{u,0}$

Subject to $I_{n,a} = 1, I_{u+1,0} = 0$ will give a solution, from which we get $f_{u,v}$ and then $P_W = \rho [f_{n-w,1}]$ we have,

$$f_{u+1,0} = \frac{n!(u+1+\rho-1)!}{0(u+1)!(n+\rho)!} \rho^{n+a-u-1-0} I_{u+1,0}$$

$\Rightarrow \qquad I_{u+1,0} = 0$ \hfill (7.45)

Also

$$\frac{dp_{n,a}}{d\tau} = -a(n+\rho) p_{n,a}$$

$$-1 + S L\,(p_{n,\,a}) = -a\,(n+\rho)\,L\,(p_{n,\,a})$$

$$(S + a\,(n+\rho))\,L\,(p_{n,\,a}) = 1$$

$$\lim_{s \to 0}\,(s + a\,(n+\rho))\,L\,(p_{n,\,a}) = 1$$

i.e.
$$a\,(n+\rho)\,\lim_{s \to 0}\,L\,(p_{n,\,a}) = 1.$$

$$a\,(n+\rho)\,f_{n,\,a} = 1$$

$$1 - a\,(n+\rho)\,f_{n,\,a} = 0. \Rightarrow f_{n,\,a} = \frac{1}{a\,(n+\rho)}$$

Next
$$f_{n,\,a} = \frac{n!\,(n+\rho-1)!\,\rho^{\,n+a-n-a}}{a\,n!\,(n+\rho)!}\,I_{n,\,a} = \frac{1}{a\,(n+\rho)}\,I_{n,\,a}$$

$$\Rightarrow \quad f_{n,\,a}\,a\,(n+\rho) = I_{n,\,a}$$

But
$$f_{n,\,a} = \frac{1}{a\,(n+\rho)} \quad \Rightarrow \quad I_{n,\,0} = 1 \tag{7.46}$$

For small values of n one can use (7.44) and (7.46) to calculate expressions for $I_{u,\,v}$ and hence for p_w (actually we need only $I_{u,\,1}$ but the method entails computations of all $I_{u,\,v}$). Bailey has given graphs of P_w for $n = 10, 20, 40$ with $a = 1$ for various values of ρ and has shown that for $\rho \geq n$, the distributions are all J-shaped with the highest point at $w = 0$, but if $\rho \leq n$, the distribution is U-shaped, so that there is either a very small no. of total cases, or a large no. of cases while the intermediate situations being rare. These results are obviously related to some kind of threshold phenomenon but with groups as small as $n = 40$ there is no sharp transition from one group to another.

7.2.3 Carrier Borne Epidemic Model: (Weiss,(1965), Biometrics)

Modern concept on epidemic models is now different from the classical epidemic models which we have discussed. Today, it is almost inconceivable that individuals affected with contageous diseases having palpable symptoms are unrestrictably showering infections to susceptibles, who are exposed to infections. On the other hand, with the appearance of visible symptoms infected individuals are segregated in any part of the civilized world. However, a class of individuals (who are either affected so mildly or the disease has a special way of progress without much visible symptoms (like AIDS etc.)) remain in the population having infections and spreading infections among susceptibles without themselves being physically visible or identifiable. They are potential danger to the population of susceptibles. They are known as carriers. More precisely following Weiss (1965), we define a carrier to be an individual who does not have overt disease symptoms but nevertheless is able to communicate the disease to others. Under this category, we may not only include such human carriers, but also inanimate sources of infection say, a polluted stream for Malarial infection which may be used by a large population.

Below we present an outline of the Stochastic Version of a Carrier borne model due to Weiss:

Weiss's Carrier Borne Model: (Random Walk Approach)
Let

(i) $\pi_k(m, n) = P$ [that a population of m susceptibles are reduced to k survivors because of an epidemic initiated by n carriers]

(ii) $\beta r \, \delta t = P$ [that a Carrier is eliminated during $(t, t + \delta t) \,|\,$ Population size of Carrier $= n$ time t]

(iii) $\alpha rs \, dt \equiv P$ [a carrier infecting a susceptible during $(t, t + \delta t) \,|\,$ size of Carrier population is 'r' at time t, size of susceptible population is 's' at time t]

Further we assume that a susceptible while being infected will not be a further source of infection as he will be identified and removed from the population.

Thus the problem reduces to random walk on a lattice with permissible transitions $(m, n) \to (m, n - 1)$ and $(m, n) \to (m - 1, n)$ for $m, n \neq 0$ where m, n are the initial number of susceptible and carriers at the beginning of the epidemic.
Let

$$m(t) = \text{\# susceptible at time } t$$

$$n(t) = \text{\# carriers at time } t$$

A carrier is eliminated at the rate $\beta n(t)$ and a susceptible is infected at the rate $\alpha m(t) \, n(t)$

$$m'(t) = -\alpha m(t) \, n(t) \tag{7.47}$$

and $\qquad n'(t) = -\beta n(t) \tag{7.48}$

$\therefore \qquad \dfrac{dm(t)}{dt} = -\alpha m(t) \, n(t)$

$\Rightarrow \qquad \dfrac{dm(t)}{m(t)} = -\alpha n(t) \, dt$

$\Rightarrow \qquad \log \dfrac{m(t)}{m(0)} = -\alpha \int\limits_{0}^{t} n(\tau) \, d\tau$

$$m(t) = m(0) \, e^{-\alpha \int\limits_{0}^{t} n(\tau) \, d\tau} \tag{7.49}$$

Similarly, $\qquad \log \dfrac{n(t)}{n(0)} = -\beta t$

$$\Rightarrow \qquad n(t) = \overset{\bullet}{n}(0)\, e^{-\beta t} \qquad (7.50)$$

$$\therefore \qquad m(t) = m(0)\, e^{-\alpha n(0)\int_0^t e^{-\beta \tau}\, d\tau}$$

$$= m(0)\, e^{-\alpha n(0)\frac{(1-e^{-\beta t})}{\beta}}$$

$$= m(0)\, e^{-\left[\frac{\alpha}{\beta} n(0)\, (1-e^{-\beta t})\right]}$$

$$\therefore \qquad m(t) = m(0)\, e^{-\sigma n(0)\, (1-e^{-\beta t})} \text{ where } \frac{\alpha}{\beta} = \sigma \qquad (7.51)$$

$$\Rightarrow \qquad m(\infty) = m(0)\, e^{-\sigma n(0)}$$

Further
$$P\{(m, n)_t \ \to \ (m-1, n)_{t+\delta t}\}$$

$$= \frac{(\alpha mn)\, \delta t}{\alpha mn\, \delta t + \beta n\, \delta t} = \frac{\alpha m}{\beta + \alpha m} = \frac{\dfrac{\alpha}{\beta} m}{1 + \dfrac{\alpha}{\beta} m}$$

$$= \frac{\sigma m}{1+\sigma m} \text{ where } \frac{\alpha}{\beta} = \sigma \qquad (7.52)$$

Similarly
$$P\{(m, n)_t \ \to \ (m, n-1)_{t+\delta t}\}$$

$$= \frac{\beta n\, \delta t}{\beta n\, \delta t + \alpha mn\, \delta t} = \frac{1}{1 + \dfrac{\alpha}{\beta} m} = \frac{1}{1+\sigma m} \qquad (7.53)$$

steady state solution of two state Markov chain on continuous time, (vide Art 2.10)

$$\therefore \qquad \pi_k(m, n) = \pi_k(m, n-1)\left[\frac{1}{1+\sigma m}\right] + \pi_k(m-1, n)\left[\frac{\sigma m}{1+\sigma m}\right]$$

$$\text{(using 7.52 \& 7.53)}$$

Also
$$\pi_k(m, 0) = \delta_{km} \qquad (7.54)$$

where δ_{km} is Kronekar's delta.

Now putting $m = k$ and $n = 1$ in (7.54), we get

$$\pi_k(k, 1) = \pi_k(k, 0)\cdot\frac{1}{1+\sigma k} + \pi_k(k-1, 1)\left[\frac{\sigma k}{1+\sigma k}\right]$$

$$\Rightarrow \qquad \pi_k(k, 1) = \frac{1}{1+\sigma k} \quad (\because \pi_k(k-1, 1) = 0]$$

Similarly,

$$\pi_k\,(k,2) \;=\; \frac{\pi_k\,(k,1)+\sigma\,m\,\pi_k\,(k-1,2)}{1+\sigma k} \;=\; \frac{1}{(1+\sigma k)^2} \quad (\because \pi_k\,(k-1,2)=0)$$

Similarly,
$$\pi_k\,(k,n) \;=\; \frac{1}{(1+\sigma k)^n} \quad n\neq 0 \tag{7.55}$$

For p.g. f of $\pi_k\,(m,n)$, we have

$$\sum_{n=1}^{\infty}\pi_k\,(m,n)\,s^n \;=\; \left[\,s\sum_{n=1}^{\infty}\pi_k\,(n,n-1)\,s^{n-1}+\sigma m\sum_{n=1}^{\infty}\pi_k\,(m-1,n)\,s^n\,\right]\Big/(1+\sigma m)$$
$$\tag{7.56}$$

In order to solve equation (7.55), we employ the generating function

$$\eta_{k,\,m}\,(s) \;=\; \sum_{n=1}^{\infty}\pi_k\,(m,n)\,s^n$$

$$\eta_{k,\,m}\,(s) \;=\; 0 \text{ for } m<k$$

$$\therefore \qquad \eta_{k,\,m}\,(s) \;=\; \frac{s\,\eta_{k,m}\,(s)+\sigma m\,\eta_{k,m-1}\,(s)}{1+\sigma m} \quad \text{from (7.56)}$$

$$\Rightarrow \qquad \eta_{k,\,m}\,(s)\,(1+\sigma m)-s\,\eta_{k,\,m}\,(s) \;=\; \sigma m\,\eta_{k,\,m-1}\,(s)$$
$$\eta_{k,\,m}\,(s)\,(1+\sigma m-s) \;=\; \sigma m\,\eta_{k,\,m-1}\,(s)$$

$$\Rightarrow \qquad \eta_{k,\,m}\,(s) \;=\; \frac{\sigma m}{1+\sigma m-s}\,\eta_{k,m-1}\,(s) \tag{7.57}$$

Also
$$\eta_{k,\,k}\,(s) \;=\; \sum_{n=1}^{\infty}\pi_k\,(k,n)\,s^n \;=\; \sum_{n=1}^{\infty}\frac{s^n}{(1+\sigma k)^n} \quad \text{from (7.55)}$$

$$= \frac{s}{1+k\sigma}\cdot\frac{1}{1-\dfrac{s}{1+\sigma k}} \;=\; \frac{s}{1+k\sigma}\cdot\frac{1+\sigma k}{1+\sigma k-s} \;=\; \frac{s}{1+\sigma k-s} \tag{7.58}$$

Putting $m=k+1$ in (7.57)

$$\eta_{k,\,k+1}\,(s) \;=\; \frac{\sigma\,(k+1)}{1+\sigma\,(k+1)-s}\,\eta_{k,k}\,(s)$$

$$= \frac{\sigma\,(k+1)}{1+\sigma\,(k+1)-s}\cdot\frac{s}{1+\sigma k-s} \quad \text{from (7.58)}$$

Similarly

$$\eta_{k,\,k+2}\,(s) \;=\; \frac{\sigma\,(k+2)\,s}{1+\sigma\,(k+2)-s}\,\eta_{k,k+1}\,(s)$$

$$= \frac{\sigma^2 (k+2)(k+1) s}{[1+\sigma(k+2)-s][1+\sigma(k+1)-s]} \frac{1}{(1+\sigma k - s)}$$

Proceeding in this way,

$$\eta_{k, k+(m-k)}(s) = \frac{\sigma^{m-k} [(k+m-k)\dots m \text{ factors}] s}{(1+(\sigma_m - s))\dots(1+\sigma k - s)}$$

$$= \sigma^{m-k} \frac{m!}{k!} \frac{s}{\displaystyle\prod_{j=k}^{m}(1+j\sigma-s)} \tag{7.59}$$

$$= \frac{m!}{k!} s \frac{\sigma^{m-k+1}}{\sigma} \frac{1}{\displaystyle\prod_{j=k}^{m}(1+j\sigma-s)}$$

$$= \frac{m!}{k!} \frac{s}{\sigma} \frac{1}{\displaystyle\prod_{j=k}^{m}\left(\frac{1-s}{\sigma}+j\right)}$$

$$= \frac{m!}{k!} \frac{s}{\sigma} \frac{\left[\left(\frac{1-s}{\sigma}+k-1\right)\dots 1\right]}{\left(\frac{1-s}{\sigma}+k\right)\left(\frac{1-s}{\sigma}+k+1\right)\dots\left(\frac{1-s}{\sigma}+m\right)\left[\left(\frac{1-s}{\sigma}+k-1\right)\dots 1\right]}$$

$$= \frac{m!}{k!} \frac{s}{\sigma} \frac{\left(\frac{1-s}{\sigma}+k-1\right)!}{\left(\frac{1-s}{\sigma}+m\right)!}$$

$$= \frac{m!}{k!} \frac{s}{\sigma} \frac{\Gamma\left(\frac{1-s}{\sigma}+k\right)}{\Gamma\left(\frac{1-s}{\sigma}+m+1\right)} \tag{7.60}$$

Also let $\displaystyle\prod_{j=k}^{m} \frac{1}{j\sigma+1-s}$ be expressed as sums given by

$$= \frac{A_k}{k\sigma+1-s} + \frac{A_{k+1}}{(k+1)\sigma+1-s} + \dots + \frac{A_m}{m\sigma+1-s}$$

$$\Rightarrow \lim_{s\to k\sigma+1} \frac{1}{k\sigma+1-s}\left[A_k + \frac{A_{k+1}(k\sigma+1-s)}{(k+1)\sigma+1-s} + \dots + \frac{A_m(k\sigma+1-s)}{m\sigma+1-s}\right]$$

$$= \lim_{s \to k\sigma + 1} \prod_{j=k}^{m} \frac{1}{j\sigma + 1 - s}$$

$$\Rightarrow \qquad A_k = \prod_{j=k+1}^{m} \frac{1}{j\sigma + 1 - 1 - k\sigma} = \prod_{j=k+1}^{m} \frac{1}{(j-k)\sigma}$$

$$= \frac{1}{[(k+1-k)\dots(m-k)] \underbrace{[\sigma\dots\sigma]}_{(m-k)\ \text{terms}}}$$

$$= \frac{1}{(m-k)!\,\sigma^{m-k}} \qquad\qquad (7.61)$$

Precisely in a similar way

$$\lim_{s \to (k+1)\sigma + 1} \left[\frac{A_k\,[(k+1)\sigma + 1 - s]}{k\sigma + 1 - s} + A_{k+1} + A_{k+2}\,\frac{(k+1)\sigma + 1 + s}{(k+2)\sigma + 1 - s} + \right.$$

$$\left. \dots + \frac{A_m\,(k+1)^{(k+1)\sigma + 1 + S}}{m\sigma + 1 - S} \right]$$

$$\lim_{s \to (k+1)\sigma + 1} \left[\frac{1}{k\sigma + 1 - s}\,\frac{1}{(k+2)\sigma + 1 - s} \dots \frac{1}{m\sigma + 1 - s} \right]$$

$$\Rightarrow \qquad A_{k+1} = \frac{1}{(-1)\sigma}\,\frac{1}{2\sigma \dots (m - k - 1)\sigma}$$

$$= \frac{(-1)}{\sigma^{m-k}\,(m - k - 1)!}$$

In general,

$$A_j = \frac{(-1)^{j-k}}{(j-k)!\,(m-j)!}\,\frac{1}{\sigma^{m-k}}, \quad j = k, k+1\dots$$

$$= (-1)^{j-k} \binom{m-k}{j-k} \frac{1}{(m-k)!}\,\frac{1}{\sigma^{m-k}} \qquad (7.62)$$

$$\eta_{k,m}(s) = \frac{m!}{k!}\,s\,\sigma^{m-k} \sum_{j=k}^{m} \left(\frac{A_j}{(1 + j\sigma - s)} \right) \quad \text{(from (7.59))}$$

Putting (7.62) in the above

$$= \frac{m!}{k!}\,s\,\sigma^{m-k} \sum_{j=k}^{m} (-1)^{j-k} \binom{m-k}{j-k} \frac{1}{(m-k)!}\,\frac{1}{\sigma^{m-k}}\,\frac{1}{1 + j\sigma - s}$$

$$= \frac{m!}{k!(m-k)!} \sum_{j=k}^{m} (-1)^{j-k} \binom{m-k}{j-k} \frac{s}{1+j\sigma - s}$$

$$\therefore \quad \eta_{k,m}(s) = \binom{m}{k} \sum_{j=k}^{m} (-1)^{j-k} \binom{m-k}{j-k} \frac{s}{1+j\sigma - s} \tag{7.63}$$

Again equating coefficients of s^n on both sides of (7.63), we get

$$\pi_{k,m} \text{ as } \binom{m}{k} \sum_{j=k}^{m} (-1)^{j-k} \binom{m-k}{j-k} \frac{s}{(1+j\sigma)^m} \tag{7.64}$$

$$\left(\because \frac{s}{1+j\sigma - s} = \sum_{n=1}^{\infty} \frac{s^n}{(1+j\sigma)^n} \right)$$

Also the term $\dfrac{1}{1+j\sigma - s}$ in (7.63) $= \displaystyle\int_{0}^{\infty} e^{-(j\sigma + 1 - s)t} \, dt$

$(7.63) \Rightarrow$

$$\eta_{k,m}(s) = s \binom{m}{k} \sum_{j=k}^{m} (-1)^{j-k} \binom{m-k}{j-k} \int_{0}^{\infty} e^{-(j\sigma + 1 - s)t} \, dt$$

$$= s \binom{m}{k} \int_{0}^{\infty} e^{-(1-s)t} \sum_{j=k}^{m} \binom{m-k}{j-k} (-1)^{j-k} e^{-j\sigma t} dt \tag{7.65}$$

Now the $(j-k)^{\text{th}}$ term in the expansion of $(1-e^{-\sigma t})^{m-k}$

$$= \binom{m-k}{j-k} (-1)^{j-k} e^{-\sigma t(j-k)} \text{ where } k \le j \le m$$

$$= e^{k\sigma t} \binom{m-k}{j-k} (-1)^{j-k} e^{-\sigma t j}$$

$$\Rightarrow \quad \left[\sum_{j=1}^{m} \binom{m-k}{j-k} (-1)^{j-k} e^{-\sigma jt} \right] e^{k\sigma t} = (1-e^{-\sigma t})^{m-k}$$

$$\Rightarrow \quad (1-e^{-\sigma t})^{m-k} e^{-k\sigma t} = \sum_{j=k}^{m} \binom{m-k}{j-k} (-1)^{j-k} e^{-\sigma jt} \tag{7.66}$$

Putting (7.66) in (7.65), we get

$$\eta_{k,m}(s) = s\binom{m}{k}\int_0^\infty e^{-(1-s)t}(1-e^{-\sigma t})^{m-k}e^{-k\sigma t}\,dt$$

$$= s\binom{m}{k}\int_0^\infty e^{-(1-s+k\sigma)t}(1-e^{-\sigma t})^{m-k}\,dt \tag{7.67}$$

Also $\pi_{k,m}$ is obtainable from the coefficient of s^n of above

$$\eta_{k,m}(s) = s\binom{m}{k}\int_0^\infty e^{-(1+k\sigma)t}\left(1+st+\ldots+\frac{s^{n-1}t^{n-1}}{(n-1)!}+\ldots\right)(1-e^{-\sigma t})^{m-k}\,dt$$

$$\pi_k(m,n) = \binom{m}{k}\int_0^\infty e^{-(1+k\sigma)t}\frac{t^{n-1}}{(n-1)!}(1-e^{-\sigma t})^{m-k}\,dt$$

$$= \frac{1}{(n-1)!}\binom{m}{k}\int_0^\infty e^{-(1+k\sigma)t}t^{n-1}(1-e^{-\sigma t})^{m-k}\,dt \tag{7.68}$$

$$\sum_{k=1}^m \pi_k(m,n) = 1.$$

From this we can estimate $E(K)$ and $\mathrm{Var}(K)$ as follows:

$$E(K) = \frac{1}{(n-1)!}\int_0^\infty \left\{\sum_{k=0}^m \binom{m}{k}e^{-k\sigma t}(1-e^{-\sigma t})^{m-k}e^{-t}t^{n-1}\right\}dt$$

$$= \frac{m}{(n-1)!}\int_0^\infty \sum_{k=1}^m \frac{(m-1)!}{(k-1)!(m-k)!}e^{-(k-1)\sigma t}(1-e^{-\sigma t})^{m-k}e^{-\sigma t}e^{-t}t^{n-1}\,dt$$

$$= \frac{m}{(n-1)!}\int_0^\infty (1-e^{-\sigma t}+e^{-\sigma t})^{m-1}e^{-t(1+\sigma)}t^{n-1}\,dt$$

$$= \frac{m}{(n-1)!}\int_0^\infty e^{-t(1+\sigma)}t^{n-1}\,dt$$

$$= \frac{m}{(n-1)!}\frac{(n)}{(1+\sigma)^n} = \frac{m}{(1+\sigma)^n}. \tag{7.69}$$

Again

$$\mathrm{Var}(K) = E(k^2)-[E(k)]^2$$

$$E(K^2) = \frac{1}{(n-1)!} \int_0^\infty \left\{ \sum_{k=0}^m k^2 \binom{m}{k} e^{-k\sigma t} (1-e^{-\sigma t})^{m-k} e^{-t} t^{n-1} dt \right\}$$

$$= \frac{1}{(n-1)!} \int_0^\infty \sum_{k=0}^m (k(k-1)+k) \binom{m}{k} e^{-k\sigma t} (1-e^{-\sigma t})^{m-k} e^{-t} t^{n-1} dt$$

$$= \frac{m(m-1)!}{(n-1)!} \left[\int_0^\infty \sum_{k=0}^m \frac{(m-2)!}{(k-2)!(m-k)!} e^{-k\sigma t} (1-e^{-\sigma t})^{m-k} e^{-t} t^{n-1} dt \right]$$

$$+ \frac{1}{(n-1)!} \int_0^\infty \sum_{k=2}^m k \binom{m}{k} e^{-k\sigma t} (1-e^{-\sigma t})^{m-k} e^{-t} t^{n-1} dt$$

$$= \frac{m(m-1)}{(n-1)!} \int_0^\infty \sum_{k=0}^m \binom{m-2}{k-2} e^{-(k-1)\sigma t} (1-e^{-\sigma t})^{(m-2)-(k-2)}$$

$$e^{-2\sigma t} e^{-t} t^{n-1} dt + \frac{m}{(1+\sigma)^n}$$

$$= \frac{m(m-1)}{(n-1)!} \int_0^\infty (1-e^{-\sigma t} + e^{-\sigma t})^{n-2} e^{-t(1+2\sigma)} t^{n-1} dt + \frac{m}{(1+\sigma)^n}$$

$$= \frac{m(m-1)}{(n-1)!} \int_0^\infty e^{-t(1+2\sigma)} t^{n-1} dt + \frac{m}{(1+\sigma)^n}$$

$$= \frac{m(m-1)}{(n-1)!} \frac{\Gamma(n)}{(1+2\sigma)^n} + \frac{m}{(1+\sigma)^n} = \frac{m(m-1)}{(1+2\sigma)^n} + \frac{m}{(1+\sigma)^n} \qquad (7.70)$$

Thus

$$\text{Var}(K) = \frac{m(m-1)}{(1+2\sigma)^n} + \frac{m}{(1+\sigma)^n} - \frac{m^2}{(1+\sigma)^{2n}}. \qquad (7.71)$$

It may be noted that Stochastic mean $= \dfrac{m}{(1+\sigma)^n}$, whereas the Deterministic mean is

$$m(0) e^{-\sigma} = m(1 - n\sigma + ...)$$

For small n,

$$(1+\sigma)^{-n} > e^{-\sigma n}$$

Hence deterministic model underestimates the size of the epidemic.

Weiss (1965) has also evolved a time dependent solution for $n = 1$. It is assumed that the epidemic can be terminated either by elimination of carriers, or when the entire population contracts the disease before the carrier is eliminated. Defining $p_j(t)$ to be the probability that the population is reduced from m to j in time t subject to the condition of the presence of a carrier at time t the expression for $p_j(t)$ is shown in the section (7.76) and the probability density function of the duration of epidemic is shown in (7.78). For obtaining the moments, Laplace transform technique has been used. The mean duration of the carrier borne epidemic has been worked out in (7.79) and (7.80).

Time dependence on the evolution of a carrier
We have

$$p_m(t + \Delta t) = p_m(t)(1 - (m\alpha \, \Delta t + o(\Delta t))$$

$$\Rightarrow \qquad \lim_{\Delta t \to 0} \frac{p_m(t + \Delta t) - p_m(t)}{\Delta t} = -m\alpha \, p_m(t)$$

$$\therefore \qquad p'_m(t) = -m\alpha \, p_m(t)$$

$$-1 + SL(p_m(t)) = -m\alpha \, L(p_m(t))$$

$$\Rightarrow \qquad (S + m\alpha)L(p_m(t)) = 1$$

$$L(p_m(t)) = \frac{1}{s + m\alpha}$$

$$\Rightarrow \qquad p_m(t) = e^{-m\alpha t} \qquad\qquad (7.72)$$

Also, $\quad p_{m-1}(t + \Delta t) = p_m(t)(m\alpha \, \Delta t + o(\Delta t))$

$$+ p_{m-1}(t)(1 - ((m-1)\alpha \, \Delta t + o(\Delta t))$$

$$\lim_{\Delta t \to 0} \frac{p_{m-1}(t + \Delta t) - p_{m-1}(t)}{\Delta t} = -(m-1)p_{m-1}(t)\alpha + p_m(t)m\alpha + \lim_{t \to 0} \frac{o(\Delta t)}{t}$$

$$\therefore \qquad p'_{m-1}(t) = -\alpha(m-1)p_{m-1}(t) + \alpha m \, p_m(t) \qquad\qquad (7.73)$$

Taking Laplace transform on both sides of (7.73),

$$L(p'_{m-1}(t)) = -\alpha(m-1)L(p_{m-1}(t)) + \alpha mL(p_m(t))$$

$$\Rightarrow \qquad SL(p_{m-1}(t)) + \alpha(m-1)L(p_{m-1}(t)) = \alpha mL(p_m(t))$$

$$\Rightarrow \qquad [S + \alpha(m-1)]L(p_{m-1}(t)) = \alpha mL(p_m(t))$$

$$\Rightarrow \qquad L(p_{m-1}(t)) = \frac{\alpha mL(p_m(t))}{(S + \alpha(m-1))}$$

$$= \frac{\alpha m}{s + \alpha(m-1)} \frac{1}{s + m\alpha} \qquad\qquad (7.74)$$

$$= m\left[\frac{1}{s + (m-1)\alpha} - \frac{1}{(s + m\alpha)}\right]$$

$$p_{m-1}(t) = m\left[e^{-(m-1)\alpha t} - e^{-m\alpha t}\right]$$

$$= \binom{m}{1} e^{-(m-1)\alpha t} (1-e^{-\alpha t}) \qquad (7.75)$$

$$P_{m-2}(t + \Delta t) = P_{m-1}(t)((m-1)\,\alpha\,\Delta t + 0(\Delta t))$$
$$+ P_{m-2}(t)(1-(m-2)\,\alpha\,\Delta t + 0(\Delta t))$$

$$\Rightarrow \qquad \lim_{\Delta t \to 0} \frac{P_{m-2}(t + \Delta t) - P_{m-2}(t)}{\Delta t} = -\alpha\,(m-2)\,P_{m-2}(t)$$

$$+ (m-1)\,\alpha\,P_{m-1}(t) + \lim_{t \to 0} \frac{0(\Delta t)}{\Delta t}$$

$$\Rightarrow \qquad P'_{m-2}(t) = -\alpha\,(m-2).P_{m-2}(t) + (m-1)\,\alpha\,P_{m-1}(t)$$

$$SL\,(P_{m-2}(t) + \alpha\,(m-2)\,L\,(P_{m-2}(t)) = (m-1)\,\alpha\,L\,(P_{m-1}(t))$$

$$\Rightarrow \qquad (s + \alpha\,(m-2))\,L\,(P_{m-2}(t)) = (m-1)\,\alpha\,L\,(P_{m-1}(t))$$

$$\Rightarrow \quad L\,(P_{m-2}(t)) = \frac{(m-1)\alpha}{s+\alpha(m-2)}\,L\,(P_{m-1}(t))$$

$$= \frac{(m-1)\alpha}{s+\alpha(m-2)} \cdot \frac{\alpha\,m}{s+\alpha(m-1)} \cdot \frac{1}{s+m\alpha} \qquad \text{(from (7.74))}$$

Proceeding in this way, we get,

$$L\,(P_{m-j}(t)) = \frac{(m-j+1)\alpha \cdot \alpha\,(m-j)\dots 1}{(s+\alpha(m-j))\,(s+\alpha(m-j+1))\dots(s+\alpha(m))}; \quad j=0,1,2,..$$

Thus

$$P_j(t) = \binom{m}{j} e^{-\alpha t j} (1-e^{-\alpha t})^{m-j}, \qquad (7.76)$$

which corresponds to a pure death process.

$\therefore$ P [at any time $\exists\, j$ susceptibles and the carrier is present]

$$= \binom{m}{j} e^{-\alpha t j} (1-e^{-\alpha t})^{m-j} e^{-\beta t} \qquad (7.77)$$

Probability density of the duration of the Epidemic is obtainable by considering either all the m susceptibles which are infected with rate α and the carrier remains which is

$$\left[(1-e^{-\alpha t})^{m-1} \binom{m}{1} (\alpha e^{-\alpha t})(e^{-\beta t}) \right] = \frac{m!}{(m-1)!} (1-e^{-\alpha t})^{m-1} \alpha e^{-\alpha t} e^{-\beta t}$$

Now

$$\int_0^\infty e^{-st}\, p(t)\, dt = \int_0^\infty \left[1 - st + \frac{s^2 t^2}{2!} \dots \right] p(t)\, dt$$

$$L(p(t)) = \int_0^\infty p(t)\,dt - s\int_0^\infty t\,p(t)\,dt + \frac{s^2}{2!}\int_0^\infty t^2\,p(t)\,dt - \frac{dL(p(t))}{ds}\Big|_{s=0}$$

$$= -s\int_0^\infty t\,p(t)\,dt - \frac{dL(p(t))}{ds}\Big|_{s=0} = \mu_1'$$

$$\frac{d^2L(p(t))}{ds^2}\Big|_{s=0} = \mu_2'$$

$$\frac{d^2L(p(t))}{ds^2}\Big|_{s=0} = (-)^r\,\mu_r$$

Or the probability that Carrier is off with rate β at time t while some of the susceptibles are infected

$$= \beta\,e^{-\beta t}[1 - (1 - e^{-\alpha t})^m]$$

Hence the probability density function of the epidemic is given by

$$\Psi(t) = [(1 - e^{-\alpha t})^{m-1}\,m\,\alpha\,e^{-(\alpha+\beta)\,t}] + [1 - (1 - e^{-\alpha t})^m]\,\beta\,e^{-\beta t} \quad (7.78)$$

Here the first term represents the probability density assuming termination of epidemic by elimination of carriers, while the second term represents the termination of the infection of the entire population before the carrier is eliminated.

Now, $\displaystyle\int_0^\infty e^{-st}\,\Psi(t)\,dt = L(\Psi(t))$

$$= \frac{\beta}{\beta+s} + \frac{s}{\alpha}\frac{m!\,\Gamma\left(\dfrac{s}{\alpha}+\dfrac{1}{\sigma}\right)}{\Gamma\left(\dfrac{s}{\alpha}+\dfrac{1}{\sigma}+m+1\right)}$$

$$\frac{dL(\Psi(t))}{ds}\Big|_{s=0} = -\frac{\beta}{(\beta+s)^2}\Big|_{s=0} + \frac{1}{\alpha}\frac{d}{ds}\left[\frac{m!\left(\dfrac{s}{\alpha}+\dfrac{1}{\sigma}-1\right)!}{\left(\dfrac{s}{\alpha}+\dfrac{1}{\sigma}+m\right)!}\right]_{s=0}$$

$$-\frac{dL(\Psi(t))}{ds}\Big|_{s=0} = -\frac{\beta}{(\beta+s)^2}\Big|_{s=0}\frac{1}{\alpha}\frac{\Gamma\left(\dfrac{1}{\sigma}\right)m!}{\Gamma\left(\dfrac{1}{\sigma}+m+1\right)}$$

$$\because \frac{\alpha}{\beta} = \sigma,\ \alpha = \beta\sigma$$

$$= \frac{1}{\beta} - \frac{1}{\beta}\frac{1}{\sigma}\Gamma\left(\frac{1}{\sigma}\right)\left(\Gamma\left(\frac{1}{\sigma}+m+1\right)\right)$$

$$\therefore \text{ Mean duration of epidemic } = \frac{1}{\beta}\left[1 - \frac{m!\,\Gamma\left(1 + \frac{1}{\sigma}\right)}{\sigma\,\Gamma\left(\frac{1}{\sigma} + m + 1\right)}\right] \tag{7.79}$$

By using Sterling's approximation to factorials the mean duration is given by,

$$T \sim \frac{1}{\beta}\left[1 - \frac{\Gamma\left(\frac{1}{\sigma} + 1\right)}{\sigma\,m^{\frac{1}{\sigma}}}\right] \tag{7.80}$$

7.2.4 A Martingale Based Carrier Borne Epidemic Model (Picard (1980))

Picard has considered Downton's (1968) bivariate birth and death process (X_t, Y_t) given as follows

$$P(X_{t+\Delta t} = r,\ Y_{t+\Delta t} = s \mid X_t = i, Y_t = j)$$

$$= \alpha\pi ij\,\Delta t + 0(\Delta t),$$

$$\qquad\qquad \text{if } r = i - 1,\ s = j + 1$$

$$= \alpha(1 - \pi)\,ij\,\Delta t + o(\Delta t),$$

$$\qquad\qquad \text{if } r = i - 1,\ s = j$$

$$= \beta j\,\Delta t + 0(\Delta t), \tag{7.81}$$

$$\qquad\qquad \text{if } r = i,\ s = j - 1$$

$$= 1 - (\alpha i + \beta)\,j\,\Delta t + o(\Delta t)$$

$$\qquad\qquad \text{if } r = i,\ s = j$$

$$= o(\Delta t) \text{ in all other cases,}$$

$$\alpha > 0,\ \beta > 0, 0 \le \pi \le 1$$

and (x_0, y_0) being the initial population of susceptibles and carriers respectively. We have

$$0 \le i + j \le x_0 + y_0$$

Now, (x_t, y_t) represent the sizes of susceptible and carriers respectively. Here he has developed a model under the assumption that any susceptible may be removed by infection or changed into a carrier. While a carrier may only be removed by death (or isolation).

Note that $\quad \pi = 0 \Rightarrow$ Weiss's model
and $\quad\quad \pi = 1 \Rightarrow$ Classical general epidemic model as discussed under sections 7.2.3 and 7.2.2 respectively.

We shall illustrate Picard's method of building up a class of Martingales on the bivariate birth and death process. Picard build up a class of Martingales of the form V_t given by

$$V_t = a(X_t, Y_t) e^{-Z_t}$$

$$\text{and} \quad Z_t = \int_0^t h(X_u, Y_u)\, du \qquad \Bigg\} \qquad (7.82)$$

where 'a' and 'h' are arbitrary functions to be fixed in a way so that $V_{t,n}$ for different class is a class of Martingales.

We set
$$E\{|a(X_t, Y_t) e^{-Z_t}|\} < \infty \qquad (7.83)$$

and
$$E[a(X_t, Y_t) e^{-Z_t} | \mathcal{F}_{t0}] = a(X_{t_0}, Y_{t_0}) e^{-Z_{t_0}}, \quad 0 \le t_0 \le t \qquad (7.84)$$

Again
$$0 \le a(X_t, Y_t) \le \max a(i, j).$$

$\therefore$ the condition (7.83) is obviously satisfied.

Again we have

$$(e^{-Z_{t+\Delta t}}) = \left(e^{-\int_0^{t+\Delta t} h(X_u, Y_u)\, du} \right)$$

$$= \left(e^{-\left[\int_0^t h(X_u, Y_u)\, du \quad \int_t^{t+\Delta t} h(X_u, Y_u)\, du \right]} \right)$$

$$= \left(e^{-\int_0^t h(X_u, Y_u)\, du} \; e^{-\Delta t(h(X_t, Y_t))} \right)$$

$$= e^{-Z_t} (1 - h(X_t, Y_t)\, \Delta t + o(\Delta t))$$

$$\left(\because \int_0^t h(X_u, Y_u)\, du = z_t, \quad e^{-\Delta t}(h(X_t, Y_t)) = 1 - \Delta t\, h(X_0, Y_0) + o(\Delta t) \right)$$

We have
$$m(t) = E_{t_0}\left[a(X_t, Y_t) e^{-z_t} | \mathcal{F}_{t_0} \right]; \quad 0 \le t_0 < t$$

$\therefore$
$$m(t + \Delta t) = E_{t_0}\left[a(X_{t+\Delta t}, Y_{t+\Delta t}) e^{-z_{t+\Delta t}} | \mathcal{F}_{t0} \right]$$

and
$$m(t + \Delta t) - m(t) = E_{t_0}\left(a(X_{t+\Delta t}, Y_{t+\Delta t}) e^{z_{t+\Delta t}} - a(X_t, Y_t) e^{-z_t} | \mathcal{F}_{t0} \right)$$

$$= E_{t_0}\left[\{ a(X_t - 1, Y_t + 1) \pi a X_t Y_t\, \Delta t \right.$$

$$+ a(X_t - 1, Y_t)\, \alpha (1 - \pi) X_t Y_t \Delta t$$

$$+ a(X_t, Y_{t-1}) \beta_a Y_t\, \Delta t \} e^{-z_t + \Delta t}$$

$$+ a(X_t, Y_t)\, [(1 - (\alpha X_t + \beta) Y_t\, \Delta t)\, e^{z+\Delta t} - e^{zt}] + o(\Delta t).$$

from Downton's Bivariate birth and death process (7.81)

Let $\Delta t \to 0$

$$\Rightarrow \lim_{t \to 0} \frac{m(t + \Delta t) - m(t)}{\Delta t}$$

$$= E_{t_0} \left[\{a (X_t - 1, Y_t + 1) \pi \alpha\, X_t\, Y_t + a (X_t - 1, Y_t) \alpha\, (1 - \pi) X_t\, Y_t \right.$$

$$+ a (X_t, Y_t - 1)\, \beta\, Y_t \} \, e^{-z_{t + \Delta T}} + a(X_t, Y_t) \left[1 - (\alpha\, X_t + \beta)\, Y_t \right.$$

$$\left. - a (X_t, Y_t)\, e^{-zt} \right] (1 - h (X_t, Y_t))\, e^{-zt} \Big] \Big] + \lim_{\Delta t \to 0} \frac{o(\Delta t)}{t} \qquad (7.85)$$

This, by using Lebesgue theorem gives us the derivative at right of m i.e. $m'_r(t)$

$$m'_r(t) = E_{t_0} \{ [a (X_t - 1, Y_t + 1) - a (X_t, Y_t))\, (\pi \alpha\, X_t\, Y_t) + (a(X_t - 1, Y_t)$$

$$- a(X_t, Y_t))\, \alpha\, (1 - \pi)\, X_t\, Y_t + (a (X_t, Y_t - 1) - a(X_t, Y_t)) \cdot \beta\, Y_t$$

$$- a(X_t, Y_t) - h(X_t, Y_t)] e^{-zt} \} + \frac{o(\Delta t)}{\Delta t}$$

$$= E_{t_0} (a(X_t, Y_t)\, e^{-zt}).$$

Similarly one can, see that the derivative at the left also acquires this value. Hence $m'_r(t) = m'_l(t) = m'(t)$. In other words $m'(t)$ exists. Putting $X_t = i$ and $Y_t = j \Rightarrow$

$$[a(i - 1, j + 1) - a(i, j)\,]\, \pi \alpha_{ij} + [a(i - 1, j) - a(i, j)](1 - \pi)\, \alpha_{ij}$$

$$+ [a(i, j - 1) - a(i, j)]\, \beta_j - a(i, j)\, h\, (i, j) = 0 \,\forall\, (i, j) \in D \qquad (7.86)$$

D the domain of all non-negative integers.
since we want to $m'(t) = 0$

$$(dm(t) \,|\, \mathcal{F}_t) = 0 \Rightarrow m(t) \text{ is a Martingale.}$$

Putting
$$\left. \begin{array}{l} h(i, j) = (Ai + B)j \\[2mm] a (i, j) = (c_i \lambda^j) \end{array} \right\} \qquad (7.87)$$

$$\Rightarrow \qquad (c_{i-1} \lambda^{j+1} - c_i \lambda^j)\, \pi \alpha_{ij} + [c_{i-1} \lambda^j - c_i \lambda^j]\, (1 - \pi)\, \alpha_{ij}$$

$$+ (c_i \lambda^{j-1} - c_i \lambda^j)\, \beta j = (c_i \lambda^j)\, (A_i + B)j$$

$$\Rightarrow \qquad \lambda^j (c_{i-1} \lambda - c_i)\, \pi \alpha i j + (c_{i-1} - c_i)\, \lambda^j (i - \pi)\, \alpha i j + c_i \left(\frac{1}{\lambda} - 1 \right) \lambda^j \beta_j$$

$$= c_i\, \lambda^j\, (Ai + B)j, \qquad (7.88)$$

which is an identity for $j = 0$.
For $j > 0$ we have

$$((A + \alpha)\, i + B + (\beta - \frac{\beta}{\lambda})\, c_i = \alpha i\, (1 - \pi + \lambda \pi)\, c_{i-1}. \qquad (7.89)$$

We put

$$\lambda = \lambda_n = \frac{\beta}{(A+\alpha)n + B + \beta}$$

$$= \frac{\beta}{\lambda} = (A + \alpha) n + \beta + B. \Rightarrow B + \beta - \frac{\beta}{\lambda} = n (A + \alpha) \quad (7.90)$$

Further by taking $c_1 = c_2 = \dots = c_{n-1} = 0$

$$\Rightarrow \quad c_n = n! \left(\frac{\alpha}{A+\alpha} (1 - \pi + \lambda_n \pi) \right)^n \quad \text{in which case (7.89) holds} \quad (7.91)$$

Also by putting (7.90) in (7.89) $\Rightarrow$

$$(A + \alpha) (i - n) c_i = \alpha i (1 - \pi + \lambda_n \pi) c_{i-1}.$$

$$\Rightarrow \qquad c_i = \frac{i\alpha}{(i-n)} \frac{1}{(A+\alpha)} (1 - \pi + \lambda_n \pi) c_{i-1} \qquad (7.92)$$

Putting $i = n + 1$ in (7.92) we get

$$(A + \alpha)c_{n+1} = \alpha (n + 1)(1 - \pi + \lambda_n \pi) c_n$$

$$c_{n+1} = \frac{\alpha (n+1) (1 - \pi + \lambda_n \pi)}{(A+\alpha)} c_n. \qquad (7.93)$$

Putting c_n from (7.91) in (7.93),

$$C_{n+1} = \frac{\alpha^{n+1} (1 + \lambda_n \pi - \pi)^{n+1}}{(A+\alpha)^{n+1}} \cdot (n+1)!$$

$$\Rightarrow \qquad C_{n+2} = \frac{\alpha^{n+2} (1 + \lambda_n \pi - \pi)^{n+2}}{(A+\alpha)^{n+2}} \cdot (n+2)! \qquad (7.94)$$

Finally, replacing $(n + 1)$ by i, for $i \geq n$

$$C_i = \left(\frac{\alpha (1 + \lambda_n \pi - \pi)}{(A + \alpha)} \right)^i i!, \quad \text{for } i \geq n$$

$$= \left(\frac{\alpha (1 + \lambda_n \pi - \pi)}{A + \alpha} \right)^i (i)_n \qquad (7.95)$$

where $(i)_n = i (i - 1) \dots (i - n + 1)$.

And $\qquad\qquad a (i, j) = C_i \lambda^j.$

Therefore from one V_t we can construct a class of Martingales $V_{t,n}$ given by

$$V_{t,n} = a_n (X_t, Y_t) e^{-\int_0^t h(X_u, Y_u) du} \quad ; \quad \text{for } n = 1, 2, 3, \dots$$

$$= C_{X_t} \lambda_n^{Y_t} e^{-\int_0^t (AX_u + B) Y_u du}$$

$$= (X_t)_n \left(\frac{\alpha}{A+\alpha}(1-\pi+\lambda_n\pi) \right)^{X_t} \lambda_n^{Y_t} \left\{ e^{-\int_0^t (AX_u + B)Y_u \, du} \right\} \quad (7.96)$$

is a Martingale over $\mathcal{F}_{t,\,n}$. Further, $\mathrm{Sup}\, t \geq 0 \,|\, V_{t,\,n}\,| < \infty$ holds $\forall\, X_t > n$.

It may be noted that for $n > x_0$ as $0 \leq X_t \leq x_0$, $(X_t)_n = 0$.

Hence $V_{t,\,n}$ gives only $(x_0 + 1)$ distinct martingales of interest. Picard used three stopping rules for the above epidemic given by

$$T_0 = \inf\{t : Y_t = r\} \text{ for an integer } r \ni 0 \leq r \leq y_0$$

$$T_1 = \inf\{t : X_t + Y_t = r\},\, r \text{ being an integer } \ni x_0 \leq r \leq x_0 + y_0 \quad (7.97)$$

$$T_2 = \inf\{t : 2X_t + Y_t = r\} \text{ for an integer } r \ni 2x_0 \leq r \leq 2x_0 + y_0.$$

Further T_2 has been defined only for $\pi = 1$. Because there exist $(x, y) \in N$(set of +ve integers) $\ni 2x + y = (r + 1)$ for $\pi < 1$, i.e., there are sample paths going through $(x, y) \rightarrow (x - 1, y)$ for which T_2 is not defined because $2X_t + Y_t$ never reaches the value of r. The three stopping rules may be combined in one as

$$T = \inf\{t : \varepsilon X_t + Y_t = r\} \quad (7.98)$$

$$\varepsilon\, x_0 \leq r < x_0 + y_0$$

with $\varepsilon = 0, 1, 2$.

Now from optional stopping theorem $\Rightarrow$

$$E(V_{t,\,n}) = V_{0,\,n} \quad (7.99)$$

but as

$$X_t + Y_t = r \Rightarrow Y_T = r - \varepsilon X_T \text{ holds for stopping time } T_\varepsilon$$

$$\varepsilon = 0, 1, 2, \dots.$$

we may eliminate Y_T by $r - \varepsilon X_T$ from (7.99)

Therefore the following result holds:

$$T_\varepsilon = \inf\{t : \varepsilon X_t + Y_t = r\}\, ;\, \varepsilon\, x_0 < r < \varepsilon\, x_0 + y_0$$

and $\varepsilon = 0, 1, 2, \dots$

$$\therefore\ E\left[(X_{T_\varepsilon})_n \left(\frac{\alpha}{A+\alpha}(1-\pi+\pi\lambda_n) \right)^{X_t} \varepsilon\, \lambda_n^{-\varepsilon X_t}\, \varepsilon\, e^{-\int_0^{T_\varepsilon}(AX_u+B)Y_u\,du} \right]$$

$$= (X_0)_n \left(\frac{\alpha}{A+\alpha}(1-\pi+\pi\lambda_0) \right)^{X_0} \lambda_n^{-\varepsilon X_0} \quad (7.100)$$

where

$$\lambda_n = \frac{\beta}{(A+\alpha)^n + B + \beta}.$$

Now, by making use of optional sampling theorem and using Gontacharoff's polynomials as in Daniels (1967). Picard has obtained the joint Laplace forms of

$$X_{T_\varepsilon}, \ \int_0^{T_\varepsilon} X_u \, du, \ \int_0^{T_\varepsilon} X_u \, Y_u \, du$$

and obtains

$$E \int_0^{T_\varepsilon} X_u \, Y_u \, du \ = \ \frac{x_0 - E(X_{T_\varepsilon})}{\alpha} \tag{7.101}$$

$$E\left[\int_0^{T_\varepsilon} Y_u \, du\right] \ = \ \frac{\pi x_0 \, y_0 - r + (\varepsilon - \pi) \, E(X_{T_\varepsilon})}{\beta} \tag{7.102}$$

Further by using

$$\varepsilon E(X_{T_\varepsilon}) + E(Y_{T_\varepsilon}) \ = \ r$$

we have

$$E\left[\int_0^{T_\varepsilon} Y_u \, du\right] = \frac{\pi(x_0 - E(X_{T_\varepsilon})) + y_0 - E(X_{T_\varepsilon})}{\beta} \tag{7.103}$$

The right hand side of (7.102) and (7.103) have simple interpretations because $(1 - \pi)(x_0 - E(X_{T_\varepsilon}))$ is the expected number of susceptibles that have been removed and $\pi(x_0 - E(X_{T_\varepsilon}))$, the expected number that have been changed to carriers. He showed that these results are also true for T_ε changed to T_ε^* given by

$$T_\varepsilon^* \ = \ \inf\{t : X_t = 0 \text{ or } \varepsilon X_t + Y_t = r.\}$$

Further, Picard has shown that

$$E\left[\alpha \int_0^{T_\varepsilon} X_u \, Y_u \, du\right]$$

$=$ Expected # susceptibles involved in the Epidemic between 0 and T_ε (removed or changed to carriers)

and $\quad E\left[\beta \int_0^{T_\varepsilon} Y_u \, du\right] =$ Expected # detected and eliminated

carriers during $(0, T_\varepsilon)$.

Picard suggested

$$N \ = \ \frac{1}{X_0 - X T_\varepsilon} \int_0^{T_\varepsilon} X_u \, Y_u \, du \tag{7.104}$$

$$M = \frac{1}{\pi\,(X_0 - XT_\varepsilon) + y_0 - Y_{T_\varepsilon}} \int\limits_0^{T_\varepsilon} Y_u\,du \qquad (7.105)$$

as estimators of $\frac{1}{\alpha}$ and $\frac{1}{\beta}$ when π is known; since direct estimates from (7.101) and (7.102) are biased.

i.e.
$$\frac{1}{\alpha} = \frac{E\left[\int\limits_0^{T_\varepsilon} X_u\,Y_u\,du\right]}{E\,[X_0 - XT_\varepsilon]}$$

and
$$\frac{1}{\beta} = \frac{E\left[\int\limits_0^{T_\varepsilon} Y_u\,du\right]}{\pi\,X_0 + Y_0 - r + (\varepsilon - \pi)\,E\,(X_{T_\varepsilon})}$$

Because $p\,(X_{T_\varepsilon} = x_0) < 0, E\,(N) = +\infty$

but if $t = \pi$, we have $E\,(M) = \frac{1}{\beta}$.

Further if we use the stopping time T_1 for the general epidemic (i.e. $\pi = 1$) we get unbiased estimates of $\frac{1}{\beta}$.

Also he has shown that N is M.L.E. of $\frac{1}{\alpha}$ while M is such only if $\pi = 0$ or 1.

Several other useful results on carrier borne epidemic models, based on Downton's bivariate birth and death process (constant parameter, density dependent, time dependent) have been given. For details the reader is referred to Picard's papers (1980, 1981, 1984).

References

1. Bailey N.T.J. (1957): *The Mathematical theory of Epidemics*. Griffin, London.
2. Bailey N.T.J. (1963): *Elements of Stochastic Processes with applications to Natural Sciences*: John Wiley & Sons. London, New York, Sydney.
3. Bartlett M.S. (1960): Stochastic Population models in Ecology and Epidemiology. *Mathuen Monograph on Applied Probability and Statistics*. London, New York.
4. Becker Neils G (1989) *Analysis of Infectious Disease Data*. Chapman and Hall. London, New York.
5. Biswas S (1988): *Stochastic Processes in Demography and Applications*. Wiley Eastern and John Wiley & Sons.
6. Chiang C.L. (1968): *Introduction to Stochastic Processes in Biostatistics*. John Wiley & Sons. New York.
7. Chiang C.L. (1980): *An Introduction to Stochastic Processes and their Applications*. Kreiger. New York.
8. Daniells H.E. (1967): The distribution of the total size of an epidemic. *Proceedings 5th Berkeley Symposium, Math. Statistics and Probability*. Vol. 4, page 281–293.

9. Downton F (1968): The ultimate size of carrier borne epidemics. *Biometrika*, Vol. 55, page 277–89.

10. Gani J (1965): On a partial differential equation of the epidemic theory. *Biometrika*, Vol. 52, page 617–622.

11. Kendall D.G.(1950): An artificial realisation of a simple birth and death process. *Journal of Royal Stat. Soc.* Series B, No. 12, page 116–19.

12. Kendall D.G. (1980): Birth and Death processes in the theory of Carcinogenesis. *Biometrika*, Vol. 47, No 1, page 2–13.

13. Picard P (1980): Applications of Martingales theory to some epidemic models. *Journal of Applied Probability.* Vol. 17, page 583–99.

14. Picard P (1981): Applications of Martingales theory to some epidemic models with time dependent parameters. *Math. Biosciences*, Vol. 55, page 205–229.

15. Picard P (1984): Applications of Martingales Theory to some epidemic models II, *Journal of Applied Probability*, Vol. 21, page 677–84.

16. Siskind V(1965): A solution of the general Stochastic epidemic. *Biometrika*, Vol. 52, page 613–616.

17. Tan Wai-Yuan (1991): *Stochastic models of Carcinogenesis.* Marcel Dekker Inc. New York, Bassel, Hongkong.

18. Weiss G.H. (1965): On the spread of epidemics by carriers. *Biometrics* Vol. 21, page 481–90.

Stochastic Processes of Clinical Drug Trials

8.0 Introduction

Suppose we have two competing treatments say A and B for a certain therapy. In practice to test the superiority of A to B (or B to A) without having any prior knowledge about any of them one cannot have the usual random allocation of two treatments in two groups of patients. Because there is an ethical consideration behind such Drug trials viz. minimum number of patients should be exposed to Inferior Drug during experimentation. The idea is to start with a particular treatment (A or B) and then continue with the treatment or change the treatment by an alternative treatment depending upon the success of the last treatment. However, there are two basic difficulties while facing the problem of allocation of drugs:

  (i)   smallness in the size of the patients and staggering of the entries of the patients.
  (ii) delay in getting the information of success (or failure) of a treatment till then a treatment whose effect may be adverse, is continued.

We shall present the outline of two types of sequential clinical trials, the allocation rules of which are known as:

  (i)   Two Armed Bandit Rule
  (ii) Play the Winner Rule.

8.1 Two Armed Bandit Rule (Robbins (1952))

Rule: Choose A or B at random (say A has been chosen). Then if for the ith trial ($i = 1, 2, ...$) in which A has been applied leads to a success then apply A again in the $(i + 1)^{th}$ trial. If, however, the application of A on the ith trial (or i^{th} patient) leads to a failure then in the $(i + 1)^{th}$ trial we switch over to a different treatment B.

Let α and β stand for the probabilities of success by treatments A and B respectively. Then the outcome of successive trials represent a simple Markov chain with four states (A, S) (A, F) (B, S) and (B, F) (S = success, F = failure). The transition matrix of the two state Markov chain is as follows:

i^{th} trial	$(i+1)^{th}$ trial (A, S)	(A, F)	(B, S)	(B, F)	Total
(A, S)	α	$(1-\alpha)$	0	0	1
(A, F)	0	0	β	$1-\beta$	1
(B, S)	0	0	β	$1-\beta$	1
(B, F)	α	$1-\alpha$	0	0	1

Let p_i be the probability that a success occurs at the ith trial $(i = 1, 2, 3, ...)$. Then the following recurrence relation holds

$$p_{i+1} = (\alpha + \beta - 1)p_i + (\alpha + \beta - 2\alpha\beta). \tag{8.1}$$

Proof: $p_{i+1} = P(A, S)_{i+1} + P(B, S)_{i+1}$

$$= P(A, S)_i \alpha + P(B, F)_i \alpha + P(B, S)_i \beta + P(A, F)_i \beta. \tag{8.2}$$

$$= \alpha [P(A, S)_i + P(B, F)_i] + \beta [P(B, S)_i + P(A, F)_i]. \tag{8.3}$$

where $P(A, S)_i = P$ [of success at the ith trial due to treatment A]
$\qquad\quad P(B, S)_i = P$ [of success at the ith trial due to treatment B]
We have

$$\left. \begin{array}{c} P(A, S)_i + P(B, S)_i = p_i \\[2mm] P(A, F)_i + P(B, F)_i = 1 - p_i \end{array} \right\} \tag{8.4}$$

$$\Rightarrow \qquad \frac{P(A, S)_i}{P(A, F)_i} = \frac{\alpha}{1-\alpha} \quad \text{and} \quad \frac{P(B, S)_i}{P(B, F)_i} = \frac{\beta}{1-\beta}$$

$$\Rightarrow \qquad \left. \begin{array}{c} P(A, F)_i = \dfrac{1-\alpha}{\alpha} P(A, S)_i \\[4mm] P(B, F)_i = \dfrac{1-\beta}{\beta} P(A, S)_i \end{array} \right\} \tag{8.5}$$

and

$\therefore$ (8.3), (8.4) and (8.5) $\Rightarrow$

$$p_{i+1} = (\alpha + \beta - 1) p_i + (\alpha + \beta - 2\alpha\beta)$$

which proves the result.
Next to find the limiting probability of p_i, let us put

$$\alpha + \beta - 1 = C$$

and

$$\alpha + \beta - 2\alpha\beta = D.$$

$\therefore$ (8.1) $\Rightarrow \quad p_{i+1} = C p_i + D$

$$= C^2 p_{i-1} + D(1 + C)$$

$$= C^3 p_{i-2} + D(1 + C + C^2 + ...)$$

$$= C^i p_1 + D(1 + C + C^2 + ...)$$

where $|C| < 1$,

$$= (\alpha + \beta - 1)^i \, p_1 + \frac{D}{1 - C}$$

$$= (\alpha + \beta - 1)^i \, p_1 + \frac{\alpha + \beta - 2\alpha\beta}{1 - (\alpha + \beta - 1)}$$

$$= (\alpha + \beta - 1)^i \, p_1 + \frac{\alpha + \beta - 2\alpha\beta}{2 - (\alpha + \beta)}$$

Since $|\alpha + \beta - 1| < 1$, therefore,

$$\lim_{i \to \infty} p_i = \frac{\alpha + \beta - 2\alpha\beta}{2 - (\alpha + \beta)} \qquad (8.6)$$

So we can write (8.4) using (8.5) as

$$\left. \begin{aligned} P(A,S)_i + P(B,S)_i &= p_i \\[2mm] \frac{1 - \alpha}{\alpha} P(A,S)_i + \frac{1 - \beta}{\beta} P(B,S)_i &= 1 - p_i \end{aligned} \right\} \qquad (8.7)$$

Multiplying 1st equation of (8.7) by $\dfrac{1 - \alpha}{\alpha}$ and then subtracting from the 2nd equation we get

$$\frac{1 - \beta}{\beta} P(B,S)_i - \frac{1 - \alpha}{\alpha} P(B,S)_i = \left(\frac{\alpha - p_i}{\alpha} \right)$$

$$\Rightarrow \qquad \frac{p_i - \alpha}{\alpha} = \frac{\beta - \alpha}{\alpha\beta} P(B,S)_i$$

or $$P(B,S)_i = \frac{\beta(\alpha - p_i)}{(\alpha - \beta)} \qquad (8.8)$$

Similarly, $$P(A,S)_i = \frac{\alpha(p_i - \beta)}{\alpha - \beta} \qquad (8.9)$$

Putting (8.8) and (8.9) in (8.3) $\Rightarrow$

$$p_{i+1} = \alpha \left[\frac{\alpha(p_i - \beta)}{\alpha - \beta} + \frac{1 - \beta}{\beta} \frac{\beta(\alpha - p_i)}{\alpha - \beta} \right]$$

$$+ \beta \left[\frac{\beta(\alpha - p_i)}{(\alpha - \beta)} + \frac{1 - \alpha}{\alpha} \frac{\alpha(p_i - \beta)}{\alpha - \beta} \right]$$

$$= \frac{1}{\alpha - \beta} \, p_i \left(\alpha^2 + \alpha\beta - \alpha - \beta^2 + \beta - \alpha\beta + (\alpha^2 - \alpha^2\beta - \alpha^2\beta + \alpha\beta^2 - \beta^2 + \alpha\beta^2) \right)$$

$$= \frac{1}{\alpha - \beta} \left[p_i \, (\alpha^2 - \beta^2) - (\alpha - \beta) + \alpha^2 - \beta^2 + 2\alpha\beta(\beta - \alpha) \right] \qquad (8.10)$$

From (8.6)

$$\lim_{i \to \infty} p_i = \frac{\alpha + \beta - 2\alpha\beta}{2 - (\alpha + \beta)} = \frac{(\alpha + \beta)/2 - \alpha\beta}{1 - (\alpha + \beta)/2} = \frac{v - \alpha\beta}{1 - v} \qquad (8.11)$$

where
$$v = \frac{\alpha + \beta}{2}$$

Now let $\delta^2 = \frac{1}{4}(\alpha - \beta)^2 \Rightarrow 4\delta^2 = (\alpha - \beta)^2 = (\alpha + \beta)^2 - 4\alpha\beta \qquad (8.12)$

$$\Rightarrow \qquad \alpha\beta = -\delta^2 + v^2 = v^2 - \delta^2$$

$$\therefore \qquad \lim_{i \to \infty} p_i = \frac{v - v^2 + \delta^2}{1 - v} = v + \frac{\delta^2}{1 - v} \qquad (8.13)$$

where
$$v = \frac{\alpha + \beta}{2}$$

$$\delta = |\alpha - \beta|/2.$$

The result is based on the allocation rule known as 'Two armed bandit rule'. This rule is asymptotic or valid for large sample which is a drawback of the rule for application in sequential drug trials for which the sample size is relatively small.

Let the outcome of every trial be denoted by a r.v., say X_i, for i^{th} trial ($i = 1, 2, \ldots n$)

Let, $\qquad X_i = 1$, if i^{th} trial results in success with probability p_i

$\qquad = 0$, if otherwise with probability $(1 - p_i)$

Then X_i is a Bernoullian variate. X_i's are assumed to be i.d.r.v.s

$$\therefore \qquad \left. \begin{array}{l} E(X_i) = p_i, \forall\ i = 1, 2, \ldots, n \\[3mm] S_n = X_1 + X_2 + \ldots + X_n \end{array} \right\} \qquad (8.14)$$

Let,

Then $\qquad E\left(\frac{S_n}{n}\right) = E\left(\frac{\sum X_i}{n}\right) = \frac{1}{n}\sum_i E(X_i) = \frac{1}{n}\sum p_i. \qquad (8.15)$

Also $\qquad \lim_{n \to \infty} E\left(\frac{S_n}{n}\right) = v + \frac{\delta^2}{1 - v}.$ (from 8.3)

This is an ideal condition which can be achieved if the sampling is continued indefinitely.

Let us consider the situation where we know which one of the treatments (A or B) is better. If we know, then we apply that treatment only. In that case,

$$E\left(\frac{S_n}{n}\right) = \text{Max}\,(\alpha, \beta)$$

$$= \frac{\alpha + \beta}{2} + \frac{\alpha - \beta}{2} = \alpha, \text{ if } \alpha > \beta$$

$$= \frac{\alpha + \beta}{2} + \frac{\beta - \alpha}{2} = \beta, \text{ if } \beta > \alpha, \qquad (8.16)$$

i.e.
$$E\left(\frac{S_n}{n}\right) = \frac{\alpha + \beta}{2} + \frac{|\alpha - \beta|}{2} = v + \delta \qquad (8.17)$$

Let us designate the Two Armed Bandit Rule of allocation of treatment by R_1 and $L\,(A, B, R_1)$ = loss of applying treatment as per R_1 when the better of the two treatments is not known.

$$L\,(A, B, R_1) = v + \delta - v - \frac{\delta^2}{1 - v} = \delta\left(1 - \frac{\delta}{1 - v}\right) \qquad (8.18)$$

Now
$$\frac{\delta}{1 - v} = \frac{|\alpha - \beta|}{2}\,\frac{2}{2 - (\alpha + \beta)} = \frac{|\alpha - \beta|}{2 - (\alpha + \beta)} < 1$$

$$(\because |\alpha + \beta - 1| < 1)$$

$$\therefore \quad L\,(A, B, R_1) = \delta\left(1 - \frac{\delta}{1 - v}\right) \geq 0 \qquad (8.19)$$

is a measure of the asymptotic loss per trial by a treatment who uses R_1 due to ignorance of efficacy of treatment A. It can be shown that Max $L_1 = 0.172$.

Further, let R_0 be a rule dictating a random allocation of either of the treatments A or B at start and thereafter stick to random allocation of treatments irrespective of the result of the previous trial. Hence

$$p_i = \frac{\alpha + \beta}{2} = v \quad \text{and} \quad p_1 = p_2 = ... = p_n$$

Again
$$\lim_{n \to \infty} E\left(\frac{S_n}{n}\right) = v.$$

$$\therefore \qquad L_0 = L\,(A, B, R_0) = v + \delta - v = \delta \qquad (8.20)$$

Also,
$$\text{Max } L_0 = 0.5 > \text{Max } L_1 = 0.172$$

This clearly shows R_1 is better than R_0.

We conclude that it is better to have a memory of the previous event (and utilize the information) rather than to have no information of the previous events. By induction, it would be more advisable to base our choice on several events and not on the same event in the recent past.

In fact, this procedure of keeping the records of the earlier treatments and then adopting a strategy for the allocation of treatment is indeed feasible in medical trails.

Therefore, consider a general rule R, in which the choice for the ith patient will depend not only on the result of the treatment of the preceding patients but on all the earlier results.

For such a rule R,

$$L_n(A, B, R) = \text{Max}(\alpha, \beta) - E\left(\frac{S_n}{n}\right) \tag{8.21}$$

Let

$$M_n(R) = \underset{\alpha, \beta}{\text{Max}}\, L_n(A, B, R)$$

$$\phi(n) = \underset{R}{\text{Min}}\,[M_n(R)]$$

If for any rule R, $\phi(n)$ is attained it, will be called the Minimax rule. Robbins (1952) suggested that one may look for the rules $\overline{R}$ whose asymptotic loss function is Zero in the absence of the explicit description of a minimax rule.

Let $A(i = 1)$ and $B(i = 1)$ be arbitrary population for which expectations exist. Further let $\{a_n\}$ and $\{b_n\}$ be two increasing sequences of fixed positive integers; both the sequences being disjoint and $a = 1, b_1 = 2$. Let their density function be Zero; i.e. the population of integers, $i = 1, 2, \dots n$ which are either a's or b's tend to zero as $n \to \infty$.

8.2   Sampling Design of the Rule *R*:

If the integer i ($i = 1, 2, 3, \dots$) is one of the a's i.e. from $\{a_n\}$ then take the ith unit from population-1 (i.e. A). If it belongs to $\{b_n\}$ then the ith unit is to be taken from 2 (i.e. B).

If it does not belong to any of the above $\{a_n\}$ or $\{b_n\}$ then take i^{th} unit X_i from $1(A)$ or $2(B)$; according as the arithmetic mean of all previous observations from A exceed or does not exceed from that of all previous observations from B. Then by strong law of large numbers (SLLN)

$$\lim_{n \to \infty} E\left(\frac{S_n}{n}\right) = \text{Max}(\alpha, \beta)$$

so that

$$\lim_{n \to \infty} L_n(A, B, R) = \text{Max}(\alpha, \beta) - \lim_{n \to \infty} E\left(\frac{S_n}{n}\right) = 0 \tag{8.22}$$

As a matter of fact, Robbins (1956), has shown that the asymptotic loss function is zero for a rule R which makes use of all previous observations.

Robbins (1956) has considered a Markov chain of order r ($r > 1$), the family of rules with a finite memory of the 'r' preceding trials. Then a rule R_r is defined as follows:

(i)   Choose either of the treatment A ($i = 1$) or B ($i = 2$) at random at start.

(ii) If the first patient dies or do not get the relief, change the treatment or switch over to an alternate treatment.

(iii) Other wise, continue till a run of 'r' failures is obtained; after a run of r failures again change the treatment.

The first lot of patients under one treatment forms a block, say first block. Block size may be one or more. Again second block begins with the alternative treatment.

Let $\quad X_i$ = 1, if there is a success at i^{th} treatment ($i = 1, 2$)

$\qquad\qquad$ = 0, otherwise.

Then

$$\lim_{n \to \infty} \left[\frac{\text{No. of successes in the 1st } n \text{ trials}}{n} \right] = \text{Max}\,(p_1, p_2) \qquad (8.23)$$

where $p_i \equiv$ probability of success due to i^{th} treatment ($i = 1, 2$).

To prove that

$$\lim_{n \to \infty} \left[\frac{\text{No. of successes in the 1st } n \text{ trials}}{n} \right] \equiv \frac{p_1 q_2^r + p_2 q_1^r}{q_1^r + q_2^r} \qquad (8.24)$$

Proof: First we show that the number of trials required to have failures (due to treatment) of run length r is given by

$$\frac{(1 - q_i^r)}{p_i q_i^r}; \quad i = 1, 2$$

For the ith treatment,

let $\quad$ V = Expected number of trials

$\qquad$ G = Expected number of failure sequences encountered before getting a run of r number of successes (i.e. failures due to i^{th} treatment)

$\qquad$ H = Expected number of units in a failure sequence (in which the last unit is a failure)

Let p be the probability of success (i.e. the probability that the ith treatment is a failure).

Expected no. of treatments in a failure sequence; the last unit n being a failure '$n = 1, 2....$) is given by

$$H = \frac{1 \cdot p_i + 2 \cdot q_i p_i + ... + r q_i^{r-1} p_i}{p_i + q_i p_i + ... + q_i^{r-1} p_i}$$

$$H = \sum_{n=1}^{r} \frac{n q_i^{n-1} p_i}{\displaystyle\sum_{n=1}^{r} q_i^{n-1} p_i} \qquad (8.25)$$

Now $\qquad H = \sum_{n=1}^{r} \frac{n q_i^{n-1} p_i}{(p_i) \dfrac{(1 - q_i^r)}{(1 - q_i)}} = \frac{p_i \displaystyle\sum_{n=1}^{r} \dfrac{d}{dq_i}(q_i^n)}{1 - q_i^r}$

$$= \frac{p_i}{1-q_i^r}\left[\frac{d}{dq_i}\sum_{n=1}^{r}(q_i^n)\right]$$

$$= \frac{p_i}{1-q_i^r}[(1-q_i)(1-(r+1)q_i^r)-q_i(1-q_i^r)(-1)]/(1-q_i)^2$$

$$= \frac{p_i}{1-q_i^r}\frac{[1-(r+1)q_i^r-q_i+(r+1)q_i^{r+1}+q_i-q_i^{r+1}]}{(1-q_i)^2}$$

$$= \frac{p_i}{1-q_i^r}\cdot\frac{1}{p_i^2}[1-q_i^r(1+rp_i)] \tag{8.26}$$

Let P be the probability of a terminal failure sequence) i.e. the sequence in which all trials lead to success but the last one i.e. $(n-1)^{th}$—one is a failure (i.e. i^{th}, treatment of success) and $Q = 1 - P$.
Then

$$G = \text{the expected number of failure sequences}$$

$$= \sum_{j=0}^{\infty} j\, p^j\, Q = PQ(1-P)^{-2} = \frac{P}{Q}$$

$$= [1-q_i^r/q_i^r] \tag{8.27}$$

where $\qquad P = 1-q_i^r$ and $Q_i = q_i^r$

Therefore the expected number of trials for a run of length r number of consecutive failure

$$GH + r = \frac{(1-q_i^r)}{q_i^r}\cdot\frac{1}{p_i(1-q_i^r)}[1-q_i^r(1+rp_i)]+r$$

$$= \frac{1-q_i^r-rp_iq_i^r+rp_iq_i^r}{p_iq_i^r}$$

$$= \frac{1-q_i^r}{p_iq_i^r} \tag{8.28}$$

Next suppose the allocation of the ith treatment repeatedly with the following stopping rule:
Stop if the first trial is a failure other wise continue till we get a succession (or run) of r number of consecutive failures. Then the expected number of trials necessary

$$= q_i\cdot 1 + p_i\left(1+\frac{1-q_i^r}{p_iq_i^r}\right) = \frac{1}{q_i^r} \tag{8.29}$$

Next we prove the following:

Define the rule R_r as follows. Start trial with treatment 1. Stop if the first trial is a failure; otherwise continue the trial until first run of r consecutive failures occur and then stop.

This defines the first block of trials with the treatment A. Now start trials with treatment B and apply the same rule. Obtain the first block of trials with treatment B.

Then start again with treatment A and apply the same giving the second block of trials with treatment A. Similarly the second block of trials with treatment B. Thus the process continues.

$$\text{Then it follows } \lim_{n \to \infty} \frac{\text{Successes in the first } n \text{ trials}}{n} = \lim_{n \to \infty} \frac{W_n}{n}$$

$$= \frac{p_1 q_2^r + p_2 q_1^r}{q_1^r + q_2^r} \tag{8.30}$$

To show this let $\dfrac{x_i}{y_i}$ denote the length of the ith block of trials with treatment 1 (or 2) ($1 \equiv A$, $2 \equiv B$).

The process of trials generated with probability 1 is an infinite sequence of i.d.r.v.s.

$$(x_1, y_1);\ (x_2, y_2);\ (x_3, y_3)\ \dots;\ (x_n, y_n)$$

The proportion of times that the treatment A(or 1) is used, is

$$= \frac{x_1 + x_2 + \dots + x_n}{(x_1 + y_1) + (x_2 + y_2) + \dots + (x_n + y_n)}$$

$$= \frac{(x_1 + x_2 + \dots + x_n)/n}{\dfrac{x_1 + x_2 + \dots + x_n}{n} + \dfrac{y_1 + y_2 + \dots + y_n}{n}} \tag{8.31}$$

$\therefore$ Expected number of trials with treatment 1

$$= q_1 \cdot 1 + p_1 \left(1 + \frac{1 - q_1^r}{p_1 q_1^r} \right) = \frac{1}{q_1^r} \tag{8.32}$$

$$\lim_{n \to \infty} \frac{\left[\dfrac{1}{q_1^r} + \dfrac{1}{q_1^r} + \dots + \dfrac{1}{q_1^r} \right] / n}{\dfrac{\dfrac{1}{q_1^r} + \dots + \dfrac{1}{q_1^r} + \dots + \dfrac{1}{q_1^r}}{n} + \dfrac{\dfrac{1}{q_2^r} + \dots + \dfrac{1}{q_2^r}}{n}}$$

$$= \frac{\dfrac{1}{q_1^r}}{\dfrac{1}{q_1^r} + \dfrac{1}{q_2^r}} = \frac{q_2^r}{q_1^r + q_2^r} \dots \tag{8.33}$$

Let $\quad u_n \equiv$ [Proportion of successes in $2n$ blocks of trials]

$\quad\quad\quad = $ [Proportion of times treatment 1 is used in $2n$ blocks]

$\quad\quad\quad \times$ [proportion of times successes occur with treatment 1]

$\quad\quad\quad +$ [Proportion of times treatment 2 is used in first $2n$ blocks]

$\quad\quad\quad \times$ [Proportion of times successes occur with treatment 2]

$$= \frac{q_2^r}{q_1^r + q_2^r} \, p_1 + \frac{q_1^r}{q_1^r + q_2^r} \cdot p_2$$

$$= \frac{p_1 q_2^r + p_2 q_1^r}{q_1^r + q_2^r}$$

Let N be any positive integer and define the random integer $N = N(n)$ so that

$$(x_1 + y_1) + (x_2 + y_2) + \ldots + (x_N + y_N) \leq n$$

$$\leq (x_1 + y_1) + (x_2 + y_2) + \ldots + (x_{N+1} + y_{N+1}) \tag{8.34}$$

Denote by w_n the # of successes among the first n trials. We have, the number of successes in the first $2N$ blocks $\leq$ number of successes in n blocks $= w_n \leq$ number of successes in $(2N + 2)$ blocks

i.e. $\quad\quad\quad u_n (x_1 + y_1 + x_2 + y_2 + \ldots + x_N + y_N)$

$$\leq w_n \leq u_{n+1} (x_1 + y_1 + x_2 + y_2 + \ldots + x_{N+1} + y_{N+1}) \tag{8.35}$$

Using the rule of rates and proportion

i.e. if $\quad\quad\quad\quad\quad a \leq b \leq c$

$$a' \leq b' \leq c' \quad \text{hold}$$

$\Rightarrow \quad\quad\quad\quad\quad \dfrac{a'}{c} \leq \dfrac{b'}{b} \leq \dfrac{c'}{a}$

we have from (8.34) and (8.35)

$$\frac{u_n (x_1 + y_1 + \ldots + x_N + y_N)}{x_1 + y_1 + x_2 + y_2 + \ldots + x_{N+1} + y_{N+1}} \leq \frac{w_n}{n}$$

$$\leq u_{n+1} \frac{(x_1 + x_2 + \ldots + x_{N+1} + y_1 + y_2 + \ldots + y_{N+1})}{x_1 + y_1 + x_2 + y_2 + \ldots + x_N + y_N}$$

As $n \to \infty$, $\dfrac{x_{N+1} + y_{N+1}}{N} \to 0$ with probability 1, therefore, it follows that

$$\lim_{n \to \infty} \frac{x_1 + y_1 + \ldots + x_{N+1} + y_{N+1}}{x_1 + y_1 + \ldots + x_N + y_N} = 1$$

As
$$n \to \infty, \quad u_n \le \frac{w_n}{n} \le u_{n+1}$$

$$\lim_{n \to \infty} u_n = \frac{w_n}{n}$$

Also
$$\lim_{n \to \infty} u_n = \frac{p_1 q_2^r + p_2 q_1^r}{q_1^r + q_2^r} \tag{8.36}$$

Hence
$$\lim_{n \to \infty} \frac{w_n}{n} = \frac{p_2 q_1^r + p_1 q_2^r}{q_1^r + q_2^r} \tag{8.37}$$

Thus the result is proved.

8.3 Play the Winner Sampling Rule (Sobel and Weiss (1970))

As pointed out earlier by Robbins' (1952) two armed bandit rule for allocation of treatments although, theoretically, is more efficient than random allocation, it is not practically feasible because the results are asymptotically true. The problem is to evolve a rule which will give a definite conclusion as to suitability of either of the treatments in an infinite patient horizon even sometimes with staggering entry.

A finite termination rule leading to a conclusion in favour of treatment A (or B) with an infinite patient horizon is given by (Sobel and Weiss (1970) known as 'Play the Winner rule', an outline of which is given as follows:

Let A and B be two treatments and

$S_A \equiv$ successes due to treatment A

$S_B \equiv$ successes due to treatment B

$p =$ probability of success with treatment A

$p' =$ probability of success with treatment B.

The trial design will be such that the probability of selection $P\,(CS)$ is made to satisfy

$$P\,(CS) \ge p^*$$

where
$$p - p' > \Delta^* \tag{8.38}$$

where the parameter Δ^* furnished by the experimenter will be referred to as the 'discrimination level' and p^* which is also fixed, denotes the power.

The termination rule is as follows:

'*Stop when the absolute difference in the number of successes for the two treatments first reaches a pre-determined integer s*'.

A sampling rule R_s is as follows:

Let A is superior to B i.e.

$$p - p' = \Delta > 0 \tag{8.39}$$

Then R_s declares treatment i to be better when $S_i - S_j = r$ where j is the other treatment.

The integer 'r' is chosen to be the smallest for which (8.38) is satisfied, i.e. r is to be chosen such that

$$P(CS) \geq p^*$$

Let $NT = A$ denote that the next treatment to be chosen is A.

$$P_n = P\{A \text{ is selected as better} \mid S_A - S_B = n, NT = A\} \tag{8.40}$$

$$Q_n = P\{A \text{ is selected as better} \mid S_A - S_B = n, NT = B\} \tag{8.41}$$

Here "$NT = A$", means either treatment A has been used in the last trial and it was successful, or treatment B has been applied in the last trial and it was failure. Similarly for "$NT = B$".

Since the first treatment is chosen at random, $P(CS)$ at termination is given by

$$P(CS) = \frac{1}{2} \cdot P_0 + \frac{1}{2} \cdot Q_0$$

$$= \frac{1}{2}(P_0 + Q_0) \tag{8.42}$$

Now consider the outcomes of the events (8.40) and (8.41)

(i) If a failure occurs on A, then

$$S_A - S_B = n \quad \text{and} \quad NT = B$$

(ii) If a success occurs on A, then

$$S_A - S_B = n + 1 \quad \text{and} \quad NT = A$$

Therefore,

$$P_n = p P_{n+1} + q Q_n, \quad q = 1 - p \tag{8.43}$$

Similarly for (8.41),

(i) If B fails then

$$S_A - S_B = n, \quad NT = A$$

(ii) If a success occurs on B, then

$$S_A - S_B = n - 1; \quad NT = B$$

and

$$Q_n = p' Q_{n-1} + q' P_n \tag{8.44}$$

Thus we have

$$P_n = p P_{n+1} + q Q_n$$

$$Q_n = p' Q_{n-1} + q' P_n$$

with boundary conditions

$$P_r = 1 \quad \text{and} \quad Q_{-r} = 0 \tag{8.45}$$

From (8.44)

$$P_n = \frac{Q_n - p' Q_{n-1}}{q'}$$

Substituting in (8.43)

$$\frac{Q_n - p' Q_{n-1}}{q'} = q Q_n + \frac{p[Q_{n+1} - p' Q_n]}{q'}$$

$$\Rightarrow \qquad Q_n - p' Q_{n-1} = qq' Q_n + p Q_{n+1} - pp' Q_n$$

$$\Rightarrow \qquad p Q_{n+1} - (pp' - qq' - 1) Q_n + p' Q_{n-1} = 0 \qquad (8.46)$$

Let $Q_n = \lambda^n$ then (8.46) becomes

$$p \lambda^2 - (pp' - qq' + 1) \lambda + p' = 0$$

$$\Rightarrow \qquad p \lambda^2 - [pp' - (1-p)(1-p') + 1] \lambda + p' = 0$$

$$\Rightarrow \qquad p \lambda^2 - (p + p') \lambda + p' = 0$$

$$\Rightarrow \qquad p \lambda (\lambda - 1) - p' (\lambda - 1) = 0$$

$\Rightarrow \lambda_1 = 1$ and $\lambda_2 = p'/p$ are the two roots of the quadratic equation.

The general solution of Q_n is

$$Q_n = C_1 \lambda_1^n + C_2 \lambda_2^n$$
$$= C_1 + C_2 (p'/p)^n$$

i.e. $\qquad Q_n = C_1 + C_2 \lambda^n$ where $\lambda = \dfrac{p'}{p} = \lambda_2 \qquad (8.47)$

To find C_1 and C_2, consider $P_r = 1$

Then from $\qquad Q_r = q' P_r + p' Q_{r-1}$ we get

$$Q_r = q' + p' Q_{r-1}$$

or $\qquad Q_r - p' Q_{r-1} = q' \qquad (8.48)$

Again $\qquad Q_{-r} = 0$

$$\Rightarrow Q_{-r} = C_1 + C_2 \lambda^{-r} = 0$$

$$\Rightarrow \quad C_2 = -C_1 \lambda^r \qquad (8.49)$$

Again $\qquad Q_r = C_1 + C_2 \lambda^r$

and $\qquad Q_{r-1} = C_1 + C_2 \lambda^{r-1}$

Using (8.48) we get

$$Q_r - p' Q_{r-1} = (C_1 + C_2 \lambda^r) - p' (C_1 + C_2 \lambda^{r-1}) = q'$$

$$\Rightarrow \qquad C_2 \lambda^{r-1} (-p' + \lambda) + C_1 (1 - p') = q'$$

$$\Rightarrow \qquad (-C_1 \lambda) \lambda^{r-1} (\lambda - p') + C_1 q' = q'$$

$$\Rightarrow \qquad -C_1 \lambda^{2r} + C_1 \lambda^{2r-1} p' + C_1 q' = q'$$

$$\Rightarrow \qquad C_1(q' - \lambda^{2r} + p'\lambda^{2r-1}) = q'$$

$$\Rightarrow \qquad C_1\left[\lambda^{2r}\left(\frac{p'}{\lambda} - 1\right) + q'\right] = q'$$

$$\because \left[\frac{p'}{\lambda} = \frac{p'}{p'/p} = p\right]$$

$$\Rightarrow \qquad C_1(\lambda^{2r}(p-1) + q') = q'$$

$$\Rightarrow \qquad C_1(q' - q\lambda^{2r}) = q'$$

$$\Rightarrow \qquad C_1 = \frac{q'}{q' - q\lambda^{2r}} \tag{8.50}$$

$$C_2 = -C_1\lambda^r \Rightarrow C_2 = \frac{-\lambda^r q'}{q' - q\lambda^{2r}} \tag{8.51}$$

Then
$$Q_n = C_1 + C_2\lambda^n$$

$$= \frac{q'}{q' - q\lambda^{2r}} + \frac{-q'\lambda^r\lambda^n}{q' - q\lambda^{2r}}$$

$$\therefore \qquad Q_n = \frac{q'(1 - \lambda^{r+n})}{q' - q\lambda^{2r}} \tag{8.52}$$

Again from (8.47) we have

$$Q_n = p'Q_{n-1} + q'P_n$$

$$\Rightarrow \qquad q'P_n = Q_n - p'Q_{n-1}$$

or
$$q'P_n = \frac{q'(1 - \lambda^{r+n})}{q' - q\lambda^{2r}} - \frac{p'q'(1 - \lambda^{r+n-1})}{q' - q\lambda^{2r}}$$

or
$$P_n = \frac{1 - \lambda^{r+n}}{q' - q\lambda^{2r}} - \frac{p'(1 - \lambda^{r+n-1})}{q' - q\lambda^{2r}}$$

$$= \frac{1 - \lambda^{r+n} - p' + p'\lambda^{r+n-1}}{q' - q\lambda^{2r}}$$

$$P_n = \frac{q' - \lambda^{r+n}(1 - p'/\lambda)}{q' - q\lambda^{2r}}$$

$$= \frac{q' - q\lambda^{r+n}}{q' - q\lambda^{2r}} \qquad \left(\because \frac{p'}{\lambda} = p\right)$$

Thus
$$P_n = \frac{q' - q\lambda^{r+n}}{q' - q\lambda^{2r}} \tag{8.53}$$

$$Q_n = \frac{q'(1 - \lambda^{r+n})}{q' - q\lambda^{2r}} \tag{8.54}$$

$$\therefore \qquad P(CS) = \frac{1}{2}(P_0 + Q_0)$$

$$= \frac{1}{2}\left[\frac{q' - q\lambda^r}{q' - q\lambda^{2r}} + \frac{q' - q'\lambda^r}{q' - q\lambda^{2r}}\right]$$

$$= \frac{1}{2}\left[\frac{2q' - \lambda^r(q + q')}{q' - q\lambda^{2r}}\right]$$

$$P(CS) = \frac{q'}{q' - q\lambda^{2r}} - \frac{(q + q')\lambda^r}{2(q' - q\lambda^{2r})} \tag{8.55}$$

Suppose we get

$$P(CS) = p^* \quad \text{then}$$

$$p^*(q' - q\lambda^{2r}) = q' - \left(\frac{q + q'}{2}\right)\lambda^r$$

$$\Rightarrow \qquad qp^*\lambda^{2r} - \left(\frac{q + q'}{2}\right)\lambda^r - (p^*q' - q') = 0$$

$$\Rightarrow \qquad qp^*(\lambda^r)^2 - \left(\frac{q + q'}{2}\right)\lambda^r - (q' - p^*q') = 0$$

$$\Rightarrow \qquad \lambda^r = \frac{(q + q')/2 \pm \sqrt{\left(\frac{q + q'}{2}\right)^2 - 4qp^*(q' - p^*q')}}{2qp^*}$$

$$\lambda^r = (2qp^*)^{-1}\left(\frac{q + q'}{2}\right) - \left\{\left(\frac{q + q'}{2}\right)^2 - 4qp^*(q' - p^*q')\right\}^{1/2} \tag{8.56}$$

Since p and p' are not known, a numerical solution of (8.56) is not obtained.

Rule Rs′

We now regard the situation after each *vector* of observation as one stage and our analysis takes from a stage to the succeeding one. We take a vector of two patients; one given a treatment A and the other treatment B.

Let
$$P_n = P[CS \mid S_A - S_B = n] \tag{8.57}$$

and let us terminate the procedure as soon as

$$|S_A - S_B| = S$$

i.e.
$$P_S = 1 \text{ and } P_{-S} = 0 \tag{8.58}$$

where S_A = successes due to treatment A

S_B = successes due to treatment B

and p and p' are corresponding probabilities of successes and $q = 1 - p$, $q' = 1 - p'$.

Then at the ith stage, there are four probabilities;

	(S, S)	(F, F)	(S, F)	(F, S)
$S_A - S_B$	n	n	$n + 1$	$n - 1$
Probability	pp'	qq'	pq'	qp'

Thus

$$P_n = P[CS \mid S_A - S_B = n]$$

$$= pp'P_n + qq'P_n + pq'P_{n+1} + qp'P_{n-1}$$

$$= (pp' + qq')P_n + pq'P_{n+1} + qp'P_{n-1}$$

$\Rightarrow$

$$pq'P_{n+1} + (pp' + qq' - 1)P_n + qp'P_{n-1} = 0 \qquad (8.59)$$

Putting

$$P_n = \lambda^n,$$

$\Rightarrow$

$$pq'\lambda^2 + (pp' + qq' - 1)\lambda + qp' = 0 \qquad (8.60)$$

Let λ_1 and λ_2 be two roots of the above equation.

Then

$$\lambda_1 + \lambda_2 = \frac{pp' + qq' - 1}{pq'}$$

and

$$\lambda_1 \lambda_2 = \frac{qp'}{pq'}$$

$\Rightarrow$

$$\lambda_1 = 1 \quad \text{and} \quad \lambda_2 = \frac{qp'}{pq'}$$

$\therefore$

$$P_n = C_1 \lambda_1^n + C_2 \lambda_2^n$$

i.e.

$$P_n = C_1 + C_2 \left(\frac{qp'}{pq'}\right)^n \qquad (8.61)$$

where C_1 and C_2 are to be determined such that

$$P_s = 1 \quad \text{and} \quad P_{-s} = 0$$

Now

$$P_n = pp'P_n + qq'P_n + pq'P_{n+1} + qp'(P_{n-1})$$

$$= (pp' + qq')P_{n-1} pq'P_{n+1} + qp'P_{n-1}$$

and

$$P_n = C_1 + C_2\lambda^n; \quad \lambda = \frac{qp'}{pq'} = \delta, \text{ say}$$

$\therefore$

$$P_n = C_1 + C_2 \delta^n, \text{ where } \delta = \frac{qp'}{pq'}$$

$$P_s = 1 \Rightarrow C_1 + C_2 \delta^s = 1$$

$$\Rightarrow C_1 = 1 - C_2 \delta^s \qquad (8.62)$$

$$P_{-s} = 0 \Rightarrow C_1 + C_2 \delta^{-s} = 0$$

$$\Rightarrow C_1 = -C_2 \delta^{-s} \tag{8.63}$$

$$\therefore \quad 1 - C_2 \delta^s = -C_2 \delta^{-s}$$

$$\Rightarrow \quad C_2(\delta^s - \delta^{-s}) = 1$$

$$\Rightarrow \quad C_2(\delta^{2s} - 1) = \delta^s$$

$$\Rightarrow \quad C_2 = \frac{\delta^s}{\delta^{2s} - 1} = \frac{-\delta^s}{1 - \delta^{2s}}$$

$$C_1 = \frac{\delta^s}{1 - \delta^{2s}} \delta^{-s} = \frac{1}{1 - \delta^{2s}}$$

$$\therefore \quad C_1 = \frac{1}{1 - \delta^{2s}}; \quad C_2 = \frac{-\delta^s}{1 - \delta^{2s}} \tag{8.64}$$

$$\Rightarrow \quad P_n = \frac{1}{1 - \delta^{2s}} - \frac{\delta^s}{1 - \delta^{2s}} \cdot \delta^n$$

$$\therefore \quad P_n = \frac{1 - \delta^{s+n}}{1 - \delta^{2s}} \tag{8.65}$$

where $\quad \delta = qp'/pq'$

Again $\quad p_0 = \dfrac{1 - \delta^s}{1 - \delta^{2s}}$

$$\Rightarrow \quad 1 - p_0 = 1 - \frac{1 - \delta^s}{1 - \delta^{2s}}$$

$$= \frac{1 - \delta^{2s} - 1 + \delta^s}{1 - \delta^{2s}}$$

$$= \frac{\delta^s(1 - \delta^s)}{1 - \delta^{2s}} \tag{8.66}$$

$$\frac{1 - p_0}{p_0} = \frac{\delta^s(1 - \delta^s)}{1 - \delta^{2s}} \frac{(1 - \delta^{2s})}{(1 - \delta^s)} = \delta^s$$

By setting $\Delta = \Delta^*$,

$$p = \frac{1}{2}(1 + \Delta^*) \Rightarrow q = \left(\frac{1}{2} - \frac{\Delta^*}{2}\right)$$

and min $p_0 = p^*$

But
$$p' = p - \Delta^* \quad \text{(from 8.38)}$$
$$= \frac{1}{2}(1 + \Delta^*) - \Delta^*$$
$$= \frac{1 - \Delta^*}{2}$$

$$\Rightarrow \quad q' = \left(\frac{1 + \Delta^*}{2}\right)$$

$$\therefore \quad \delta = \frac{qp'}{pq'} = \frac{\left(\dfrac{1 - \Delta^*}{2}\right)\left(\dfrac{1 - \Delta^*}{2}\right)}{\left(\dfrac{1 + \Delta^*}{2}\right)\left(\dfrac{1 + \Delta^*}{2}\right)}$$

$$= \frac{(1 - \Delta^*)^2}{(1 + \Delta^*)^2}$$

$$\Rightarrow \quad \frac{1 - p^*}{p^*} = \delta^s = \left(\frac{1 - \Delta^*}{1 + \Delta^*}\right)^{2s} \tag{8.67}$$

from which we can get the required s

$$s = \frac{1}{2} \frac{\log \dfrac{1 - p^*}{p^*}}{\log \left(\dfrac{1 - \Delta^*}{1 + \Delta^*}\right)} \tag{8.68}$$

References

1. Armitage P (1960): *Sequential Medical trial*. Charles C. Thomas, Springfield III.
2. Chih-Haising Ho (1986): One sided Group Sequential procedures for Clinical trials, *Technical Report* No. 462, University of Minnesota, January, 1986.
3. Demets D.L and Ware J.H. (1980): Group sequential methods for clinical trials with a one sided hypothesis. *Biometrika*, Vol. 67, page 651–660.
4. Freedman L.S. Lowe D and Mackaskill P (1984): Stopping rules for clinical trials incorporating clinical opinion, *Biometrics*, Vol. 40, page 575–586.
5. Hopper John L and Young Greame P (1988): A random walk model for evaluating clinical trials involving serial observations. *Statistics in Medicine*, Vol. 7, page 581–590.
6. Morgan Timothy M. and Elasoff Robert M (1986): Effect of censoring on adjusting for covariates in comparison of survival times. *Communications in Statistics, Theory and Methods*, Vol. 15, No. 6, page 1837–54.
7. Robbins H (1952): Some aspects of sequential design of experiments. Bull. *American Math. Soc.* Vol. 58, page 529–532.
8. Robbins H (1956): A sequential decision problem with a finite memory–*Proceedings of the National Academy of Science*, Vol. 42, page 920–923.
9. Sobel M and Weiss G.H. (1970): Play the winner Sampling for selecting the better of two Binomial Populations. *Biometrika*, Vol. 57, page 357–365.

Techniques of Stochastic Processes in Mortality Analysis—Applications on Life Table

9.0 Introduction

Stochastic Process techniques are profitably utilised in the problems relating to sampling distributions of life table functions. These problems arise when a life table is constructed on the basis of sample. Since the columns of a traditional life table may be considered as sequences of random variables, the parameter space being the ages taken over all nonnegative integers, life tables provide wide scope of application of Stochastic Processes. A very fundamental question that arises of any life table constructed from sample is that how reliable the given life table constructed on the basis of the sample? Or what are the properties of the estimators of life table functions constructed on the basis of a sample? Since life tables occupy very formidable positions, so far their applications are concerned in Survival Analysis, Stochastic Population models, apart from its indispensability in the investigation of mortality in Demography and Medical Statistics; and often for clinical or other related data the construction of a life table may be on the basis of small data. The traditional actuarial techniques of analysing life tables (which are generally undertaken on the basis of complete enumeration of a large population) may not be quite suitable. Here the techniques of Stochastic Processes can play an important role.

Since the applications of Stochastic Processes oriented techniques require a good background of different types of life table functions, their interrelations and their significances so far as the mortality analysis is concerned, we propose to start with the same in the study of this chapter.

9.1 Life Tables

A life table is mortality table which presents the survival experience of a hypothetical cohort of a number of new born infants exposed to a particular type of mortality experience. It consists of the following columns:

(i) x = exact age in years.

(ii) l_x = number of persons surviving at exact age x.

(iii) d_x = number of persons dying while passing from x to $(x + 1)$.

(iv) $q_x = \dfrac{d_x}{l_x}$ = probability of dying within one year following the attain-

ment of the age x.

(v) m_x = probability of dying for a person whose exact age is between x to $(x + 1)$ years.

(vi) μ_x = instantaneous force of mortality or hazard rate, which means that $\mu_x\, \delta x$ is the conditional probability of dying between x to $x + \delta x$ given that the person is surviving at age x.

(vii) L_x = person-years lived by the cohort between x to $(x + 1)$ years

$$= \int_0^1 l_{x+t}\, dt$$

(viii) $T_x = \displaystyle\sum_{t=0}^{\infty} L_{x+t}$ = total person years lived by the cohort from x years

to ∞.

(ix) $\varepsilon_x^0 = \dfrac{T_x}{l_x}$ = complete expectation of life.

9.1.1 Relationship between Life Table Functions

(i) *Relationship between the mortality rates q_x and m_x*
We have

$$m_x = \frac{d_x}{L_x} \cong \frac{d_x}{l_{x+\frac{1}{2}}}$$

$$q_x = \frac{d_x}{l_x} \cong \frac{d_x}{l_{x+\frac{1}{2}} + \frac{1}{2} d_x}$$

$$\therefore \qquad q_x = \frac{d_x/d_x}{\dfrac{l_{x+\frac{1}{2}}}{d_x} + \frac{1}{2}} = \frac{1}{1 \Big/ \dfrac{d_x}{l_{x+\frac{1}{2}}} + \frac{1}{2}}$$

$$q_x = \frac{1}{\dfrac{1}{m_x} + \dfrac{1}{2}} = \frac{2m_x}{2 + m_x} \tag{9.1}$$

is approximate relation between q_x and m_x based on the linearity of l_x curve between x to $(x + 1)$.

Result
Prove that

$$m_x = \frac{2q_x}{2 - q_x}$$

Solution:
We know that

$$m_x = \frac{d_x}{L_x}$$

Also
$$L_x \cong \frac{1}{2}(l_x + l_{x+1})$$

∴
$$m_x = \frac{d_x}{\frac{1}{2}(l_x + l_{x+1})}$$

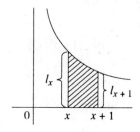

Fig. 9.1

Divide numerator and denominator by l_x

⇒
$$m_x = \frac{\dfrac{d_x}{l_x}}{\dfrac{1}{2}\left\{1 + \dfrac{l_{x+1}}{l_x}\right\}}$$

$$= \frac{q_x}{\frac{1}{2}(1 + p_x)} \qquad \left(\begin{array}{l} \because \ q_x = \dfrac{d_x}{l_x} \\[2mm] \quad p_x = \dfrac{l_{x+1}}{l_x} \end{array} \right)$$

$$= \frac{2q_x}{(1 + p_x)}$$

$$= \frac{2q_x}{2 - q_x}$$

(ii) *Relationship between q_x and μ_x*
We have $\mu_x \, \delta x$ is the conditional probability of dying between x to $x + \delta x$ years

given that the person is surviving at age x.

Then we may write

$$\mu_x \, \delta x \; = \; \frac{f(x) \, \delta x}{1 - F(x)} \tag{9.2}$$

where $1 - F(x)$ is the probability of surviving at least upto age x and $f(x) \, \delta x$ represents the probability of dying between $(x, x + \delta x)$

$$\Rightarrow \mu_x \; = \; \frac{f(x)}{R(x)} \qquad \text{where } [1 - F(x)] = R(x) \tag{9.3}$$

$R(x)$ is called the reliability or survival function at age x. Obviously $F(x)$ is the cumulative distribution function of the mortality distribution and

$$\frac{dF(x)}{dx} \; = \; f(x)$$

Therefore by integrating (9.2) we have

$$\int_0^x \mu_\tau \, d\tau \; = \; -\log_e (1 - F(x))$$

or

$$-\int_0^x \mu_\tau \, d\tau \; = \; \log_e R(x)$$

or

$$R(x) \; = \; \exp\left(-\int_0^x \mu_\tau \, d\tau \right) \tag{9.4}$$

If we put $R(x) = \dfrac{l_x}{l_0} = P$ [surviving at least upto age x] $\tag{9.5}$

Then we have

$$\frac{l_x}{l_0} \; = \; \exp\left(-\int_0^x \mu_\tau \, d\tau \right)$$

$$-\log \frac{l_x}{l_0} \; = \; \int_0^x \mu_\tau \, d\tau$$

Differentiating both sides w.r.t. x we have

$$-\frac{1}{l_x} \frac{dl_x}{dx} \; = \; \mu_x \qquad \text{or} \qquad \mu_x \; = \; -\frac{d(\log l_x)}{dx} \tag{9.6}$$

Hence $\mu_x = -\dfrac{1}{l_x} \dfrac{dl_x}{dx}$ represents the instantaneous force of mortality.

If we compare the deaths between x to $x + \delta x$ viz.

$$[l(x) - l(x + \delta x)]$$

which implies a relative decrease in the number of persons in the life table stationary population while passing from x to $(x + \delta x)$ as $\delta x \to 0$ is given by

$$\left[\frac{l(x) - l(x + \delta x)}{l(x)}\right] \tag{9.7}$$

while the relative increase in age of δx (from x to $x + \delta x$) takes place.

Therefore the elasticity of the decrease in the number of persons with respect to age or age elasticity of the number of persons (analogously compared defined with the price elasticity of demand) is given by

$$\lim_{\delta x \to 0} \left[\frac{l(x) - l(x + \delta x)}{l(x)} \Big/ \delta x\right]$$

$$= -\frac{1}{l(x)} \frac{dl(x)}{dx} = \mu_x \tag{9.8}$$

Thus μ_x represents the age elasticity of the decrease in the number of persons (at the age x).

Next

$$\mu_x = -\frac{1}{l_x} \frac{dl_x}{dx}$$

$$\Rightarrow \qquad \mu_{x+t} = -\frac{1}{l_{x+t}} \frac{dl_{x+t}}{d(x+t)}$$

$$\Rightarrow \qquad \int_0^1 \mu_{x+t}\, l_{x+t}\, dt = -\int_0^1 dl_{x+t}$$

we keep x fixed and t variable in $(x + t)$

$$\Rightarrow \int_0^1 \mu_{x+t}\, l_{x+t}\, dt = -[l_{x+1} - l_x] = d_x$$

$$\therefore \qquad \frac{d_x}{l_x} = \frac{1}{l_x} \int_0^1 \mu_{x+t}\, l_{x+t}\, dt$$

$$\text{or} \qquad q_x = \frac{1}{l_x} \int_0^1 \mu_{x+t}\, l_{x+t}\, dt \tag{9.9}$$

which *provides a relationship between* q_x *and* μ_x.

(iii) *Deterministic and Stochastic formulation of ε_x^0*

$$\varepsilon_{x_\alpha}^0 = \int\limits_0^\infty \exp\left(-\int\limits_{x_\alpha}^{x_\alpha + y_\alpha} \mu(\tau)\,d\tau\right) dy_\alpha \tag{9.10}$$

$$= \int\limits_0^\infty y_\alpha \exp\left(-\int\limits_{x_\alpha}^{x_\alpha + y_\alpha} \mu(\tau)\,d\tau\right) \mu\,(x_\alpha + y_\alpha)\,dy_\alpha \tag{9.11}$$

where $\varepsilon_{x_\alpha}^0$ represents the complete expectation of life at age x_α.

Remarks

[Eqns. (9.10) and (9.11) represent the deterministic and Stochastic definition of the complete expectation of life at age x_α respectively; the result shows that both the definitions lead to one and the same result].

Proof

We have

$$E(X) = \int\limits_0^\infty x\,f(x)\,dx$$

where X is absolutely continuous r.v. and $f(x)$ represents its density function

$$E(X) = -\int\limits_0^\infty x\,\frac{d\,R(x)}{dx}\,dx$$

where $\qquad F(x) = \int\limits_0^x f(t)\,dt$

$\Rightarrow \qquad R(x) = 1 - F(x) = 1 - \int\limits_0^x f(t)\,dt$

$\Rightarrow \qquad \dfrac{d\,R(x)}{dx} = -f(x)$

Hence integrating by parts we have

$$E(X) = -[x\,R(x)]_0^\infty + \int\limits_0^\infty R(x)\,dx$$

Again $\qquad R(0) = 1,\ R(\infty) = 0$

$\therefore \qquad E(X) = \int\limits_0^\infty x\,f(x)\,dx = \int\limits_0^\infty R(x)\,dx \tag{9.12}$

Now we have shown in (9.4)

$$R(x) = \exp\left(-\int_0^x \mu(\tau)\,d\tau\right) \tag{9.13}$$

$$\therefore \quad E(X) = \int_0^\infty \exp\left(-\int_0^x \mu(\tau)\,d\tau\right)dx \tag{9.14}$$

Similarly $R(y_\alpha \mid x_\alpha) = P$ [of surviving upto $x_\alpha + y_\alpha$ | the person has survived upto x_α]

$$= \exp\left(-\int_{x_\alpha}^{x_\alpha + y_\alpha}\mu(\tau)\,d\tau\right) \tag{9.15}$$

in analogy with (9.13)

and

$$\varepsilon^0_{x_\alpha} = \int_0^\infty \exp\left(-\int_{x_\alpha}^{x_\alpha + y_\alpha}\mu(\tau)\,d\tau\right)dy_\alpha \tag{9.16}$$

in analogy with (9.14)

Further we have

$$R(y_\alpha \mid x_\alpha) = \exp\left[-\int_{x_\alpha}^{x_\alpha + y_\alpha}\mu(\tau)\,d\tau\right] \text{ from (9.15)}$$

$$F(y_\alpha \mid x_\alpha) = 1 - \exp\left(-\int_{x_\alpha}^{x_\alpha + y_\alpha}\mu(\tau)\,d\tau\right) \tag{9.17}$$

$$f(y_\alpha \mid x_\alpha) = \frac{d}{dy_\alpha}F(y_\alpha \mid x_\alpha)$$

$$= \exp\left[-\int_{x_\alpha}^{x_\alpha + y_\alpha}\mu(\tau)\,d\tau\right]\mu(x_\alpha + y_\alpha) \tag{9.18}$$

Also

$$\varepsilon^0_{x_\alpha} = \int_0^\infty y_\alpha\, f(y_\alpha \mid x_\alpha)\,dy_\alpha$$

$$= \int_0^\infty y_\alpha \exp\left[-\int_{x_\alpha}^{x_\alpha + y_\alpha}\mu(\tau)\,d\tau\right]\mu(x_\alpha + y_\alpha)\,dy_\alpha \tag{9.19}$$

Thus a comparison of (9.16) and (9.19) shows that

$$\varepsilon_{x_\alpha}^0 = \int_0^\infty \exp\left[-\int_{x_\alpha}^{x_\alpha + y_\alpha} \mu(\tau)\, d\tau\right] dy_\alpha$$

$$= \int_0^\infty y_\alpha \exp\left[-\int_{x_\alpha}^{x_\alpha + y_\alpha} \mu(\tau)\, d\tau\right] \mu\,(x_\alpha + y_\alpha)\, dy_\alpha,$$

which proves (9.10) and (9.11).

(iv)   A result on the behaviour of $x + \varepsilon_x^0$

$f(x) = x + \varepsilon_x^0$ is monotonically non-decreasing function of x.

Proof:

$$f(x + n) = (x + n) + \varepsilon_{x+n}^0 \varepsilon_{x+n}^0$$

$$f(x + n) - f(x) = n + \varepsilon_{x+n}^0 - \varepsilon_x^0, \quad (n > 0) \tag{9.20}$$

Also

$$\varepsilon_x^0 = \frac{T_x}{l_x} = \frac{\int_0^\infty l_{x+t}\, dt}{l_x}$$

$\Rightarrow$

$$\varepsilon_x^0 = \frac{\int_0^n l_{x+t}\, dt}{l_x} + \frac{\int_n^\infty l_{x+t}\, dt}{l_x}$$

$$= \frac{T_{x:\overline{n}}}{l_x} + \frac{T_{x+n}}{l_{x+n}} \frac{l_{x+n}}{l_x}$$

where $T_{x:\overline{n}}$ represents the total person year lived by the cohort between x to $(x + n)$ years

$$\varepsilon_x^0 = \varepsilon_{x:\overline{n}}^0 + \varepsilon_{x+n}^0 \left({}_n p_x\right) \tag{9.21}$$

where $\varepsilon_{x:\overline{n}}^0 = \dfrac{T_{x:\overline{n}}}{l_x}$ = complete expectation of life between x to $(x + 1)$ years.

Obviously

$$\varepsilon_{x:\overline{n}}^0 \leq n \tag{9.22}$$

and ${}_n p_x = \dfrac{l_{x+n}}{l_x}$, the probabilities of surviving from the age x to $(x + n)$.

Putting (9.21) in (9.20) $\Rightarrow$

$$f(x+n) - f(x) = n + \varepsilon^0_{x+n} - \varepsilon^0_{x\,:\,n1} - \varepsilon^0_{x+n} \, ({}_nP_x)$$

$$= n + \varepsilon^0_{x+n} \, (1 - {}_nP_x) - \varepsilon^0_{x\,:\,n1}$$

$$= (n - \varepsilon^0_{x\,:\,n1}) + \varepsilon^0_{x+n} \, (1 - {}_nP_x)$$

Since $\qquad n - \varepsilon^0_{x\,:\,n1} \geq 0$, from, (9.22)

$$\varepsilon^0_{x+n} \geq 0 \text{ and } (1 - {}_nP_x) \geq 0$$

it follows that $\qquad f(x+n) - f(x) \geq 0 \; \forall \; n > 0$

Thus $f(x)$ is a monotonically nondecreasing function of x.

(v)   *Relation between m_x and μ_x*

$$\mu_{x + 1/2} \cong m_x$$

Proof.   By definition of central rate of mortality

$$m_x = \frac{d_x}{L_x}, \; \mu_x = -\frac{1}{l_x} \frac{dl_x}{dx}$$

Also $\qquad L_x = \int_0^1 l_{x+t} \, dt$

Differentiating w.r.t. x we get

$$\frac{dL_x}{dx} = \frac{d}{dx} \int_0^1 l_{x+t} \, dt$$

$$= \int_0^1 \frac{d}{dt} (l_{x+t}) \, dt$$

(since l_{x+t} is a symmetric function of x and t)

$$= |\, l_{x+t} \,|_0^1$$

$$= (l_{x+1} - l_x) = -d_x$$

$$\Rightarrow -\frac{1}{L_x} \frac{dL_x}{d_x} = \frac{d_x}{L_x}$$

$\therefore \qquad m_x = -\frac{1}{L_x} \frac{dL_x}{d_x}$ $\qquad\qquad$ (9.23)

Since $\qquad L_x = \int_0^1 l_{x+t} \, dt = l_{x+\frac{1}{2}}$

$$\therefore \qquad m_x = -\frac{1}{l_{x+\frac{1}{2}}} \frac{dl_{x+\frac{1}{2}}}{d_x} = \mu_{x+\frac{1}{2}} \qquad (9.24)$$

(vi) An Exact Expression of L_x

$$L_x = \left(\frac{1}{m_x} - \frac{1}{q_x} + 1\right) l_x + \left(\frac{1}{q_x} - \frac{1}{m_x}\right) l_{x+1}$$

Proof:

$$L_x = \int_0^1 l_{x+t} \, dt$$

$$= f_x \, l_x + (1 - f_x) \, l_{x+1}$$

where f_x is the weight function of l_x (to be obtained)

$$L_x = f_x \, l_x + l_{x+1} - f_x \, l_{x+1}$$
$$= f_x (l_x - l_{x+1}) + l_{x+1}$$
$$= f_x \, d_x + l_{x+1}$$

$$\Rightarrow \qquad f_x = \frac{L_x - l_{x+1}}{d_x}$$

$$\Rightarrow \qquad f_x = \frac{L_x}{d_x} - \frac{l_{x+1}}{d_x}$$

$$\Rightarrow \qquad f_x = \frac{1}{m_x} - \left(\frac{l_x - d_x}{d_x}\right) \left(\text{since } m_x = \frac{d_x}{L_x}, d_x = l_x - l_{x+1}\right)$$

$$\Rightarrow \qquad f_x = \frac{1}{m_x} - \frac{l_x}{d_x}\left(1 - \frac{d_x}{l_x}\right)$$

$$\Rightarrow \qquad f_x = \frac{1}{m_x} - \frac{(1 - q_x)}{q_x} \qquad \left(\text{since } q_x = \frac{d_x}{l_x}\right)$$

$$= 1 + \frac{1}{m_x} - \frac{1}{q_x}$$

$$\therefore \qquad L_x = f_x \, l_x + (1 - f_x) \, l_{x+1}$$

$$= \left(1 + \frac{1}{m_x} - \frac{1}{q_x}\right) l_x + \left(\frac{1}{q_x} - \frac{1}{m_x}\right) l_{x+1} \qquad (9.25)$$

(vii) A relationship between ε_x^0 and ε_{x+1}^0

Prove that

$$\varepsilon_x^0 = \varepsilon_{x+1}^0 \, p_x + f_x + (1 - f_x) \, p_x$$

where
$$f_x = \left(\frac{1}{m_x} - \frac{1}{q_x} + 1\right)$$

Proof:

$$L_x = \int_0^1 l_{x+t}\, dt$$

$$= f_x\, l_x + (1 - f_x)\, l_{x+1}$$

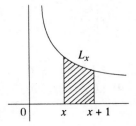

$$L_x = f_x(l_x - l_{x+1}) + l_{x+1}$$
$$L_x = f_x\, d_x + l_{x+1}$$

$$\Rightarrow f_x = \frac{L_x}{d_x} - \frac{l_{x+1}}{d_x} = \frac{1}{m_x} - \left(\frac{l_x - d_x}{d_x}\right) = \frac{1}{m_x} - \frac{1}{q_x} + 1$$

Again
$$\overset{0}{\varepsilon}_x = \frac{T_x}{l_x} = \frac{T_{x+1} + L_x}{l_x}$$

$$\Rightarrow \overset{0}{\varepsilon}_x = \frac{T_{x+1}}{l_x} + \frac{L_x}{l_x}$$

$$\Rightarrow \overset{0}{\varepsilon}_x = \frac{T_{x+1}}{l_{x+1}} \cdot \frac{l_{x+1}}{l_x} + \frac{f_x\, l_x + (1 - f_x)\, l_{x+1}}{l_x}$$

$$\Rightarrow \overset{0}{\varepsilon}_x = p_x\, \overset{0}{\varepsilon}_{x+1} + f_x + (1 - f_x)\frac{l_{x+1}}{l_x}$$

$$\Rightarrow \overset{0}{\varepsilon}_x = \overset{0}{\varepsilon}_{x+1}\, p_x + f_x + (1 - f_x)\, p_x$$

where
$$f_x = \frac{1}{m_x} - \frac{1}{q_x} + 1 \qquad (9.26)$$

(viii) *Relationship between* μ_x *and* $\overset{0}{\varepsilon}_x$

Prove that

$$\mu_x = \frac{1}{\overset{0}{\varepsilon}_x}\left[1 + \frac{d\overset{0}{\varepsilon}_x}{dx}\right]$$

Proof:

We have

$$\varepsilon_x^0 = \frac{T_x}{l_x} = \frac{\int_0^\infty l_{x+t}\, dt}{l_x}$$

Differentiating w.r.t. x, we get

$$\frac{d\varepsilon_x^0}{dx} = \frac{d}{dx}\left[\frac{1}{l_x}\int_0^\infty l_{x+t}\, dt\right]$$

$$= \left(\int_0^\infty l_{x+t}\, dt\right)\frac{d}{dx}\left(\frac{1}{l_x}\right) + \frac{1}{l_x}\left(\int_0^\infty \frac{d}{dx}(l_{x+t})\, dt\right)$$

$$= -\frac{1}{l_x^2}\frac{dl_x}{dx}\int_0^\infty l_{x+t}\, dt + \frac{1}{l_x}\int_0^\infty \frac{d}{dt}(l_{x+t})\, dt$$

$$(\because l_{x+t} \text{ is a symmetric function of } x \text{ and } t)$$

$$= \left(-\frac{1}{l_x}\frac{dl_x}{dx}\right)\left(\frac{\int_0^\infty l_{x+t}\, dt}{l_x}\right) + \frac{1}{l_x}[l_{x+t}]_0^\infty$$

$$= \mu_x \frac{T_x}{l_x} + \frac{1}{l_x}[l_\infty - l_x]$$

$$= \mu_x \varepsilon_x^0 - 1 \quad (\because l_\infty = 0)$$

$$\Rightarrow \mu_x = \frac{1}{\varepsilon_x^0}\left[\frac{d\varepsilon_x^0}{dx} + 1\right] \tag{9.27}$$

(ix) *Relationship between Complete and Curtailed (or Curtate) Expectation of life*

Let w be the last age (in integer) for survival; w is obviously a random variable (r.v.)

We denote $p_{0x} = P$ [surviving from the age 0 to x]

Therefore, probability of no person surviving upto age $(w+1)$ out of a cohort of l_0 who started life together is

$$= (1 - p_{0w+1})^{l_0}$$

Similarly out of the newly form cohort of l_0 individuals the probability that no one is alive at w is given by

$$(1 - p_{0w})^{l_0}$$

Thus the probability of at least one person surviving (out of l_0) between w and $(w + 1)$ is

$$= (1 - p_{0w+1})^{l_0} - (1 - p_{0w})^{l_0}$$

This is the discrete probability distribution of w which is given in nonnegative integer only. Thus

$$E(w) = \sum_{w=0}^{\infty} w \{(1 - p_{0w+1})^{l_0} - (1 - p_{0w})^{l_0}\}$$

which is the curtate expectation of life at the age 0 viz. ε_0. If we take $l_0 = 1$ then

$$E(w) = \sum_{w=0}^{\infty} w (p_{0w} - p_{0w+1})$$

$$= \sum_{x=1}^{\infty} p_{0x} \tag{9.28}$$

Similarly, the complete expectation of life at zero viz.

$$\varepsilon_0^0 = \frac{\int_0^{\infty} l_t \, dt}{l_0}$$

$$\cong \frac{1}{2} \frac{l_0 + l_1 + l_2 + \ldots + l_w + \ldots}{l_0}$$

$$= \frac{1}{2} + \frac{l_1}{l_0} + \ldots + \frac{l_w}{l_0} + \ldots +$$

$$= \frac{1}{2} + p_{01} + p_{02} + \ldots + p_{0w} + \ldots$$

$$= \frac{1}{2} + \sum_{x=1}^{\infty} p_{0x} \tag{9.29}$$

Thus comparing (9.28) and (9.29) we have

$$\varepsilon_0^0 \cong \frac{1}{2} + \sum_{x=1}^{\infty} p_{0x} = \frac{1}{2} + \varepsilon_0$$

$$\varepsilon_0^0 \cong \frac{1}{2} + \varepsilon_0 \tag{9.30}$$

i.e. complete expectation of life at birth

$$\cong \frac{1}{2} + \text{curtate expectation of life at birth.}$$

The result (9.30) is however true for any x, i.e.

$$\varepsilon_x^0 \cong \frac{1}{2} + \varepsilon_x \tag{9.31}$$

Proof:

In this case

$$\varepsilon_x^0 = \frac{\int_0^\infty l_{x+t} \, dt}{l_x} \cong \frac{1}{2} + \sum_{t=0}^\infty p_{x,\,x+t} \quad \text{from (9.29) and}$$

$$\varepsilon_x = \sum_{t=0}^\infty p_{x,\,x+t} \quad \text{from (9.28)}$$

Hence

$$\varepsilon_x^0 \cong \frac{1}{2} + \varepsilon_x \quad \text{which proves (9.31)}$$

We have the probability of surviving upto t years for a newly born infant

$$_t p_0 = \exp\left(-\int_0^t \mu(\tau) \, d\tau\right),$$

where $\mu(\tau)$ represents the instantaneous force of mortality or the Hazard rate at the age τ.

Therefore the probability of surviving upto the age t and then dying between $(t, t + dt)$ is given by

$$\exp\left(-\int_0^t \mu(\tau) \, d\tau\right) \mu(t) \, dt = f_0(t) \, dt$$

which represents the probability of dying between t to $(t + dt)$. Hence

$$f_0(t) = \exp\left(-\int_0^t \mu(\tau) \, d\tau\right) \mu(t)$$

represents the density function of the time of death measured from the time of birth.

∴ Expectation of life at birth

$$\varepsilon_0^0 = \int_0^\infty t \exp\left(-\int_0^t \mu(\tau) \, d\tau\right) \mu(t) \, dt \tag{9.32}$$

Also if the survival probability upto the age t is given by

$$_tP_0 = \exp\left(-\int_0^t \mu(\tau) \, d\tau\right)$$

then $\quad 1 - {}_tP_0 = 1 - \exp\left(-\int_0^t \mu(\tau) \, d\tau\right) = F_0(t) =$ c.d.f. of the failure

distribution or the distribution of the age at death.

Also $\quad \varepsilon_0^0 = \int_0^\infty t \, f_0(t) \, dt = \int_0^\infty t \frac{d}{dt} [F_0(t)] \, dt$

$$= \int_0^\infty t \frac{d}{dt} [1 - p_0(t)] \, dt$$

$$= -\int_0^\infty t \frac{dp_0(t)}{dt} \, dt$$

$$= -t \, p_0(t) \Big|_0^\infty + \int_0^\infty p_0(t) \, dt, \text{ integrating by parts}$$

$$= \int_0^\infty p_0(t) \, dt = \int_0^\infty \exp\left(-\int_0^t \mu(\tau) \, d\tau\right) dt \quad (\because p_0(\infty) = 0)(9.33)$$

Thus combining (9.32) and (9.33) we have

$$\varepsilon_0^0 = \int_0^\infty t \exp\left(-\int_0^t \mu(\tau) \, d\tau\right) \mu(t) \, dt$$

$$= \int_0^\infty \exp\left(-\int_0^t \mu(\tau) \, d\tau\right) dt \qquad (9.34)$$

Again the life table death rate is given by $\dfrac{1}{\varepsilon_0^0}$

Because $\quad \varepsilon_0^0 = \dfrac{T_0}{l_0}; \ \dfrac{1}{\varepsilon_0^0} = \dfrac{l_0}{T_0}$

$T_0 =$ Total person years lived by the cohort by l_0 persons till the end of the life table.

$$\frac{l_0}{T_0} = \frac{\text{Total number of persons}}{\text{Total person years lived by the cohort}}$$

$$\frac{1}{\varepsilon_0^0} = \frac{d_0 + d_1 + \dots + d_w + \dots}{L_0 + L_1 + \dots + L_w + \dots}$$

Thus of a life table population of L_0 (whose ages lie between 0 and 1) a number of d_0 persons died within a course of a year. Similarly out of L_1 population a number of d_1 persons died within the course of a year and so on.

$$\frac{1}{\varepsilon_0^0} = \frac{\sum\limits_{x=0}^{\infty} d_x}{\sum\limits_{x=0}^{\infty} L_x} \cong D \tag{9.35}$$

where D represents a measure of the crude death rate of a Stationary life table population.

Again $\quad \sum\limits_{x=0}^{\infty} d_x = l_0 \quad$ and $\quad \sum\limits_{x=0}^{\infty} L_x = T_0$ gives

$$\varepsilon_0^0 = \frac{T_0}{l_0} = \frac{1}{D}$$

holds for a Stationary Population.

9.2 Abridged Life Table

An abridged life table is the abridged version of a complete life table. The abridgment is made with respect to two aspects viz. (i) the age entries are in groups (often quinoquinical age groups) (ii) number of columns are abridged in complete life table. The first Indian Abridged life table was constructed based on 1941 Census data collected on the basis of 1% sample, the design of which was given by Frank Yates (1941) known as 1% Y-sample.

The following are the columns of an abridged life table.

(i) $(x - x + n) \equiv$ age groups for all x.

(ii) $l_x \equiv$ no. of persons surviving at the age x.

(iii) $_n d_x \equiv$ no. of deaths while passing from x to $(x + n)$.

(iv) $_n q_x = \dfrac{_n d_x}{l_x}$, the annual yearly mortality rate.

(v) $_n m_x = \dfrac{_n d_x}{_n L_x}$, the central rate of mortality in the age group $(x - x + n)$.

(vi) $\quad _nL_x = \int\limits_0^n l_{x+t}\, dt$

(vii) $\quad T_x =$

9.2.1 Relationship between Abridged Life Table Function $_nm_x$ and $_nq_x$

To show that $\qquad\qquad _nq_x = \dfrac{2n\,(_nm_x)}{2 + n\,(_nm_x)}$ $\qquad\qquad$ (9.36)

Proof:

We know from the definition of abridged life table functions

$$_nq_x = \frac{_nd_x}{l_x} \qquad\qquad (9.37)$$

and $\qquad\qquad _nm_x = \dfrac{_nd_x}{_nL_x}$

$$_nd_x = (_nm_x)\,_nL_x \qquad\qquad (9.38)$$

Combining (9.37) and (9.38) we get

$$_nq_x = \frac{(_nm_x)\,_nL_x}{l_x}$$

$$\cong \frac{(_nm_x)}{l_x}\left[\frac{n}{2}(l_x + l_{x+n})\right]$$

$\left(\text{assuming linearity of } l_x \text{ between } x \text{ to } x+n. \Rightarrow {}_nL_x \cong \dfrac{n}{2}(l_x + l_{x+n})\right)$

$$\cong \frac{_nm_x}{l_x}\left[\frac{n}{2}(l_x + (l_x - {}_nd_x))\right]\;(\because l_x - l_{x+n} = {}_nd_x)$$

$$= \frac{_nm_x}{l_x}\left\{\frac{n}{2}(2l_x - {}_nd_x)\right\}$$

$$\Rightarrow {}_nq_x = n\,(_nm_x)\left\{1 - \frac{_nd_x}{2l_x}\right\}$$

$$\Rightarrow {}_nq_x = n\,(_nm_x)\left\{1 - \frac{_nq_x}{2}\right\}$$

$$\Rightarrow {}_nq_x = n\,(_nm_x) - \frac{n}{2}(_nm_x)(_nq_x)$$

$$\Rightarrow \left(1 + \frac{n}{2}(_nm_x)\right)\,{}_nq_x = n\,(_nm_x)$$

$$\Rightarrow \ {}_n q_x \ = \ \frac{n\left({}_n m_x\right)}{1+\dfrac{n}{2}\left({}_n m_x\right)}$$

or

$$_n q_x \ = \ \frac{2n\left({}_n m_x\right)}{2+n\left({}_n m_x\right)}$$

This result is known as abridged Greville's formula for constructing abridged life table.

9.2.2   Greville's Formula

$$_n q_x \ = \ \frac{{}_n m_x}{\dfrac{1}{n}+\left({}_n m_x\right)\left\{\dfrac{1}{2}+\dfrac{n}{12}\left({}_n m_x - k\right)\right\}} \tag{9.39}$$

where $k = \log_e C$ and C is a parameter in the Gompertz curve given by

$$_n m_x \ \cong \ BC^{\,x}$$

(Thus B and C are parameters of Gompertz curve)

Proof:

We know that

$$_n q_x \ = \ \frac{{}_n d_x}{l_x} \ \text{ and } \ {}_n m_x \ = \ \frac{{}_n d_x}{{}_n L_x}$$

$$\Rightarrow \ {}_n q_x \ = \ {}_n m_x \left(\frac{{}_n L_x}{l_x}\right)$$

In the next place, ${}_n L_x = \displaystyle\int_0^n l_{x+t}\, dt$.

Differentiating with respect to x, we get

$$\frac{d\,{}_n L_x}{dx} \ = \ \frac{d}{dx}\left\{\int_0^n l_{x+t}\, dt\right\} \ = \ \int_0^n \frac{d}{dt}\left(l_{x+t}\right) dt$$

(since l_{x+t} is a symmetric function of x and t)

$$\Rightarrow \qquad \frac{d\,{}_n L_x}{dx} \ = \ \left|l_{x+t}\right.\big|_0^n \ = \ l_{x+n}-l_x \ = \ -\ {}_n d_x$$

$$\Rightarrow \qquad -\frac{1}{{}_n L_x}\frac{d\,{}_n L_x}{dx} \ = \ \frac{{}_n d_x}{{}_n L_x}$$

$$\Rightarrow \qquad \frac{d\,\log {}_n L_x}{dx} \ = \ -\frac{{}_n d_x}{{}_n L_x} \ = \ -\ {}_n m_x$$

$$\Rightarrow \qquad d\,\log {}_n L_x \ = \ -\left({}_n m_x\right) dx$$

Integrating $\qquad \int d \log {}_n L_x = -\int {}_n m_x \, dx + \log_e C$

$\Rightarrow \qquad\qquad {}_n L_x = C \exp\left(-\int {}_n m_x \, dx\right)$ $\qquad$ (9.40)

We have, by Euler's Maclaurin's Quadrature Formula

$$\frac{1}{h} \int\limits_a^{a+nh} f(t) \, dt = \frac{1}{2} f(a) + f(a+h) + \dots + f(a+(n-1)h)$$

$$+ \frac{1}{2} f(a+nh) - \frac{h}{12} [f'(a+nh) - f'(a)]$$

$$+ \frac{h^2}{720} [f'''(a+nh) - f'''(a)] + \dots$$

$$\Rightarrow \frac{1}{h} \int\limits_a^{a+nh} f(t) \, dt = \sum_{r=0}^{n} f(a+rh) - \frac{1}{2} [f(a)$$

$$+ f(a+nh)] - \frac{h}{12} [f'(a+nh) - f'(a)]$$

$$+ \frac{h^2}{720} [f'''(a+nh) - f'''(a)] - \dots \qquad (9.41)$$

Put $a = x$ and let $n \to \infty$ both sides of (9.41), whence we get

$$\frac{1}{h} \int\limits_x^{\infty} f(t) \, dt = \sum_{r=0}^{\infty} f(x+rh) - \frac{1}{2} [f(x) + f(x+nh)]$$

$$- \frac{h}{12} [f'(\infty) - f'(x)] + 0\left(\frac{h^2}{720}\right)$$

Again, putting $h = n$ and $f(t) = {}_n L_t$, we get

$$\frac{1}{n} \int\limits_x^{\infty} {}_n L_t \, dt = \sum_{r=0}^{\infty} {}_n L_{x+rn} - \frac{1}{2} [{}_n L_x] + \frac{n}{12} \left[\frac{d \, {}_n L_x}{dx}\right]$$

$$\Rightarrow \qquad \sum_{r=0}^{\infty} {}_n L_{x+rn} = \frac{1}{n} \int\limits_x^{\infty} {}_n L_t \, dt + \frac{1}{2} [{}_n L_x] - \frac{n}{12} \left[\frac{d \, {}_n L_x}{dx}\right] \qquad (9.42)$$

From (9.40) we have by putting

$${}_n L_x = C \exp\left(-\int {}_n m_x \, dx\right) \quad \text{in (9.42)}$$

$$\sum_{r=0}^{\infty} {}_n L_{x+rn} = \frac{1}{n} \int\limits_x^{\infty} C \exp\left(-\int {}_n m_t \, dt\right) dt + \frac{1}{2} \left[C \exp\left(-\int {}_n m_x \, dx\right)\right]$$

$$-\frac{n}{12}\left[\left(-\,_nm_x\right)C\exp\left(-\int_n m_x\,dx\right)\right]$$

$$=\frac{1}{n}\int_x^\infty C\exp\left(-\int_n m_t\,dt\right)dt+\frac{1}{2}\left[C\exp\left(-\int_n m_x\,dx\right)\right]+\frac{n}{12}\left[\left(_nm_x\right)C\exp\left(-\int_n m_x\,dx\right)\right]$$

$$(9.43)$$

Differentiating both sides of (9.43) w.r.t. x, we get

$$\frac{d}{dx}\left[\sum_{r=0}^\infty {}_nL_{x+rn}\right]=\frac{-1}{n}C\exp\left(-\int_n m_x\,dx\right)-\frac{1}{2}\left(_nm_x\right)C\exp\left(-\int_n m_x\,dx\right)$$

$$-\frac{n}{12}\left[\left(_nm_x\right)^2 C\exp\left(-\int_n m_x\,dx\right)\right]$$

$$+\frac{n}{12}\left[\frac{d}{dx}\left(_nm_x\right)C\exp\left(-\int_n m_x\,dx\right)\right]\qquad(9.44)$$

If
$$F(a,b)=\int_{a(x)}^{b(x)}f(t)\,dt$$

then using differentiation under integral sign, we get

$$\frac{d}{dx}F(a(x),b(x))=\int_a^b\left(\frac{df(t)}{dx}\right)dt+\frac{db(x)}{dx}\cdot f(b)-\frac{da(x)}{dx}\cdot f(a)$$

We put $_nm_x=BC^x$ (Gompertz curve)

$$\therefore\qquad\frac{d}{dx}\left(_nm_x\right)=BC^x\log_e C$$

$$=k\left(_nm_x\right)\text{ where }k=\log_e C\qquad(9.45)$$

Substituting (9.45) in (9.44) we get

$$\frac{d}{dx}\left[\sum_{r=0}^\infty {}_nL_{x+rn}\right]=-C\exp\left(-\int_n m_x\,dx\right)\left[\frac{1}{n}+\frac{1}{2}\left(_nm_x\right)\right.$$

$$\left.+\frac{n}{12}\left(_nm_x\right)^2-\frac{n}{12}k\left(_nm_x\right)\right]$$

or, $$-\frac{d}{dx}\left[\sum_{r=0}^\infty {}_nL_{x+rn}\right]={}_nL_x\left[\frac{1}{n}+\frac{1}{2}\left(_nm_x\right)+\frac{n}{12}\left(_nm_x\right)\left(_nm_x-k\right)\right]$$

$$={}_nL_x\left[\frac{1}{n}+\left(_nm_x\right)\left\{\frac{1}{2}+\frac{n}{12}\left(_nm_x-k\right)\right\}\right]$$

$$\Rightarrow\quad -\frac{d}{dx}\left[\sum_{r=0}^\infty {}_nL_{x+rn}\right]={}_nL_x\left[\frac{1}{n}+\frac{1}{2}\left(_nm_x\right)\left\{1+\frac{n}{6}\left(_nm_x-k\right)\right\}\right]\qquad(9.46)$$

Again

$$\frac{d}{dx}\left[\sum_{r=0}^{\infty} {}_nL_{x+rn}\right] = \frac{d}{dx}\left[{}_nL_x + {}_nL_{x+n} + {}_nL_{x+2n} + \cdots\right]$$

$$= \frac{d}{dx}\left[\int_0^n l_{x+t}\, dt + \int_n^{2n} l_{x+t}\, dt + \int_{2n}^{3n} l_{x+t}\, dt + \cdots\right]$$

$$= (l_{x+n} - l_x) + (l_{x+2n} - l_{x+n}) + (l_{x+3n} - l_{x+2n}) + \cdots$$

$$= -l_x \qquad (9.47)$$

From (9.46) and (9.47) we get

$$l_x = ({}_nL_x)\left[\frac{1}{n} + \frac{1}{2}({}_nm_x)\left(1 + \frac{n}{6}({}_nm_x - k)\right)\right]$$

or

$$\frac{1}{l_x} = \frac{1}{{}_nL_x\left[\frac{1}{n} + ({}_nm_x)\left(\frac{1}{2} + \frac{n}{12}({}_nm_x - k)\right)\right]}$$

or

$$\frac{{}_nd_x}{l_x} = \frac{\left(\frac{{}_nd_x}{{}_nL_x}\right)}{\frac{1}{n} + ({}_nm_x)\left[\frac{1}{2} + \frac{n}{12}({}_nm_x - k)\right]}$$

or

$$_nq_x = \frac{{}_nm_x}{\frac{1}{n} + ({}_nm_x)\left[\frac{1}{2} + \frac{n}{12}({}_nm_x - k)\right]}$$

where

$$_nq_x = \frac{{}_nd_x}{l_x}, \quad {}_nm_x = \frac{{}_nd_x}{{}_nL_x}$$

which establishes Greville's formula for the construction of abridged life table.

9.2.3 Reed and Merrell's Formula

$$_nq_x = 1 - \exp\left(-n\,({}_nm_x)\left[1 + \frac{k}{12}n^2\,({}_nm_x)\right]\right) \qquad (9.48)$$

The formula holds on ignoring terms involving $({}_nm_x)^k$, for $k \geq 3$.

Proof:

We have, from the Greville's Formula given in (9.39)

$$_nq_x = \frac{({}_nm_x)}{\frac{1}{n} + {}_nm_x\left[\frac{1}{2} + \frac{n}{12}({}_nm_x - k)\right]}$$

where $\quad \dfrac{d}{dx}({}_nm_x) = k\,({}_nm_x)$ and $k = \log_e C$

$$\Rightarrow {}_nP_x = 1 - {}_nq_x = 1 - \frac{({}_n m_x)}{\frac{1}{n} + {}_n m_x\left[\frac{1}{2} + \frac{n}{12}({}_n m_x - k)\right]}$$

$$= 1 - \frac{({}_n m_x)}{\frac{1}{n} + \frac{1}{2}({}_n m_x) + \frac{n}{12}\left[({}_n m_x)^2 - \frac{d}{dx}({}_n m_x)\right]}$$

$$\Rightarrow {}_nP_x = \frac{\frac{1}{n} + \frac{1}{2}({}_n m_x) + \frac{n}{12}\left[({}_n m_x)^2 - \frac{d}{dx}({}_n m_x)\right] - ({}_n m_x)}{\frac{1}{n} + \frac{1}{2}({}_n m_x) + \frac{n}{12}\left[({}_n m_x)^2 - \frac{d}{dx}({}_n m_x)\right]}$$

$$= \frac{1 + \frac{n^2}{12}\left[({}_n m_x)^2 - \frac{d}{dx}({}_n m_x)\right] - \frac{n}{2}({}_n m_x)}{1 + \frac{n}{2}({}_n m_x) + \frac{n^2}{12}\left[({}_n m_x)^2 - \frac{d}{dx}({}_n m_x)\right]}$$

$$= \frac{1 + \alpha - \beta}{1 + \alpha + \beta} \quad \text{where } \alpha = \left\{\frac{n^2}{12}({}_n m_x)^2 - \frac{d}{dx}({}_n m_x)\right\}$$

$$\beta = \frac{n}{2}({}_n m_x)$$

Taking logarithm on both sides, we get

$$\log {}_nP_x = \log_e(1 + \alpha - \beta) - \log_e(1 + \alpha + \beta)$$

$$= \left[(\alpha - \beta) - \frac{(\alpha - \beta)^2}{2} + \frac{(\alpha - \beta)^3}{3} + \cdots\right]$$

$$- \left[(\alpha + \beta) - \frac{(\alpha + \beta)^2}{2} + \frac{(\alpha + \beta)^3}{3} - \cdots\right]$$

Assuming that $|\alpha + \beta| < 1$ and $|\alpha - \beta| < 1$ ignoring higher powers of $(\alpha - \beta)^3$ and $(\alpha + \beta)^3$, we get

$$\log {}_nP_x = \left\{(\alpha - \beta) - \frac{1}{2}(\alpha^2 + \beta^2 - 2\alpha\beta) + 0(\alpha - \beta)^3\right\}$$

$$- \left\{(\alpha + \beta) - \frac{1}{2}(\alpha^2 + \beta^2 + 2\alpha\beta) + 0(\alpha + \beta)^3\right\}$$

$$= -2\beta + 2\alpha\beta$$

$$= -2\beta(1 - \alpha)$$

$$= -2 \cdot \frac{n}{2}({}_n m_x)\left\{1 - \frac{n^2}{12}\left(({}_n m_x)^2 - \frac{d}{dx}({}_n m_x)\right)\right\}$$

$$= -\left[n({}_n m_x) - \frac{n^3}{12}({}_n m_x)^3 + \frac{n^3}{12}({}_n m_x)\frac{d}{dx}({}_n m_x)\right]$$

$$= -n\left({}_n m_x\right) - \frac{n^2}{12}\left({}_n m_x\right)^2 k$$

(neglecting terms involving $\left({}_n m_x\right)^3$)

$$\Rightarrow {}_n P_x = \exp\left(-n\left({}_n m_x\right) - \frac{n^3}{12}\left({}_n m_x\right)^2 k\right)$$

$${}_n q_x = 1 - {}_n P_x = 1 - \exp\left(-n\left({}_n m_x\right) - \frac{n^3}{12}\left({}_n m_x\right)^2 k\right)$$

or $${}_n q_x = 1 - \exp\left(-n\left({}_n m_x\right)\left(1 + \frac{k}{12} n^2 \left({}_n m_x\right)\right)\right)$$

This is *Reed* and *Marrell's Formula* for the construction of abridged life table.

9.3 Sampling Distribution of Life Table Functions

Multivariate probability generating function of l_1, l_2, l_w/l_0:
The problem arises when we construct a life table from a sample as to test (i) the reliability of the life table as well as (ii) the sampling variance associated with the estimates of different life table functions. The results are due to Chiang (1968). Basically, the parameters in the life table are the survival probabilities (or intrinsic risk of mortality) viz.,

$$P_{0x} = \exp\left(-\int_0^x \mu(\tau)\,d\tau\right), \quad P_x = \exp\left(-\int_0^{x+1} \mu(\tau)\,d\tau\right)$$

and $$P_{0x} = \prod_{i=0}^{x-1} P_i = P_0\, P_1 \cdots P_{x-1} \tag{9.49}$$

Hence $\mu(x)$, P_{0x} and p_x are the parameters of the life table, where

$$P_{0x} = P\{\text{surviving from 0 to } x\}$$

$$p_x = P\{\text{of surviving from } x \text{ to } (x+1)\}$$

The sequence of r.v.'s are $\{l_x\}$ which conforms to a simple Markov Chain

i.e. $$P\{l_x = k_x \mid l_{x-1} = k_{x-1}, l_{x-2} = k_{x-2} \cdots l_0 = k_0\}$$
$$= P\{l_x = k_x \mid l_{x-1} = k_{x-1}\} \tag{9.50}$$

and $$\{dx\}, \{L_x\}, \{T_x\} \text{ and } \{\varepsilon_x^0\}, x = 0, 1, 2, \ldots$$

are also sequences of r.v.'s in the life table. To test the reliability of a life table constructed from a sample, we at the outset, consider

$$E\left[S_1^{l_1} S_2^{l_2}, \ldots S_w^{l_w} \mid l_0\right]$$

as multivariate probability generating function (p.g.f.) of $l_1, l_2, \ldots l_w$, where w

has been conceived as possibly the last age for survival. A univariate p.g.f. for a discrete r.v. is defined by

$$\phi = E(S^X), X = 0, 1, 2, 3, ..., |S| < 1$$

$$\Rightarrow \phi = p_0 + p_1 S + p_2 S^2 + ... \quad \text{where } p_i = P(X = i)$$

$$\frac{d\phi}{dS}\bigg]_{S=0} = p_1 + 2p_2 S + 3 p_2 S^2 + ...]_{S=0} = p_1$$

$$\frac{d^2\phi}{dS^2}\bigg]_{S=0} = 2p_2 \Rightarrow p_2 = \frac{1}{2!}\frac{d^2\phi}{dS^2}\bigg]_{S=0}$$

$$P(X = k) = p_k = \frac{1}{k!}\frac{d^k\phi}{dS^k}\bigg]_{S=0}$$

Extending this concept, we have

$$P[l_1 = k_1, ..., l_w = k_w | l_0]$$

$$= \frac{1}{k_1!}\frac{d^{k_1}\phi}{dS_1^{k_1}}\bigg]_{\substack{S_1=0 \\ S_i=1, i \neq 1}} \cdot \frac{1}{k_2!}\frac{d^{k_2}\phi}{dS_2^{k_2}}\bigg]_{\substack{S_2=0 \\ S_i=1, i \neq 2}} \cdots \frac{1}{k_w!}\frac{d^{k_w}\phi}{dS_w^{k_w}}\bigg]_{\substack{S_w=0 \\ S_i=1, i \neq w}} \quad (9.51)$$

where

$$\phi = E(S_1^{l_1} S_2^{l_2} ... S_w^{l_w} | l_0)$$

and

$$\{l_i\}\text{'s}, \ i = 1, 2, ... k \text{ are r.v.'s.}$$

UNIVARIATE p. g. f. of $l_x | l_0$
We assume

$$P\{l_{x=k} | l_0\} = \binom{l_0}{k}(p_{0x})^k (1 - p_{0x})^{l_0 - k}$$

i.e.

$$l_x \sim B(l_0, p_{0x})$$

(i.e. l_x is distributed as a Binomial variate with parameter l_0 and p_{0x}.)
The p.g.f. of $l_x | l_0$ is $E(S^{l_x} | l_0)$

$$= \sum_{l_x=0}^{l_0} S^{l_x}\binom{l_0}{l_x}(p_{0x})^{l_x} (1 - p_{0x})^{l_0 - l_x}$$

$$= \sum_{l_x=0}^{l_0} \binom{l_0}{l_x}(S p_{0x})^{l_x} (1 - p_{0x})^{l_0 - l_x}$$

$$= (1 - p_{0x}(1 - S))^{l_0}, \ |S| < 1 \quad (9.52)$$

Similarly the p.g.f. for conditional distribution of $l_j | l_i$ for $i < j$

$$= E(S^{l_j} | l_i)$$

$$= (1 - p_{ij} (1 - S))^{l_i}, \quad i < j \tag{9.53}$$

Precisely, in the same way it can be shown that

$$E (S_{i+1}^{l_{i+1}} | l_0, l_1 \dots l_i)$$

$$= \sum_{l_{i+1}=0}^{l_i} S_{i+1}^{l_{i+1}} \binom{l_i}{l_{i+1}} p_{i,i+1}^{l_{i+1}} (1 - p_{i,i+1})^{l_i - l_{i+1}}$$

$$= (1 - p_{i,i+1} + S_{i+1} p_{i,i+1})^{l_i}$$

$$= (1 - p_{i,i+1} (1 - S_{i+1}))^{l_i} \, E (S_{i+1}^{l_{i+1}} | l_i) \tag{9.54}$$

which shows that $\{l_i\}$ conforms to a simple Markov Chain.

9.3.1 Probability Distribution of the Number of Deaths

To obtain the joint distribution of $d_0, d_1, \dots d_w$ we note that

$$d_0 + d_1 + \dots + d_w = l_0$$

and $\qquad p_{00} q_0 + p_{01} q_1 + \dots + p_{0w} q_w = 1$ (p_{00} implies a live birth)

which shows that the joint probability distribution of d_i's is clearly multinomial. That is,

$$P [d_0 = \delta_0, d_1 = \delta_1, \dots d_w = \delta_w \mid l_0]$$

$$= \frac{l_0!}{\delta_0! \delta_1! \dots \delta_w!} (p_{00} q_0)^{\delta_0} (p_{01} q_1)^{\delta_1} \dots (p_{0w} q_w)^{\delta_w} \tag{9.55}$$

where $\qquad\qquad E (d_i \mid l_0) = l_0 p_{0i} q_i \tag{9.56}$

$$\text{Var} (d_i \mid l_0) = l_0 p_{0i} q_i (1 - p_{0i} q_i) \tag{9.57}$$

and $\qquad\qquad \text{Cov} (d_i, d_j) = -l_0 p_{0i} q_i p_{0j} q_j \tag{9.58}$

$$i, j = 0, 1, 2, \dots$$

9.3.2 Distribution of the Observed Residual Expectation of Life (given that the person has survived upto the age x_α) and the Sample Mean Length of Life

Let l_α be the number of survivors at age x_α. The future life time of l_α survivors may be regarded as a sample of l_α independent and identically distributed random variables (i.i.d.r.v.'s) $Y_{\alpha k}$, $k = 1, 2, \dots l_\alpha$.

The probability density of $Y_{\alpha k}$ is given by

$$f(y_\alpha) \, dy_\alpha = \left[\exp - \int_{x_\alpha}^{x_\alpha + y_\alpha} \mu(\tau) \, d\tau \right] \mu(x_\alpha + y_\alpha) \, dy_\alpha$$

from (9.18) for $y_\alpha \geq 0$

$$\therefore \quad \varepsilon_\alpha^0 = E(Y_\alpha) = \int_0^\infty y_\alpha \exp\left(-\int_{x_\alpha}^{x_\alpha + y_\alpha} \mu(\tau)\,d\tau\right)\mu(x_\alpha + y_\alpha)\,dy_\alpha$$

$$= \int_0^\infty \exp\left(-\int_{x_\alpha}^{x_\alpha + y_\alpha} \mu(\tau)\,d\tau\right)dy_\alpha \tag{9.59}$$

from (9.10) and (9.11) respectively.

$$\text{Var}(Y_\alpha) = \int_0^\infty (y_\alpha - \varepsilon_\alpha^0)^2\, f(y_\alpha)\,dy_\alpha = \sigma_{Y_\alpha}^2 \tag{9.60}$$

where e_α^0 and $f(y_\alpha)$ are given as above. Accordingly by the Lindberg-Levy central limit theorem of (vide appendix A - 4 (i)) of i.i.d.r.v.'s. The distribution of the sample mean

$$\overline{Y}_\alpha = \frac{1}{l_\alpha}\sum_{k=1}^{l_\alpha} Y_{\alpha k}$$

is given as $\overline{Y}_\alpha$ a.d. $N\left(\varepsilon_\alpha^0, \dfrac{\sigma_{Y_\alpha}^2}{l_\alpha}\right)$ where ε_α^0 and $\sigma_{Y_\alpha}^2$ are the same as given in (9.59) and (9.60).

It follows that the sample mean length of life is consistently asymptotic normal estimator (CAN) of the complete expectation of life following x_α.

9.4   Multivariate p.g.f. of $l_1, l_2 \dots l_w$ given l_0

To show that

$$\phi = E[S_1^{l_1}\, S_2^{l_2} \dots S_w^{l_w}\,|\,l_0] = \{1 - [p_{01}\,(1 - S_1) + p_{02}\,S_1\,(1 - S_2)$$

$$+\, p_{03}\,S_1\,S_2\,(1 - S_3) + \dots$$

$$\dots + p_{0w-1}\,S_1\,S_2 \dots S_{w-2}\,(1 - S_{w-1}) + p_{0w}\,S_1\,S_2 \dots S_{w-1}\,(1 - S_w)]\}^{l_0}$$

This result is due to Chiang , C.L.

Proof :

$$\phi = E(S_1^{l_1}\, S_2^{l_2} \dots S_w^{l_w}\,|\,l_0)$$

$$= E[S_1^{l_1}\, S_2^{l_2} \dots S_{w-1}^{l_{w-1}}\, E(s_w^{l_w}\,|\,l_{w-1})\,|\,l_0]$$

$$= E\left[S_1^{l_1}\, S_2^{l_2}\, S_{w-1}^{l_{w-1}} \sum_{l_w=0}^{l_{w-1}} \binom{l_{w-1}}{l_w} S_w^{l_w}\, p_{w-1}^{l_w}\,(1 - p_{w-1})^{l_{w-1} - l_w}\,|\,l_0\right]$$

$$= E\left[S_1^{l_1} S_2^{l_2} \dots S_{w-1}^{l_{w-1}} (1 - p_{w-1} + S_w \, p_{w-1})^{l_{w-1}} \, \middle| \, l_0\right]$$

Let us put

$$S_{w-1}(1 - p_{w-1}(1 - S_w)) = t_{w-1} \qquad (9.61)$$

$$\phi = E\left(S_1^{l_1} S_2^{l_2} \dots S_{w-2}^{l_{w-2}} \, t_{w-1}^{l_{w-1}} \, \middle| \, l_0\right)$$

We will prove the result by the Principle of Mathematical induction.

For $w = 1$
$$E\left(S_1^{l_1} \middle| l_0\right) = \sum_{l_1=0}^{l_1} S_1^{l_1} \binom{l_0}{l_1} p_{01}^{l_1} (1 - p_{01})^{l_0 - l_1}$$

$$= (1 - p_{01} + S_1 \, p_{01})^{l_1 + l_0 - l_1}$$

$$= (1 - p_{01} + S_1 \, p_{01})^{l_0}$$

Hence the result is true for $w = 1$.

Assuming that the result is true for $(w - 1)$ r.v.'s viz. $l_1, l_2, \dots l_{w-1}$, we may write

$$\phi = E\left[S_1^{l_1} S_2^{l_2} \dots S_{w-2}^{l_{w-2}} \, t_{w-1}^{l_{w-1}} \, \middle| \, l_0\right]$$

$$= [1 - \{ p_{01}(1 - S_1) + p_{02} S_1 (1 - S_2) + \dots +$$

$$+ p_{0w-1} S_1 S_2 \dots S_{w-2}(1 - t_{w-1})\}]^{l_0}$$

From (9.61) we have

$$1 - t_{w-1} = 1 - S_{w-1} + S_{w-1} p_{w-1} - S_{w-1} S_w p_{w-1}$$

$$= (1 - S_{w-1}) + p_{w-1}(S_{w-1} - S_{w-1} S_w) \qquad (9.62)$$

$\therefore$ From (9.62)

$$\phi = [1 - \{ p_{01}(1 - S_1) + p_{02}(1 - S_2) S_1 + \dots + p_{0w-1}$$

$$S_1 S_2 \dots S_{w-2} p_{w-1} S_{w-1})(1 - S_2)\}]^{l_0}$$

But $\quad p_{0w-1} p_{w-1} = p_{0w}$

$\Rightarrow$
$$\phi = E\left(S_1^{l_1} S_2^{l_2} \dots S_{w-1}^{l_{w-1}} S_w^{l_w} \middle| l_0\right)$$

$$= [1 - \{ p_{01}(1 - S_1) + p_{02} S_1 (1 - S_2) + \dots$$

$$+ p_{0w} S_1 S_2 \dots S_{w-1}(1 - S_w)\}]^{l_0} \qquad (9.63)$$

which proves the result for w number of r.v.'s viz. $l_1, l_2, \dots l_w$.

Hence by mathematical induction the result is true for any finite number of r.v.'s.

9.4.1 Application of the Result

We have,
$$\phi = [1 - \{ p_{01}(1 - S_1) + p_{02} S_1 (1 - S_2) + \dots$$

$$+ p_{0w} S_1 S_2 \dots S_{w-1} (1 - S_w) \}]^{l_0}$$

$$\therefore \quad \frac{\partial \phi}{\partial S_1} = l_0 [1 - \{ p_{01} (1 - S_1) + p_{02} S_1 (1 - S_2) + \dots$$

$$+ p_{0w} S_1 S_2 \dots S_{w-1} (1 - S_w) \}]^{l_0 - 1}$$

$$\times [p_{01} - p_{02} (1 - S_2) - \dots - p_{0w} S_2 \dots S_{w-1} (1 - S_w)]$$

$$\frac{\partial \phi}{\partial S_1} \bigg]_{S_i = 1} = l_0 p_{01} = E (l_1 \mid l_0) \tag{9.64}$$

Similarly

$$\frac{\partial \phi}{\partial S_i} \bigg]_{\substack{S_i = 1 \\ S_j = 1 \\ j \neq i}} = E (l_i) = l_0 p_{0i} \tag{9.65}$$

$$\therefore \quad \frac{\partial \phi}{\partial S_i} \bigg]_{\substack{S_i = 1 \\ S_j = 1 \\ j \neq i}} + \frac{\partial^2 \phi}{\partial S_i^2} \bigg]_{S_i = 1} = l_0 (l_0 - 1) p_{0i}^2 + l_0 p_{0i}$$

$$\Rightarrow \quad \text{Var} (l_i) = l_0 (l_0 - 1) p_{0i}^2 + l_0 p_{0i} - l_0^2 p_{0i}^2$$

$$= l_0 p_{0i} (1 - p_{0i}).$$

Now $\quad \dfrac{\partial^2 \phi}{\partial S_1 \partial S_2} = l_0 (l_0 - 1) [1 - \{ p_{01} (1 - S_1) + p_{02} S_1 (1 - S_2) + \dots$

$$+ p_{0w} S_1 S_2 \dots S_{w-1} (1 - S_w) \}]^{l_0 - 2}$$

$$\times [p_{01} - p_{02} (1 - S_2) - \dots p_{0w} S_2 \dots S_{w-1} (1 - S_w)]$$

$$[p_{02} S_1 - p_{03} S_1 (1 - S_3) \dots - p_{0w} S_1 S_3 \dots S_{w-1} (1 - S_w)]$$

$$+ l_0 [1 - \{ p_{01} (1 - S_1) + p_{02} S_1 (1 - S_2) + \dots$$

$$+ p_{0w} S_1 S_2 \dots S_{w-1} (1 - S_w) \}]^{l_0 - 1}$$

$$\times [p_{02} - p_{03} (1 - S_3) - \dots - p_{0w} S_3 S_4 \dots S_{w-1} (1 - S_w)] \tag{9.66}$$

$$\therefore \quad \frac{\partial^2 \phi}{\partial S_1 \partial S_2} \bigg]_{\substack{S_1 = S_2 = 0 \\ S_i = 1 \\ i \neq 1, 2}} = l_0 (l_0 - 1) p_{01} p_{02} + l_0 p_{02}$$

$$= E (l_1 l_2 \mid l_0)$$

$$\text{Cov} (l_1, l_2 \mid l_0) = E (l_1 l_2 \mid l_0) - E (l_1 \mid l_0) E (l_2 \mid l_0)$$

$$= l_0 (l_0 - 1) p_{01} p_{02} + l_0 p_{02} - (l_0 p_{01}) (l_0 p_{02})$$

$$= l_0 p_{02} - l_0 p_{01} p_{02} = l_0 p_{02} (1 - p_{01}) \tag{9.67}$$

Generalizing,

$$\text{Cov}\,(l_i, l_j \mid l_0) \; = \; l_0 p_{0j}(1 - p_{0i}) \text{ for } i < j \tag{9.68}$$

$$\text{Corr.}\,(l_i, l_j \mid l_0) \; = \; \frac{l_0\, p_{0j}\,(1 - p_{0i})}{\sqrt{l_0\, p_{0i}\,(1 - p_{0i})}\,\sqrt{l_0\, p_{0j}\,(1 - p_{0j})}}$$

$$= \; \sqrt{\frac{p_{0j}\,(1 - p_{0i})}{p_{0i}\,(1 - p_{0j})}} \;\; \forall \; i < j \tag{9.69}$$

9.5 Estimation of Survival Probability $p_j (\, j = 1, 2,..)$ by the Method of Maximum Likelihood

Following Chiang (1968), let us define a random variable $\varepsilon_{i\beta}$ as follows:

$$\left.\begin{aligned} \varepsilon_{i\beta} \; &= \; 1 \text{ if } \beta \text{th person dies between } i \text{ to } i + 1 \\[4pt] &= \; 0, \text{otherwise}, \beta = 1, 2, \dots l_0, i = 0, 1, 2, .. \end{aligned}\right\} \tag{9.70}$$

Then the likelihood function L of the sample is given by

$$L \; = \; \prod_{i=0}^{\infty} \prod_{\beta=1}^{l_0} [p_{0i}\,(1 - p_i)]^{\varepsilon_{i\beta}}$$

$$= \; \prod_{i=0}^{\infty} [p_{0i}\,(1 - p_i)]^{\sum_{\beta=1}^{l_0} \varepsilon_{i\beta}} \tag{9.71}$$

Again

$$\sum_{\beta=1}^{l_0} \varepsilon_{i\beta} \; = \; d_i \tag{9.72}$$

Putting (9.72) in (9.71), we get

$$L \; = \; \prod_{i=0}^{\infty} [p_{0i}\,(1 - p_i)]^{d_i} \tag{9.73}$$

$$\Rightarrow \log_e L \; = \; \sum_{i=0}^{\infty} d_i\, [\log_e p_{0i} + \log_e (1 - p_i)]$$

$$\log_e L \; = \; d_0\, [\log_e p_{00} + \log_e (1 - p_0)] + d_1 [\log_e p_{01} + \log_e (1 - p_1)]$$

$$+ \; \dots + d_j\, \log_e p_{0j} + d_j\, \log_e (1 - p_j) + d_{j+1}\, \log_e p_{0j+1}$$

$$+ \; d_{j+1}\, \log_e (1 - p_{j+1}) + d_{j+2}\, \log_e p_{0j+2} + d_{j+2}\, \log_e (1 - p_{j+2}) + \dots$$

Setting $\dfrac{\partial \log_e L}{\partial p_j} = 0$, we have

$$-\frac{d_j}{1-p_j}+\frac{d_{j+1}}{p_j}+\frac{d_{j+2}}{p_j}+\frac{d_{j+3}}{p_j}+\dots=0$$

$$\Rightarrow \quad -\frac{d_j}{1-p_j}+\frac{\displaystyle\sum_{i=j+1}^{\infty}d_i}{p_j}=0$$

$$\Rightarrow \quad \frac{\displaystyle\sum_{i=j+1}^{\infty}d_i}{p_j}=\frac{d_j}{1-p_j}$$

$$\Rightarrow \quad \frac{d_j}{\displaystyle\sum_{i=j+1}^{\infty}d_i}=\frac{1-p_j}{p_j}$$

$$\Rightarrow \quad \frac{\displaystyle\sum_{i=j}^{\infty}d_i}{\displaystyle\sum_{i=j+1}^{\infty}d_i}=\frac{1}{p_j}$$

$$\Rightarrow \quad \hat{p}_j=\frac{\displaystyle\sum_{i=j+1}^{\infty}d_i}{\displaystyle\sum_{i=j}^{\infty}d_i}$$

or

$$\hat{p}_j=\frac{l_{j+1}}{l_j} \tag{9.74}$$

$\therefore\ \hat{p}_j=\dfrac{l_{j+1}}{l_j}$ is the *maximum likelihood estimator* (m.l.e.) of p_j.

To prove $\hat{p}_j$ is unbiased for p_j.

$$E(\hat{p}_j)=E\left(\frac{l_{j+1}}{l_j}\right)$$

$$=E\left(\frac{1}{l_j}E(l_{j+1}|l_j)\right)$$

$$=E\left(\frac{1}{l_j}\cdot l_j\,p_j\right) \qquad (\because E(l_{j+1}|l_j)=l_j\,p_j)$$

$$= E(p_j) = p_j$$

$$\Rightarrow E(\hat{p}_j) = p_j \tag{9.75}$$

$\Rightarrow \hat{p}_j$ is unbiased for p_j.

Consider

$$E(\hat{p}_j^2) = E\left[\frac{l_{j+1}^2}{l_j^2}\right]$$

$$= E\left[\frac{1}{l_j} E\left(\frac{l_{j+1}^2}{l_j}\bigg| l_j\right)\right]$$

$$= E\left[\frac{1}{l_j^2} E(l_{j+1}^2 | l_j)\right]$$

$$= E\left[\frac{1}{l_j^2}\{l_j\, p_j\,(1-p_j)+l_j^2\, p_j^2\}\right]$$

$$= E\left[\frac{1}{l_j}\, p_j\,(1-p_j)+p_j^2\right]$$

Since

$$E(l_{j+1} | l_j) = l_j p_j$$

$$V(l_{j+1} | l_j) = l_j p_j\,(1-p_j)$$

$$E(l_{j+1}^2 | l_j) = l_j\, p_j\,(1-p_j)+l_j^2\, p_j^2$$

$$\text{Var}(\hat{p}_j) = E(\hat{p}_j^2)-[E(\hat{p}_j)]^2$$

$$= E\left(\frac{1}{l_j}\right) p_j\,(1-p_j)+p_j^2-p_j^2$$

$$= E\left(\frac{1}{l_j}\right) p_j\,(1-p_j) \tag{9.76}$$

$\therefore$

$$\left.\begin{array}{l} E(\hat{p}_j) = p_j \\[2mm] \text{Var}(\hat{p}_j) = E\left(\dfrac{1}{l_j}\right) p_j\,(1-p_j) \end{array}\right\} \tag{9.77}$$

9.5.1 Crámer-Rao Lower Bound for the Estimator of p_j

We have

$$\frac{\partial \log_e L}{\partial p_j} = \frac{-d_j}{1-p_j}+\frac{\displaystyle\sum_{i=j+1}^{\infty} d_i}{p_j}$$

$$\frac{\partial^2 \log_e L}{\partial p_j^2} = \frac{-d_j}{(1-p_j)^2} - \frac{\sum_{i=j+1}^{\infty} d_i}{p_j^2}$$

$$= \frac{-d_j}{(1-p_j)^2} - \frac{l_{j+1}}{p_j^2} \left(\because \sum_{i=j+1}^{\infty} d_i = l_{j+1} \right)$$

or $\quad -\dfrac{\partial^2 \log_e L}{\partial p_j^2} = \dfrac{d_j}{(1-p_j)^2} + \dfrac{l_{j+1}}{p_j^2}$

$$= \frac{l_j - l_{j+1}}{(1-p_j)^2} + \frac{l_{j+1}}{p_j^2} \quad (\because d_j = l_j - l_{j+1})$$

$\Rightarrow \quad E\left[\dfrac{-\partial^2 \log_e L}{\partial p_j^2} \right] = \dfrac{E(l_j) - E(l_{j+1})}{(1-p_j)^2} + \dfrac{E(l_{j+1})}{p_j^2}$

$$= \frac{l_0 \, p_{0j} - l_0 \, p_{0j+1}}{(1-p_j)^2} + \frac{l_0 \, p_{0j+1}}{p_j^2}$$

$$= \frac{l_0 \, p_{0j} - l_0 \, p_{0j} \, p_j}{(1-p_j)^2} + \frac{l_0 \, p_{0j}}{p_j} \quad (\because p_{0j+1} = p_{0j} \, p_j)$$

$$= \frac{l_0 \, p_{0j}}{(1-p_j)} + \frac{l_0 \, p_{0j}}{p_j}$$

$$= \frac{l_0 \, p_{0j} \, (p_j + 1 - p_j)}{p_j \, (1-p_j)}$$

$\Rightarrow \quad E\left[\dfrac{-\partial^2 \log_e L}{\partial p_j^2} \right] = \dfrac{l_0 \, p_{0j}}{p_j \, (1-p_j)}$ \hfill (9.78)

Crámer Rao Lower Bound (vide Appendix A–2) is given by

$$\frac{1}{E\left(\dfrac{-\partial^2 \log_e L}{\partial p_j^2} \right)} = \frac{p_j \, (1-p_j)}{l_0 \, p_{0j}}$$

$$= \frac{p_j \, (1-p_j)}{E(l_j)}$$

Variance of m.l.e. $\hat{p}_j = E\left(\dfrac{1}{l_j} \right) p_j \, (1 - p_j)$

Since
$$E\left(\frac{1}{l_j}\right) \geq \frac{1}{E\,(l_j)} \tag{9.79}$$

$$\mathrm{Var}\,(\hat{p}_j) = E\left(\frac{1}{l_j}\right) p_j\,(1-p_j) \geq \frac{p_j\,(1-p_j)}{E\,(l_j)} \tag{9.80}$$

∴ Variance of m.l.e. $\hat{p}_j \geq$ Crámer Rao lower bound
and Relative efficiency of m.l.e. $\hat{p}_j$

$$= \frac{(E\,(l_j))^{-1}\,p_j\,(1-p_j)}{E\left(\dfrac{1}{l_j}\right) p_j\,(1-p_j)} \tag{9.81}$$

We shall now show that the maximum likelihood estimate provides the minimum variance estimator of p_j and as such the Crámer-Rao Lower bound which is less than the maximum likelihood estimate is not attained.

9.5.2 Sufficiency of $\hat{p}_j$

We reconsider the likelihood function of the sample

$$L = \prod_{i=0}^{\infty} \prod_{\beta=1}^{l_0} \{p_{0j}\,(1-p_i)\}^{\varepsilon_{i\beta}}$$

where $\varepsilon_{i\beta}$ is a random variable taking value 1 when the βth person ($\beta = 1, 2, ...l_0$) dies in between i to $(i + 1)$ $(i = 0, 1, 2,..)$ and takes value zero when he does not die.

Also
$$\sum_{\beta=1}^{l_0} \varepsilon_{i\beta} = d_i \quad (i = 0, 1, 2, ...)$$

$$L = \prod_{i=0}^{\infty} \{p_{0i}\,(1-p_i)\}^{\sum\limits_{\beta=1}^{l_0}\varepsilon_{i\beta}}$$

$$= \prod_{i=0}^{\infty} \{p_{0i}\,(1-p_i)\}^{d_i}$$

$$= (p_{00}^{d_0}\,(1-p_0)^{d_0})\,(p_{01}^{d_1}\,(1-p_1)^{d_1})\,p_{02}^{d_2}\,(1-p_2)^{d_2}\,p_{03}^{d_3}\,(1-p_3)^{d_3}\,...$$

$$= (p_{00}\,p_0)^{d_1}\,(p_{00}\,p_0\,p_1)^{d_2}\,(p_{00}\,p_0\,p_1\,p_2)^{d_2}\,(1-p_0)^{d_0}$$
$$(1-p_1)^{d_1}\,(1-p_2)^{d_2}\,(1-p_3)^{d_3}\,...\,\text{(assuming } p_{00} = 1)$$

$$= (p_{01})^{\sum\limits_{i=1}^{\infty} d_i} (p_1)^{\sum\limits_{i=2}^{\infty} d_i} (p_2)^{\sum\limits_{i=3}^{\infty} d_i} \ldots (1-p_0)^{d_0} (1-p_1)^{d_1} (1-p_2)^{d_2} \ldots$$

$$= (p_0^{l_1} p_1^{l_2} p_2^{l_3} \ldots)(1-p_0)^{d_0} (1-p_1)^{d_1} (1-p_2)^{d_2} \ldots$$

$$\left(\because \sum_{i=1}^{\infty} d_i = l_1, \ \sum_{i=2}^{\infty} d_i = l_2 \text{ etc.} \right)$$

$$= \prod_{i=0}^{\infty} p_i^{l_{i+1}} (1-p_i)^{d_i}$$

$$= \prod_{i=0}^{\infty} p_i^{l_{i+1}} (1-p_i)^{l_i - l_{i+1}} \quad (\because d_i = l_i - l_{i+1}) \tag{9.82}$$

Note that l_i's are random variables and p_i's are set of parameters in the joint distribution probability function.

Therefore (9.82) can be written as

$$L = \prod_{i=0}^{\infty} \frac{l_i!}{d_i!(l_{i+1})!} p_i^{l_{i+1}} (1-p_i)^{l_i - l_{i+1}} \prod_{i=0}^{\infty} \frac{d_i!(l_{i+1})!}{l_i!}$$

$$= \prod_{i=0}^{\infty} \frac{l_i! \, p_i^{l_{i+1}}}{(l_i - l_{i+1})! \, l_{i+1}!} (1-p_i)^{l_i - l_{i+1}} \prod_{i=0}^{\infty} \frac{\left(\sum\limits_{\beta=1}^{l_0} \varepsilon_{i\beta}\right)! \left(l_i - \sum\limits_{\beta=1}^{\infty} \varepsilon_{i\beta}\right)!}{l_i!}$$

$$\left(\because d_i = l_i - l_{i+1} = \sum_{\beta=1}^{l_0} \varepsilon_{i\beta} \right) \tag{9.83}$$

Therefore 'L' is capable of being factorised into the distinct products. The first being a function of the parameters p_i and (l_i, l_{i+1}) and the second being independent of p_i (a function of r.v.'s only). By Neyman's Factorisation Criterion (vide Appendix A–3) it shows that l_i and l_{i+1} are jointly sufficient for $\tilde{p}_i$. Also if $\tilde{p}_j$ be any other estimate of p_j then $E(\tilde{p}_j \mid l_j, l_{j+1})$ must be independent of p_j.

Let

$$E(\tilde{p}_j \mid l_j, l_{j+1}) = f(l_j, l_{j+1}) \tag{9.84}$$

Let $\tilde{p}_j$ be unbiased for p_j, i.e.

$$E[\tilde{p}_j] = p_j \tag{9.85}$$

Now we shall compare the estimator of $\tilde{p}_j$ with the maximum likelihood estimator (m.l.e) $\hat{p}_j$.

Since

$$E(\hat{p}_j) = p_j$$

$$E(\bar{p}_j) = E[E(\tilde{p}_j | l_j, l_{j+1})]$$

$$= E[f(l_j, l_{j+1})] \tag{9.86}$$

To obtain $E(f(l_j, l_{i+1}))$, we require joint probability density function of l_j and l_{j+1}. The joint p.d.f. of l_j and l_{j+1} subject to the condition $l_j \neq 0$ is given by

$$\frac{\binom{l_0}{l_j} p_{0j}^{l_j} (1 - p_{0j})^{l_0 - l_j} \binom{l_j}{l_{j+1}} p_j^{l_{j+1}} (1 - p_j)^{l_j - l_{j+1}}}{[1 - (1 - p_{0j})^{l_0}]} \tag{9.87}$$

$$(\text{Since } P[l_j \neq 0] = 1 - (1 - p_{0j})^{l_0})$$

From (9.86) and (9.87), it follows that

$$p_j = E(\bar{p}_j)$$

$$= \frac{\sum\limits_{l_j=0}^{l_0} \sum\limits_{l_{j+1}=0}^{l_0} f(l_j, l_{j+1}) \binom{l_0}{l_j} p_{0j}^{l_j} (1 - p_{0j})^{l_0 - l_j} \binom{l_j}{l_{j+1}} p_j^{l_{j+1}} (1 - p_j)^{l_j - l_{j+1}}}{[1 - (1 - p_{0j})^{l_0}]}$$

$$\tag{9.88}$$

$$\Rightarrow \sum\limits_{l_j=0}^{l_0} \sum\limits_{l_{j+1}=0}^{l_0} f(l_j, l_{j+1}) p(l_j, l_{j+1}) = p_j \tag{9.89}$$

Also we have shown that

$$E(\hat{p}_j) = E\left(\frac{l_{j+1}}{l_j}\right) = p_j$$

$$\Rightarrow \sum\limits_{l_j=0}^{l_0} \sum\limits_{l_{j+1}=0}^{l_j} \left(\frac{l_{j+1}}{l_j}\right) p(l_j, l_{j+1}) = p_j \tag{9.90}$$

Since (9.89) is an identity in p_j it has unique solution

$$f(l_j, l_{j+1}) = \frac{l_{j+1}}{l_j} = \hat{p}_j$$

$$\Rightarrow E(\bar{p}_j | l_j, l_{j+1}) = \hat{p}_j \tag{9.91}$$

By Rao-Blackwell Theorem (vide appendix A–3), it is therefore shown that $\hat{p}_j = \dfrac{l_{j+1}}{l_j}$ is MVUE (Minimum Variance Unbiased Estimator).
Also

$$\text{Var}(\bar{p}_j) = E(\bar{p}_j - p_j)^2$$

$$= E\,(\tilde{p}_j - \hat{p}_j + \hat{p}_j - p_j)^2$$

$$= E\,(\tilde{p}_j - \hat{p}_j)^2 + E\,(\hat{p}_j - p_j)^2 + 2E\,(\tilde{p}_j - \hat{p}_j)(\hat{p}_j - p_j)$$

$$= \text{Var}\,(\hat{p}_j) + E\,(\tilde{p}_j - \hat{p}_j)^2 \qquad (9.92)$$

Since $\qquad E\,(\tilde{p}_j - \hat{p}_j)(\hat{p}_j - p_j)$

$$= E\,[E\,(\tilde{p}_j - \hat{p}_j)(\hat{p}_j - p_j)|l_j, l_{j+1}]$$

$$= E\,[E\,(\tilde{p}_j - \hat{p}_j)|(l_j, l_{j+1})(\hat{p}_j - p_j)]$$

But $\qquad E\,(\tilde{p}_j\,|l_j, l_{j+1}) = \hat{p}_j = \dfrac{l_{j+1}}{l_j}$

Hence $\qquad E\,(\tilde{p}_j - \hat{p}_j)(\hat{p}_j - p_j) = 0$

$\therefore \qquad \text{Var}\,(\tilde{p}_j) = \text{Var}\,(\hat{p}_j) + E\,(\tilde{p}_j - \hat{p}_j)^2$

$$\geq \text{Var}\,(\hat{p}_j) \quad (\because E\,(\tilde{p}_j - \hat{p}_j)^2 \geq 0) \qquad (9.93)$$

Hence there exists no unbiased estimate whose variance is lower than that of $\hat{p}_j$, the m.l.e. of p_j. Therefore $\hat{p}_j$ provides Minimum Variance Unbiased Estimate (MVUE) of p_j and Crámer Rao Lower bound is not realized.

Example 9.1 Show that

$$\hat{q}_i = \frac{n_i\,M_i}{[1 + (1 - a_i)\,n_i\,M_i]} \qquad (9.94)$$

where a_i is the average number of years lived in the interval $(x_i, x_i + 1)$ for those who die in it; and hence obtain the sampling variances of $\hat{q}_i, \hat{p}_{ij}$ in respect of a (i) Current life table (ii) Generation (cohort) life table.

Solution:

We may note the sampling variance of $\hat{q}_i$ viz.

$$\sigma_{\hat{q}_i}^2 = \text{Var}\,(\hat{q}_i) = \text{Var}\,(1 - \hat{p}_i) = \text{Var}\,(\hat{p}_i) = \sigma_{\hat{p}_i}^2 \qquad (9.95)$$

The sample variance denoted by $S_{\hat{q}_i}^2$ and $S_{\hat{p}_i}^2$ will also be consequently equal viz $S_{\hat{q}_i}^2 = S_{\hat{p}_i}^2$.

First we consider a current life table. For a current life table

$$\hat{q}_i = \frac{D_i}{N_i}$$

$D_i \equiv$ number of deaths in (x_i, x_{i+1}) given that N_i persons are alive at age x_i.

Note that if we assume that all individuals in the group of population N_i have the same probability of dying on the interval. Then D_i is a binomial variate and $\hat{q}_i$ is the estimated binomial proportion. We have,

$$S_{D_i}^2 = N_i\, \hat{q}_i\, (1 - \hat{q}_i)$$

and

$$S_{\hat{q}_i}^2 = \frac{\hat{q}_i\,(1 - \hat{q}_i)}{N_i}$$

Also

$$N_i = \frac{D_i}{\hat{q}_i}$$

$$\Rightarrow \qquad S_{\hat{q}_i}^2 = \frac{\hat{q}_i^2\,(1 - \hat{q}_i)}{D_i} \tag{9.96}$$

where

$$\hat{q}_i = \frac{D_i}{N_i} \tag{9.97}$$

and

$$M_i = \frac{D_i}{(N_i - D_i)\, n_i + a_i\, n_i\, D_i} \tag{9.98}$$

Eliminating N_i from (9.97) and (9.98), we have

$$\hat{q}_i = \frac{n_i\, M_i}{1 + (1 - a_i)\, n_i\, M_i} \quad \text{which proves (9.94)} \tag{9.99}$$

Now substituting (9.99) in (9.97) $\Rightarrow$

$$S_{\hat{q}_i}^2 = \frac{n_i^2\, M_i^2 \left(1 - \dfrac{n_i\, M_i}{1 + (1 - a_i)\, n_i\, M_i}\right)}{\left[1 + (1 - a_i)\, n_i\, M_i\right]^2 D_i}$$

$$= \frac{n_i^2\, M_i^2\, [1 - a_i\, n_i\, M_i]}{[1 + (1 - a_i)\, n_i\, M_i]^3\, D_i} = \frac{n_i^2\, M_i\, [1 - a_i\, n_i\, M_i]}{[1 + (1 - a_i)\, n_i\, M_i]^3 \left(\dfrac{D_i}{M_i}\right)}$$

$$\left(\because M_i = \frac{D_i}{P_i} \right)$$

$$= n_i^2\, M_i\, [1 - a_i\, n_i\, M_i] / [1 + (1 - a_i)\, n_i\, M_i]^3\, P_i \tag{9.100}$$

If we take the length of the age interval as unity i.e. $n_i = 1$, the subscripts 'i' in the relation (9.100) is replaced by x for obtaining formula for single year of age

$$S_{\hat{q}_x}^2 = \frac{M_x\,(1 - a_x'\, M_x)}{P_x\,[1 + (1 - a_x')\, M_x]^3} \tag{9.101}$$

while replacing a_i by a_x'.

Taking $a_x' \cong \dfrac{1}{2}$, we have

$$S_{\hat{q}_x}^2 = \frac{M_x\,(1 - \frac{1}{2}\, M_x)}{P_x\,[1 + \frac{1}{2}\, M_x]^3}$$

$$= \frac{4M_x(2-M_x)}{P_x(2+M_x)^3} \tag{9.102}$$

which provides an expression of the sampling variance of $\hat{q}_x$.

Next to obtain the estimated sampling variance $S^2_{\hat{P}_i}$ we proceed as follows

$$\hat{P}_{ij} = \hat{P}_i\,\hat{P}_{i+1}\cdots\hat{P}_{j-1} \;(j>i)\;(i,j=0,1,2,..) \tag{9.103}$$

where the sample proportions

$$\hat{P}_i = 1 - \hat{q}_i,\; (i=0,1,2,..) \tag{9.104}$$

are based on the estimates of the age specific death rates and $\hat{p}_i$ and $\hat{p}_k$ $(i \ne k)$ are assumed to be independent of each other. Therefore, by the formula of large sample variance of a function of an estimate

$$\hat{S}^2_{\hat{P}_i} = \sum_{k=1}^{j-1}\left[\frac{\partial(\hat{P}_{ij})}{\partial\hat{P}_k}\right]^2_E S^2_{P_k} \tag{9.105}$$

(vide appendix A –1)

$$\hat{P}_{ij} = \hat{P}_i\,\hat{P}_{i+1}\cdots\hat{P}_{j-1}\;;\,j-1\ge i$$

$$\log\hat{P}_{ij} = \log\hat{P}_i + \log\hat{P}_{i+1} + ... + \log\dot{P}_{j-1}$$

$$\frac{\partial\log\hat{P}_{ij}}{\partial\hat{P}_k} = \frac{1}{\hat{P}_k}$$

$$\frac{\partial\hat{P}_{ij}}{\partial\hat{P}_k} = \frac{\partial\hat{P}_{ij}}{\partial\log\hat{P}_{ij}}\cdot\frac{\partial\log\hat{P}_{ij}}{\partial\hat{P}_k}$$

$$= \hat{P}_{ij}\cdot\frac{1}{\hat{P}_k}$$

$$\left(\frac{\partial(\hat{P}_{ij})}{\partial\hat{P}_k}\right)_E = P^2_{ij}\sum_k\frac{1}{P^2_k}$$

$$\therefore\; S^2_{\hat{P}_i} = P^2_{ij}\sum_{k=1}^{j-1}(P_k)^{-2}\,S^2_{P_k} \tag{9.106}$$

Cohort life table:

In generation life table

$$\hat{q}_i = \frac{d_i}{l_i} \;\text{ and }\; \hat{p}_i = \frac{l_{i+1}}{l_i}$$

These are ordinary binomial proportions with sampling variances

$$S^2_{\hat{q}_i} = S^2_{\hat{P}_i} = \frac{1}{l_i}\,\hat{p}_i\,\hat{q}_i \tag{9.107}$$

Again as before, as in current life table, we have also for a cohort life table

$$\hat{p}_{ij} = \hat{p}_i \, \hat{p}_{i+1} \cdots \hat{p}_{j-1} \tag{9.108}$$

$$j - 1 \geq i$$

$$i, j = 0, 1, 2, \ldots$$

[Note that $\hat{p}_{ij}$ is a general form of $\hat{p}_i$] For example if we put $j = i + 1$.

$$\hat{p}_{i,i+1} = \hat{p}_i$$

$$\therefore \qquad \mathrm{Var}\,(\hat{p}_{ij}) = S^2_{\hat{p}_{ij}} = \frac{1}{l_i} \, \hat{p}_{ij} \, (1 - \hat{p}_{ij}) \tag{9.109}$$

$$i < j, i \quad i, j = 0, 1, 2 \ldots$$

In view of (9.108) and $\hat{p}_i$'s being linearly uncorrelated we have the result (9.106) viz.

$$S^2_{\hat{p}_{ij}} = \hat{p}^2_{ij} \sum_{k=1}^{j-1} p_k^{-2} S^2_{\hat{p}_k} \tag{9.110}$$

$$\Rightarrow \qquad S^2_{\hat{p}_{ij}} = \hat{p}^2_{ij} \sum_{k=1}^{j-1} p_k^{-2} \frac{1}{l_k} \, \hat{p}_k \, (1 - \hat{p}_k) \tag{9.111}$$

from (9.107)

Further, it is immediately verified that

$$\frac{1}{l_i} \, \hat{p}_{ij} \, (1 - \hat{p}_{ij}) = \hat{p}^2_{ij} \sum_{k=1}^{j-1} p_k^{-2} \frac{1}{l_k} \, \hat{p}_k \, (1 - \hat{p}_k)$$

$$= S^2_{\hat{p}_{ij}} \tag{9.112}$$

Thus the results of the cohort life table remain the same as that of a current life table (vide Chiang C.L. (1968)).

References

1. Berclay G.W. (1968): *Techniques of Population Analysis*, John Wiley & Sons, New York.
2. Chiang C.L. (1968): *Introduction to Stochastic Processes in Biostatistics*, John Wiley & Sons, New York.
3. Cox C.R. (1976): *Demography*, Cambridge University Press.
4. Jafee A.J. (1966): *Handbook of Statistical Methods for Demographers*. U.S. Govt. Printing Press, Washington.
5. Keyfitz Nathan (1966): A life table that agrees with the data, *Journal of American Statistical Association*, Vol. 61, page 305-12.
6. Keyfitz Nathan (1968): A life table that agrees with the data II, *Journal of American*

Statistical Association, Vol. 63, page 1253-68.

7. Keyfitz N (1977): *Introduction to the Mathematics in Population with revisions*, Addison-Wesley, London.

8. Logan W.P. (1953): *The measurement of Infant Mortality, Population Bulletin of the United Nations*, No. 3, page 30-55, New York.

9. Makeham W.H. (1960): On the law of Mortality and the construction of Annuity tables. *Journal of the Institute of Actuaries*, Vol. 8.

10. Mathen K.K. and Poti S.J. (1955): An adjustment for the changing of Birth rates on Infant Mortality rates, *Sañkhya* Vol. 15, Part IV, page 417-22.

11. Poti S.J. and Biswas S (1963): A Study of Child Health during the first year of life, *Sankhya*, Vol. 26, Series I and II, page 35-120.

12. Rogers A (1975): *Introduction to Multiregional Demography*, John Wiley & Sons, New York.

13. Spiegelman M (1980): *Introduction to Demography* (revised edition). Harvard University Press, Cambridge University Press, Cambridge.

14. Tetly R. (1950): *Actuarial Statistics*, Vol. 3, Statistics and Graduation, The Institute of Actuaries and the Faculty of Actuaries, Cambridge.

15. Wolfenden H.H. (1954): *Population Statistics and their compilation*, The University of Chicago Press, Chicago.

16. United Nations (1955): *Age sex patterns of Mortality model*. Life tables for Under developed countries, New York.

Techniques of Stochastic Processes in Fertility Analysis

10.0 Introduction

Fertility corresponds to the additive component of the growth of a population. The inherent biological capacity to produce births is known as "Fecundity"; whereas the realized level of producing births under a given socio-economic and cultural set up is known as 'Fertility'. Unlike mortality fertility measures are estimable mostly through Female Population who are exposed to the risk of child bearing. Again certain measures of fertility as 'Net Reproduction Rate' are associated with the mortality condition in the sense to what extent mortality in the child-bearing period affects total fertility performance of the mothers. Below we present a few measures of fertility most of which, can be taken up independent of mortality situation.

10.1 Indices of Fertility Measures

1. Crude Birth Rate (CBR)
2. General Fertility Rate (GFR)
3. Age-Sex-Specific Fertility Rate (ASSFR)
4. Total Fertility Rate (TFR)
5. Gross Reproduction Rate (GRR)
6. Net Reproduction Rate (NRR)

10.1.1 Crude Birth Rate (CBR)

In a particular locality in the period Z (calender year), let the total number of births be B^Z and the mean population measured in the middle of the calender year Z be P^Z. Then a fertility index is given by

$$\text{CBR} = \frac{B^Z}{P^Z} \times k \qquad (10.1)$$

where $k = 10^3$ usually
B^Z = Number of births in the Calender year Z.

$$P^Z = \text{Mid-year population of the Calender year } Z.$$

Remarks :

(1) It is a very crude measure of fertility but easily understandable index and easy to calculate. It is not necessary to have the births in different sectors of the population.

(2) It is not a probability rate since whole population is not exposed to the risk of giving births. The whole of male population and the part of the female population not exposed to the risk of child bearing should have been excluded.

(3) CBR may not be a suitable index for comparing the fertility status of two communities. For example, two populations may have the same CBR. This does not mean that the two populations have the same fertility status. The age distribution of the female population or the distortion or imbalance in the sex ratios in the population may mask the actual fertility picture presented in the form of CBR.

10.1.2 General Fertility Rate (GFR)

It is defined as

$$\text{GFR} = \frac{\text{Total number of live births in a given period in a given region}}{\substack{\text{Mid-year female population in the child bearing age-group in} \\ Z\text{th calender year}}}$$

Symbolically,
$$\text{GFR} = \frac{B^Z}{{}_r W^Z} \times k \tag{10.2}$$

where B^Z = number of live births in the calender year Z

${}_r W^Z$ = female population in the reproductive age group in the calender year Z

$k = 10^3$ (usually)

Remarks:

(i) It is a probability rate because the denominator corresponds to the female population who are exposed to the risk of child bearing.

(ii) It does not enable us to study the variation of fertility by ages as it overlooks the age composition of the female population.

(iii) It has also a defect of non-comparability in respect of time and place. It may so happen that for two places 'A' and 'B', general fertility rates are the same. Even in that case we can hardly say that the two places have the same fertility status; for the proportion of young females (who are under greater risk of giving birth) for 'A' may be different than from B.

10.1.3 Age-specific Fertility Rate (ASFR)

Age-specific fertility rate between the age-group x to $(x + n)$ is given by

$$\text{ASFR} = \frac{{}_nB_x^Z}{{}_nW_x^Z} \times k \qquad (10.3)$$

where ${}_nB_x^Z$ = number of live births to women of age x to $x + n$ in the calender year Z

${}_nW_x^Z$ = number of women of age x to $x + n$ in the calender year Z

n = 5 (usually)

k = 1000 (usually)

Remarks :
- (i)   It is a probability rate and is not a heterogeneous figure.
- (ii)   It enables us to compare the fertility status of two different populations corresponding to the same age group.
- (iii)   The fertility status of two populations cannot be compared on the basis of two sets of age specific fertility rates unless there is a consolidated fertility index for each one of them.

10.1.4   Total Fertility Rate (TFR)

It is defined as

$$\text{TFR} = k \sum_{x=w_1}^{w_2} n \text{ (age specific fertility rate between ages } x \text{ to}$$

$(x + n)$ where w_1 and w_2 are the lower and upper bounds of the child bearing are groups)

$$\text{TFR} = k \sum_n n \left\{ \frac{\text{no. of births to women from } x \text{ to } (x + n)}{\text{no. of female in ages } x \text{ to } (x + n)} \right\}$$

$$= k \sum_x \left[\frac{{}_nB_x^Z}{{}_nW_x^Z} \right] \cdot n = k \sum_{x=w_1}^{w_2} n_n f_x$$

where ${}_n f_x = \frac{{}_nB_x^Z}{{}_nW_x^Z}$

Physical interpretation of total fertility rate:
Let $k = 10^3$, usually and suppose 1,000 females are born at the same time and none of them die before reaching the end of the child bearing age interval. Then the total fertility rate (TFR) can be interpreted as the number of babies born to the cohort of 1,000 women (all born at the same period), assuming that at each group (in the child bearing ages) they are subject to the fertility condition given by the observed age specific fertility rates.

Pearl's Vital Index:
Pearl's Vital Index (PVI) is defined by

$$PVI = \frac{\text{No. of births in a given period of time}}{\text{No. of deaths in a given period of time}} \times k$$

where $k = 10^3$ usually.

From the above, it is possible to employ PVI to find out whether the population is increasing or decreasing. As PVI is very simple to calculate, Pearl has suggested the index for measuring Population growth. But as may be found in the following lines it has some serious defects.

(1)   PVI, no doubt, gives a measure, whether births exceed deaths or not. But our object is not merely to know this. We may, as well, want to know whether a Population has a tendency to increase or decrease and to measure the trend in population growth. But this index certainly fails to provide such measures.

(2)   For comparative purpose PVI is defective. Suppose for two countries I and II births BI = BII holds but DI > DII in respect of deaths. Then PVI for I < PVI for II. But this does not imply that the country II has a greater tendency to grow faster. The snag is obvious for we do not consider the age composition of either of the two countries.

(3)   In Pearls Vital Index we take into consideration of the whole Population; but population can increase through females only, so that we should consider only female population and their births and deaths.

The last point in (3) is employed to construct two important fertility growth indices as (i) Gross and (ii) Net Reproduction rates.

10.1.5   Gross Reproduction Rate (GRR)

To explain the computation of Gross Reproduction rate, the following table is constructed:

$(x - x + n)$	$_nW_x$	$_nB_x^f$	$[_nB_x^f /_nW_x] \times k$
(1)	(2)	(3)	(4)

(1) $(x - x + n)$ represents the age groups of female population.

(2) $_nW_x$ = no. of females in the age group $(x - x + n)$

(3) $_nB_x^f$ = no. of female births to mothers in the age group $(x - x + n)$ in a year (calender year)

(4) $k = 10^3$ usually.

Then the gross reproduction rate (GRR) is defined as

$$GRR = \left\{ \sum n \frac{_nB_x^f}{_nW_x} \right\} k$$

Physical interpretation of GRR :

GRR gives the number of female babies that would be born to k ($k = 10^3$ usually) women (all born at the same time) and if none of these women die before reaching the end of child bearing age interval and if all of them are subject to the risk of giving births under the given fertility schedule. A value of GRR = $k \Rightarrow$ under the existing condition of fertility, k women would replace themselves in the next generation to k^2 women so that the GRR can be taken as the measure of population growth under the hypothesis of no mortality during the entire fertility span.

10.1.6 Net Reproduction Rate (NRR)

Net reproduction rate (NRR) improves Gross reproduction rate (GRR) not only by considering the extent of replacement of one female cohort of newly born babies to another cohort formed of the female children born to these female cohort of the newly born babies (who were potential mothers); but also by considering the ability to produce children subject to the condition of survival of the mothers during the fertility span. More precisely, if $f_x = P$ [of a female aged x to produce a female child during the course of a calender year Z under the given fertility condition] and $p_x = P$ [that a newly born female baby will survive up to the age x under the given mortality schedule] then $\int_{w_1}^{w_2} p_x f_x \, dx$ is defined to be the net reproduction rate per women. In other words.

$$\text{NRR per women} \;=\; \int_{w_1}^{w_2} p_x f_x \, dx$$

where w_1 and w_2 are the lower and the upper bounds of the fertility span. If we require to estimate NRR from grouped data, then $\sum_{x=w_1}^{w_2} \left[\dfrac{{}_n^Z B_x^f}{{}_n W_x} \right] \left(\dfrac{{}_n L_x}{l_0} \right) . \, k = \text{NRR}$

per k women where $k = 10^3$ generally and ${}_n^Z B_x^f$ = female births to women in the age group $(x - x + n)$ in the calender year Z

$${}_n W_x = \text{women in the age group } (x - x + n)$$

Further, if we assume that of the k newly born infants at a particular time, $k \left(\dfrac{{}_n L_x}{l_0} \right)$ will represent the female population in the age group $(x - x + n)$ over which the probability of having a female baby $\dfrac{{}_n^Z B_x^f}{{}_n W_x}$ will be applied.

Thus the NRR per k women will imply the total number of female babies born by k number of newly born female infants during their fertility span subject to their survival at the present mortality schedule throughout the fertility span. This is the physical interpretation of the NRR.

Thus NRR can be taken as a growth Index. If NRR $> k$, we can conclude that the population is increasing and NRR $< k$ we may conclude that the population is decreasing. In other words, k number of potential mothers will fail to produce sufficient number of future mothers to maintain the same level.

Hence, for rough work NRR may be used as the measure of replacement index of the population.

The tacit assumptions underlying the calculation of NRR:

(1) The same survival factor (i.e. mortality) will be applicable for the entire newly born babies throughout their life time i.e.

$$\left(\frac{l_x}{l_0}\right) = \frac{l_x:t}{l_0:t} = \text{constant and independent of } t$$

(2) Same fertility pattern will be applicable for the newly born babies throughout their life time.

(3) The sex ratio $= \dfrac{\text{number of males}}{\text{total population}}$ should remain the same over all the years to come. But the above assumptions are not always justified.

If the mortality rate has a decreasing trend then the actual $NRR \geq$ calculated NRR on the basis of the above assumption and if the mortality is increasing then a reverse conclusion holds good.

10.2 Relationship Between Crude Birth Rate (CBR), General Fertility Rate (GFR) and the Total Fertility Rate (TFR)

Let $c(x; t) \equiv$ observed proportion of females in the age group $(x - x + 1]$ at time t.

$f(x; t) \equiv$ observed proportion of females giving birth to female children in the age group $(x - x + 1]$ at time t

$$\int_{\alpha}^{\beta} f(x, t)\, dx \equiv \text{estimated total fertility rate } = \hat{T}_f(t), \text{ say at time } t$$

$$\int_{\alpha}^{\beta} c(x; t) f(x; t)\, dx \equiv \text{estimated female birth rate at time } t = \hat{B}_f(t)$$

where α and β represent the lower and the upper bound of the child bearing ages. The correlation between $c(x; t)$ and $f(x; t)$ at a given t is given by

$$\text{Cor}(c(x; t), f(x; t) \mid t)$$

$$= \frac{[E(c(x; t) f(x; t) \mid t] - E(c(x; t \mid t) E(f(x; t) \mid t)}{\sqrt{\text{Var}(c(x; t) \mid t)}\sqrt{\text{Var}(f(x; t) \mid t)}}$$

Also

$$E(c(x; t) f(x; t) \mid t) = \int_{\alpha}^{\beta} [\phi_x c(x; t) f(x; t) \mid t]\, dx$$

where ϕ_x is the probability distribution of the r.v. X which denotes the age. We have

$$E\,(c\,(x;t)\,|\,t) := \int_\alpha^\beta [\phi_x\,c\,(x;t)\,|\,t]\,dx$$

$$E\,(f\,(x;t)\,|\,t) = \int_\alpha^\beta [\phi_x\,f\,(x;t)\,|\,t]\,dx$$

$$\mathrm{Var}\,(c\,(x;t)\,|\,t) = \int_\alpha^\beta [\phi_x\,c^2\,(x;t)\,|\,t]\,dx - \left[\int_\alpha^\beta (\phi_x\,c\,(x;t)\,|\,t)\,dx\right]^2$$

$$\mathrm{Var}\,[f\,(x;t)\,|\,t] = \int_\alpha^\beta [\phi_x\,f^2\,(x;t)\,|\,t]\,dx - \left[\int_\alpha^\beta (\phi_x\,f\,(x;t)\,|\,t)\,dx\right]^2$$

Let us assume that X is uniformly distributed in (α, β) i.e.,

$$\phi_x = \frac{1}{\beta - \alpha};\ \alpha \le x \le \beta$$

$$= 0 \text{ otherwise}$$

$$\mathrm{Cor}\,(c(x;t), f(x;t)\,|\,t) = \frac{1}{\beta - \alpha} \int_\alpha^\beta [c\,(x;t)\,f\,(x;t)\,|\,t]\,dx$$

$$- \frac{1}{\beta - \alpha} \int_\alpha^\beta [c\,(x;t)\,|\,t]\,dx\, \frac{1}{\beta - \alpha} \int_\alpha^\beta [f\,(x;t)\,|\,t]\,dx$$

$$\div \left[\sqrt{\frac{1}{\beta - \alpha} \int_\alpha^\beta [c\,(x;t)\,|\,t]^2\,dx - \frac{1}{\beta - \alpha} \left[\int_\alpha^\beta [c\,(x;t)\,|\,t]\,dx\right]^2}\right]$$

$$\left\{\sqrt{\frac{1}{\beta - \alpha} \int_\alpha^\beta [f\,(x;t)\,|\,t]^2\,dx - \frac{1}{\beta - \alpha} \left[\int_\alpha^\beta [f\,(x;t)\,|\,t]\,dx\right]^2}\right\}$$

$$= \left\{\frac{1}{\beta - \alpha} \int_\alpha^\beta [c\,(x;t)\,f\,(x;t)\,|\,t]\,dx - \frac{1}{(\beta - \alpha)^2} \int_\alpha^\beta [c\,(x;t)\,|\,t]\,dx\right.$$

$$\left. \int_\alpha^\beta [f\,(x;t)\,|\,t]\,dx\right\} \Big/ \sigma_c \cdot \sigma_f$$

where σ_c^2 and σ_f^2 stand for the variances of $[c\,(x;t)\,|\,t]$ and $[f(x;t)\,|\,t]$ as given in the expression of correlation between $[c\,(x;t)\,|\,t]$ and $[f(x;t)\,|\,t]$,

$$\Rightarrow [\hat{r}_{c,f}\,|_t]\,\sigma_c\,\sigma_f \;=\; \frac{1}{\beta - \alpha}\int_\alpha^\beta [c\,(x;t)\,f\,(x;t)\,|\,t]\,dx$$

$$-\frac{1}{(\beta - \alpha)^2}\int_\alpha^\beta [c\,(x;t)\,|\,t]\,dx\int_\alpha^\beta [f\,(x;t)\,|\,t]\,dx$$

$$=\frac{1}{\beta - \alpha}\,\hat{B}_f\,(t) - \frac{\hat{T}_f\,(t)}{(\beta - \alpha)^2}\int_\alpha^\beta [c\,(x;t)\,|\,t]\,dx \tag{10.4}$$

where $\hat{r}_{c,f}\,|_t$ represents the estimated product moment correlation coefficient between c and f given t.

Again, estimated General Fertility rate;

$$\hat{G}_F\,(t) \;=\; \frac{N\int_\alpha^\beta [f\,(x;t)\,|\,t]\,dx}{N\int_\alpha^\beta [c\,(x;t)\,|\,t]\,dx}$$

$$\Rightarrow \int_\alpha^\beta [c\,(x;t)\,|\,t]\,dx \;=\; \frac{\hat{T}_f\,(t)}{\hat{G}_F\,(t)} \tag{10.5}$$

Putting (10.5) in (10.4) $\Rightarrow$

$$[\hat{r}_{c,f}\,|_t]\,\sigma_c\,\sigma_f \;=\; \frac{\hat{B}_f(t)}{\beta - \alpha} - \frac{\hat{T}_f\,(t)}{(\beta - \alpha)^2}\,\frac{\hat{T}_f\,(t)}{\hat{G}_F\,(t)}$$

$$\Rightarrow (\beta - \alpha)\,\hat{r}_{c,f}\,|_t\,\hat{\sigma}_c\,\hat{\sigma}_f \;=\; \hat{B}_f\,(t) - \frac{(\hat{T}_f\,(t))^2}{\beta - \alpha}\,\frac{1}{\hat{G}_F\,(t)}$$

where $\hat{\sigma}_c$ and $\hat{\sigma}_f$ are the observed standard deviations of c and f respectively.

$$\Rightarrow \hat{B}_f\,(t) \;=\; \left[\hat{r}_{c,f}\,|_t\,\sigma_c\,\sigma_f\,(\beta - \alpha) + \frac{(\hat{T}_f\,(t))^2}{(\beta - \alpha)}\,\frac{1}{\hat{G}_F\,(t)}\right] \tag{10.6}$$

which establishes a relation between birth rate, total fertility rate and general fertility rate.

A special case:

If $\hat{r}_{c,f}\,|_t = 0 \Rightarrow \hat{B}_f\,(t) = \dfrac{(\hat{T}_f\,(t))^2}{(\beta - \alpha)}\,\dfrac{1}{\hat{G}_F\,(t)}$ \tag{10.7}

10.3 An Approach to the Net Reproduction Rate (NRR) from Branching Process Point of View

Let Z_n be the # female children in the nth generation

$$Z_{n+1} = X_{1,n} + X_{2,n} + \ldots + X_{Z_n}, n$$

where $X_{i,n}$ = # female children born to the ith female in the nth generation $(i = 1, 2 \ldots Z_n)$ where $X_{i,n}$'s $(i = 1, 2, \ldots Z_n)$ are i.i.d., r.v.s. for given n.

From art 1.37, we have

$$E\,(Z_{n+1} \mid Z_n = k_n) = k_n E\,(X_{i,n}) = k_n\,\alpha_n$$

where $E\,(X_{i,n}) = \alpha_n$

$$E\,(Z_{n+1}) = \alpha_n E\,(Z_n) = \alpha_n\,\alpha_{n-1}\,E\,(Z_{n-1})$$

$$\Rightarrow \qquad E\,(Z_n) = \alpha_n\,\alpha_{n-1}\,\alpha_{n-2}\,E\,(Z_{n-2})$$

$$= [\alpha_n\,\alpha_{n-1} \ldots \alpha_0]\,E\,(Z_0)$$

Suppose given $(Z_0) = \pi$, a fixed non-negative integral valued quantity.

Then

$$E\,(Z_{n+1}) = \left\{ \pi \prod_{i=0}^{n} \alpha_i \right\} \qquad \text{for } n = 0, 1, 2, \ldots$$

We thus get a stochastic analogue of the NRR as

$$E\,(Z_{n+1})/E\,(Z_n)$$

Also

$$\left[\frac{E\,(Z_{n+1})}{\pi} \right] = \prod_{i=1}^{n} \alpha_i \qquad (10.8)$$

$$\text{For } n = 0, 1, 2, \ldots$$

we get the mean # female children per female child in the first, second... and $(n + 1)$th generations respectively.

If $\dfrac{E\,(Z_{n+1})}{\pi}$ is an increasing function over n then we say that the population growth has an increasing behaviour whereas if it is consistently decreasing function of n then we say that the population growth has a declining trend.

Further from example ref. 4.1 we have for $\alpha_1 = \alpha_2 = \ldots = \alpha_n = \alpha$;

$$Y_n = (\alpha)^{-n}\,Z_n \text{ is a Martingale.}$$

Denoting S_T as stopping time in terms of a given Z_{S_T}

$\Rightarrow E\,(Y_{S_r}) = E\,[(\alpha)^{-S_T}\,(Z_{S_r})] = E\,(Y_0) = Z_0$ under the regularity conditions of optional sampling theorem of Martingales (from article 4.8.2). Given Z_0 and the stopping time S_T

$$\Rightarrow E\,(Y_{S_r}) = E\,[(\alpha)^{-S_T}]\,Z_{S_T} = Z_0 \qquad (10.9)$$

Eq. (10.8) may be utilized to obtain the expected size of the female population reaching a level given by a stopping time S_T.

10.4 Stochastic Models On Fertility and Human Reproductive Process

10.4.1 Dandekar's Modified Binomial and Poisson Distribution

The systematic origin of the development of Stochastic models in India, representing the fertility behaviour under uncontrolled condition of fertility or a description of the Human reproductive process, perhaps had its foundation with the work of Dandekar (1955). Dandekar while formulating an appropriate probability model of the number of birth in a given marital exposure of duration t developed special interrupted probability distributions known as 'Modified' Binomial and Poisson distributions based on the following assumptions.

Assumptions

    (i)   Probability of a conception (vis-a-vis a birth, assuming a one-to-one correspondence between a conception and a birth) is p in every trial (every trial is assumed to be of duration of one month approximately which is the interval between two consecutive ovulatory cycles).

    (ii) Given that there is a success (a conception leading to a birth, the probability of a further success in another π number of trials (π is an integer) inclusive of the trial in which a success took place, is zero.

Notations

    (i) $X \equiv$ number of successes (conceptions or births) (a.r.v.)

    (ii) $n \equiv$ number of trials

    (iii) $p \equiv$ probability of success in a trial; $q = 1 - p$

    (iv) $P(x; n) \equiv$ probability of exactly x successes in n trials,

    (v) $F(x, n) \equiv$ probability of not more than x successes in n trials, i.e.

$$P[X \leq x \,|\, n]$$

Development of the Model

Consider a sequence of n trials. If there is no success in $(n - x\pi)$ number of trials then obviously in the remaining $n - (n - x\pi) = x\pi$ number of trials at most x number of successes can occur. Probability of more than x successes in the remaining $x\pi$ number of trials given that no successes occurred upto the first $(n - x\pi)$ number of trials is zero. In other words, given that the $(n - x\pi)$ leads to failures the probability of at most x number of successes in the remaining $x\pi$ number of trials is one.

    Again when the first success occurs after S number of consecutive failure $(S < n - x\pi - 1)$, a case distinct from the earlier one where in the remaining $(n - \pi - S)$ effective trials at most $(x - 1)$ successes occur with probability $F(x - 1; n - \pi - S)$.

Thus

$$F(x\,;n) \;=\; F(0\,;n-x\pi)+ \sum_{S=0}^{n-x\pi-1} pq^{S}\,F(x-1\,;n-\pi-S) \qquad (10.10)$$

where x and π are integers.

Obviously $\qquad\qquad F(0\,;n) \;=\; P(0,n) \;=\; q^{n} \qquad\qquad\qquad (10.11)$

Putting $x = 1$

$$F(1\,;n) \;=\; F(0\,;n-\pi)+ \sum_{S=0}^{n-\pi-1} pq^{S}\,F(0\,;n-\pi-S)$$

$$=\; q^{n-\pi} + \sum_{S=0}^{n-\pi-1} pq^{s}\,q^{n-\pi-s}$$

$$=\; q^{n-\pi} + pq^{n-\pi}\,(n-\pi)$$

$$=\; q^{n-\pi}\,[1 + p\,(n-\pi)]$$

$$\therefore \qquad F(1,n) \;=\; q^{n-\pi}\,[1 + (n-\pi)p] \qquad\qquad (10.12)$$

which is generalised Binomial distribution.

With $\pi = 0$, it gives Binomial distribution.

For $x = 2$

$$F(2,n) \;=\; F(0,n-2\pi)+ \sum_{S=0}^{n-2\pi+1} q^{S}p\,F(1,n-\pi-S)$$

$$=\; q^{n-2\pi} + \sum_{S=0}^{n-2\pi+1} q^{S}p\,q^{(n-\pi-S)-\pi}\,[1 + ((n-\pi-S)-\pi)\,p]$$

$$\qquad\qquad\qquad\qquad\qquad\qquad \text{from (10.12)}$$

$$=\; q^{n-2\pi} + \sum_{S=0}^{n-2\pi+1} q^{S}p\,q^{n-\pi-S-\pi}$$

$$+\; \sum_{S=0}^{n-2\pi+1} q^{S}p\,q^{n-\pi-S-\pi}\,(n-\pi-S-\pi)\,p$$

$$=\; q^{n-2\pi} + (n-2\pi)\,pq^{n-2\pi} + p^{2}q^{n-2\pi}\left(\sum_{S=0}^{n-2\pi+1}(n-2\pi-S)\right)$$

$$=\; q^{n-2\pi} + (n-2\pi)\,pq^{n-2\pi} + p^{2}\,q^{n-2\pi}\,(n-2\pi)^{2}$$

$$- \frac{p^2 q^{n-2\pi} (n-2\pi)(n-2\pi-1)}{1 \cdot 2}$$

$$= q^{n-2\pi} \left\{ 1 + (n-2\pi)p + p^2(n-2\pi)^2 - \frac{p^2 (n-2\pi)(n-2\pi-1)}{2} \right\}$$

$$= q^{n-2\pi} \left[1 + (n-2\pi)p + \frac{(n-2\pi)(n-2\pi+1)}{2} \cdot p^2 \right] \tag{10.13}$$

Proceeding in this way

$$F(3,n) = q^{n-3\pi} \left[1 + (n-3\pi)p + \frac{(n-3\pi)(n-3\pi+1)}{2} p^2 \right.$$

$$\left. + \frac{(n-3\pi)(n-3\pi+1)(n-3\pi+2)}{1 \cdot 2 \cdot 3} p^3 \right]$$

$$\times \qquad \times \qquad \times \qquad \times \qquad \times$$

$$F(x,n) = q^{n-x\pi} \left[1 + p(n-x\pi) + \frac{(n-x\pi)(n-x\pi+1)}{2!} p^2 \right.$$

$$\left. + \ldots + \frac{(n-x\pi)(n-x\pi+1)\ldots(n-x\pi+x-1)}{x!} p^x \right] \tag{10.14}$$

= the first $(x+1)$ terms in the expansion of

$$q^{n-x\pi} (1-p)^{-(n-x\pi)}$$

The relation (10.14) is true for all integral values of x for which

$$n - x\pi \geq 0$$

i.e.
$$\frac{n}{\pi} \geq x.$$

For large values of x for which $n - x\pi < 0$. Clearly

$$F(x;n) = 1$$

From this the required probability $P(x, n)$ is given by

$$P(x;n) = F(x;n) - F(x-1;n) \tag{10.15}$$

Dandekar's Modified Poisson Distribution

Dandekar (1955) took the limiting case of his modified Binomial distribution and obtained the modified Poisson distribution as obtained in (10.14) as follows:

Putting $np = \lambda t; \dfrac{\pi}{n} = \dfrac{\theta}{t}$ in (10.14), we have

$$F(x,n) = (1-p)^{n-\pi x} \left[1 + (n-\pi x)p + \frac{(n-\pi x)(n-\pi x+1)}{2!} p^2 \right.$$

$$\left. + \ldots + \frac{(n-\pi x)(n-\pi x+1)\ldots(n-\pi x+x-1)}{x!} p^x \right]$$

and $P(x, n) = F(x, n) - F(x - 1, n)$.

Also we can put

$$F(x, n) = (1 - p)^{n\left(1 - \frac{\pi}{n}x\right)} \left[1 + np\left(1 - \frac{\pi}{n}x\right)\right.$$

$$+ n^2 p^2 \frac{\left(1 - \frac{\pi}{n}x\right)\left(1 - \frac{\pi}{n}x + \frac{1}{n}\right)}{2!}$$

$$\left. + ... + \frac{(np)^x}{x!}\left(1 - \frac{\pi}{n}x\right)\left(1 - \frac{\pi}{n}x + \frac{1}{n}\right)...\left(1 - \frac{\pi}{n}x + \frac{x-1}{n}\right)\right]$$

$$= \left(1 - \frac{\lambda t}{n}\right)^{n\left(1 - \frac{\theta x}{t}\right)} \left[1 + \lambda t\left(1 - \frac{\theta x}{t}\right)\right.$$

$$+ \frac{(\lambda t)^2}{2!}\left(1 - \frac{\theta}{t}x\right)\left(1 - \frac{\theta}{t}x + \frac{1}{n}\right)$$

$$\left. + ... + \frac{(\lambda t)^x}{x!}\left(1 - \frac{\theta}{t}x\right)\left(1 - \frac{\theta}{t}x + \frac{1}{n}\right)...\left(1 - \frac{\theta}{t}x + \frac{x}{n} - \frac{1}{n}\right)\right]$$

As $n \to \infty$, the above reduces to

$$= \exp(-\lambda(t - \theta x)) \sum_{r=0}^{x} \frac{(\lambda t)^r}{r!} \frac{(t - \theta x)^r}{t^r}$$

$$\therefore \quad F(x) = \exp(-\lambda(t - \theta x)) \sum_{r=0}^{x} \frac{[\lambda(t - \theta x)]^r}{r!} \tag{10.16}$$

This is the modified Poisson Process of Dandekar which gives the probability distribution of the number of conceptions of births in time $(0, t]$ subject to the condition that each conception with intensity λ is followed by an infecundable exposure (duration) θ.

From (10.16) we have

$$P(X = x) = F(x) - F(x - 1)$$

$$= \exp(-\lambda(t - \theta x)) \sum_{r=0}^{x} \frac{(\lambda(t - \theta x))^r}{r!}$$

$$- \exp(-\lambda(t - (x - 1)\theta)) \sum_{r=0}^{x-1} \frac{[\lambda(t - (x - 1)\theta)]^r}{r!} \tag{10.17}$$

However, the modified Poisson Process derived using differnece equation (10.10) seems to lack mathematical rigour, presicely due to switching from discrete process to continuous by taking $n \to \infty$. π and n both being positive integers, $\dfrac{\pi}{n}$ is a rational number which cannot essentially be equal to $\dfrac{\theta}{t}$, where both θ and t are continuous variables admitting irrational values also. (For more rigorous derivation vide (5.12) and (5.13))

10.4.2 William Brass Model

William Brass (1958) generalised the model given by (10.17) further by assuming "λ" to conform to a probability distribution of the form

$$f(\lambda) = \frac{a^k\, e^{-a\lambda}\, \lambda^{k-1}}{\Gamma(k)}\,;\ 0 \le \lambda < \infty \tag{10.18}$$

where λ = fecundability parameter in fertility analysis or intensity of sickness in morbidity studies. This assumptions be a fairly plausible one in representing the differential risk in human fertility or differential susceptibility in communicable diseases or accident proneness.

Writing

$$\left.\begin{aligned} F(t, x \mid \lambda) &= \sum_{r=0}^{x} e^{-\lambda(t - \theta x)}\, \frac{[\lambda\,[t - \theta x)]^r}{r!} \quad \text{if } t > \theta \\ &= 0 \quad \text{otherwise} \end{aligned}\right\} \tag{10.19}$$

Therefore the interrupted Poisson Process in view (10.10) is thus given by

$$F(t, x) = \int_{0}^{\infty} F(t, x \mid \lambda)\, f(\lambda)\, d\lambda$$

$$= \int_{0}^{\infty} \sum_{r=0}^{x} e^{-\lambda(t - \theta x)}\, \frac{[\lambda\,[t - \theta x)]^r}{r!} \cdot \frac{a^k\, e^{-a\lambda}\, \lambda^{k-1}}{\Gamma(k)}\, d\lambda$$

$$= \sum_{r=0}^{x} \int_{0}^{\infty} e^{-\lambda(t - \theta x)}\, \frac{[\lambda\,[t - \theta x)]^r}{r!} \cdot \frac{a^k\, e^{-a\lambda}\, \lambda^{k-1}}{\Gamma(k)}\, d\lambda$$

$$= \sum_{r=0}^{x} \frac{a^k}{\Gamma(k)} \cdot \frac{(t - \theta x)^r}{r!} \int_{0}^{\infty} e^{-\lambda(a + t - \theta x)}\, \lambda^{r+k-1}\, d\lambda$$

$$= \sum_{r=0}^{x} \frac{a^k}{\Gamma(k)} \cdot \frac{(t - \theta x)^r}{r!} \cdot \frac{\Gamma(r + k)}{(a + t - \theta x)^{r+k}} \tag{10.20}$$

In view of (10.20), we have

$$P(t, x) = F(t, x) - F(t, x - 1)$$

$$= \sum_{r=0}^{x} \frac{a^k}{\Gamma(k)} \cdot \frac{(t - \theta x)^r}{r!} \cdot \frac{\Gamma(r+k)}{(a+t-\theta x)^{r+k}}$$

$$- \sum_{r=0}^{x-1} \frac{a^k}{\Gamma(k)} \frac{(t-(x-1)\theta)^r \Gamma(r+k)}{r!(a+t-(x-1)\theta)^{r+k}} \tag{10.21}$$

$$= \frac{a^k}{\Gamma(k)} \frac{(t-\theta x)^r}{r!} \cdot \frac{\Gamma(x+k)}{(a+t-\theta x)^{x+k}} + o\left(\frac{\pi}{t}\right) \tag{10.21'}$$

The result (10.20) and (10.21) are the cumulative distribution function (c.d.f) and probability distribution function of the model of William Brass which he used for obtaining the probability distribution of births in $(0, t]$ (in a given marriage duration), clearly the probability distribution of the number of births in time $(0, t]$ given by $P(t, x)$ in (10.21) for an integer k and x ($a > 0, k > 0, x > 0$) is a negative Binomial Distribution with a, k and θ as parameters.

$$P(t; x) = \frac{\Gamma(x+k)}{\Gamma(k)x!} \left(1 - \frac{a}{a+t-\theta x}\right)^x \left(\frac{a}{a+t-\theta x}\right)^k \tag{10.21''}$$

$$= \binom{x+k-1}{k-1} \left(\frac{a}{a+t-\theta x}\right)^{k-1} \left(1 - \frac{a}{a+t-\theta x}\right)^x \left(\frac{a}{a+t-\theta x}\right)$$

Putting $p' = \dfrac{a}{a+t-\theta x} \geq 0$ and $p' \leq 1$ when $t = \theta\, x \Rightarrow x \leq \left[\dfrac{t}{\theta}\right]$ where $\left[\dfrac{t}{\alpha}\right]$ contains the greatest integer in $\dfrac{t}{a}$

$$q' = 1 - \frac{a}{t+a-\theta x} \quad \text{where } p' + q' = 1$$

It follows that

$$P(t; x) = \binom{x+k-1}{k-1} (p')^{k-1} (q')^x \cdot p' \tag{10.22}$$

represents the probability of precisely k number of successes in $(x + k)$ trials subject to the condition that the last trial leads to a success.

Brass (1958) fitted the Distribution in the data of empirical distribution of births to women in the United States with completed marital span while ignoring θ, the period of infecundable exposure corresponding to every birth. This gave rise to the following model

$$P(t, x) = \frac{\Gamma(x+k)}{\Gamma(k)x!} \left(\frac{t}{a+t}\right)^x \left(\frac{a}{a+t}\right)^k \quad (x = 0, 1, 2, \ldots) \tag{10.23}$$

As the fit was not found to be good, especially for the mothers with no births in the entire marital span (which might have been either due to Biological

Sterility or due to the preference of the mothers to have no children) which could have been due to a heterogeneous combination of two different groups of women. Brass therefore, excluded mothers with zero number of births and considered a zero truncated model of (10.23) given by the probability distribution.

$$P(t; x) = \frac{\Gamma(x+k)}{\Gamma(k) \, x!} \left(\frac{t}{a+t}\right)^x \left(\frac{a}{a+t}\right)^k \Bigg/ \left\{ 1 - \left(\frac{a}{a+t}\right)^k \right\} \qquad (10.24)$$

$$x = 1, 2, 3, \dots$$

Under the above, Brass obtained good fit of (10.24) using the empirical distribution of births.

The raw moments of the distribution (from the origin) can easily be obtained, for the fitting of the model to the data, as follows:

$$\mu_1' = \frac{kt}{a} \Bigg/ \left\{ 1 - \left(\frac{a}{a+t}\right)^k \right\} \qquad (10.25)$$

$$\mu_2' = \frac{kt}{a} + \frac{k(k+1)\,t^2}{a^2} \Bigg/ \left[1 - \left(\frac{a}{a+t}\right)^k \right] \qquad (10.26)$$

$$\mu_3' = \frac{kt}{a} + \frac{3k(k+1)\,t^2}{a^2} + \frac{k(k+1)(k+2)\,t^3}{a^3} \Bigg/ \left[1 - \left(\frac{a}{a+t}\right)^k \right] \qquad (10.27)$$

using the above three equations, the parameters a, k and t are estimable. However, Brass model suffered from an inherent defect by ignoring the infecundable exposure corresponding to every live birth as pointed out by Biswas (1973).

Biswas (1973) generalized Brass model (10.23) by (i) introducing $(1 - \alpha)$ proportion of Biologically sterile mothers and (ii) assuming the following each live birth there is an infecundable period of constant length θ so that for x # births, $\left(\frac{x\theta}{t}\right)^n = o(1) \; \forall \; n \geq 2$ subject to max $x = 6$ and t being the total martial exposure.

Starting from (10.17) under assumption of (i) and (ii)

$$f(t - x\theta \,|\, \lambda) = f\!\left(t \left(1 - \frac{x\theta}{t}\right) \,|\, \lambda\right) = p[x \,; t]$$

$$= \left[e^{-\lambda t} \frac{(\lambda t)^x}{x!} - x\theta\lambda \left(e^{-\lambda t} \frac{(\lambda t)^{x-1}}{(x-1)!} - e^{-\lambda t} \frac{(\lambda t)^x}{\lambda!} \right) + \frac{x\theta}{t} e^{-\lambda t} \frac{(\lambda t)^x}{x!} \right] + o\left(\frac{x\theta\lambda}{t}\right)$$

$$(10.28)$$

By using Taylor expansion of the series about t

$$f(t - x\theta \,|\, \lambda) = \sum_{r=0}^{x} e^{-\lambda(t - \theta x)} \frac{\lambda(t - \theta x)^r}{r!}$$

$$-\sum_{r=0}^{x-1} e^{-\lambda(t-(x-1)\theta)} \frac{\lambda(t-(x-1)\theta)^r}{r!} \tag{10.29}$$

the probability of zero # births in $(0, t]$ as given by (10.28) as $\alpha e^{-\lambda t}$ where α is the proportion of biologically non-sterile mothers in a population and $(1-\alpha)$ is the proportion of sterile mothers in the population.

The probability of x # births in $(0, t]$ is given by

$$(1-\alpha) f \left(t \left(1 - \frac{x\theta}{t}\right) | \lambda \right)$$

Further assuming the probability distribution of λ (i.e., variation of fecundability) as:

$$\phi(\lambda) = \frac{ak}{\Gamma(k)} e^{-a\lambda} \lambda^{k-1}; \; 0 \leq \lambda < \infty, \; a, k > 0 \tag{10.30}$$

the probability of a biologically fecund mother to have x # births $(x = 1, 2, 3...)$ is given by

$$(1-\alpha) \int_0^{\infty} f\left[t\left(1 - \frac{x\theta}{t}\right) | \lambda \right] \phi(\lambda) \, d\lambda$$

$$= (1-\alpha) \left[\left(\frac{a}{a+t}\right)^k \frac{\Gamma(x+k)}{\Gamma(k) x!} \left(\frac{t}{a+t}\right)^x \right.$$

$$+ \frac{k(k+1)\theta t}{a^2} \left(\frac{a}{a+t}\right)^{k+2} \frac{\Gamma(k+2+x-1)}{\Gamma(k+2)(x-1)!} \left(\frac{t}{a+t}\right)^{x-1}$$

$$\left. - \frac{k(k+1)\theta t}{a^2} \left(\frac{a}{a+t}\right)^{k+2} \frac{\Gamma(k+2+x-2)}{\Gamma(k+2)(x-2)!} \left(\frac{t}{a+t}\right)^{x-2} \right] \tag{10.31}$$

and that the probability of zero number of births is given by

$$(1-\alpha) + \alpha \left(\frac{a}{a+t}\right)^k$$

Remarks

Now the probability of having x # births in $(0, t]$ given that (i) the probability of a conception (or a birth) in an infinitesimal interval of length δt is $\lambda \, \delta t + o(\delta t)$ where λ is a constant and (ii) given that a renewal (conception or birth) while taking place at time t, for a fixed time π following t, there cannot occur another success, is really a problem of Geiger Muller Counter model type I with a fixed dead time π. Hence the Dandekar's model derived in (10.19) can be derived in a much more rigorous way using Counter theory as in art 5.1. One may compare (10.29) with that of (5.13) for $n = x$ and $2 = \pi$. Again starting from (5.13) one can derive William Brass model given in (10.31).

Using (10.20) we have by replacing π by θ

$$F(t, n) = \sum_{r=0}^{n} \frac{a^k}{\Gamma(k)} \frac{(t - \theta n)^r}{r!} \frac{\Gamma(r + k)}{(a + t - \theta n)^{r+k}}$$

$$= P[X \leq n \mid t] \Rightarrow 1 - F(t, n) = P[X > n \mid t]$$

$$\Rightarrow 1 - F(t, n) = P[T_n \leq t] \tag{10.32}$$

where T_n represents the waiting time for the nth renewal (conception or birth)

Therefore

$$P[t \leq T_n \leq t + \delta t \mid t] = \frac{d}{dt}[1 - F(t, n)] \tag{10.33}$$

$$\Rightarrow f_n(t_n) = \frac{d}{dt}\left[1 - \sum_{r=0}^{n} \frac{a^k}{\Gamma(k)} \frac{(t - \theta n)^r}{r!} \frac{\Gamma(r + k)}{(a + t - \theta n)^{r+k}}\right] \tag{10.34}$$

Putting $n = 0$ we have the waiting time distribution for the first conception (or first birth using a one-to-one correspondence between a conception and a birth) we have

$$f_0(t_0) = \frac{d}{dt}\left[1 - \frac{a^k}{\Gamma(k)} \frac{\Gamma(k)}{(a + t)^k}\right]$$

$$= \frac{ka^k}{(a + t)^{k+1}} \tag{10.35}$$

This is the model used by Singh (1964) for obtaining the waiting time distribution of the first birth from marriage.

10.4.3   Some Modifications of the Singh's (1964) Result

The model for the waiting time of first conception given in (10.35) is based on the premises (i) that the waiting time distribution from marriage to first conception is given by

$$f(t \mid \lambda) = \lambda e^{-\lambda t} \quad 0 \leq \lambda < \infty$$

(ii) given that the conception rate remains fixed at λ. However, given that the fecundity varies from woman to woman and the distribution of λ follow a Gamma distribution given by

$$\phi(\lambda) = \frac{a^k}{\Gamma(k)} e^{-a\lambda} \lambda^{k-1}; \quad \lambda > 0 \ a, k > 0$$

We have the unconditional waiting time distribution for first conception, given by

$$\int_0^\infty f(t \mid \lambda) \phi(\lambda) d\lambda = \frac{ka^k}{(a + t)^{k+1}} \tag{10.36}$$

One major objection in the model (10.35) is that marital exposure for first conception need not be infinite. Suppose,

$$f(t_1|\lambda) = \frac{\lambda \exp(-\lambda t_1)}{1 - \exp(-\lambda T')}, \quad \lambda > 0, 0 \le t_1 < T'$$

which is a negative exponential right truncated at T', may be taken as the distribution of reasonable exposure for the first conception given λ.

Then the unconditional waiting time distribution for first conception is given by

$$\int_0^\infty f(t_1|\lambda)\,\phi(\lambda)\,d\lambda = \int_0^\infty \frac{\lambda \exp(-\lambda t_1)}{1 - \exp(-\lambda T')} \frac{a^k}{\Gamma(k)} e^{-a\lambda}\,\lambda^{k-1}\,d\lambda$$

$$= \frac{a^k}{\Gamma(k)} \int_0^\infty \lambda^k \exp(-\lambda(t_1 + a))(1 - \exp(-\lambda T'))^{-1}\,d\lambda$$

$$= \frac{a^k}{\Gamma(k)} \int_0^\infty [\exp(-\lambda(t_1 + a)) + \exp(-\lambda(t_1 + a + T')) + \ldots]\lambda^k\,d\lambda$$

$$= \frac{a^k}{\Gamma(k)} \int_0^\infty \sum_{k'=0}^\infty \exp[-\lambda(t_1 + a + k'T')]\lambda^k\,d\lambda$$

$$= \frac{a^k}{\Gamma(k)} \sum_{k'=0}^\infty \frac{\Gamma(k+1)}{(t_1 + a + k'T')^{k+1}} \tag{10.37}$$

which is the modified distribution for the waiting time for first conception due to Biswas and Shrestha (1985)

It is seen that

$$E(T_1) = \frac{a}{k-1}\left[1 - \left(\frac{a}{a+T'}\right)^{k-1}\right] \tag{10.38}$$

after some routine calculation and

$$\text{Var}(T_1) = \frac{a}{(k-1)^2}\left[\frac{ka}{k-2} - \left(\frac{a}{a+T'}\right)^{k-1}\right]\left\{\left(\frac{kT'(k-1)+2a}{k-2}\right) + a\left(\frac{a}{a+T'}\right)^{k-1}\right\} \tag{10.39}$$

The parameters of the distribution (10.36) are estimable by the method of moments by using (10.38) and (10.39).

10.4.4 Models for the Waiting Time of Conception of Various Orders

We have by putting $n = 1$ in (10.34)

$$f_1(t_1) = \frac{d}{dt}\left[1 - \sum_{r=0}^1 \frac{a^k}{\Gamma(k)} \frac{(t-\theta)^r}{r!} \cdot \frac{\Gamma(k+r)}{(a+t-\theta)^{r+k}}\right]$$

$$= \frac{d}{dt}[1 - F,(t;1)] = \frac{d}{dt}(P[T_1 \le t])$$

$$\Rightarrow P[t \le T_1 \le t + dt] = f_1(t_1)\, dt_1$$

$$= \frac{ka^k\,(k+1)\,(t_1-\theta)}{(a+t_1-\theta)^{k+2}};\ \theta \le t_1 < \infty \tag{10.40}$$

which is the waiting time distribution of the second conception or birth from the marriage.

Putting $n = 2$ in (10.34)

$$f_2(t_2) = \frac{d}{dt}\left[1 - \sum_{r=0}^{2}\frac{a^k}{\Gamma(k)}\frac{(t-2\theta)^r}{r!}\frac{\Gamma(k+r)}{(a+t-2\theta)^{r+k}}\right]$$

$$= \frac{a^k\,k\,(k+1)\,(k+2)\,(t_2-2\theta)^2}{2!\,(a+t_2-2\theta)^{k+3}};\ 2\theta \le t_2 < \infty \tag{10.41}$$

which is the waiting time distribution of the third conception (or birth) from marriage.

Finally, the waiting time distribution of the nth conception or birth from the marriage is given by

$$f_{n-1}(t_{n-1}) = \frac{k\,(k+1)\ldots(k+n-1)}{(n-1)!}\frac{a^k\,(t_{n-1}-(n-1)!\theta)}{(a+t_{n-1}-(n-1)\theta)^{k+n}};$$

$$(n-1)\,\theta \le t_{n-1} < \infty \tag{10.42}$$

It is seen that $(t_{n-1}(n-1)\,\theta)$ conforms to a Beta distribution of type- II. The results are due to Biswas and Nauhria (1980). Another generalization of Dandekar's model which is based on practical consideration is due to Biswas (1980). This generalization is based on taking the parity specific conception rate (hazard rate) $\lambda_1, \lambda_2, \lambda_3, \ldots, \lambda_n$ instead of one conception rate λ for all order of conceptions. Here keeping correspondence with empirical situation $\{\lambda_i\}$'s are assumed to conform to a decreasing sequence i.e.,

$$\lambda_1 = \lambda,\ \lambda_2 = \lambda e^{-\delta},\ \lambda_3 = \lambda e^{-2\delta},\ \ldots,\ \lambda_k = \lambda e^{-(k-1)\delta}\ \text{where}\ \delta > 0 \tag{10.43}$$

Again λ_i's are Poisson intensities leading to ith registration of an event (conception vis a vis a birth) during a free time t_i after which there follows an infecundable of dead period π. The convolutions of $t_1, t_2, \ldots, t_n$ given by

$$S_n = t_1 + t_2 + \ldots + t_n \tag{10.44}$$

are assumed to be independently but not identically distributed, because t_1 is preceded with no dead time whereas $t_2, \ldots, t_n$ are preceded by dead time with length θ (say) . Then

$$f_n(S_n \mid \lambda_1, \lambda_2, \ldots \lambda_n) = \sum_{i=1}^{n}\prod_{j \ne i}^{n}\frac{\lambda_j}{\lambda_j - \lambda_i}\lambda_i \exp(-\lambda_i S_n)$$

where
$$S_n = \sum_{i=1}^{n} t_i \tag{10.45}$$

Again following Brass if we assume a probability distribution of λ given by

$$\phi(\lambda) = \frac{a^k e^{-a\lambda} \lambda^{k-1}}{\Gamma(k)}; 0 < \lambda < \infty \text{ and } a, k > 0 \qquad \text{(vide (10.18))}$$

the distribution of the waiting time for the nth registration (or conception or birth) is given by

$$f_n(S_n) = \frac{a^k}{\Gamma(k)} \int_0^{\infty} \sum_{i=1}^{n} \prod_{j \neq i} \frac{\exp(-(j-1)\delta - (i-1)\delta)}{[\exp(-(j-1)\delta) - \exp(-(i-1)\delta)]}$$

$$\times [-\exp(-\lambda^{(\exp(-(j-1)\delta)}S_n + \delta)] \, d\lambda$$

$$= \frac{ka^k}{\Gamma(k)} \sum_{j \neq i}^{n} \frac{e^{-\delta(i+j-\delta)}}{[e^{-(j-1)\delta} - e^{-(i-1)\delta}]} \frac{1}{[a + e^{-(i-1)\delta} S_n]^{k+1}} \tag{10.46}$$

Finally, incorporating a dead time π (infecundable period) following every registration (conception or birth) which in Demographic Jargon implies that the total infecundable exposure following every conception is fixed but the conceptions occur with gradually decreasing intensities with differential fecundities given by $\lambda_i = \lambda \exp(-(i-1)\delta)$ and the distribution of λ is given by (10. 16). Then the nth registration time

$$\xi_n = S_n + (n-1)\theta \tag{10.47}$$

Hence the distribution of ξ_n is given by Biswas (1960)

$$f_n(\xi_n) = ka^k \sum_{i=1}^{n} \prod_{j \neq i} \frac{\exp(-\delta(i+j-2))}{[\exp(-(j-1)\delta) - \exp(-(i-1)\delta)]}$$

$$\cdot \frac{1}{[a + \exp(-(i-1)\delta)(\xi_n - (n-1)\theta)]^{k+1}} \tag{10.48}$$

which is the waiting time distribution of the nth arrival in a counter model type I with fixed dead time θ.

Moments of the distribution

$$\mu_r' = E(S_n^r) = \sum_{i=1}^{n} \prod_{j \neq i} \frac{\exp(-\delta(i+j-2))}{[\exp(-(j-1)\delta) - \exp(-(i-1)\delta)]} ka^k$$

$$\int_0^{\infty} \frac{S_n^r \, dS_n}{(a + \exp(-(i-1)\delta) S_n)^{k+1}}$$

$$= \sum_{i=1}^{n} \prod_{j \neq i} \frac{\exp(-\delta(i+j-2))}{[\exp(-(j-1)\delta) - \exp(-(i-1)\delta)]}$$

$$\exp(-\delta(i-1)(r+1)) a^{r-k} \beta(r+1, k-r) \qquad (10.49)$$

which gives the rth moment of the uninterrupted distribution. The corresponding moment of the interrupted distribution is given by

$$v_r'^{(n)} = E[S_n + (n-1)\theta]^r$$

$$= E(\xi_n')$$

$$= \mu_r' + \binom{r}{1}\mu_{r-1}'(n-1)\theta + \binom{r}{2}\mu_{r-2}'(n-1)\theta^2 + ... + ((n-1)\theta)^r$$

$$r = 1, 2, ... n \qquad (10.50)$$

In fact, to estimate a, k and δ in (10.48) it is sufficient to consider the estimating equation $v_1'^{(n)}$, $v_2'^{(n)}$ and $v_3'^{(n)}$.

A Special Case: Let

$$\lambda_1 = \lambda_2 = ... = \lambda_n = \lambda$$

In this case

$$L(f_n(s_n | \lambda)) = L[f(t_i | \lambda)]^n$$

Since $s_n = \sum_{i=1}^{n} t_i$ and T_i's are independent random variables

$$L(f_n(S_n | \lambda)) = E[\exp(-S(T_1 + T_2 + ... + T_n)) | \lambda]$$

$$= \left[\lambda \int_0^\infty e^{-St} e^{-\lambda t} dt \right]^n$$

$$= \frac{\lambda^n}{(\lambda + s)^n}$$

$$\Rightarrow \qquad f_n(S_n | \lambda) = L^{-1}\left(\frac{\lambda^n}{(\lambda + s)^n} \right) = \frac{\lambda^n \exp(-\lambda S_n) S_n^{n-1}}{\Gamma(n)} \qquad (10.51)$$

If λ conforms to the distribution (10.18)

$$\Rightarrow \qquad f_n(S_n) = \int_0^\infty f_n(S_n | \lambda) \phi(\lambda) d\lambda$$

$$= \frac{a^k}{\Gamma(k)} \int_0^\infty \exp(-\lambda(a + S_n))(S_n)^{n-1} \lambda^{n+k-1} d\lambda$$

Put $\lambda (a + S_n) = \xi$

$$= \frac{a^k}{\Gamma(k)} \int_0^\infty e^{-\xi} \left(\frac{\xi}{a + S_n}\right)^{n+k-1} \frac{d\xi}{a + S_r} (S_n)^{n-1}$$

$$f_n(S_n) = \frac{a^k}{\Gamma(k)} \frac{S_n^{n-1}}{(a + S_n)^{n+k}} \frac{\Gamma(n+k)}{\Gamma(n)} \qquad (10.52)$$

Finally, incorporating a dead time θ corresponding to every registration

$$\xi_n = S_n + (n - 1) \theta$$

we have

$$f_n(\xi_n) = \frac{a^k k (k+1) \dots (k+n-1) (\xi_n - (n-1)\theta)^{n-1}}{1.2.3 \dots (n-1) (a + \xi_n - (n-1)\theta)^{k+n}};$$

$$(n - 1) \theta \le \xi_n < \infty \qquad (10.53)$$

we observe that $(\xi_n - (n - 1) \theta)$ conforms to a Beta distribution of type II and also

$$v_r'^{(n)} = E (\xi_n - (n - 1) \pi)^r; \quad r = 1, 2, 3, \dots$$

$$= \frac{k (k+1) \dots (k+n-1) a^r \Gamma(k-r) \Gamma(r+n)}{\Gamma(k+n)} \qquad (10.54)$$

(10.54) has also been alternatively derived by Biswas and Nauharia (1978) by using Dandekar's modified distribution.

10.5 Problems for the Development of the Model of the Interarrival Waiting Time Distribution

Introduction

Let $(T_i - T_{i-1})$ be the waiting time for the ith order of conception measured from the data of $(i-1)$th order of conception $(i = 2, 3, \dots)$; or $(T_i - T_{i-1})$ may be called 'Inter Conception' interval. If we assume that the conception rate λ varies from individual to individual even within the same interconception interval conforming to some probability distribution say Gamma distribution as in the earlier section; then because of weighting of λ we shall see

  (i)   Renewal intervals $(T_i - T_{i-1}) \forall i = 2, 3, \dots$ will cease to become i.i.d. r.v.s. (or i.d.r.v.s) contrary to the traditional assumption.
  (ii) The # renewals even in two non-overlapping intervals will be correlated. The extent of correlation will be proportional to the variance of λ.

As a result of the correlation because of the weighting of the process, the renewal structure is completed destroyed leading to the process to conform to infinitely divisible distributions with dependent increments.

The problem is then to obtain the interarrival distribution of such dependent processes.

10.5.1 Correlation between T_i and $T_{j-i} = T_j - T_i$ for $j > i$

Let us assume the probability distribution of λ to be

$$\phi(\lambda) = \frac{a^k}{\Gamma(k)} \exp(-a\lambda) \lambda^{k-1}; 0 \le \lambda < \infty; a, k > 0$$

We have

$$\text{Cov}(T_i, T_j) = E_\lambda(E(T_i T_j | \lambda) - E_\lambda(E(T_i | \lambda)) E_\lambda(E(T_j | \lambda)$$
$$(E(T_i T_j | \lambda) = E(T_i(T_i + T_{j-i}) | \lambda))$$
$$= E(T_i^2 | \lambda) + [\text{Cov}(T_i, T_{j-i} | \lambda) + E(T_i | \lambda) E(T_{j-i} | \lambda)]$$

Now
$$\text{Cov}(T_i, T_{j-i} | \lambda) = 0$$

since T_i and T_{j-i} being two non overlapping intervals.

$$E(T_i^2 | \lambda) = \frac{i(i+1)}{\lambda^2}, \quad E(T_i | \lambda) = \frac{i}{\lambda}$$

and
$$E(T_{j-i} | \lambda) = \frac{j-i}{\lambda}$$

Therefore
$$E(T_i T_j | \lambda) = i(j+1) E\left(\frac{1}{\lambda^2}\right)$$

$$E(T_i T_j) = \frac{i(j+1) a^2}{(k-1)(k-2)}$$

$$\left(\because E\left(\frac{1}{\lambda^2}\right) = \frac{a^2}{(k-1)(k-2)}\right)$$

$$\therefore \quad \text{Cov}(T_i, T_j) = \frac{a^2 i(j+k-1)}{(k-1)^2(k-2)}, \quad k > 2$$

$$\text{Var}(T_i) = \frac{a^2 i(i+k-1)}{(k-1)^2(k-2)}, \quad k > 2 \tag{10.55}$$

Next consider $\text{Cov}(T_i, T_{j-1})$
We note

$$\text{Cov}(T_i, T_j) = \text{Cov}(T_i, T_i + T_{j-i})$$
$$= E(T_i, (T_i + T_{j-i})) - E(T_i) E(T_i + T_{j-i})$$
$$= E(T_i^2) + E(T_i T_{j-i}) - [E(T_i)]^2 - E(T_i) E(T_{j-i})$$
$$= \text{Var}(T_i) + \text{Cov}(T_i, T_{j-i})$$

$$\therefore \quad \text{Cov}(T_i, T_{j-i}) = \text{Cov}(T_i T_j) - \text{Var}(T_i)$$

$$= \frac{a^2 \, i \, (j + k - 1)}{(k - 1)^2 \, (k - 2)} - \frac{a^2 \, i \, (i + k - 1)}{(k - 1)^2 \, (k - 2)}$$

$$= \frac{a^2 \, i \, (j - i)}{(k - 1)^2 \, (k - 2)}, \; k > 2 \tag{10.56}$$

Also

$$\text{Var} \, (T_{j-i}) = \text{Var} \, (T_j) + \text{Var} \, (T_i) - 2 \, \text{Cov} \, (T_i, T_j)$$

$$= \frac{a^2 \, j \, (j + k - 1)}{(k - 1)^2 \, (k - 2)} + \frac{a^2 \, i \, (i + k - 1)}{(k - 1)^2 \, (k - 2)} - \frac{a^2 \, i \, (j + k - 1)}{(k - 1)^2 \, (k - 2)}$$

$$= \frac{a^2 \, (j - i) \, (j - i + k - 1)}{(k - 1)^2 \, (k - 2)} \tag{10.57}$$

$$\text{Cor} \, (T_i, T_{j-i}) = \frac{\text{Cov} \, (T_i, T_{j-i})}{\sqrt{\text{Var} \, (T_i)} \, \sqrt{\text{Var} \, (T_{j-i})}} \tag{10.58}$$

Putting (10.55), (10.56) and (10.57) in (10.58) $\Rightarrow$

$$\text{Cor} \, (T_i, T_{j-i}) = \frac{i \, (j - i)}{\sqrt{i \, (i + k - 1)} \, \sqrt{(j - 1) \, (j - i + k - 1)}} \tag{10.59}$$

which is independent of a.

As a special case if $i = 1$ and $j = 2$

$$\text{Cor} \, (T_1, T_2 - T_1) = \frac{1}{k} \tag{10.59'}$$

The result is due to Biswas and Sehgal (1986).

10.5.2 Correlation Between the # Events in Two Non-overlapping Intervals

Let us take two non-overlapping intervals $(0, s)$ and (s, t); $t > s$

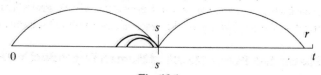

Fig. 10.1

Let $\quad X(s)$ denote the # conceptions in $(0, s)$

and $\quad [X(t) - X(s)]$ the # conceptions in (s, t)

Then

$\text{Cov} \, (X(s), X(t) - X(s))$

$= E_\lambda \, [E \, (X(s)) \, (X(t) - X(s) \, | \, \lambda)] - E_\lambda \, [E \, (X(s)) \, | \, \lambda] \, E_\lambda \, [E(X(t) - X(s)) \, | \, \lambda]$

$$= E_\lambda [\lambda s \, \lambda \, (t - s)] - E_\lambda [s\lambda] \, E_\lambda [\lambda \, (t - s)]$$

$$= s(t - s) \, E_\lambda \, (\lambda^2) - s(t - s) \, [E_\lambda \, (\lambda)]^2$$

$$= s(t - s) \, \text{Var} \, (\lambda) \tag{10.60}$$

The result is due to Biswas and Pachal (1983).

10.5.3 Discussion and the Role of Palm Probability

Because of (10.58), (10.59) and (10.60) it shows that the waiting time distribution between $(1 - 2)$th order of conceptions cannot be taken independent $(0-1)$st order of conceptions (where '0' may stand for the date of effective marriage). Also the interval from $(0-2)$ cannot be imagined as the convolution $(0-1)$ and $(1-2)$ since T_{01} and T_{12} will be correlated because of the weighting of the process. Further, since the renewal structure is destroyed because of the weighting of the process, therefore, the problem of obtaining the waiting time distribution for the ith conceptive delay T_i given that $(i-1)$th conceptive delay has occured at some $T_{i-1} = t_{i-1}$ naturally arises. We have shown in 6.2 the application of Palm Probability in solving the problems of above type because of the weighting of the process. However, the motivation of using Palm Probabilistic technique while obtaining the distribution of Inter conceptive delays of a weighted process from above is very clear. The illustrations of the same are in art 6.2.

10.5.4 Remarks

The result (10.60) shows that

$$\text{Cov} \, (X(s), X(t) - X(s)) \, = \, s \, (t - s) \, \text{Var} \, (\lambda) \quad \text{for} \quad t > s$$

As $\quad s \to 0 \Rightarrow \text{Cor} \, (X(s), X(t) - X(s)) \to 0.$

By the condition of orderliness, as $s \to 0$ the infinitesimal interval cannot have more than one renewal. As $s \to 0$ the same event, under the limiting condition, at most one event may occur at the beginning of the interval (s, t) (See figure (10.1), in which case we get the Palm Probability has been defined as a limiting probability of having a given # renewals in (s, t); given that a renewal has occurred at the beginning of the interval, as a natural sequence.

10.6 Sheps and Perrin Model of Human Reproductive Process

Here we discuss a model of Perrin and Sheps (1964) dealing with various states of the human reproductive process; thus describing the intrinsic fertility status of a woman by a semi-markov process (or Markov renewal process). Semi-markov processes are generalizations of both Markov-chain (discrete) and markov-processes (continuous) with countable number of state spaces. They were independently introduced by P. Levy, W.L. Smith and L. Takac's in 1954. Roughly speaking, a semi-Markov process is a stochastic process, under which movement occurs from one state to another (of at most a countable number of states) with the visit of successive states in the pattern of Markov chain; while the stay in a given state is for a random length of time (unlike Markov-process), the distribution of which may depend on the present state as well as on the state to be visited next. If we denote the process by $\{X_n; t_n\}$ $(n = 1, 2, 3, ...)$ with the above condition then the process is called a Markov-renewal process and then $X(t)$ is a semi-Markov process.

The formulation of Sheps and Perrin model starts with the modelling viz a woman at $T = 0$ being in the state of S_0 which is the non-pregnant fecundable state. After a random period of time the woman may enter into the state of pregnancy denoted by S_1 from which she may enter the post-partum non-susceptible (or non-fecundable state) following a live birth (state S_4) or the state of still birth (state S_3 or the state of early foetal death (state S_2). After a stay for a variable duration of time in any of these states, it is assumed that the first phase of the sojurn is over while the woman comes back to state S_0. However, the renewal of the state S_0 of the process continuous indefinitely, under the model which is not the actual situation in reality.

A diagrammatic representation of Perrin and Sheps Model is presented below;

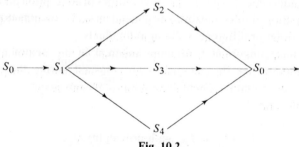

Fig. 10.2

The length of stay in each state is viewed as a random variable as in the case of Markov-Renewal Process. The details for the development of the model are presented below:

S_0 = Non-pregnant fecundable state.

S_1 = Pregnant state.

S_2 = Post-partum infecundable State associated with abortion or early foetal loss.

S_3 = Post-partum infecundable period associated with still birth.

S_4 = Post-partum infecundable period associated with live birth.

ρ = fecundability (probability of passage from S_0 to S_1).

For i, j = 0, 1, 2, 3, 4:

T_{ij} = the random time (in months) required for passage from S_i to S_j.

$$\mu_{ij} = E\,(T_{ij}) = \text{mean passage time from } S_i \text{ to } S_j$$

$$\sigma_{ij}^2 = \text{Var}\,(T_{ij}) = \text{Variance of passage time from } S_i \text{ to } S_j$$

For $i = 2, 3, 4:$

$$\theta_i = \text{Probability of direct passage from } S_i \text{ to } S_j$$

and $v_i, \xi_i^2 = $ the mean and variance respectively, of the length of the stay in S_i given that S_i is the next state period from S_1

$$\eta_i = v_i + \mu_{i0} = \text{mean duration of gestation plus post-partum in-} \\ \text{fecundable for outcome } i \qquad (10.61)$$

$$\lambda_i^2 = \xi_i^2 + \sigma_{i0}^2 = \text{Variance of duration of gestation plus post-} \\ \text{partum infecundable period for outcome } i. \ (10.62)$$

Assuming for the present that the relevant parameters remain constant during the period of observation it may be shown that the resulting Stochastic Process is a Markov renewal process.

Restrictions of the Model
   (i)   The model does not permit the probabilities of conception or foetal loss or duration of infecundability to depend on such salient factors as age, birth order or differences among individuals.
   (ii)   The most important restrictions inherent in the treatment are the assumption that the parameters are constant during the period of observation and the model does not provide any scope.

Now under this model,

$$T_{00} = \begin{cases} T_{01} + T_{12}^* + T_{20} & \text{with probability } \theta_2\,; \\ T_{01} + T_{13}^* + T_{30} & \text{with probability } \theta_3\,; \\ T_{01} + T_{14}^* + T_{40} & \text{with probability } \theta_4\,; \end{cases} \qquad (10.63)$$

$\therefore$ $E\,[T_{01}] = \mu_{01} = \dfrac{(1-\rho)}{\rho}$ (being the mean of geometric

distribution with parameter ρ)   (10.64)

and $\text{Var}\,[T_{01}] = \sigma_{01}^2 = \dfrac{(1-\rho)}{\rho^2}$ (being the variance of geometric

distribution with parameter ρ)   (10.65)

Thus the expected value of T_{00}, the recurrence time for the fecundable non-pregnant state, is a weighted mean of the expected time spent in each of three paths leading to recurrence.
In particular, if we let

$$\eta_i = v_i + \mu_{i0}; \ i = 1, 2, 3, 4$$

$$\mu_{00} = E(T_{00}) = \sum_{i=2}^{4} \theta_i\, E\,[T_{01} + T_{1i}^* + T_{i0}]$$

$$= (\theta_2 + \theta_3 + \theta_4)\, E\,(T_{01}) + \sum_{i=2}^{4} \theta_i\, E\,[T_{1i}^* + T_{i0}]$$

$$= E\,(T_{01}) + \sum_{i=2}^{4} \theta_i\,(\nu_i + \mu_{i0})$$

$$(\because\ \theta_2 + \theta_3 + \theta_4 = 1,\ E(T_{1i}^*) = \nu_i, E(T_{i0}) = \mu_{i0})$$

$$\therefore \qquad \mu_{00} = E(T_{00}) = \frac{(1-\rho)}{\rho} + \sum_{i=2}^{4} \theta_i\, \eta_i \qquad\qquad (10.66)$$

Similarly,

$$E(T_{00}^2) = \sum_{i=2}^{4} \theta_i\, E\,(T_{01} + T_{1i}^* + T_{i0})^2$$

$$= \sum_{i=2}^{4} \theta_i\, E\,(T_{01}^2) + \sum_{i=2}^{4} \theta_i\, E\,[T_{1i}^* + T_{i0}]^2 + 2\sum_{i=2}^{4} \theta_i\, E\,[T_{0i}\,(T_{1i}^* + T_{i0})]$$

$$= E\,(T_{01}^2) + \sum_{i=2}^{4} \theta_i\, E\,[T_{1i}^* + T_{i0}]^2 + 2\sum_{i=2}^{4} \theta_i\, E\,(T_{01})\,(T_{1i}^* + T_{i0}) \tag{10.67}$$

Similarly,

$$\{E\,(T_{00})\}^2 = \left\{ \sum_{i=2}^{4} \theta_i\, E\,(T_{01} + T_{1i}^* + T_{i0}) \right\}^2$$

$$= \left\{ \sum_{i=2}^{4} \theta_i\, E\,(T_{01}) \right\}^2 \left\{ \sum_{i=2}^{4} \theta_i\, E\,(T_{1i}^* + T_{i0}) \right\}^2$$

$$+ 2E\,(T_{01}) \sum_{i=2}^{4} \theta_i\, E\,(T_{1i}^* + T_{i0})$$

$$= \{E\,(T_{01})\}^2 + \sum_{i=2}^{4} \theta_i^2\, [E\,(T_{1i}^* + T_{i0})]^2$$

$$+ 2 \sum_{i<j} \theta_i \theta_j \left[E\left(T_{1i}^* + T_{i0}\right) E\left(T_{1j}^* + T_{j0}\right) \right]$$

$$+ 2 E\left(T_{01}\right) \sum_{i=2}^{4} \theta_i E\left(T_{1i}^* + T_{i0}\right)$$

$$= \left[E\left(T_{01}\right) \right]^2 + \left(\sum_{i=2}^{4} \theta_i \right) \left(\sum_{i=2}^{4} \theta_i \{ E\left(T_{1i}^* + T_{i0}\right) \}^2 \right)$$

$$- \sum_{\substack{i=2 \\ i<j}}^{4} \theta_i \theta_j \{ E\left(T_{1i}^* + T_{i0}\right) \}^2 - \sum_{\substack{i=2 \\ i<j}}^{4} \theta_i \theta_j \{ E\left(T_{1j}^* + T_{j0}\right) \}^2$$

$$+ 2 \sum_{\substack{i=2 \\ i<j}}^{4} \theta_i \theta_j E\left(T_{1i}^* + T_{i0}\right) E\left(T_{1j}^* + T_{j0}\right)$$

$$+ 2 E\left(T_{01}\right) \sum_{i=2}^{4} \theta_i E\left(T_{1i}^* + T_{i0}\right)$$

$$= \left[E\left(T_{01}\right) \right]^2 + \sum_{i=2}^{4} \theta_i \left[E\left(T_{1i}^* + T_{i0}\right) \right]^2$$

$$- \sum_{\substack{i,j=2 \\ i<j}}^{4} \theta_i \theta_j \left[E\left(T_{1i}^* + T_{i0}\right) \right]^2 - \sum_{\substack{i,j=2 \\ i<j}}^{4} \theta_i \theta_j \left[E\left(T_{1j}^* + T_{j0}\right) \right]^2$$

$$+ 2 \sum_{\substack{i,j=2 \\ i<j}}^{4} \theta_i \theta_j E\left(T_{1i}^* + T_{i0}\right) E\left(T_{1j}^* + T_{j0}\right)$$

$$+ 2 \sum_{i=2}^{4} \theta_i E\left(T_{01}\right) E\left(T_{1i}^* + T_{i0}\right) \tag{10.68}$$

Subtracting (10.68) from (10.67), we get

$$\text{Var}\left(T_{00}\right) = E\left(T_{00}^2\right) - \left[E\left(T_{00}\right) \right]^2$$

$$= \{ E\left(T_{01}^2\right) - \left[E\left(T_{01}\right) \right]^2 \} + \sum_{i=2}^{4} \theta_i E\left(T_{1i}^* + T_{i0}\right)$$

$$+ 2 \sum_{i=2}^{4} \theta_i E\left(T_{01}\right) E\left(T_{1i}^* + T_{i0}\right) - \sum_{i=2}^{4} \theta_i \left[E\left(T_{1i}^* + T_{i0}\right) \right]^2$$

$$+ \sum_{\substack{i,j=2 \\ i<j}}^{4} \theta_i \, \theta_j \, [E \, (T_{1i}^* + T_{i0})]^2 - 2 \sum_{\substack{i<j}}^{4} \theta_i \, \theta_j \, E \, (T_{1i}^* + T_{i0}) \, E \, (T_{1j}^* + T_{j0})$$

$$+ \sum \theta_i \, \theta_j \, [E \, (T_{1j}^* + T_{j0}))^2 - 2 \sum_{i=2}^{4} \theta_i \, E \, (T_{01}) \, E \, (T_{1i}^* + T_{i0})$$

$$= \operatorname{Var} (T_{01}) + \sum_{i=2}^{4} \theta_i \, \{[E \, (T_{1i}^* + T_{i0})]^2 - E \, (T_{1i}^* + T_{i0})^2\}$$

$$+ \sum_{i<j}^{4} \theta_i \, \theta_j \, \{E \, (T_{1i}^* + T_{i0}) - E \, (T_{1j}^* + T_{j0})\}^2$$

$$= \operatorname{Var} (T_{01}) + \sum_{i=2}^{4} \theta_i \, \operatorname{Var} (T_{1i}^* + T_{i0}) + \sum_{i<j}^{4} \theta_i \, \theta_j \, \{E \, (T_{1i}^* + T_{i0})$$

$$- E \, (T_{1j}^* + T_{j0})\}^2 \tag{10.69}$$

Now

$$\operatorname{Var} (T_{01}) = \frac{(1-\rho)}{\rho^2}, \quad \operatorname{Var} (T_{1i}^*) = \xi_i^2 \text{ and } \operatorname{Var} (T_{i0}) = \sigma_{i0}^2$$

$$i = 2, 3, 4.$$

and $\operatorname{Cov} (T_{1i}^*, T_{i0}) = 0$ (this assumption is not strictly justified in this model)

Putting
$$\lambda_i^2 = \xi_i^2 + \sigma_{i0}^2; \; i = 2, 3, 4$$

$$\eta_i = \nu_i + \mu_{i0}$$

where
$$\nu_i = E \, (T_{1i}^*) \text{ and } \mu_{i0} = E \, (T_{i0}).$$

Therefore (10.69) becomes

$$\operatorname{Var} (T_{00}) = \frac{(1-\rho)}{\rho^2} + \sum_{i=2}^{4} \theta_i \, \lambda_i^2 + \sum_{i<j}^{4} \theta_i \theta_j \, (\eta_i - \eta_j)^2 \tag{10.70}$$

A quantity of special interest is T_{44}, the waiting time between successive live births per female. This time can be represented as follows:

$$T_{44} = T_{40} + T_{01} + T_{11.4}^{(1)} + T_{11.4}^{(2)} + \ldots + T_{11.4}^{(N)} + T_{14}^* \tag{10.71}$$

where $T_{11.4}^{(k)}$ represents the time spent in the kth consecutive recurrence of state S_1 accomplished without passage through state S_4. The total number of times, N, which the female cycles between pregnancies without the occurrence of a live birth is a random variable and is easily seen to have (in this model) a geometric distribution with mean $\dfrac{\theta_2 + \theta_3}{\theta_4}$ and variance $\dfrac{(\theta_2 + \theta_3)}{\theta_4^2}$

Now

$$T_{11.\bar{4}} = \begin{cases} T_{12}^* + T_{20} + T_{01} \text{ with probability } \dfrac{\theta_2}{\theta_2 + \theta_3} \\[3mm] T_{13}^* + T_{30} + T_{01} \text{ with probability } \dfrac{\theta_3}{\theta_2 + \theta_3} \end{cases} \tag{10.72}$$

$$E\,(T_{11.\bar{4}}) = \frac{1}{\theta_2 + \theta_3} \sum_{i=2}^{3} \theta_i\, E\,(T_{1i}^* + T_{i0} + T_{01})$$

$$= \frac{\theta_2}{\theta_2 + \theta_3}\, E\,(T_{12}^* + T_{20} + T_{01}) + \frac{\theta_3}{\theta_2 + \theta_3}\, E\,(T_{13}^* + T_{30} + T_{01})$$

$$= \frac{\theta_2}{\theta_2 + \theta_3}\, E\,(T_{12}^* + T_{20}) + \frac{\theta_3}{\theta_2 + \theta_3}\, E\,(T_{13}^* + T_{30}) + \frac{(\theta_2 + \theta_3)}{(\theta_2 + \theta_3)}\, E\,(T_{01})$$

$$= \frac{\theta_2}{\theta_2 + \theta_3}\, (\nu_2 + \mu_{20}) + \frac{\theta_3}{\theta_2 + \theta_3}\, (\nu_3 + \mu_{30}) + E\,(T_{01})$$

$$(\because E\,(T_{i0}) = \mu_{i0},\ E\,(T_{1i}^*) = \nu_i)$$

$$= \frac{\theta_2}{\theta_2 + \theta_3}\, \eta_2 + \frac{\theta_3}{\theta_2 + \theta_3}\, \eta_3 + \frac{(1-\rho)}{\rho} \tag{10.73}$$

and $\quad \text{Var}\,(T_{11.\bar{4}}) = E\,(T_{11.\bar{4}}^2) - [E\,(T_{11.\bar{4}})]^2$

Now $\quad E\,(T_{11.\bar{4}}^2) = \dfrac{1}{\theta_2 + \theta_3} \sum\limits_{i=2}^{3} \theta_i\, E\,(T_{1i}^* + T_{i0} + T_{01})^2$

$$= \frac{1}{\theta_2 + \theta_3} \left\{ \sum_{i=2}^{3} \theta_i\, [E\,(T_{1i}^* + T_{i0})^2 + E\,(T_{01})^2 \right.$$

$$\left. + 2\, E\,(T_{01})\, E\,(T_{1i}^* + T_{i0})] \right\}$$

$$= \frac{1}{\theta_2 + \theta_3} \left\{ \sum_{i=2}^{3} \theta_i\, E\,(T_{01})^2 + \sum_{i=2}^{3} \theta_i\, E\,(T_{1i}^* + T_{i0})^2 \right.$$

$$\left. + 2\, E\,(T_{01}) \sum_{i=2}^{3} \theta_i\, E\,(T_{1i}^* + T_{i0})] \right\}$$

$$= \frac{\left(\sum\limits_{i=2}^{3} \theta_i\right)}{\theta_2 + \theta_3} E(T_{01}^2) + \frac{1}{\theta_2 + \theta_3} \sum\limits_{i=2}^{3} \theta_i E(T_{1i}^* + T_{i0})^2$$

$$+ \frac{2 E(T_{01})}{\theta_2 + \theta_3} \sum\limits_{i=2}^{3} \theta_i E(T_{1i}^* + T_{i0})$$

or $\quad E(T_{11.4}^2) = E(T_{01}^2) + \dfrac{1}{\theta_2 + \theta_3} \sum\limits_{i=2}^{3} \theta_i E(T_{1i}^* + T_{i0})^2$

$$+ \frac{2 E(T_{01})}{\theta_2 + \theta_3} \sum\limits_{i=2}^{3} \theta_i E(T_{1i}^* + T_{i0}) \tag{10.74}$$

Similarly,

$$[E(T_{11.4})]^2 = \left\{ \frac{1}{\theta_2 + \theta_3} \sum\limits_{i=2}^{3} \theta_i E(T_{1i}^* + T_{i0} + T_{01}) \right\}^2$$

$$= \frac{1}{(\theta_2 + \theta_3)^2} \left\{ \sum\limits_{i=2}^{3} \theta_i E(T_{1i}^* + T_{i0}) + \sum\limits_{i=2}^{3} \theta_i E(T_{01}) \right\}^2$$

$$= \frac{1}{(\theta_2 + \theta_3)^2} \left\{ \left[\sum\limits_{i=2}^{3} \theta_i E(T_{1i}^* + T_{i0}) \right]^2 + \left(\sum\limits_{i=2}^{3} \theta_i \right)^2 (E(T_{01}))^2 \right.$$

$$+ 2 E(T_{01}) \left(\sum\limits_{i=2}^{3} \theta_i \right) \sum\limits_{i=2}^{3} \theta_i E(T_{1i}^* + T_{i0}) \Bigg\}$$

$$= \frac{1}{(\theta_2 + \theta_3)^2} \sum\limits_{i=2}^{3} \theta_i^2 [E(T_{1i}^* + T_{i0})]^2$$

$$+ \frac{2}{(\theta_2 + \theta_3)^2} \cdot \theta_2 \, \theta_3 \, E(T_{12}^* + T_{20})(T_{13}^* + T_{30}) + [E(T_{01})]^2$$

$$+ \frac{2 E(T_{01})}{(\theta_2 + \theta_3)^2} \sum\limits_{i=2}^{3} \theta_i E(T_{i1}^* + T_{i0})$$

$$= \frac{1}{(\theta_2 + \theta_3)^2} \left\{ \left(\sum\limits_{i=2}^{3} \theta_i \right) \sum\limits_{i=2}^{3} \theta_i [E(T_{1i}^* + T_{i0})]^2 \right.$$

$$- \theta_2 \theta_3 E \left(T_{12}^* + T_{20} \right)^2 - \theta_2 \theta_3 E \left(T_{13}^* + T_{30} \right) \}$$

$$+ \frac{2 \theta_2 \theta_3}{(\theta_2 + \theta_3)^2} E \left(T_{12}^* + T_{20} \right) E \left(T_{13}^* + T_{30} \right) + \left[E \left(T_{01} \right) \right]^2$$

$$+ \frac{2 E \left(T_{01} \right)}{(\theta_2 + \theta_3)} \sum_{i=2}^{3} \theta_i E \left(T_{1i}^* + T_{i0} \right) \tag{10.75}$$

Subtracting (10.75) from (10.74), we get

$$\text{Var} \left(T_{11.\overline{4}} \right) = E \left(T_{11.\overline{4}}^2 \right) - \left[E \left(T_{11.\overline{4}} \right) \right]^2$$

$$= E \left(T_{01}^2 \right) - \left[E \left(T_{01} \right) \right]^2 + \frac{1}{\theta_2 + \theta_3} \sum_{i=2}^{3} \theta_i \{ E \left(T_{1i}^* + T_{i0} \right)^2$$

$$- E \left(T_{1i}^* + T_{i0} \right)]^2 \} + \frac{\theta_2 \theta_3}{(\theta_2 + \theta_3)^2} \{ E \left(T_{12}^* + T_{20} \right) - E \left(T_{13}^* + T_{30} \right) \}^2$$

$$= \text{Var} \left(T_{01} \right) + \frac{1}{(\theta_2 + \theta_3)} \sum_{i=2}^{3} \theta_i \text{ Var} \left(T_{1i}^* + T_{i0} \right)$$

$$+ \frac{\theta_2 \theta_3}{(\theta_2 + \theta_3)^2} \left[(\nu_2 + \mu_{20}) - (\nu_3 + \mu_{30}) \right]^2$$

$$= \frac{(1 - \rho)}{\rho^2} + \frac{1}{(\theta_2 + \theta_3)} \sum_{i=2}^{3} \theta_i \lambda_i^2 + \frac{\theta_2 \theta_3}{(\theta_2 + \theta_3)^2} \left(\eta_2 - \eta_3 \right)^2 \tag{10.76}$$

Also from (10.71) we have

$$E \left(T_{44} \right) = E \left(T_{40} \right) + E \left(T_{01} \right) + E \left(T_{11.\overline{4}}^{(1)} \right) + E \left(T_{11.\overline{4}}^{(2)} \right) + \ldots + E \left(T_{11.\overline{4}}^{(N)} \right) + E \left(T_{14}^* \right)$$

$$= \mu_{40} + \frac{(1 - \rho)}{\rho} + \underbrace{E \left(T_{11.\overline{4}} \right) + E \left(T_{11.\overline{4}} \right) + \ldots + E \left(T_{11.\overline{4}} \right)}_{\text{upto some terms } N, \text{ which is a random variable}} + \nu_4$$

$$= \mu_{40} + \frac{(1 - \rho)}{\rho} + E \left(N \right) E \left(T_{11.\overline{4}} \right) + \nu_4 \tag{10.77}$$

We can also write

$$E \left(T_{44} \right) = \mu_{40} + \frac{(1 - \rho)}{\rho} + \left\{ \underbrace{E \left(T_{11.\overline{4}} \right) + E \left(T_{11.\overline{4}} \right) + \ldots + E \left(T_{11.\overline{4}} \right)}_{N \text{ terms}} - N E \left(T_{11.\overline{4}} \right) \right\}$$

$$+ E \left(N \right) E \left(T_{11.\overline{4}} \right) + \nu_4$$

$$= \mu_{40} + \frac{(1-\rho)}{\rho} + E(N)\, E(T_{11.\overline{4}}) + \nu_4$$

$$= \mu_{04} + \frac{(1-\rho)}{\rho} + E(N)\left\{ \frac{\displaystyle\sum_{i=2}^{3}\theta_i\,\eta_i}{\theta_2 + \theta_3} + \frac{(1-\rho)}{\rho} \right\} + \nu_4 \quad \text{from (10.73)}$$

Also we know that

$$\left. \begin{aligned} E(N) &= \frac{\theta_2 + \theta_3}{\theta_4} \\[2mm] \mathrm{Var}(N) &= \frac{\theta_2 + \theta_3}{\theta_4^2} \end{aligned} \right\} \tag{10.78}$$

$$\therefore \quad E(T_{44}) = \left[\mu_{40} + \frac{(1-\rho)}{\rho} \right] + \left[\frac{\displaystyle\sum_{i=2}^{3}\theta_i\,\eta_i}{\theta_2 + \theta_3} + \frac{(1-\rho)}{\rho} \right] \cdot \frac{\theta_2 + \theta_3}{\theta_4} + \nu_4$$

$$= \mu_{40} + \frac{(1-\rho)}{\rho} + \frac{\displaystyle\sum_{i=2}^{3}\theta_i\,\eta_i}{\theta_4} + \frac{(1-\rho)}{\rho}\cdot\frac{(\theta_2+\theta_3)}{\theta_4} + \nu_4$$

$$= \nu_4 + \mu_{40} + \frac{(1-\rho)}{\rho}\left\{ 1 + \frac{(\theta_2+\theta_3)}{\theta_4} \right\} + \frac{\displaystyle\sum_{i=2}^{3}\theta_i\,\eta_i}{\theta_4}$$

$$= \eta_4 + \frac{(1-\rho)}{\rho}\left\{ \frac{(\theta_2+\theta_3+\theta_4)}{\theta_4} \right\} + \frac{\displaystyle\sum_{i=2}^{3}\theta_i\,\eta_i}{\theta_4}$$

$$= \frac{\theta_4\,\eta_4}{\theta_4} + \frac{(1-\rho)}{\rho}\cdot\frac{1}{\theta_4} + \frac{\displaystyle\sum_{i=2}^{3}\theta_i\,\eta_i}{\theta_4}$$

or $\quad E(T_{44}) = \dfrac{1}{\theta_4}\left\{ \dfrac{(1-\rho)}{\rho} + \displaystyle\sum_{i=2}^{4}\theta_i\,\eta_i \right\} \tag{10.79}$

Precisely in a similar way,

$$E(T_{22}) = \mu_{22} = \text{Average waiting time between mis-carriages}$$

$$= \frac{1}{\theta_2}\left\{\frac{(1-\rho)}{\rho} + \sum_{i=2}^{4}\theta_i\,\eta_i\right\} \tag{10.80}$$

$$E(T_{33}) = \mu_{33} = \text{Average waiting time between still births}$$

$$= \frac{1}{\theta_3}\left\{\frac{(1-\rho)}{\rho} + \sum_{i=2}^{4}\theta_i\,\eta_i\right\} \tag{10.81}$$

Now to find the variance of T_{44}, the waiting time between live births

$$\text{Var}(T_{44}) = E(T_{44}^2) - [E(T_{44})]^2$$

First consider,

$$T_{44} - E(T_{44}) = T_{44} - \left[\mu_{40} + \frac{(1-\rho)}{\rho} + E(N)E(T_{11.4}) + v_4\right]$$

$$= (T_{40} - \mu_{40}) + \left(T_0 - \frac{(1-\rho)}{\rho}\right) + (T_{14}^* - v_4)$$

$$+ (T_{11.4}^{(1)} - E(T_{11.4})) + \dots + [T_{11.4}^{(N)} - E(T_{11.4})] + NE(T_{11.4}) - E(N)E(T_{11.4}) \tag{11.82}$$

Squaring both sides and taking expected values, we get

$$\text{Var}(T_{44}) = \sigma_{44}^2 = E(T_{44} - E(T_{44})]^2$$

$$= E(T_{40} - \mu_{40})^2 + E(T_{01} + \mu_{01})^2 + E(T_{14}^* - v_4)^2$$

$$+ E(N)E(T_{11.4} - E(T_{11.4}))^2 + [E(T_{11.4})]^2 E(N - E(N))^2$$

$$= \lambda_4^2 + \sigma_{01}^2 + E(N)\text{Var}(T_{11.4}) + [E(T_{11.4})]^2\text{Var}(N)$$

$$(\because E(T_{40} - E(T_{40}))^2 = \sigma_{40}^2,\ E(T_{14}^* - v_4)^2 = \xi_4^2 \text{ and } \xi_4^2 + \sigma_{40}^2 = \lambda_4^2)$$

$$\therefore\ \text{Var}(T_{44}) = \lambda_4^2 + \frac{(1-\rho)}{\rho^2} + \frac{(\theta_2 + \theta_3)}{\theta_4}\text{Var}(T_{11.4}) + \frac{(\theta_2 + \theta_3)}{\theta_4^2}\{E(T_{11.4})\}^2$$

$$= \lambda_4^2 + \frac{(1-\rho)}{\rho^2} + \frac{(\theta_2 + \theta_3)}{\theta_4}\left\{\frac{(1-\rho)}{\rho^2} + \frac{1}{(\theta_2 + \theta_3)}\sum_{i=2}^{3}\theta_i\,\lambda_i^2\right.$$

$$+ \frac{\theta_2\,\theta_3}{(\theta_2 + \theta_3)^2}(\eta_2 - \eta_3)^2\bigg\} + \frac{(\theta_2 + \theta_3)}{\theta_4^2}\left\{\frac{\theta_2\,\eta_2}{\theta_2 + \theta_3}\right.$$

$$\left.+ \frac{\theta_3\,\eta_3}{\theta_2 + \theta_3} + \frac{(1-\rho)}{\rho}\right\}^2 \quad \text{(using (10.73) and (10.76))}$$

$$= \lambda_4^2 + \frac{(1-\rho)}{\rho^2}\left\{1 + \frac{\theta_2 + \theta_3}{\theta_4}\right\} + \frac{1}{\theta_4}\sum_{i=2}^{3}\theta_i\,\lambda_i^2$$

$$+ \frac{\theta_2\,\theta_3}{(\theta_2 + \theta_3)\,\theta_4}\cdot(\eta_2 - \eta_3)^2 + \frac{(\theta_2 + \theta_3)}{\theta_4^2}\left\{\frac{\theta_2^2\,\eta_2^2}{(\theta_2 + \theta_3)^2}\right.$$

$$+ \frac{\theta_3^2\,\eta_3^2}{(\theta_2 + \theta_3)^2} + \frac{(1-\rho)^2}{\rho^2} + \frac{2\,\theta_2\,\theta_3}{(\theta_2 + \theta_3)^2}\,\eta_2\,\eta_3$$

$$+ \frac{2\,\theta_2\,\eta_2}{(\theta_2 + \theta_3)}\,\frac{(1-\rho)}{\rho} + \left.\frac{2\,\theta_3\,\eta_3}{(\theta_2 + \theta_3)}\,\frac{(1-\rho)}{\rho}\right\}$$

$$= \frac{\theta_4\,\lambda_4^2}{\theta_4} + \frac{(1-\rho)}{\rho^2}\frac{1}{\theta_4} + \frac{1}{\theta_4}\sum_{i=2}^{3}\theta_i\,\lambda_i^2 + \frac{\theta_2\,\theta_3}{\theta_4\,(\theta_2 + \theta_3)}\,(\eta_2 - \eta_3)^2$$

$$+ \frac{\theta_2^2}{\theta_4^2\,(\theta_2 + \theta_3)}\,\eta_2^2 + \frac{\theta_3^2}{\theta_4^2\,(\theta_2 + \theta_3)}\,\eta_3^2$$

$$+ \frac{(\theta_2 + \theta_3)}{\theta_4^2}\cdot\frac{(1-\rho)^2}{\rho^2} + \frac{2\,\theta_2\,\theta_3}{\theta_4^2\,(\theta_2 + \theta_3)}\,\eta_2\,\eta_3$$

$$+ \frac{2\,\theta_2\,\eta_2}{\theta_4^2}\cdot\frac{(1-\rho)}{\rho} + \frac{2\,\theta_3\,\eta_3}{\theta_4^2}\cdot\frac{(1-\rho)}{\rho}$$

$$= \frac{(1-\rho)}{\theta_4\cdot\rho} + \frac{1}{\theta_4}\sum_{i=2}^{4}\theta_i\,\lambda_i^2 + \frac{\theta_2\,\theta_3}{\theta_4\,(\theta_2 + \theta_3)}\,(\eta_2^2 + \eta_3^2 - 2\eta_2\eta_3)$$

$$+ \frac{\theta_2^2\,\eta_2^2}{\theta_4^2\,(\theta_2 + \theta_3)} + \frac{\theta_3^2\,\eta_3^2}{\theta_4^2\,(\theta_2 + \theta_3)} + \frac{(\theta_2 + \theta_3)}{\theta_4^2}\frac{(1-\rho)^2}{\rho^2}$$

$$+ \frac{2\,\theta_2\,\theta_3}{\theta_4^2\,(\theta_2 + \theta_3)}\,\eta_2\,\eta_3 + \frac{2\,\theta_2\,\eta_2}{\theta_4^2}\cdot\frac{(1-\rho)}{\rho} + \frac{2\,\theta_3\,\eta_3}{\theta_4^2}\frac{(1-\rho)}{\rho}$$

$$= \frac{1}{\theta_4}\cdot\frac{(1-\rho)}{\rho} + \frac{1}{\theta_4}\sum_{i=2}^{4}\theta_i\,\lambda_i^2 + \frac{(\theta_2 + \theta_3)}{\theta_4^2}\frac{(1-\rho)^2}{\rho^2}$$

$$+ \frac{2}{\theta_4}\frac{(1-\rho)}{\rho}\sum_{i=2}^{3}\theta_i\,\eta_i + \frac{\theta_2\,\theta_3}{\theta_4\,(\theta_2 + \theta_3)}\,(\eta_2^2 + \eta_3^2 - 2\eta_2\eta_3)$$

$$+ \frac{1}{\theta_4^2\,(\theta_2 + \theta_3)}\,\{\theta_2^2\,\eta_2^2 + \theta_3^2\,\eta_3^2 + 2\theta_2\theta_3\eta_2\eta_3\}$$

$$= \frac{1}{\theta_4} \cdot \frac{(1-\rho)}{\rho} + \frac{1}{\theta_4} \sum_{i=2}^{4} \theta_i \, \lambda_i^2 + \frac{(\theta_2 + \theta_3)}{\theta_4^2} \frac{(1-\rho)^2}{\rho^2}$$

$$+ \frac{2}{\theta_4} \frac{(1-\rho)}{\rho} \sum_{i=2}^{3} \theta_i \, \eta_i + \frac{\theta_2 \, \theta_3}{\theta_4 \, (\theta_2 + \theta_3)} (\eta_2 - \eta_3)^2$$

$$+ \frac{1}{\theta_4^2 \, (\theta_2 + \theta_3)} \{\theta_2 \, \eta_2 + \theta_3 \, \eta_3\}^2 \tag{10.83}$$

It is also of interest to determine the expected length and variance of the waiting time T_{04} between the onset of marriage and the first live birth. This is easily done since $T_{44} = T_{40} + T_{04}$

and therefore $\qquad \mu_{04} = E\,(T_{04}) = \mu_{44} - \mu_{40}$

$$= \nu_4 + \mu_{40} + \frac{1}{\theta_4} \left(\theta_2 \eta_2 + \theta_3 \eta_3 + \frac{(1-\rho)}{\rho} \right) - \mu_{40}$$

$$= \nu_4 + \frac{1}{\theta_4} \left(\theta_2 \eta_2 + \theta_3 \eta_3 + \frac{(1-\rho)}{\rho} \right)$$

And $\qquad \sigma_{04}^2 = \sigma_{44}^2 - \sigma_{40}^2$

$$= \xi_4^2 + \frac{(1-\rho)}{\rho^2} + \frac{(\theta_2 + \theta_3)}{\theta_4} \cdot \text{Var}\,(T_{11.\overline{4}}) + \frac{(\theta_2 + \theta_3)}{\theta_4^2} [E\,(T_{11.\overline{4}})]^2$$

$$= \xi_4^2 + \frac{(1-\rho)}{\rho^2} \cdot \frac{1}{\theta_4} + \frac{1}{\theta_4} \sum_{i=2}^{3} \theta_i \, (\lambda_i^2 + \eta_i^2) + \frac{(\theta_2 + \theta_3)}{\theta_4} \cdot \frac{(1-\rho)^2}{\rho^2}$$

$$+ \frac{2}{\theta_4} \cdot \left(\frac{1-\rho}{\rho} \right) \sum_{i=2}^{3} \theta_i \, \eta_i$$

The analogous expression for the first moment of the waiting time between marriage and the first mis-carriage or still birth (i.e. of T_{02} and T_{03}) may similarly be derived. Application of the above results to the study of intervals to the first live birth and between consecutive births also have been described.

References

1. Berclay, G.W. (1958): *Techniques of population analysis*, John Wiley & Sons Inc. New York.
2. Brass William (1958): The distribution of births in Human population, *Population studies*; Vol. 12, page 51–72.
3. Biswas, S. (1973): A note on the generalization of William Brass model: *Demography*, August, 1973 Vol. 10, No. 3 page 450–452.
4. Biswas, S. (1975): On a more generalised probability model of the waiting time of conception based on censored sampling from a mixed population: *Sankhyā*, series

B, Vol. 37 Part 3, August, 1975, page 343–354.

5. Biswas, S. (1980): On the extension of some results of counter models with Poisson inputs and their applications; *Journal of Indian Statistical Association.* Vol. 18 page 45–53.

6. Biswas, S. and Nauhria Indu (1980): A note on the development of some interrupted waiting time distribution; *Pure and Applied mathematica Sciences* Vol. XI.

7. Biswas S and Pachal, T.K (1981): On the application of Palm probability of obtaining inter-arrival time distribution in weighted Poisson Process; *Calcutta Statistical Association bulletin.* Vol. 32, May-June, 1981, No. 125–126, page 111–123.

8. Biswas, S and Sehgal, V.K. (1986): On the correlation between inter-arrival delays of shocks in weighted Poisson Process; Under publication in Micro electronics and Reliability.

9. Biswas, S and Srestha Ganga (1984): Waiting time distribution for first conception leading to a live birth-International report IC/85/98, *International centre of Theoretical Physics,* Trieste, Italy.

10. Biswas, S and Srestha Ganga (1985): A probability model of the waiting time distribution between consecutive conceptions based on the data of live birth Metron, vol XLIV, Nos, 1-4 page 195–206.

11. Chiang, C.L (1971): A stochastic model on Human fertility, multiple transition probabilities; *Biometrics,* Vol. 27 pp. 345–356.

12. Dandekar, V.M. (1955): Certain modified forms of Binomial and Poisson distribution-*Sankhyā*, Series B, Vol 15, page 237–250.

13. Dharmadhikari, S.W (1964): A generalization of a stochastic model considered by V.M. Dandekar; *Sankhyā*, Series A, Vol. 25, page 31–38.

14. Iosufescu, M and Tautu P (1973): Stochastic Processes and applications in Biology and Medicine II Models: *Biomathematics* Vol. 4. Springer-Verlag, Berlin; Helberg, New York

15. Keyfitz, N (1977): *Introduction to the mathematics of population - with revision-* Addison Wesley & Co. London.

16. Neyman, J(1949): On the problem of estimating the number of schools of fish., University of California, Publication in statistics 1, page 21–32.

17. Neyman, J(1949): Contribution to the theory of the χ^2-test; *Proceedings of Berkley Symposium in Mathematics and Statistics,* University of California Press, page 239–273.

18. Pathak, K.B (1966): A probability distribution for number of conceptions: *Sankhyā*, Series B, Vol. 28, page 213–218.

19. Perrin, E.B. and Sheps, M.C. (1964): *Human reproduction A Stochastic Process*: Biometrics, Vol. 20. Page 28–45.

20. Poti S.J. and Biswas, S. (1963) A study of child health during the first year of life, *Sankhyā*, Vol. 26, Series B, part I and II. p, 35–120.

21. Potter, R.G. and Parker, M.P. (1964): Predicting the time required to conceive: *Population Studies,* Vol XVIII.

22. Rogers, A (1975): *Introduction to multiregional Demography,* John Wiley and Sons. New York.

23. Sheps, M.C. and Perrin, E.B. (1966): Further results from Human Fertility model with a variety of pregnancy outcomes: *Human Biology,* Vol. 38, page 180–193.

24. Sheps, M.C., Menken, J.A. and Radick, A.P. (1969): Probability models for family building, An analytical review; *Demography*, Vol. 8, page 161–183.

25. Singh S.N. (1964): On the time of first birth, *Sankhyā*, Series B, Part I and II, page 95–102.

26. Singh. S.N. (1964): A probability model for couple fertility. *Sankhyā*, Series B, Vol. 21, p, 89–94.

27. Takac's L. (1960): *Stochastic Processes (Problems and Solutions)*: Translated by P. Zador, John Wiley & Sons Inc., New York, Mathem of Co. Ltd., London, Butler & Tanner Ltd., Rome and London.

Techniques of Stochastic Processes for Demographic Analysis— Population Growth Indices

11.0 Measurement of Population Growth

Given the indices of mortality and fertility for a natural population, a question that naturally arises is whether the tendency of the given population is to increase, to decrease or to remain more or less stationary over time. Therefore, before any idea about the growth of a population is taken, it is necessary to evolve proper indices of population growth. We have seen in chapter ten , that a fairly good index of population growth is given by "Net Reproduction Rate". But the validity of the Net Reproduction Rate is again subject to the limitations of fertility and mortality conditions maintaining 'status quo'. Besides that Net Reproduction Rate essentially reflects the growth in female population. The parallel index for male is difficult to construct. Thus to estimate growth of a population, a good approach would be to evolve certain theoretical models (deterministic or stochastic) describing the growth of populations. Growth parameter may often be estimable from the model itself. Let us take the simplest situation in which we have a population of N individuals and we assume that the population remains closed for Migration.

Now as a matter of fact, modelling of Natural Populations have three basic characterizations as follows:

(i) The population over time may show the average density of population being maintained at a constant level over long period of time, unless there is a major environmental change. In otherwords, populations under such set up neither die out nor explode. (This phenomenon is called " Balancing " by Nicholson (1954).

(ii) In the next place, we have a second characterization under which the growth of a population need not necessarily remain at a constant level; but the same may fluctuate around a constant mean value randomly.

(iii) Finally a third characterization is given by superimposition of a random cycle of oscillation on the type of random variation, already considered in the second type of characterization. Thus a third type of characterization is again a generalization of the second type.

With this set up we propose to develop a simple deterministic model. Here we ignore such factors as environmental conditions etc. while developing a deterministic model. Further, as in the deterministic model the probabilistic consideration relating to the variation of N (which is an integer) is ignored, one can reasonably assume N to be a continuous variable (vide Moran (1961)). In other words, we structure the deterministic model as

$$\frac{dN}{dt} = N(B - D) \tag{11.1}$$

where B and D are the instantaneous birth and death rates at time t. Note that here the birth rate is defined with respect to the whole population since we are ignoring the distinction between male and female populations.

11.1   A Density Dependent Growth Model

Again we may note the assumption that B and D are independent of N is not valid. This otherwise implies that the reason of a Population being at a given level must be explained by the density of the population. In view of the same, we rewrite (11.1) as

$$\frac{dN}{dt} = N(B(N) - D(N)) \tag{11. 2}$$

In this set up we may again note that Population movements do not have any oscillatory character. For, had it been so one could find two points, t_1 and t_2 on the time scale at which $N(t)$ would have been the same; but $\frac{dN}{dt}$ most probably be different at these two points. This is clearly impossible under (11.2). Of course, oscillations could have been possible even under (11.2) had the birth and death rates being age specific, But again such oscillations are damped out relative to the size of the population.

11.2   A Logistic Growth Model as Density Dependent Model

Returning to (11.2) if we write $(B(N)) - D(N))$ as a linear function of N

i.e. $$[B(N) - D(N)] = \alpha - \beta N \tag{11. 3}$$

$$\Rightarrow \frac{dN}{dt} = N(\alpha - \beta N)$$

where α and β are positive constants.

$$\Rightarrow \frac{dN}{(\alpha - \beta N) N} = dt$$

$$\Rightarrow \frac{1}{\alpha}\left[\frac{1}{N} + \frac{\beta}{\alpha - \beta N}\right] dN = dt$$

$$\Rightarrow \frac{1}{\alpha}\left[\frac{1}{N} - \frac{\beta}{\beta\left(N - \frac{\alpha}{\beta}\right)}\right] dN = dt$$

$$\Rightarrow \left\{\frac{1}{N} - \frac{1}{N - \frac{\alpha}{\beta}}\right\} dN = \alpha dt$$

Integrating both sides, we get

$$\log_e \frac{N}{\left(N - \frac{\alpha}{\beta}\right)} = \alpha t + \alpha \log_e c$$

or

$$\frac{N}{N - \frac{\alpha}{\beta}} = c^\alpha e^{\alpha t}$$

$$\Rightarrow \quad N = \left(N - \frac{\alpha}{\beta}\right) c^\alpha e^{\alpha t}$$

$$\Rightarrow \quad N \cdot (1 - c^\alpha e^{\alpha t}) = -\frac{\alpha}{\beta} c^\alpha e^{\alpha t}$$

$$\Rightarrow \quad N = -\frac{\frac{\alpha}{\beta} c^\alpha e^{\alpha t}}{1 - c^\alpha e^{\alpha t}} = \frac{\frac{\alpha}{\beta}}{1 - \frac{1}{c^\alpha} e^{-\alpha t}}$$

$$N = \frac{\frac{\alpha}{\beta}}{1 + c' e^{-\alpha t}} \text{ where } c' = -\frac{1}{c^\alpha}. \tag{11.4}$$

is solution of the equation.

Here c' is a constant which is positive or negative according as the initial value at $t_0 = 0$ i.e. N_0

$$N_0 = (\alpha/\beta)/(1 + c') \tag{11.5}$$

Also the limiting value of N converges as $t \to \infty$

$$\text{viz } \lim_{t \to \infty} N = \frac{\alpha}{\beta}, \ N_0 < N \text{ if } 1 + c' > 1$$

$$\Rightarrow c' > 0 \Rightarrow -\frac{1}{c^\alpha} > 0 \tag{11.6}$$

Also $N_0 \geq N \Rightarrow c' < 0$

11.2.1 A Critical Review of Logistic Model

The result (11.4) is known as Logistic law of growth and when $N_0 < \frac{\alpha}{\beta}$ gives an S-shaped curve of growth which has often been used in the past to give close fit to observations relating to experimental and natural populations. As a matter of fact, the Logistic model with three parameters often gives a closed fit to the population data taken over periods of time. But as cautioned by Feller (1940) closed fit of the data with the logistic does not necessarily give any evidence that the underlying process of growth follows a Logistic pattern. As such the population projection by fitting of logistic curve remains often a very questionable issue. A result due to Leslie (1948) in this respect may appear to be quite interesting. Leslie has shown that if the age specific fertility and mortality rates are also density dependent i.e. they depend on the size of the Population then the Population growth rate will acquire a logistic pattern provided the initial age distribution from which we start is stable or stationary. The implication of this result is that if the initial age distribution is not stable then the growth rate will not acquire a logistic form. But Indian Population which is far from having a stable age distribution has given quite good fit with the Logistic. This is quite misleading demographically. It is, therefore, advisable to examine critically the theoretical premises in support of a model relating to a Population growth curve, (especially logistic) before the same is employed for the measurement of the Population growth or projection purpose.

11.2.2 Properties of Logistic Model of Population Growth

We have discussed the generation of the *Logistic Growth Model* in section 11.2. Verhulst (1838) evolved the Logistic Curve while experimenting with the growth of the insects under controlled environmental condition. Also Logistic Curve derived by other similar assumptions as given below has in past given good fit to the Population data of several countries as well as to other kinds of data. It is true that a logistic model for predicting the growth of a Population has all the defects and limitations which a single equation model has while

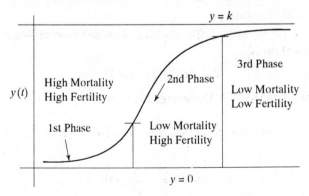

Fig. 11.1

predicting the course of a Population. Let the two asymptotes be $y = 0$ and $y = k$.

Asymptotes for the logistic curve

at $\qquad t \to -\infty \qquad y(t) = 0$

at $\qquad t \to \infty \qquad y(t) = k$

$$\left.\begin{aligned} &\frac{dy(t)}{dt} > 0 ; \; \frac{dy}{dt} = 0 \text{ at } y = 0, y = k \\ &\frac{d^2 y(t)}{dt^2} > 0 \quad \text{if } y < \frac{k}{2} \\ &\qquad < 0 \quad \text{if } y > \frac{k}{2} \end{aligned}\right\}$$

$$\frac{d^2 y}{dt^2} = 0; \quad y = \frac{k}{2} \quad \text{(the point of inflexion)} \tag{11.7}$$

We have, therefore, the differential equation governing the generation of Logistic Curve is given by

$$\frac{dy(t)}{dt} = \frac{y(k-y)}{k} \phi'(t) \tag{11.8}$$

where $\phi(t)$ represents the time trend of the population.

$$\Rightarrow \qquad k \int \frac{dy}{y(k-y)} = \int \phi'(t) \, dt + c$$

$$\Rightarrow \qquad \int \left(\frac{1}{y} + \frac{1}{k-y} \right) dy = \phi(t) + c$$

$$\Rightarrow \qquad \log y - \log (k-y) = \phi(t) + c$$

$$\Rightarrow \qquad \frac{y}{k-y} = \exp(\phi(t) + c)$$

$$\Rightarrow \qquad \frac{(k-y)}{y} = \frac{1}{\exp(\phi(t) + c)}$$

$$\Rightarrow \qquad \frac{k}{y} - 1 = \frac{1}{\exp(\phi(t) + c)}$$

$$\Rightarrow \qquad \frac{k}{y} = \frac{1}{\exp(\phi(t) + c)} + 1$$

or $\qquad \frac{k}{y} = \frac{1}{e^c e^{\phi(t)}} + 1$

or $\qquad \frac{k}{y} = \frac{1}{b' e^{\phi(t)}} + 1 \quad \text{where } b' = e^c$

or
$$(k/y) = \frac{(1+b' \, e^{\phi(t)})}{b' \, e^{\phi(t)}}$$

or
$$\frac{y}{k} = \frac{b' \, e^{\phi(t)}}{1+b' \, e^{\phi(t)}}$$

or
$$y = k \left(\frac{b' \, e^{\phi(t)}}{1+b' \, e^{\phi(t)}} \right)$$

or
$$y = \frac{k}{1+\dfrac{1}{b'} e^{-\phi(t)}}$$

or
$$y = \frac{k}{1+be^{-\phi(t)}} \quad \text{where } b = \frac{1}{b'}$$

∴
$$y(t) = \frac{k}{1+be^{-\phi(t)}} \tag{11.9}$$

is the general form of the *Logistic Curve*
Let

$$\phi(t) = a_n t^n + a_{n-1} t^{n-1} + \ldots + a_0$$

$$= t^n \left(a_n + \frac{a_{n-1}}{t} + \ldots + \frac{a_0}{t^n} \right)$$

For very large t

$$\phi(t) = a_n t^n + 0 \left(\frac{1}{|t|} \right)$$

Case I $a_n > 0$, n is even
Case II $a_n < 0$, n is even
Case III $a_n > 0$, n is odd
Case IV $a_n < 0$, n is odd

Case I: $\phi(-\infty) = +\infty, \quad \phi(+\infty) = \infty$

$$\Rightarrow \quad y(-\infty) = k, \quad y(\infty) = k, \quad \phi(0) = a_0$$

∴ Case I is ruled out
Similarly Case II and Case IV are ruled out

Case III $\phi(-\infty) = -\infty \qquad \phi(+\infty) = \infty$

$$y(-\infty) = 0 \qquad y(\infty) = k, \quad \phi(0) = a_0$$

which satisfy our assumptions.

If follows that $\phi(t)$ is a polynomial of odd degree. The simplest of which corresponds to $n = 1$ i.e.

$$\phi(t) = a_0 + a_1 t$$

Then the equation of the Logistic Curve is given by

$$y_t = \frac{k}{1 + b' \exp(-(a_0 + a_1 t))}; \quad a_0 \text{ and } a_1 \text{ are arbitrary constants}$$

$$= \frac{k}{1 + b' e^{-a_0} e^{-a_1 t}} = \frac{k}{1 + b e^{-at}},$$

for arbitrary constant b'

where $\quad a = a_1, \quad b = b' e^{-a_0}$

$$\Rightarrow \quad \frac{dy}{dt} = \frac{kba e^{-at}}{(1 + b e^{-at})^2}$$

$$= a \cdot k \cdot \frac{y}{k}\left(1 - \frac{y}{k}\right)$$

$$\Rightarrow \quad \frac{dy}{dt} = ay\left(1 - \frac{y}{k}\right)$$

$$\frac{d^2 y}{dt^2} = a\frac{dy}{dt}\left(1 - \frac{y}{k}\right) + ay\left(-1/k\right)\frac{dy}{dt}$$

$$= a\frac{dy}{dt} - a\frac{dy}{dt}\cdot\frac{y}{k} - a\frac{y}{k}\frac{dy}{dt}$$

$$= a\frac{dy}{dt} - 2a\frac{y}{k}\frac{dy}{dt}$$

$$= a\left(1 - \frac{2y}{k}\right)\frac{dy}{dt}$$

$$= a\left(1 - \frac{2y}{k}\right)ay\left(1 - \frac{y}{k}\right)$$

$$= a^2 y\left(1 - \frac{y}{k}\right)\left(1 - \frac{2y}{k}\right)$$

$$\frac{dy}{dt} = ay\left(1 - \frac{y}{k}\right)$$

$$\frac{dy}{dt} = 0 \text{ at } y = 0 \text{ and at } y = k$$

i.e., at $t = -\infty$ and $t = \infty$

$$\frac{dy}{dt} > 0 \text{ when } 1 - \frac{y}{k} > 0; y > k$$

$$\frac{d^2 y}{dt^2} = 0 \text{ if } \frac{2y}{k} = 1 \Rightarrow y = \frac{k}{2}$$

$$\Rightarrow y = \frac{k}{2} \text{ is the point of inflexion.}$$

11.3   Another Approach to the Logistic Model

Malthus (1798) while giving the first qualitative formulation of population growth observed that the increase of population follows a Geometric Progression in contrast to its means of subsistence which tend to grow in arithmetical progression.

Under condition of unlimited resources, Malthusian Geometric law can be expressed as

$$\frac{d\,N(t)}{dt} = r\,N(t) \tag{11.10}$$

where $N(t) \equiv$ Population at any time t and r is the constant of proportionality.

$$\Rightarrow \qquad N(t) = N(0)\,e^{rt} \tag{11.11}$$

$N(0)$ refers to the initial size of the population.

Malthus did not take into account of the fact that in any given environment the growth of the population may stop due to the density of the population which the environment can sustain.

However, Verhulst (1838) took account of this limitation of Malthus. He postulated that the rate of the population growth was jointly proportional to the existing population. If π is the maximum population that a given amount of food can support, then according to Verhulst

$$\frac{d\,N(t)}{dt} = r\,N(t)\left(1 - \frac{N(t)}{\pi}\right) \tag{11.12}$$

(Note that $N(t) << \pi \Rightarrow \dfrac{dN(t)}{dt} \cong r\,N(t)$, which is the Malthusian law; on the

other hand $N(t) = \pi \Rightarrow \dfrac{dN(t)}{dt} = 0$)

$$\Rightarrow \frac{d\,N(t)}{N(t)\,(\pi - N(t))} = \frac{r}{\pi}\,dt$$

$$\Rightarrow \log_e \frac{N(t)}{\pi - N(t)} = rt + c, c \text{ being the constant of integration.}$$

$$\Rightarrow \frac{N(t)}{\pi - N(t)} = e^{rt + c}, \text{ Putting } \frac{N(t)}{\pi} = f(t), \frac{N(0)}{\pi} = f(0) \tag{11.13}$$

$$\Rightarrow \frac{f(t)}{1 - f(t)} = \frac{f(0)}{1 - f(0)}\,e^{rt}$$

$$\Rightarrow f(t) = \frac{f(0)\,e^{rt}}{(1 - f(0)) + f(0)\,e^{rt}}$$

$$\Rightarrow f(t) = f(0)[f(0) + (1 - f(0)) e^{-rt}]^{-1} \tag{11.14}$$

is the equation of the logistic curve which can be taken as the generalization of Malthus's model.

11.3.1 A Generalization

A generalization of (11.12) may be given by

$$\frac{d N(t)}{dt} = r N(t) \left[1 - \left(\frac{N(t)}{\pi}\right)^{\alpha}\right] \Big/ \alpha \tag{11.15}$$

which is reduced to $\dfrac{d N(t)}{dt} = r N(t)\left[1 - \dfrac{N(t)}{\pi}\right]$ when $\alpha = 1$

and for $\alpha \to 0$ this reduces to

$$\frac{d N(t)}{dt} = -r N(t) \log_e \left(\frac{N(t)}{\pi}\right) \tag{11.16}$$

which in the literature of Demography and Actuarial Science is known as Gompertz Law and is used in the mortality analysis of the elderly persons
Now the solution of (11.16) is given by

$$\frac{d}{dt}\left(\log_e \frac{N(t)}{\pi}\right) = -r \log_e \left(\frac{N(t)}{\pi}\right)$$

$$\Rightarrow \qquad \log_e \frac{N(t)}{\pi} = e^{-rt} \log_e \left(\frac{N(0)}{\pi}\right)$$

Putting $\dfrac{N(t)}{\pi} = f(t)$ and $\dfrac{N(0)}{\pi} = f(0)$

The solution of (11.16) is obtained as:

$$f(t) = \frac{N(t)}{\pi} = f(0)\{(f(t))^{\alpha} + e^{-rt}(1 - (f(0)^{\alpha})^{-1/\alpha}\} \tag{11.17}$$

It may be noted that for $r < 0$, a population of size N $(0 < N < \pi)$ goes on increasing as long as $N < \pi$ and stays stationary at level π. Similarly a population starting with $N > \pi$ evidently decreases to a value of π and then stays there. The problem of extinction arises if the net growth rate per individual for small sized population is negative.

11.4 Other Population Growth Models

Defining
$$R_k = \int_0^{\infty} x^k \, p(x) i(x) \, dx \tag{11.18}$$

$$(k = 0, 1, 2 \dots)$$

we have, the Net reproduction rate (N.R.R.) per woman is given by putting

$k = 0$ in (11.18) as:

$p(x)$ = probability of surviving upto age x

$i(x)$ = probability of giving birth in the age group $(x - x + 1)$

w = upper bound of the age of child bearing and can reasonably be replaced by '∞'.

The mean age of giving birth or child bearing

$$\mu'_1 = E(X) = \frac{\int\limits_0^\infty x\, p(x)\, i(x)\, dx}{\int\limits_0^\infty p(x)\, i(x)\, dx} = \frac{R_1}{R_0} \qquad (11.19)$$

The r. v. X has the distribution

$$\phi(x) = \frac{p(x)\, i(x)}{\int\limits_0^\infty p(x)\, i(x)\, dx} \text{ where } X \text{ represents the age of giving birth.}$$

$p(x) \cdot i(x) = \psi(x)$ is called the 'Net maternity function'.
The kth moment of X is given by

$$\mu'_k = \int\limits_0^\infty x^k\, p(x)\, i(x)\, dx \Big/ \int\limits_0^\infty p(x)\, i(x)\, dx \qquad (11.20)$$

The variance of X is given by

$$\mu_2 = \mu'_2 - {\mu'_1}^2$$

$$= \frac{\int\limits_0^\infty x^2\, p(x)\, i(x)\, dx}{\int\limits_0^\infty p(x) \cdot i(x)\, dx} - \left[\frac{\int\limits_0^\infty x\, p(x)\, i(x)\, dx}{\int\limits_0^\infty p(x) \cdot i(x)\, dx} \right]^2$$

$$= \frac{R_2}{R_0} - \left(\frac{R_1}{R_0} \right)^2$$

$$= \sigma_x^2, \text{ say} \qquad (11.21)$$

11.4.1 Compound Interest Law of Population Growth with Annual Conversion

The analysis is made on the basis of female population. We have

$$P_M = P_0 (1 + r)^M \qquad (11.22)$$

where $\quad M \equiv$ mean length of generation

$\qquad P_M \equiv$ Population after the length of one generation, measured from the origin of the earlier generation.

$\qquad P_0 \equiv$ Initial size of the population $= k$, say (i.e. earlier generation)

$\qquad r \equiv$ Annual growth rate per individual per year.

Also we have $P_M = kR_0$ where $R_0 = \int_0^\infty p(x)\, i(x)\, dx$

$$= \text{N. R. R. per woman}$$

Thus $(11.22) \Rightarrow$

$$kR_0 = k (1 + r)^M$$

$$R_0^{1/M} = 1 + r$$

$$\Rightarrow \qquad \hat{r} = [R_0^{1/M} - 1] \qquad (11.23)$$

M is approximately given by $\dfrac{R_1}{R_0}$, the mean age of child bearing (details of

the relationship between M and $\dfrac{R_1}{R_0}$ will be discussed in the next section of

stable population analysis).

Hence $\qquad\qquad \hat{r} \cong [R_0^{R_0/R_1} - 1] \qquad (11.24)$

$\hat{r}$ represents the estimated population growth rate.

11.4.2 Compound Interest Law with Instantaneous Rate of Conversion

We assume a law where population is compounded instantaneously i.e., Rate of change of population is proportional to population size

$$\Rightarrow \qquad \frac{dP(t)}{dt} = rP(t) \qquad (11.25)$$

$$\Rightarrow \qquad \frac{dP(t)}{P(t)} = rdt$$

$$\Rightarrow \qquad P(t) = P(0)\, e^{rt} \text{ where } P(0) = P(t)\big|_{t=0} = k$$

Putting $\qquad\qquad t = M$, the mean length of generation

and $\qquad\qquad P(0) = k$

$$P(M) = ke^{rM} \qquad (11.26)$$

Also $\qquad\qquad P(M) = kR_0 \text{ where } R_0 = \text{N.R.R.}$

Therefore $\qquad kR_0 = ke^{rM} \Rightarrow R_0 = e^{rM}$

$\Rightarrow \qquad\qquad \log R_0 = rM$

$\Rightarrow \qquad\qquad\qquad \hat{r} = \dfrac{1}{M}\log_e R_0 \qquad\qquad\qquad (11.27)$

11.5 Lotka and Dublin's Model for Stable Population Analysis

Lotka and Dublin's stable population analysis is based on the consideration that (i) the population growth is independent of time when both fertility and the mortality rates are also time independent and (ii) the population is closed to migratory movement.

Below we present the mathematical modelling of the stable population analysis:

Mathematical Background of Stable Population Analysis:

As stated earlier Lotka and Dublin's stable population theory consists in assuming that in the third phase of the population growth, the structural form of the population will be characterized by the following:

  (i)   Birth rate is independent of t.
  (ii) Death rate is independent of t.
  (iii) The age distribution between ages $(x, x + \delta x)$ is independent of t. ((iii) $\Leftarrow$ (i) and (ii))
  (iv) The population is closed to migration. (This is a special assumption)

Notations:

  (i) $C(x, t)\,\delta x$ = The proportion of population in the age group

$\qquad\qquad\qquad (x, x + \delta x)$ at time $T = t$

  (ii) $B(t)$ = Total number of births at time $T = t$
  (iii) $P(t)$ = Population at time $T = t$

  (iv) $b(t) = \dfrac{B(t)}{P(t)}$ = Birth rate per individual

  (v) $p(x) = $ Prob {of surviving upto age x}
$\qquad\qquad\qquad$ (independent of t)

  (vi) $i(x)\,\delta x = $ Prob {of giving a birth between age $(x, x + \delta x)$}
$\qquad\qquad\qquad$ (independent of t)

  (vii) $d(x)\,\delta x = $ Prob {of dying between $(x, x + \delta x)$}

The entire analysis is restricted to female cohorts only.

We have the basic identity

$$P(t)\,C(x, t)\,\delta x = B(t - x)\,p(x)\,\delta x$$

$\Rightarrow \qquad\qquad P(t)\,C(x, t) = B(t - x)\,p(x) \qquad\qquad (11.28)$

Multiplying both sides by $i(x)$ and integrating in $(0, \infty)$, we have

$$\int_0^\infty P(t)\,C(x, t)\,i(x)\,dx = \int_0^\infty B(t - x)\,p(x)\,i(x)\,dx \qquad (11.29)$$

Note that the right hand side of (11.29) represents the births at time $T = t$.
Hence, we can write

$$B(t) = \int_0^\infty B(t - x) \, p(x) \, i(x) \, dx \tag{11.30}$$

This is an integral equation with lag x.

Lotka and Dublin assumed a trial solution of the form

$$B(t) = \sum_{n=0}^\infty Q_n \, e^{r_n t} \tag{11.31}$$

where $Q_0, Q_1 \dots$ are the populations at the beginning of each year under consideration (treated here as constants) and $r_0 \, r_1 \dots$ are different rates of growth over time.

Substituting (11.31) in (11.30), we get

$$\sum_{n=0}^\infty Q_n \exp(r_n t) = \int_0^\infty \sum_{n=0}^\infty Q_n \exp(r_n(t - x)) \, p(x) \, i(x) \, dx$$

$$\Rightarrow \quad \sum_{n=0}^\infty Q_n \exp(r_n t) = \sum_{n=0}^\infty Q_n \exp(r_n t) \left(\int_0^\infty \exp(-r_n x) \, p(x) \, i(x) \, dx \right) \tag{11.32}$$

It appears that $r_0, r_1, r_2, \dots r_n$ corresponds to the roots of the integral equation

$$\int_0^\infty e^{-rx} \, p(x) \, i(x) \, dx = 1 \tag{11.33}$$

This is known as **Lotka's Integral Equation.**
Again

$$B(t) = Q_0 \exp(r_0 t) + \sum_{n=1}^\infty Q_n \exp(r_n t) \tag{11.34}$$

It has been shown by Lotka that of infinite number of roots $r_0, r_1, \dots r_n \dots$ of the integral equation (11.33), only one is real and rest of all are complex.
Now (11.33) $\Rightarrow$

$$\int_0^\infty e^{-rx} \, \phi(x) \, dx = 1 \tag{11.35}$$

where $\phi(x) = p(x) \cdot i(x)$ is the net maternity function.

We shall show that the real part of any complex root must be less than the only real root r_0. This can be shown as below:

If $r_k = \alpha_k + i\beta_k$, is a complex root of the Lotka's integral equation, $k = 1, 2, 3, \ldots$; then from (11. 35), we have

$$\int_0^\infty \exp(-(\alpha_k + i\beta_k)x)\,\phi(x)\,dx = 1$$

$$\Rightarrow \qquad \int_0^\infty \exp(-\alpha_k x)(\cos \beta_k x - i \sin \beta_k x)\,\phi(x)\,dx = 1$$

Equating real and imaginary parts, we have

$$\int_0^\infty \exp(-\alpha_k x)[\cos \beta_k x]\,\phi(x)\,dx = 1 \qquad (11.\ 36)$$

This equation when compared with Lotka's integral equation with real root r_0 satisfying the same, gives rise to

$$\int_0^\infty \exp(-r_0 x)\,\phi(x)\,dx = 1 \qquad (11.\ 37)$$

which shows that $\alpha_k \leq r_0$
Since $\cos \beta_k x \leq 1$

$$\therefore \qquad \exp(-\alpha_k x) \geq \exp(-r_0 x)$$

$$\Rightarrow \alpha_k \leq r_0$$

Now $B(t) = Q_0 \exp(r_0 t) + \sum_{k=1}^\infty Q_k \exp(r_k t)$, following the representation of (11.34)

Since $r_0 \geq \alpha_k \; \forall \; k = 1, 2, \ldots$, it follows that,

$$|\exp(\alpha_k t)[\cos \beta_k t + i \sin \beta_k t]| = |\exp(\alpha_k t)| \leq |\exp(r_0 t)| \qquad (11.38)$$

and as $t \to \infty$, $\exp(\alpha_k t)$ will be negligible in comparison with $\exp(r_0 t)$ for $k = 1, 2, \ldots$.

In other words, as $t \to \infty$

$$B(t) \cong Q_0 \exp(r_0 t),$$

while the contribution of the terms under the summation

$$\sum_{k=1}^\infty Q_k \exp(\alpha_k t)[\cos \beta_k t + i \sin \beta_k t]$$

becomes infinitely small in comparison to $Q_0 \exp(r_0 t)$.

Thus $B(t) \cong Q_0 \exp(r_0 t)$ for a large t assuming that the process has started since a very long time.

$$\Rightarrow \qquad B(t - x) = Q_0 \exp(r_0(t - x))$$

$$B(t) = B(t-x) \exp(r_0 x) \qquad (11.39)$$

Dropping the suffix, we have

$$B(t) = B(t-x) e^{rx} \qquad (11.40)$$

where 'r' stands for the real root of Lotka's integral equation (11.33).

Now to obtain the real root of the Lotka's integral equation, we put

$$y = \int_0^\infty e^{-rx} p(x) i(x) dx \qquad (11.41)$$

$$\Rightarrow \qquad \frac{dy}{dr} = \int_0^\infty \frac{d}{dr}(e^{-rx}) p(x) i(x) dx$$

$$\Rightarrow \qquad \frac{dy}{dr} = -\int_0^\infty xe^{-rx} p(x) i(x) dx$$

$$= -\left\{ \frac{\int_0^\infty xe^{-rx} p(x) i(x) dx}{\int_0^\infty e^{-rx} p(x) i(x) dx} \right\} \underbrace{\int_0^\infty e^{-rx} p(x) i(x) dx}_{= y}$$

$$\underbrace{\qquad\qquad\qquad\qquad}_{\text{function of } r = A(r) \text{ (say)}}$$

$$= -y A(r)$$

where

$$A(r) = \frac{\int_0^\infty xe^{-rx} p(x) i(x) dx}{\int_0^\infty e^{-rx} p(x) i(x) dx} \qquad (11.42)$$

We have the differential equation

$$\frac{dy}{dr} = -y A(r)$$

$$\Rightarrow \qquad \frac{dy}{y} = -A(r) dr \qquad (11.43)$$

$$\log_e y = \log_e y_0 - \int A(r) dr$$

where $\log_e y_0$ is constant of integration

$$\Rightarrow y = y_0 \exp\left(-\int A(r) dr\right) \qquad (11.44)$$

Since $y = 1$, we have

$$y_0 = \exp\left(\int A(r)\, dr\right) \tag{11.45}$$

Next we note that $y = y_0$, when $r = 0$
$(\because$ At r = 0,

$$y = y_0\, e^{-\int A(0)\, dr\,|_{r=0}}$$

$$= y_0 \cdot e^{-A(0)\, r\,|_{r=0}}$$

$$= y_0\,)$$

$$y_0 = \int\limits_0^\infty e^{-rx}\, p(x)\, i(x)\, dx \,\bigg|_{r=0}$$

$$= \int\limits_0^\infty p(x)\, i(x)\, dx = R_0 = NRR \text{ per woman} \tag{11.45'}$$

$$\log y_0 = \log_e R_0 = \int A(r)\, dr \tag{11.45''}$$

To obtain $A(r)$:

$$A(r) = \frac{\displaystyle\int\limits_0^\infty x\, e^{-rx}\, p(x)\, i(x)\, dx}{\displaystyle\int\limits_0^\infty e^{-rx}\, p(x)\, i(x)\, dx} \tag{11.46}$$

$$= \frac{\displaystyle\int\limits_0^\infty x \sum_{j=0}^\infty \frac{(-1)^j\, (rx)^j}{j!}\, p(x)\, i(x)\, dx}{\displaystyle\int\limits_0^\infty \sum_{j=0}^\infty \frac{(-1)^j\, (rx)^j}{j!}\, p(x)\, i(x)\, dx}$$

$$\left(\because e^T = \sum_{x=0}^\infty \frac{T^x}{x!}\right)$$

$$= \frac{\displaystyle\sum_{j=0}^\infty \frac{(-1)^j\, r^j}{j!} \int\limits_0^\infty x^{j+1}\, p(x)\, i(x)\, dx}{\displaystyle\sum_{j=0}^\infty \frac{(-1)^j\, r^j}{j!} \int\limits_0^\infty x^j\, p(x)\, i(x)\, dx}$$

$$= \frac{\sum_{j=0}^{\infty} \frac{(-1)^j r^j}{j!} R_{j+1}}{\sum_{j=0}^{\infty} \frac{(-1)^j r^j}{j!} R_j} \qquad (11.47)$$

where

$$R_j = \int_0^{\infty} x^j \, p(x) \, i(x) \, dx$$

$$R_{j+1} = \int_0^{\infty} x^{j+1} \, p(x) \, i(x) \, dx \qquad (11.48)$$

Let

$$A(r) = \alpha + \beta r + \gamma r^2 + \delta r^3 + \dots , \text{ by actual division,}$$

where

$$\alpha = \frac{R_1}{R_0}, \ \beta = \left(\frac{R_1}{R_0}\right)^2 - \frac{R_2}{R_0} \text{ etc.}$$

$$\gamma = \left[\frac{R_3}{R_0} - \frac{3R_1 R_2}{R_0^2} + 2\left(\frac{R_1}{R_0}\right)^3\right] \Big/ 2 \qquad (11.49)$$

$$A(r) \cong \alpha + \beta r \text{ (neglecting terms involving higher power of } r)$$

$$\because \ \mu_3 = \mu_3' - 3\mu_2'\mu_1' + 2(\mu_1')^3$$

$$\therefore \quad \log_e R_0 = \int A(r) \, dr$$

$$= \int (\alpha + \beta r) \, dr$$

$$= \alpha r + \beta \frac{r^2}{2} \qquad (11.50)$$

or

$$\beta \frac{r^2}{2} + \alpha r - \log_e R_0 = 0$$

or

$$\beta r^2 + 2\alpha r - 2\log_e R_0 = 0$$

or

$$r = \frac{-2\alpha \pm \sqrt{4\alpha^2 - 4\beta(-2\log_e R_0)}}{2\beta}$$

or

$$r = \frac{-2\frac{R_1}{R_0} \pm \sqrt{4\left(\frac{R_1}{R_0}\right)^2 + 8\left[\left(\frac{R_1}{R_0}\right)^2 - \frac{R_2}{R_0}\right] \log_e R_0}}{2\left[\left(\frac{R_1}{R_0}\right)^2 - \frac{R_2}{R_0}\right]} \qquad (11.51)$$

where r is the Growth parameter of the stable population and R_0, R_1 and R_2 are estimated as

$$\hat{R}_0 = \sum p(x) i(x) = \text{N.R.R.} \tag{11.51'}$$

$$\hat{R}_1 = \frac{\sum x p(x) i(x)}{\sum p(x) i(x)} = \text{Mean age of child bearing} \tag{11.51''}$$

$$\hat{R}_2 = \frac{\sum x^2 p(x) i(x)}{\sum p(x) i(x)} \tag{11.51'''}$$

Then substituting estimates for α, β and R_0, R_1, R_2 in (11.51), two real roots of r will be obtained of which one is positive, other is negative.

Conclusion: Certain results in stable population analysis:
   (i)   Both birth and death rates are independent of time.
  (ii) $C(x, t) = C(x)$ i.e., age distribution is independent of "t",
   i.e., Birth and death rate as well as age distribution may undergo changes but these changes are only of random nature.

11.5.1   Certain Important Deductions of Stable Population Analysis

(i)   To obtain $b(t)$ and $C(x ; t)$ in stable population analysis
We have

$$P(t) C(x ; t) \, \delta x = B(t - x) p(x) \, \delta x$$

Also
$$B(t) = B(t - x) e^{rx}$$

$$\Rightarrow C(x ; t) = \frac{B(t - x) p(x)}{P(t)}$$

$$= \frac{B(t) e^{-rx} p(x)}{P(t)}$$

$$= b(t) e^{-rx} p(x) \quad \left(\because b(t) = \frac{B(t)}{P(t)} \right)$$

Also
$$\int_0^\infty C(x ; t) \, dx = 1 \Rightarrow b(t) \int_0^\infty e^{-rx} p(x) \, dx = 1$$

$$\Rightarrow b(t) = \frac{1}{\left[\int_0^\infty e^{-rx} p(x) \, dx \right]} \tag{11.52}$$

Also
$$C(x ; t) \, \delta x = b(t) e^{-rx} p(x) \, dx$$

$$\Rightarrow C(x ; t) = \frac{e^{-rx} p(x)}{\int_0^\infty e^{-rx} p(x) \, dx} \tag{11.53}$$

(11.52) and (11.53) shows that the birth rate and the age distribution are independent of time.

$$\text{Death rate} = d(t) = [r - b(t)] = \left[r - \frac{1}{\int_0^\infty e^{-rx} p(x)\, dx} \right] \tag{11.54}$$

is also independent of time.

(ii) Death rate in a stationary population

A population is stationary when $r = 0$ in which the stationary age distribution is given by

$$C(x) = \frac{p(x)}{\int_0^\infty p(x)\, dx} \tag{11.55}$$

In a life-table population, which is stationary

$$p(x) = \frac{l(x)}{l_0}$$

$$C(x) = \frac{l(x)}{\int_0^\infty l(x)\, dx} \tag{11.56}$$

which represents the life table age distribution.

Again for a stationary population $b = d$

i.e.,

$$d = \frac{1}{\int_0^\infty p(x)\, dx} = \frac{l_0}{\int_0^\infty l(x)\, dx}$$

$$= \frac{l_0}{T_0} = \frac{1}{\varepsilon_0^0} \tag{11.57}$$

where ε_0^0 is the complete expectation of life at birth

$$\therefore \qquad d = \frac{1}{\varepsilon_0^0}$$

Thus the life table death rate is usually denoted by $\dfrac{1}{\varepsilon_0^0}$

(iii) Determination of Q_0 under stable population modelling

$$B(t) = Q_0 \exp(r_0 t) + \sum_{n=0}^{\infty} Q_n \exp(r_n t)$$

we have from (11.30)

$$B(t) = \int_0^\infty B(t-x)\, p(x)\, i(x)\, dx$$

$$\Rightarrow \qquad \int_0^\infty B(t-x)\, \phi(x)\, dx = B(t),$$

where $\phi(x) = p(x)\, i(x)$, the net maternity function at age x. Lotka and Dublin proposed a trial solution of the form (11.31).

Now $B(t) = \int_0^\infty B(t-x)\, \phi(x)\, dx$ is a homogeneous equation with lag x. Therefore, the solution multiplied by an arbitrary constant still remains a solution. Further, if $B_1(t)$ and $B_2(t)$ are two solutions then $B_1(t) + B_2(t)$ is also a solution. In view of these two properties, if e^{rt} is a solution for $r = r_0, r_1\, r_2, \ldots$ then

$$B(t) = Q_0 \exp(r_0 t) + Q_1 \exp(r_1 t) + \ldots + Q_n \exp(r_n t) + \ldots$$

$$= \sum_{n=0}^\infty Q_n \exp(r_n t)$$

is also a solution for all arbitrary Q_r's provided $\sum\limits_{n=0}^\infty Q_n \exp(r_n t)$ converges:

$$\sum_{n=0}^\infty Q_n \exp(r_n t) = \int_0^\infty \sum_{n=0}^\infty Q_n \exp(r_n (t-x))\, p(x)\, i(x)\, dx$$

$$\sum_{n=0}^\infty Q_n \exp(r_n t) = \sum_{n=0}^\infty Q_n \exp(r_n t). \int_0^\infty \exp(-r_n x)\, p(x)\, i(x)\, dx$$

$$= \int_0^\infty \exp(-r_n x)\, p(x)\, i(x)\, dx = 1$$

and $i(x) = 0$, outside (α, β)
where α and β are the lower and upper bounds of the reproductive period.

We have assumed that r_0 is real and $r_1, r_2, \ldots, r_n, \ldots$ are all complex roots.

$$\Rightarrow B(t) = \sum_{n=0}^\infty Q_n \exp(r_n t) = Q_0 \exp(r_0 t) + \sum_{n=1}^\infty Q_n \exp(r_n t)$$

We are thus left with determination of Q_n's.

Consider a general formulation

$$B(t) = G(t) + \int_0^t B(t-x) \, p(x) \, i(x) \, dx \qquad (11.58)$$

where $G(t)$ births to women already born at time $t = 0$.

The term $G(t) = 0$ for $t \geq \beta$ i.e., no births can occur to women aged $\geq \beta$

$$\Rightarrow \qquad G(t) = B(t) - \int_0^t B(t-x) \, p(x) \, i(x) \, dx \qquad (11.59)$$

Taking Laplace transform of (11. 59) on both sides

$$\Rightarrow \qquad L(G(t)) = \int_0^\infty \exp(-r_0 t) \, G(t) \, dt = \int_0^\beta \exp(-r_0 t) \, G(t) \, dt$$

$$(\because G(t) = 0 \,\forall\, t \geq \beta)$$

$$= \int_0^\beta \exp(-r_0 t) \left[B(t) - \int_0^t B(t-x) \, \phi(x) \, dx \right] dt$$

$$= \int_0^\beta \exp(-r_0 t) \left[Q_0 \exp(r_0 t) + \sum_{n=1}^\infty Q_n \exp(r_n t) \right.$$

$$\left. - \int_0^t \left(Q_0 \exp[r_0(t-x)] + \sum_{n=1}^\infty Q_n \cdot \exp[r_n(t-x)] \right) \phi(x) \, dx \right] dt$$

$$= \int_0^\beta \exp(-r_0 t) \left[Q_0 \exp(r_0 t) + \sum_{n=1}^\infty Q_n \exp(r_n t) \right.$$

$$\left. - \int_0^t Q_0 \exp[r_0(t-x)] \phi(x) \, dx + \int_0^t \sum_{n=1}^\infty Q_n \exp[r_n(t-x)] \phi(x) \, dx \right] dt$$

$$= \int_0^\beta \exp(-r_0 t) \left[Q_0 \exp(r_0 t) - \int_0^t Q_0 \exp[r_0(t-x)] \phi(x) \, dx \right] dt + R_0'$$

$$(11.60)$$

where

$$R_0' = \int_0^\beta \exp(-r_0 t) \left[\sum_{n=1}^\infty Q_n \exp(r_n t) - \int_0^t \left(\sum_{n=1}^\infty Q_n \exp[r_n(t-x)] \phi(x) \, dx \right) \right] dt$$

$$(11.61)$$

Therefore, $L\,(G\,(t))$

$$= \int_0^\beta \left\{ Q_0 - \int_0^t Q_0 \exp\left(-r_0\,x\right)\phi(x)\,dx \right\} dt + R_0'$$

$$= Q_0 \int_0^\beta \left\{ 1 - \int_0^t \exp\left(-r_0\,x\right)\phi(x)\,dx \right\} dt + R_0' \tag{11.62}$$

Since r_0 is the real root of the equation,

$$\therefore \qquad \int_\alpha^\beta e^{-rx}\,p(x)\,i(x)\,dx = \int_0^\infty e^{-rx}\,p(x)\,i(x)\,dx = 1$$

or

$$\int_0^\beta e^{-rx}\,\phi(x)\,dx = 1$$

We may write

$$L\,(G\,(t)) = Q_0 \int_0^\beta \left\{ \int_0^\beta \exp\left(-r_0\,x\right)\phi(x)\,dx - \int_0^t \exp\left(-r_0\,x\right)\phi(x)\,dx \right\} dt + R_0'$$

$$= Q_0 \int_0^\beta \int_t^\beta \exp\left(-r_0\,x\right)\phi(x)\,dx\,dt + R_0' \tag{11.63}$$

Next we shall change the order of integration in the above double integral. Note that in the case given above sums of the type of vertical strips given in the figure 11.2 were taken. Now we shall take sums of the horizontal type of strips. The

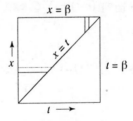

Fig. 11.2

relevant limits are thus

$$0 \le x \le \beta \text{ and } 0 \le t \le x$$

Hence

$$L(G(t)) = Q_0 \int_0^\beta \int_0^x \exp\left(-r_0\,x\right)\phi(x)\,dt\,dx + R_0'$$

$$= Q_0 \int_0^\beta \left\{ \exp\left(-r_0\,x\right)\phi(x)\int_0^x dt \right\} dx + R_0'$$

$$= Q_0 \int_0^\beta \exp\left(- r_0\, x\right) \phi(x) \cdot x\, dx + R_0' \qquad (11.64)$$

$$\Rightarrow \qquad Q_0 = \frac{L\left(G\left(t\right)\right)}{\displaystyle\int_0^\beta x \exp\left(- r_0\, x\right) \phi(x)\, dx}$$

$$= \frac{\displaystyle\int_0^\beta \exp\left(- r_0\, t\right) G(t)\, dt}{\displaystyle\int_0^\beta x \exp\left(- r_0\, x\right) \phi(x)\, dx} \qquad (11.65)$$

provided $R_0' = 0$.

Now to show that $R_0' = 0$, we see that R_0' is made up of sums of such terms as:

$$R_{0u}' = Q_u \int_0^\beta \exp\left(- r_0\, t\right) \left[\exp r_u\, t - \int_0^t \exp\left[r_u\,(t - x)\right] \phi(x)\, dx \right] dt,\ u \neq 0$$

$$= Q_u \int_0^\beta \left[\exp\left[(r_n - r_0)t\right] - \int_0^t \exp\left[(r_n - r_0)t\right] \exp\left(- r_n\, x\right) \phi(x)\, dx \right] dt$$

$$= Q_u \int_0^\beta \exp\left[(r_n - r_0)t\right] \left\{1 - \int_0^t \exp\left(- r_n\, x\right) \phi(x)\, dx \right\} dt$$

We put
$$\int_0^\beta \exp\left(- r_n\, x\right) \phi(x)\, dx = 1 \qquad (11.66)$$

Again since r_n is also a root of the Lotka's integral equation, we get

$$R_{0u}' = Q_u \int_0^\beta \exp\left[(r_u - r_0)t\right] \left\{\int_0^\beta \exp\left(- r_n\, x\right) \phi(x)\, dx \right.$$

$$\left. - \int_0^t \exp\left(- r_n\, x\right) \phi(x)\, dx \right\} dt$$

$$= Q_u \int_0^\beta \exp\left[(r_u - r_0)t\right] \left\{\int_t^\beta \exp\left(- r_n\, x\right) \phi(x)\, dx \right\} dt \qquad (11.67)$$

Again changing the order of integration

$$R'_{0u} = Q_u \int\limits_0^\beta \exp(-r_n x) \phi(x) \int\limits_0^x \exp[(r_u - r_0)t] \, dt \, dx$$

$$= \frac{Q_u}{r_u - r_0} \int\limits_0^\beta \exp(-r_n x) \phi(x) [\exp\{(r_u - r_0)\}x - 1] \, dx$$

$$= \frac{Q_u}{r_u - r_0} \left\{ \int\limits_0^\beta \exp(-r_0 x) \phi(x) \, dx - \int\limits_0^\beta \exp(-r_u x) \phi(x) \, dx \right\}$$

$$\hspace{11cm} (11.68)$$

$$= 0,$$

because r_u and r_0 are both roots of Lotka's integral equation

$$\int\limits_0^\beta \exp(-r_0 x) \phi(x) \, dx = \int\limits_0^\beta \exp(-r_u x) \phi(x) \, dx = 1$$

Again $R'_0 = \sum\limits_{u=1}^\infty R_{0u}$ and each $R_{0u} = 0$

Therefore
$$R'_0 = 0 \hspace{6cm} (11.69)$$

We have
$$Q_0 = \frac{\int\limits_0^\beta \exp(-r_0 t) G(t) \, dt}{\int\limits_0^\beta x \exp(-r_0 x) \phi(x) \, dx} \hspace{3cm} (11.70)$$

If we know $G(t)$ and the rate of survivorship and fertility $[\phi(x) = p(x) \cdot i(x)]$ then one can calculate Q_0 from r_0.

A special case
Let us take the daughters of B_0 births occured at $t = 0$.

Then
$$G(t) = B_0(t) p(t) i(t) = B_0 \phi(t)$$

In this case,
$$Q_0 = \frac{B_0 \int\limits_0^\beta \exp(-r_0 t) \phi(t) \, dt}{\int\limits_0^\beta x \exp(-r_0 x) \phi(x) \, dx}$$

$$= \frac{B_0}{\int\limits_0^\beta x \exp(-r_0 x) \phi(x) \, dx} \hspace{3cm} (11.71)$$

(iv) **A few results on stable population analysis on growth parameter and relationship between the mean length of generation and age of child bearing :**

If $T(r)$ and T_0 are the mean length of generation and the mean age of child bearing in a stable population then

$$\text{(a)} \qquad\qquad r = \frac{\log R_0}{T(r)} \qquad\qquad (11.72)$$

and $\qquad$ (b) $\qquad\qquad T_0 \cong T + \frac{r\sigma^2}{2} \qquad\qquad (11.73)$

where R_0 is the net reproduction rate and r is the intrinsic growth parameter of a stable population; σ^2 being the variance of the age distribution of the stable population.

Proof of (a)

We have from (11.41)

$$y = \int_0^\infty e^{-rx}\, p(x)\, i(x)\, dx$$

$$\Rightarrow \qquad \frac{dy}{dr} = -\frac{\int_0^\infty x\, e^{-rx}\, p(x)\, i(x)\, dx}{\int_0^\infty e^{-rx}\, p(x)\, i(x)\, dx} \left[\int_0^\infty e^{-rx}\, p(x)\, i(x)\, dx\right]$$

$$= -A(r)\, y,$$

where $\qquad A(r) = \dfrac{\displaystyle\int_0^\infty x\, e^{-rx}\, p(x)\, i(x)\, dx}{\displaystyle\int_0^\infty e^{-rx}\, p(x)\, i(x)\, dx}$

$$\Rightarrow \qquad \int \frac{dy}{y} = -\int A(r)\, dr + c$$

where c is the constant of integration.

$$\Rightarrow \qquad\qquad \log_e y = -Tr + c \qquad\qquad (11.74)$$

where we express $\int A(r)\, dr = Tr$, T being the mean length of the generation.

Also $\qquad \log_e y\big|_{r=0} = \log R_0 \quad \because y\big|_{r=0} = R_0$

$$\Rightarrow \qquad\qquad \log R_0 = c \qquad\qquad (11.75)$$

Putting (11.75) in the right hand side of (11.74) we have

$$\log_e y = -Tr + \log R_0$$

But $$\log_e y = \log_e 1 = 0 \Rightarrow \log R_0 = T_r$$

$$\Rightarrow \qquad r = \frac{\log R_0}{T}$$

which proves (11. 72)

Proof of (b)

Again

$$A(r) = \frac{\int_0^\infty x \, e^{-rx} \, p(x) \, i(x) \, dx}{\int_0^\infty e^{-rx} \, p(x) \, i(x) \, dx}$$

$$= \alpha + \beta r + \gamma r^2 + \delta r^3 + ..., \text{ say} \qquad (11.76)$$

where

$$\alpha = \frac{R_1}{R_0}, \quad \beta = \left\{ \left(\frac{R_1}{R_0} \right)^2 - \frac{R_2}{R_0} \right\}$$

Also

$$\int A(r) \, dr = Tr = \alpha + \frac{\beta r^2}{2} + \frac{\gamma r^3}{3} + \frac{\delta r^4}{4} + ...$$

$$\Rightarrow T = \alpha + \frac{\beta r}{2} + \frac{\gamma r^2}{3} \cong \alpha + \frac{\beta r}{2}$$

$$\text{(since coeffs. of } \gamma\text{'s, } \delta\text{'s are small)}$$

$$\Rightarrow T \cong \frac{R_1}{R_0} - \frac{r}{2} \left[\frac{R_2}{R_0} - \left(\frac{R_1}{R_0} \right)^2 \right] \qquad (11.77)$$

Also $\frac{R_1}{R_0} = T_0 = E(X)$, the mean age of child bearing (vide 11.19)

and

$$\frac{R_2}{R_0} - \left(\frac{R_1}{R_0} \right)^2 = \sigma_X^2 \text{ (vide 11.21)}$$

Hence

$$T_0 \cong T + r \frac{\sigma_X^2}{2},$$

a relation holding between the mean age of child bearing and the mean length of the generation. If $r = 0$, (for a stationary population). $T_0 = T$ holds exactly for a stationary Population.

11.6 Leslie Matrix Technique for Population Projection

We have the population as stable with birth and death parameters independent of time. Here   child bearing age is taken from 15 to 45 for the sake of convenience.

We denote by

(i) $_nP_x^{(t)} \equiv$ Female Population in the age sector $(x - x + n)$ at time t.

(ii) $_nF_x \equiv$ Age specific fertility rate in the age x to $x + n$ independent of time t.

With the above assumptions, the following linear equations for Projections are obtained:

$$\frac{1}{2}\left({}_5P_{15}^{(t)} + {}_5P_{15}^{(t+5)}\right){}_5F_{15}\left(\frac{{}_5L_0}{l_0}\right) + \frac{1}{2}\left({}_5P_{20}^{(t)} + {}_5P_{20}^{(t+5)}\right){}_5F_{20}\left(\frac{{}_5L_0}{l_0}\right) + \dots$$

$$\dots + \frac{1}{2}\left({}_5P_{40}^{(t)} + {}_5P_{40}^{(t+5)}\right){}_5F_{40}\left(\frac{{}_5L_0}{l_0}\right) = {}_5P_0^{(t+5)} \tag{11.78}$$

$${}_5P_0^{(t)}\left(\frac{{}_5L_5}{{}_5L_0}\right) = {}_5P_5^{(t+5)} \tag{11.79}$$

$${}_5P_5^{(t)}\left(\frac{{}_5L_{10}}{{}_5L_5}\right) = {}_5P_{10}^{(t+5)} \tag{11.80}$$

$$\dots \quad \dots \quad \dots$$
$$\dots \quad \dots \quad \dots$$

$${}_5P_{35}^{(t)}\left(\frac{{}_5L_{40}}{{}_5L_{35}}\right) = {}_5P_{40}^{(t+5)} \tag{11.81}$$

$${}_5P_{40}^{(t)}\left(\frac{{}_5L_{45}}{{}_5L_{40}}\right) = {}_5P_{45}^{(t+5)} \tag{11.82}$$

The equations from (11.78) can be re written in the following way

$$\frac{1}{2}\left({}_5P_{15}^{(t)} + {}_5P_{10}^{(t)}\frac{{}_5L_{15}}{{}_5L_{10}}\right){}_5F_{15}\left(\frac{{}_5L_0}{l_0}\right) + \frac{1}{2}\left({}_5P_{20}^{(t)} + {}_5P_{15}^{(t)}\frac{{}_5L_{20}}{{}_5L_{15}}\right){}_5F_{20}\left(\frac{{}_5L_0}{l_0}\right)$$

$$+ \dots + \frac{1}{2}\left({}_5P_{40}^{(t)} + {}_5P_{35}^{(t)}\frac{{}_5L_{40}}{{}_5L_{35}}\right){}_5F_{40}\left(\frac{{}_5L_0}{l_0}\right) = {}_5P_0^{(t+5)} \tag{11.83}$$

and again (11.83) can be written as

$$\frac{1}{2}\left(\frac{{}_5L_{15}}{{}_5L_{10}}{}_5F_{15}\right)\left(\frac{{}_5L_0}{l_0}\right){}_5P_{10}^{(t)} + \left({}_5F_{15} + \frac{{}_5L_{20}}{{}_5L_{15}}{}_5F_{20}\right)\left(\frac{{}_5L_0}{2l_0}\right){}_5P_{15}^{(t)}$$

$$+ \left({}_5F_{20} + \frac{{}_5L_{25}}{{}_5L_{20}}{}_5F_{25}\right)\frac{{}_5L_0}{2l_0}{}_5P_{20}^{(t)} + \dots + \left({}_5F_{35} + \frac{{}_5L_{40}}{{}_5L_{35}}{}_5F_{40}\right)\frac{{}_5L_0}{2l_0}{}_5P_{35}^{(t)}$$

$$+ \left(\frac{1}{2}\frac{{}_5L_0}{l_0}{}_5F_{40}\right){}_5P_{40}^{(t)} = {}_5P_0^{(t+5)} \tag{11.84}$$

Putting the equations (11.78) — (11.82) in the matrix form, we have

$$
\begin{bmatrix}
_5P_0^{(t+5)} \\
_5P_5^{(t+5)} \\
\vdots \\
_5P_{40}^{(t+5)}
\end{bmatrix}
$$

$$
=
\begin{bmatrix}
0 & 0 & \dfrac{_5L_0}{2l_0}\left(\dfrac{_5L_{15}}{_5L_{10}}\,_5F_{15}\right) & \dfrac{_5L_0}{2l_0}\left(_5F_{15}+\dfrac{_5L_{20}}{_5L_{15}}\,_5F_{20}\right) & \cdots & \dfrac{_5L_0}{2l_0}\,_5F_{40} \\[2ex]
\dfrac{_5L_5}{_5L_0} & 0 & 0 & 0 & \cdots & 0 \\[2ex]
0 & \dfrac{_5L_{10}}{_5L_5} & 0 & 0 & \cdots & 0 \\[2ex]
\vdots & \vdots & \vdots & \vdots & & \vdots \\[2ex]
0 & 0 & 0 & 0 & \dfrac{_5L_{40}}{_5L_{35}} & 0
\end{bmatrix}
$$

$$
\times
\begin{bmatrix}
_5P_0^{(t)} \\
_5P_5^{(t)} \\
\vdots \\
_5P_{40}^{(t)}
\end{bmatrix}
\tag{11.85}
$$

or
$$
P^{(t+5)} = LP^{(t)} \tag{11.86}
$$

where $P^{(t+5)}$ and $P^{(t)}$ represent the population age vector in the year $(t+5)$ and t respectively and L is the Leslie Matrix (evolved by P. H. Leslie (1945), (1948) consisting of elements which are function of fertility and mortality parameters, independent of time. Thus with time independent Leslie matrix

$$
\left.
\begin{aligned}
P^{(t+5)} &= LP^{(t)} \\
P^{(t+10)} &= LP^{(t+5)} = L^2 P^{(t)} \\
P^{(t+15)} &= LP^{(t+10)} = L^3 P^{(t)} \\
&\vdots \quad \vdots \quad \vdots \quad \vdots \quad \vdots
\end{aligned}
\right\}
\tag{11.87}
$$

For integer K, $\quad P^{(t+5k)} = L^k P^{(t)}$

which shows the sequence $P^{(t)}, P^{(t+5)}, P^{(t+10)}, \dots$ constitute a simple Markov-chain.

11.7 Properties of Time Independent Leslie Matrix

We first describe the projection process for a closed population with time independent fertility and mortality parameters.

Let n_{xt} = # persons in the age sector $(x - x + 1)$ at time $t = i$

$$i = 0, 1, 2, 3, \dots m$$

where '0' represents the base year (or the starting year of Projection) and projection is made in every year.

f_x ≡ probability of giving birth between x to $(x + 1)$ years which is survived for one more year.

p_x ≡ probability of surviving between x to $(x + 1)$ years .

Both f_x and p_x are time independent parameters and relate to a stable population based on female cohorts.

Then we have

$$
\left.
\begin{aligned}
n_{01} &= \sum_{x=0}^{m} f_x \, n_{x0} \\[2mm]
n_{11} &= p_0 \, n_{00} \\[2mm]
n_{21} &= p_1 \, n_{10} \\
&\ \ \vdots \\
n_{m1} &= p_{m-1} \, n_{(m-1)0}
\end{aligned}
\right\}
$$

where m is the upper bound of the age for survival Matrix notation. we have

$$
\begin{bmatrix} n_{01} \\ n_{11} \\ \vdots \\ n_{m1} \end{bmatrix}_{(m+1) \times 1}
=
\begin{bmatrix}
f_0 & f_1 & \cdots & f_{m-1} & f_m \\
p_0 & 0 & \cdots & 0 & 0 \\
\vdots & \vdots & & \vdots & \vdots \\
0 & 0 & \cdots & p_{m-1} & 0
\end{bmatrix}_{(m+1) \times (m+1)}
\begin{bmatrix} n_{00} \\ n_{10} \\ \vdots \\ n_{(m-1)0} \\ n_{m0} \end{bmatrix}_{(m+1) \times 1}
$$

$$(11.88)$$

or $\qquad\qquad n_1 = L \, n_0$

where $\qquad\qquad n_1 = (n_{01}, n_{11}, \dots n_{m1})'$

$$n_0 = (n_{00}, n_{01}, \dots)'$$

and L being the time independent Leslie Matrix. Precisely, in the same way

$$
\left.
\begin{aligned}
n_2 &= L \, n_1 = L^2 \, n_0 \\
n_3 &= L^3 \, n_0 \\
&\ \ \vdots \\
n_t &= L^t \, n_0
\end{aligned}
\right\}
\qquad (11.89)
$$

In the next place, we condense $(L)_{(m+1) \times (m+1)}$ to a shorter matrix

$(L)_{(K+1)\times(K+1)}$ where K is the upper bound of the child bearing age interval and we concentrate our attention only on $n_{0t} n_{1t} \dots n_{Kt}$ for $t = 0, 1, 2, \dots$ Denoting the condensed matrix $(L)_{K\times K}$ by L (as we shall be dealing with the sum only) and noting that $(L)_{K\times K}$ is nonsingular with none of the characteristics roots zeros we prove the following results:

Result I

Leslie Matrix L has only one positive eigen value and the rest of the eigen values are either complex or negative. Further, if λ_1 is the positive eigen value of L then $|\lambda_i| \le \lambda_1$ where $\lambda_i (i = 2, 3 \dots)$ are complex or negative eigen values of L. In other words λ_1 is most dominant.

Proof: We have

$$n_t = (L)'_{(K+1)\times(K+1)} (n_0) \tag{11.90}$$

where $(L)_{(K+1)\times(K+1)}$ is the non singular condensed Leslie Matrix, $n_t \equiv$ Projected population after t years and $n_0 \equiv$ Population at the base year.

If X is the characteristic vector corresponding to the characteristic root λ, then the characteristic polynomial $|L - \lambda I| = 0$ can be written as

$$\lambda^{k+1} - f_0 \lambda^k - f_1 p_0 \lambda^{k-1} - f_2 p_0 p_1 \lambda^{k-2} \dots - f_k p_0 p_1 p_2 \cdots p_{k-1} = 0 \tag{11.91}$$

$$\Rightarrow \quad \phi(\lambda) = \frac{f_0}{\lambda} + \frac{f_1 p_0}{\lambda^2} + \frac{f_2 p_0 p_1}{\lambda^3} + \dots + \frac{f_K p_0 \cdots p_{k-1}}{\lambda^{k+1}} = 1 \tag{11.92}$$

It can be immediately seen that # changes of sign of the equation in (11.91) is only one and by Descarte's rule of signs in theory of equations, we find that there can at most be one real root of λ. The remaining roots are either complex or negative.

Again as $\phi(\lambda)$ in (11.92) is a continuous function of λ in $(0 \le \lambda < \infty)$ and

$$\phi(0) = \infty \quad \text{and} \quad \phi(\infty) = 0$$

$\exists$ at least one real positive root λ of (11.91) or (11.92).

Hence in view of two considerations viz Descarte's rules of sign and the continuity of $\phi(\lambda)$ in $0 < \lambda < \infty$, it follows, that there exists one and only one positive root of the equation (11.91).

Denoting the positive root by λ_1 and other kind of roots by λ_i for $i = 2, 3 \dots (k+1)$
we can write

$$\lambda_i^{-1} = e^\alpha (\cos\beta + i \sin\beta)$$

$$\Rightarrow \quad \lambda_i^{-r} = e^{r\alpha} (\cos r\beta + i \sin r\beta), \quad i = 2, 3 \dots (k+1)$$

(11.92) reduces to

$$\phi(\lambda_i) = \frac{f_0}{\lambda_i} + \frac{f_1 p_0}{\lambda_i^2} + \dots + \frac{f_K p_0 p_1 \cdots p_{K-1}}{\lambda_i^{k+1}} = 1$$

$$\Rightarrow \quad f_0\, e^\alpha\, (\cos \beta + i \sin \beta) + f_1\, p_0\, e^{2\alpha}\, (\cos 2\beta + i \sin 2\beta) + \ldots$$
$$+ f_k p_0 p_1 \cdots p_{k-1}\, e^{(k+1)\alpha}(\cos (k+1)\, \beta + i \sin (k+1)\, \beta) \;=\; 1$$

Equating real and imaginary parts $\Rightarrow$

$$f_0\, e^\alpha \cos \beta + f_1\, p_0\, e^{2\alpha} \cos 2\beta + \ldots + f_K p_0 p_1 \cdots p_{K-1}\, e^{(k+1)\alpha} \cos (k+1)\beta = 1$$

Since $\qquad\qquad |\cos r\beta| \le 1 \quad \forall\ r = 1, 2, 3 \ldots ,$

it follows that

$$f_0\, e^\alpha + f_1\, p_0\, e^{2\alpha} + \ldots + f_k p_0 \cdots p_{k-1}\, e^{(k+1)\alpha} \ge 1 \qquad\qquad (11.93)$$

Also $\qquad\qquad e^{r\alpha} = |\lambda_i|^{-r},\ i = 2, 3 \ldots (k+1)$

$$\Rightarrow \quad f_0\, |\lambda_i|^{-1} + f_1 p_0\, |\lambda_i|^{-2} + \ldots + f_k p_0 \cdots p_{K-1}\, |\lambda_i|^{-(k+1)} \ge 1$$
$$i = 2, 3 \ldots (k+1) \qquad\qquad (11.94)$$

A comparison of (11.92) and (11.94) $\Rightarrow$

$$\frac{1}{|\lambda_i|} \ge \frac{1}{\lambda_1}$$

$$\Rightarrow \qquad\qquad \lambda_1 \ge |\lambda_i| \quad \forall\ i = 2, 3 \ldots (k+1)$$

$\Rightarrow \lambda_1$ is most dominant.

Result II

$$n_t = \lambda_1\, n_{t-1} \text{ holds asymptotically for large } t.$$

Proof: Since $(L)_{(k+1)\times(k+1)}$ is non singular, $\exists$ a non singular matrix $P \ni$

$$P^{-1} L P = \Lambda = \text{diag}\,(\lambda_1, \lambda_2 \ldots \lambda_{k+1})$$
$$L = P\Lambda P^{-1}$$
$$L^2 = P\Lambda^2 P^{-1}$$
$$\vdots \qquad \vdots$$
$$L^t = P\Lambda^t P^{-1} \qquad\qquad (11.95)$$

Now $\qquad\qquad n_t = L^t\, n_0$

$$\Rightarrow \qquad\qquad n_t = P\Lambda^t P^{-1}\, n_0$$

$$\Rightarrow \qquad\qquad \frac{n_t}{\lambda_1\, t} = P\,\frac{\Lambda^t}{\lambda_1\, t}\, P^{-1}\, n_0 \qquad\qquad (11.96)$$

$(P^{-1})\,(n_0)_{(k+1)\times 1}$ is a $(k+1)$ rowed column vector.

Let $\qquad\qquad P^{-1} n_0 = \begin{pmatrix} C \\ * \\ \vdots \\ * \end{pmatrix}_{(k+1)\times 1}$

i.e. first element of $P^{-1}\, n_0$ is C whereas other elements may be arbitrary and denoted by $*$

$$(11.96) \Rightarrow \quad \frac{n_t}{\lambda_1 t} = P \begin{bmatrix} \frac{\lambda_1^t}{\lambda_1 t} & 0 & \cdots & 0 \\ 0 & \frac{\lambda_2^t}{\lambda_1 t} & \cdots & 0 \\ 0 & 0 & \cdots & \frac{\lambda_{k+1}^t}{\lambda_1 t} \end{bmatrix} \begin{pmatrix} C \\ * \\ \vdots \\ * \end{pmatrix}$$

$$\lim_{t \to \infty} \frac{n_t}{\lambda_1 t} = P \operatorname{diag}(1, 0 \ldots 0) \begin{pmatrix} C \\ * \\ \vdots \\ * \end{pmatrix}$$

Since $\lambda_1 \geq |\lambda_i|$, $i = 2, 3 \ldots (k+1)$

and $\qquad\qquad \dfrac{\lambda_i^t}{\lambda_1^t} \to 0$ as $t \to \infty$ $\;\forall\; i = 2, 3 \ldots (k+1)$

$\therefore \qquad\qquad \dfrac{n_t}{\lambda_1 t} = P \begin{pmatrix} C \\ 0 \\ \vdots \\ 0 \end{pmatrix}$ holds for $t \to \infty$

Similarly $\qquad \dfrac{n_{t-1}}{\lambda_1^{t-1}} = P \begin{pmatrix} C \\ 0 \\ \vdots \\ 0 \end{pmatrix}$ holds for $t \to \infty$

$\Rightarrow \qquad\qquad n_t = \lambda_1 n_{t-1}$ holds for very large t.

$$\Rightarrow \qquad\qquad \begin{bmatrix} n_{0t} \\ n_{1t} \\ \vdots \\ n_{kt} \end{bmatrix} = \lambda_1 \begin{bmatrix} n_{0t-1} \\ n_{1t-1} \\ \vdots \\ n_{k(t-1)} \end{bmatrix} \qquad\qquad (11.97)$$

It shows that age distribution of a stable population

$$\frac{n_{xt}}{K} = \frac{n_{x(t-1)}}{K} \quad \text{holds} \;\forall\; x = 0, 1, 2, \ldots \text{ and for large } t$$
$$\sum_{x=0} n_{xt} \quad \sum_{x=0} n_{x(t-1)}$$

i.e., Age distribution remains independent of t.

Result III

Non zero vector solution of $LX_1 = \lambda_1 X_1$ where λ_1 is the positive eigen value of L is given by

$$X_1 = \begin{bmatrix} 1 \\ p_0/\lambda_1 \\ p_0 \, p_1/\lambda_1^2 \\ \vdots \\ p_0 \, p_1 \cdots p_{k-1}/\lambda_1^{k+1} \end{bmatrix} \qquad\qquad (11.98)$$

Proof: We have

$$(L - \lambda I) X_1 = 0$$

$$\Rightarrow \quad \begin{bmatrix} f_0 & f_1 & \cdots & & f_k \\ p_0 & 0 & \cdots & & 0 \\ 0 & p_1 & \cdots & 0 & \cdots & 0 \\ \vdots & \vdots & & \vdots & & \vdots \\ 0 & 0 & \cdots & p_{k-1} & 0 \end{bmatrix} \begin{bmatrix} x_0 \\ x_1 \\ \vdots \\ x_k \end{bmatrix} = \lambda_1 \begin{bmatrix} x_0 \\ x_1 \\ \vdots \\ x_k \end{bmatrix}$$

We have

$$f_0 x_0 + f_1 x_1 + \dots + f_k x_k = \lambda_1 x_0$$

$$p_0 x_0 = \lambda_1 x_1 \Rightarrow x_1 = p_0 x_0/\lambda_1$$

$$p_1 x_1 = \lambda_1 x_2 \Rightarrow x_2 = \frac{p_1 x_1}{\lambda_1} = \frac{p_1 p_0 x_0}{\lambda_1^2}$$

Similarly

$$x_3 = \frac{p_0 p_1 p_2}{\lambda_1^3} x_0, \dots, x_{K-1} = \frac{p_0 p_1 \cdots p_{k-2}}{\lambda_1^{k-1}} x_0$$

Finally

$$f_0 x_1 + f_1 x_1 + \dots + f_K x_k = \lambda_1 x_0$$

$$\Rightarrow \quad x_K = \frac{p_0 p_2 \cdots p_k}{\lambda_1^k} x_0$$

Hence

$$X_1 = \begin{bmatrix} x_0 \\ p_0 x_0/\lambda_1 \\ p_0 p_1 x_0/\lambda_1^2 \\ \vdots \\ p_0 p_1 \cdots p_k x_0/\lambda_1^{k+1} \end{bmatrix}$$

also $(L - \lambda_1 I)_{(k+1) \times (k+1)} (X_1)_{(k+1) \times 1} = 0$ has one linearly independent solution since

$$\text{Rank} (L - \lambda_1 I) = k$$

Therefore, # linearly independent vector solution being one only we can write the vector solution of $LX_1 = \lambda_1 X_1$ as

$$\begin{bmatrix} 1 \\ p_0/\lambda_1 \\ p_0 p_1/\lambda_1^2 \\ \vdots \\ p_0 p_1 \cdots p_k/\lambda_1^{k+1} \end{bmatrix} \quad (11.99)$$

Result IV

A sufficient condition that λ_1 is the only positive characteristic root of a non negative matrix L is λ_1 is strictly dominant.

The result holds for any matrix L with non negative elements. For the Leslie matrix, however, we *state the condition without proof*.

Result V

If two consecutive entries f_j and f_{j+1} in the first row of L are not zeros then the positive characteristic root of L is strictly dominant.

11.7.1   Density Dependent Leslie Matrix (A Result Due to P. H. Leslie (1948))

Suppose the mortality and the fertility parameters separately in the L-matrix are also density dependent i.e., these also depend on the size of the population at that time in addition to being age dependent; then the population growth rate is Logistic.

Proof:

$$
\text{Let} \quad L' = \begin{bmatrix} q^{-1} f_0 & q^{-1} f_1 & \cdots & q^{-1} f_k \\ q^{-1} p_0 & 0 & & 0 \\ 0 & q^{-1} p_1 & & 0 \\ 0 & 0 & q^{-1} p_{k-1} & 0 \end{bmatrix} = q^{-1} L \qquad (11.100)
$$

where $q = a + bN$, N being the size of the population and L being the Leslie matrix in (11.88)

Here the birth and the death parameters are each multiplied by q^{-1} to realise the same being density dependent.

Here
$$
|L' - \lambda I| = 0 \Rightarrow
$$
$$
\lambda^{k+1} - (q^{-1} f_0) \lambda^K - (q^{-2} p_0 f_1) \lambda^{k-1} \cdots - q^{-(k+1)} f_k p_0 p_1 \cdots p_{k-1} = 0 \qquad (11.101)
$$

Now if $N \to \infty$, the growth rate essentially becomes zero and the most dominant characteristic roots of L' becomes unity and that of L becomes λ_1.

$$
N(t+1) = \lambda_1 q^{-1} N(t)
$$

$$
\therefore \qquad \frac{N(t+1) - N(t)}{N(t)} = \frac{(\lambda_1 q^{-1} - 1) N(t)}{N(t)} = \lambda_1 q^{-1} - 1
$$

As $N \to \infty$,
$$
\lambda_1 q^{-1} = 1
$$

$$
\Rightarrow \qquad q = \lambda_1
$$

$$
\Rightarrow \qquad \lambda_1' = \lambda_1 q^{-1} = 1
$$

Also the characteristic roots of L' are q^{-1} times that of L ($q^{-1} \lambda_1 = 1$ as $N \to \infty$)

Hence
$$
q = \lambda_1 \qquad (11.102)
$$

Also if $N \to 0$, $q = a + bN$, the population birth and death parameters being independent of density, while density being very low $\Rightarrow a = 1 \qquad (11.103)$

$$
\Rightarrow q = a + bN \to 1 + bN \qquad (11.104)
$$

Also as N is very large say, $N = N_0$ as $t \to \infty$ i.e. ($\lim_{t \to \infty} N = N_0$) $\qquad (11.105)$

and
$$q = \lambda_1$$

$$\Rightarrow \qquad \lambda_1 = 1 + bN_0$$

$$\Rightarrow \qquad \frac{(\lambda_1 - 1)}{b} = N_0 \qquad (11.106)$$

$$\Rightarrow \qquad b = \frac{(\lambda_1 - 1)}{N_0} \qquad (11.107)$$

$$q = a + bN$$

$$= \left[1 + \frac{(\lambda_1 - 1)}{N_0} N \right] \qquad (11.108)$$

Again if $N(t + 1)$ be the population age vector at time $(t + 1)$ then

$$N(t + 1) = L' N(t) = q^{-1} LN(t)$$

$$= q^{-1} \lambda_1 N(t) \qquad (11.109)$$

where λ_1 is the most dominant positive characteristic root of L.

Adding up over all ages we get from (11.109)

$$N(t + 1) = \frac{\lambda_1}{q} N(t) \qquad (11.110)$$

$$N(t + 1) = \frac{\lambda_1 N(t)}{1 + \frac{(\lambda_1 - 1)}{N_0} N(t)} \qquad (11.111)$$

One can easily verify that

$$N(t) = \frac{K}{1 + \rho e^{-rt}} \qquad (11.112)$$

is the solution of the difference equation (11.111) which shows that the growth rate is Logistic.

Remarks:

However, if the initial age distribution is not in stable form, $N(t)$ will not increase in the Logistic form as shown. Although this unjustifies the fitting of Logistic curve, in such a situation, often it has been found Logistic curve to produce a good fit; this is indeed misleading (Feller (1949) page 52). Similar is the case of several empirically good-fittings of Population data with Logistic.

11.8 Generalised Birth and Death Process Models

A stochastic version of the population growth model, based on time dependent birth and death parameters evolved by D. G. Kendall (1948) is presented in this section. This is a more generalised stochastic model, the particular case or the deterministic version of which may lead to Stable or Quasi Stable population models described in the earlier sections. The formulation of the general time dependent birth and death process is as follows:

Let an integer valued time dependent random variable n, measure at any time t the size of the population and suppose in an infinitesimal element of time δt, the only possible transitions are

$$n_{t+\delta t} = n + 1 \text{ with probability } n_t \, \lambda(t) \, \delta t + o(\delta t)$$

$$= n_t - 1 \text{ with probability } n_t \, \mu(t) \, \delta t + o(\delta t)$$

$$= n_t \text{ with probability } [1 - (\lambda(t) + \mu(t)) \, n_t + o(\delta t)] \quad (11.113)$$

with the initial conditions

$$P_1(0) = 1, \ P_0(0) = 0, \ P_n(0) = 0 \text{ for } n \neq 1$$

$$P_0(t) = 0, \ t \neq 0, \ P_n(t) = 0 \text{ for } n < 0 \quad (11.114)$$

where $P_n(t)$ represents the probability that the population size is n at any time t ; $n = 0, 1, 2,...$

Denoting $\lambda(t) = \lambda$ and $\mu(t) = \mu$ we have the Kolmogorov equations as

$$P_n(t + \delta t) = P_n(t) \left[1 - n \left(\lambda \, \delta t + \mu \, \delta t\right) + o(\delta t)\right]$$

$$+ P_{n-1}(t) \left[(n - 1) \lambda \, \delta t + o(\delta t)\right]$$

$$+ P_{n+1}(t) \left[(n + 1) \mu \, \delta t + o(\delta t)\right] \quad (11.115)$$

$$\Rightarrow P_n(t + \delta t) - P_n(t) = - n P_n(t) \left(\lambda + \mu\right) \delta t + (n - 1) P_{n-1}(t) \lambda \, \delta t$$

$$+ (n + 1) P_{n+1}(t) \mu \, \delta t + o(\delta t)$$

$$\Rightarrow \lim_{\delta t \to 0} \frac{P_n(t + \delta t) - P_n(t)}{\delta t} = - n \, P_n(t) \left(\lambda + \mu\right) + \lambda \left(n - 1\right) P_{n-1}(t)$$

$$+ \mu \left(n + 1\right) P_{n+1}(t)$$

$$\Rightarrow \frac{\partial P_n(t)}{\partial t} = - n \left(\lambda + \mu\right) P_n(t) + \lambda \left(n - 1\right) P_{n-1}(t) + \mu \left(n + 1\right) P_{n+1}(t)$$

$$(11.116)$$

We employ the method of probability generating function for the solution of the above differential equation. The entire analysis is given by Kendall (1948).

For
$$n = 0, \frac{\partial P_0(t)}{\partial t} = \mu \, P_0(t)$$

Let
$$\phi(t) = \sum_{n=0}^{\infty} P_n(t) z^n; \ |z| < 1$$

$$\frac{\partial \phi(t)}{\partial t} = \sum_{n=0}^{\infty} \frac{\partial P_n(t)}{\partial t} z^n \quad (11.117)$$

$$\frac{\partial \phi(t)}{\partial z} = \sum_{n=1}^{\infty} n P_n(t) z^{n-1} \qquad (11.118)$$

From (11.116) and (11.117) we have

$$\frac{\partial \phi(t)}{\partial t} = \sum_{n=1}^{\infty} [-n(\lambda + \mu) P_n(t) + \lambda(n-1) P_{n-1}(t) + \mu(n+1) P_{n+1}(t)] z^n$$

$$= -(\lambda + \mu) \sum_{n=1}^{\infty} n P_n(t) z^n + \lambda z \sum_{n=1}^{\infty} (n-1) P_{n-1}(t) z^{n-1}$$
$$+ \mu \sum_{n=1}^{\infty} (n+1) P_{n+1}(t) z^n$$

$$\frac{\partial \phi}{\partial t} = \mu \sum_{n=1}^{\infty} n P_n(t) z^{n-1} + \lambda z \sum_{n=0}^{\infty} n P_n(t) z^n - (\lambda + \mu) z \sum_{n=1}^{\infty} n P_n(t) z^{n-1}$$

$$= \mu \sum_{n=1}^{\infty} n P_n(t) z^{n-1} + \lambda z^2 \sum_{n=1}^{\infty} n P_n(t) z^{n-1} - (\lambda + \mu) z \sum_{n=1}^{\infty} n P_n(t) z^{n-1}$$

or
$$\frac{\partial \phi}{\partial t} = \mu \frac{\partial \phi}{\partial z} + \lambda z^2 \frac{\partial \phi}{\partial z} - (\lambda + \mu) z \frac{\partial \phi}{\partial z}$$

or
$$\frac{\partial \phi}{\partial t} = \{\mu + \lambda z^2 - (\lambda + \mu) z\} \frac{\partial \phi}{\partial z}$$

or
$$\frac{\partial \phi}{\partial t} = \{\lambda z(z-1) - \mu(z-1)\} \frac{\partial \phi}{\partial z}$$

or
$$\frac{\partial \phi}{\partial t} = (\lambda z - \mu)(z-1) \frac{\partial \phi}{\partial z} \qquad (11.119)$$

This is a Lagrangian type of differential equation.
  Noting

$$\frac{\partial \phi}{\partial t} = \frac{\partial \phi}{\partial z} \cdot \frac{\partial z}{\partial t},$$

the above is reduced to an ordinary differential equation

$$\frac{dz}{dt} = (\lambda z - \mu)(z-1) \qquad (11.120)$$

The solution is a homographic (Waston-Bessel function) function of the form

$$z = \frac{f_1 + Cf_2}{f_3 + Cf_4} \qquad (11.121)$$

where f_1, f_2, f_3 and f_4 are functions of t.

$$\Rightarrow \quad zf_3 + zCf_4 = f_1 + Cf_2$$

$$\Rightarrow \quad C(zf_4 - f_2) = f_1 - zf_3$$

$$\Rightarrow \quad C = \frac{zf_3 - f_1}{f_2 - zf_4} \tag{11.122}$$

Consider

$$\phi(z, t) = \Phi\left(\frac{f_1 + Cf_2}{f_3 + Cf_4}, t\right)$$

or

$$\phi(z, t) = \Phi\left(\frac{zf_3 - f_1}{f_2 - zf_4}, t\right)$$

and the solution or structure of $\phi(z ; t)$ is to be obtained. Further, if we assume the boundary conditions

$$P_1(0) = 1$$

$$P_n(0) = 0, \text{ for } n \neq 1$$

$$\phi(z, t)\big|_{t=0} = \left.\sum_{n=0}^{\infty} P_n(t) z^n\right|_{t=0} = z$$

$$\Rightarrow \quad \phi(z, 0) = \Phi\left(\frac{z[f_3]_0 - [f_1]_0}{[f_2]_0 - z[f_4]_0}\right) = z$$

where $[f_i]_0$ indicates the value of f_i at $t = 0$

Let

$$Z = \frac{z[f_3]_0 - [f_1]_0}{[f_2]_0 - z[f_4]_0}$$

$$\therefore \quad \Phi(Z) = z$$

Now

$$Z = \frac{z[f_3]_0 - [f_1]_0}{[f_2]_0 - z[f_4]_0}$$

$$Z[f_2]_0 - Z\{z[f_4]_0\} = z[f_3]_0 - [f_1]_0$$

$$z\{[f_3]_0 + Z[f_4]_0\} = [f_1]_0 + Z[f_2]_0$$

$$\Rightarrow \quad z = \frac{[f_1]_0 + Z[f_2]_0}{[f_3]_0 + Z[f_4]_0}$$

$$\Phi(Z) = z = \frac{Z[f_2]_0 + [f_1]_0}{Z[f_4]_0 + [f_3]_0} \tag{11.123}$$

To obtain the structure of $\Phi(z)$

Also

$$\phi(z, t) = \Phi\left(\frac{zf_3 - f_1}{f_2 - zf_4}\right)$$

$$= \frac{\left(\frac{zf_3 - f_1}{f_2 - zf_4}\right)[f_2]_0 + [f_1]_0}{\left(\frac{zf_3 - f_1}{f_2 - zf_4}\right)[f_4]_0 + [f_3]_0} \qquad (\text{ from } (11.123))$$

$$= \frac{(zf_3 - f_1)[f_2]_0 + (f_2 - zf_4)[f_1]_0}{(zf_3 - f_1)[f_4]_0 + (f_2 - zf_4)[f_3]_0}$$

$$= \frac{g_1(t) + zg_2(t)}{g_3(t) + zg_4(t)}$$

where

$$g_1(t) = f_2[f_1]_0 - f_1[f_2]_0$$

$$g_2(t) = f_3[f_2]_0 - f_4[f_1]_0$$

$$g_3(t) = f_2[f_3]_0 - f_1[f_4]_0$$

$$g_4(t) = f_3[f_4]_0 - f_4[f_3]_0$$

$$\therefore \qquad \phi(z, t) = [g_1(t) + zg_2(t)]\left[g_3(t)\left(1 + z\,\frac{g_4(t)}{g_3(t)}\right)\right]^{-1}$$

$$= \left[\frac{g_1(t) + zg_2(t)}{g_3(t)}\right]\left[1 + z\,\frac{g_4(t)}{g_3(t)}\right]^{-1}$$

$$= \left[\frac{g_1(t) + zg_2(t)}{g_3(t)}\right]\left[1 - z\,\frac{g_4(t)}{g_3(t)} + z^2\left(\frac{g_4(t)}{g_3(t)}\right)^2 + \dots + z^{n-1}\left(\frac{g_4(t)}{g_3(t)}\right)^{n-1}\right.$$

$$\left. - z^n\left(\frac{g_4(t)}{g_3(t)}\right)^n + \dots\right] \qquad (11.124)$$

; where n is odd and provided $\left| z\,\frac{g_4(t)}{g_3(t)} \right| < 1$

Again $\qquad \phi(z, t) = \sum_{n=0}^{\infty} P_n(t)\,z^n$

$\Rightarrow \qquad \phi(1, 0) = 0 \qquad$ for $n \neq 1$

$\qquad\qquad\qquad = 1 \qquad$ for $n = 1$

$P_0(t) = $ term without involving z

$P_n(t) = $ coefficient of z^n in the above expression

Therefore, from (11.124) on comparing coefficient of $z, z^2, \dots z^n$, we get

$$P_0(t) = \frac{g_1(t)}{g_3(t)} = \xi(t) = \xi$$

$$P_n(t) = \left\{ \frac{g_2(t)}{g_3(t)} \left(\frac{g_4(t)}{g_3(t)} \right)^{n-1} - \frac{g_1(t)}{g_3(t)} \left(\frac{g_4(t)}{g_3(t)} \right)^n \right\}$$

$$= \left[\left(\frac{g_4(t)}{g_3(t)} \right)^{n-1} \left\{ \frac{g_2(t)}{g_3(t)} - \frac{g_4(t)}{g_3(t)} \xi(t) \right\} \right] \qquad (11.125)$$

Now
$$\phi(1, t) = \frac{g_1(t) + g_2(t)}{g_3(t) + g_4(t)} = 1$$

$$\Rightarrow \qquad \frac{\dfrac{g_1(t)}{g_3(t)} + \dfrac{g_2(t)}{g_3(t)}}{1 + \dfrac{g_4(t)}{g_3(t)}} = 1$$

Put
$$\frac{g_4(t)}{g_3(t)} = -\eta \text{ and } \frac{g_1(t)}{g_3(t)} = \xi \qquad (11.126)$$

$$\therefore \qquad \frac{\xi + \dfrac{g_2(t)}{g_3(t)}}{1 - \eta} = 1$$

$$\Rightarrow \qquad \xi + \frac{g_2(t)}{g_3(t)} = 1 - \eta$$

$$\Rightarrow \qquad \frac{g_2(t)}{g_3(t)} = 1 - \eta - \xi \qquad (11.127)$$

Putting (11.126) and (11.127) in (11.125)

$$\Rightarrow \qquad P_n(t) = (-\eta)^{n-1}(1 - \eta - \xi + \eta\xi) \text{ when } n \text{ is odd}$$

$$= (-\eta)^{n-1}(1 - \eta)(1 - \xi)$$

$$= (-\eta_t)^{n-1}(1 - P_0(t), (1 - \eta_t) \qquad (11.128)$$

which provides the probability distribution   of the r.v.*n* (the size of the population) at time *t*, when *n* is odd.

Similarly one can show that

$$P_n(t) = (\eta)^{n-1}(1 - \eta - \xi + \eta\xi), \text{ when } n \text{ is even}$$

$$= (\eta)^{n-1}(1 - \eta)(1 - \xi)$$

$$= (\eta)^{n-1}(1 - P_0(t))(1 - \eta_t) \qquad (11.129)$$

and
$$P_0(t) = \xi_t$$

Combining (11.128) and (11.129) we have

$$P_n(t) = (\eta_t)^{n-1}(1 - \eta_t)(1 - \xi_t) \text{ and } P_0(t) = \xi_t$$

Next
$$\phi(z, t) = P_0(t) + zP_1(t) + z^2 P_2(t) + \dots + z^{n-1} P_{n-1}(t) + \dots$$

the p.g.f. of $P_n(t)$ ($n = 0, 1, 2\dots$)

$$= \xi + z(1 - \eta - \xi + \eta\xi) + z^2\eta(1 - \eta - \xi + \eta\xi)$$

$$+ z^3\eta^2(1 - \eta - \xi + \eta\xi) + \dots + (-\eta)^{n-1}z^n(1 - \eta - \xi + \eta\xi) + \dots$$

$$= \xi + z(1 - \eta - \xi + \eta\xi)\{1 + \eta z + \eta^2 z^2 + \dots + \eta^{n-1}z^{n-1} + \eta^n z^n + \dots\}$$

$$= \xi + z(1 - \eta - \xi + \eta\xi)\frac{1}{1 - \eta z} \text{ provided } |\eta z| < 1$$

$$= \frac{\xi(1 - \eta z) + z(1 - \eta - \xi + \eta\xi)}{1 - \eta z}$$

$$= \frac{\xi - \xi\eta z + z - \eta z - \xi z + \eta\xi z}{1 - \eta z}$$

$$\therefore \qquad \phi(z, t) = \frac{\xi - \eta z - \xi z + z}{1 - \eta z} \qquad (11.130)$$

Now we require to estimate the parameters $\xi = \xi(t)$ and $\eta = \eta(t)$

Since $\qquad \phi(z, t) = \dfrac{\xi - \eta z - \xi z + z}{1 - \eta z}$

$$\log_e \phi(z, t) = \log(\xi - \eta z - \xi z + z) - \log_e(1 - \eta z)$$

$$\frac{1}{\phi}\frac{\partial\phi}{\partial z} = \frac{1 - \eta - \xi}{(\xi - \eta z - \xi z + z)} - \frac{(-\eta)}{1 - \eta z}$$

or $\qquad \dfrac{1}{\phi}\dfrac{\partial\phi}{\partial z} = \dfrac{(1 - \eta - \xi)(1 - \eta z) + \eta(\xi - \eta z - \xi z + z)}{(1 - \eta z)(\xi - \eta z - \xi z + z)}$

$$= \frac{(1 - \xi - \eta - \eta z + \eta^2 z + \xi\eta z + \eta\xi - \eta^2 z - \eta\xi z + z\eta)}{(1 - \eta z)(\xi - \eta z - \xi z + z)}$$

$$= \frac{(1 - \eta - \xi + \eta\xi)}{(1 - \eta z)(\xi - \eta z - \xi z + z)}$$

or $\qquad \dfrac{\partial\phi}{\partial z} = \phi \cdot \dfrac{(1 - \eta - \xi + \eta\xi)}{(1 - \eta z)(\xi - \eta z - \xi z + z)}$

$$= \frac{(\xi - \eta z - \xi z + z)}{(1 - \eta z)} \cdot \frac{(1 - \eta - \xi + \eta\xi)}{(1 - \eta z)(\xi - \eta z - \xi z + z)}$$

$$\therefore \qquad \frac{\partial\phi}{\partial z} = \frac{(1 - \eta)(1 - \xi)}{(1 - \eta z)^2} \qquad (11.131)$$

Now $\qquad \dfrac{\partial\phi}{\partial t} = \dfrac{\partial\phi}{\partial z} \cdot \dfrac{\partial z}{\partial t}$

$$= \frac{(1-\xi)(1-\eta)}{(1-\eta z)^2} \cdot (\lambda z - \mu)(z-1)$$

(from (11.119) & (11..131))

$$= \frac{(1-\xi)(1-\eta)}{(1-\eta z)^2} (\lambda z^2 - (\lambda+\mu)z + \mu) \qquad (11.132)$$

Again, $\qquad \phi(z,t) = \dfrac{\xi - \eta z - \xi z + z}{1 - \eta z}$

$\log_e \phi(z,t) = \log_e(\xi - \eta z - \xi z + z) - \log_e(1 - \eta z)$

$$\frac{1}{\phi} \cdot \frac{\partial \phi}{\partial t} = \frac{\xi' - \eta' z - \xi' z}{\xi - \eta z - \xi z + z} + \frac{\eta' z}{1 - \eta z}$$

where $\qquad \xi' = \dfrac{\partial \xi}{\partial t}, \ \eta' = \dfrac{\partial \eta}{\partial t}$

$\Rightarrow \qquad \dfrac{\partial \phi}{\partial t} = \dfrac{\xi' - \eta' z - \xi' z}{1 - \eta z} + \dfrac{\eta' z (\xi - \eta z - \xi z + z)}{(1 - \eta z)^2}$

$$= \frac{(\xi' - \eta' z - \xi' z)(1 - \eta z) + \eta' z(\xi - \eta z - \xi z + z)}{(1 - \eta z)^2}$$

$$= \frac{\begin{array}{c}\xi' - \eta' z - \xi' z - \xi' \eta z + \eta \eta' z^2 + \eta \xi' z^2 + \eta' \xi z \\ - \eta \eta' z^2 - \eta' \xi z^2 + \eta' z^2\end{array}}{(1 - \eta z)^2}$$

$$= \frac{\xi' + (\eta'\xi - \xi' - \eta' - \eta\xi')z + (\eta' + \eta\xi' - \eta'\xi)z^2}{(1 - \eta z)^2} \qquad (11.133)$$

Equating terms independent of z and coefficient of z, z^2 from (11.133) and (11.132) we get

$$\left.\begin{array}{rcl} \xi' &=& \mu(1-\xi)(1-\eta) \qquad \text{(i)} \\ \eta'\xi - \xi'\eta - \xi' - \eta' &=& -(\lambda+\mu)(1-\xi)(1-\eta) \quad \text{(ii)} \\ \eta\xi' - \eta'\xi + \eta' &=& \lambda(1-\xi)(1-\eta) \qquad \text{(iii)} \end{array}\right\} \qquad (11.134)$$

In eqn. (11.134), (i) and (iii) are the estimating equations for the parameters $\xi = \xi(t)$ and $\eta = \eta(t)$.

Now put

$$U = 1 - \xi \qquad\qquad V = 1 - \eta$$
$$U' = -\xi' \qquad\qquad V' = -\eta'$$

where $\qquad U' = \dfrac{dU}{dt}, \ V' = \dfrac{dV}{dt}$

$\Rightarrow \qquad\qquad 1 - U = \xi \text{ and } 1 - V = \eta$

(i)–(iii) $\Rightarrow$ $\quad\quad$ $\xi' - \eta\xi' + \eta'\xi - \eta' = (\mu - \lambda)(1 - \xi)(1 - \eta)$

$\Rightarrow$ $\quad\quad$ $-U' - (1 - V)(-U') + (-V')(1 - U) - (-V') = (\mu - \lambda)\,UV$

or $\quad\quad$ $-U' + (1 - V)\,U' - V'(1 - U) + V' = (\mu - \lambda)\,UV$

Dividing both sides by U we get,

$$-\frac{U'}{U} + (1 - V)\frac{U'}{U} - V'\frac{(1 - U)}{U} + \frac{V'}{U} = (\mu - \lambda)V$$

or $\quad\quad$ $-\dfrac{U'}{U} + \dfrac{U'}{U} - \dfrac{U'V}{U} - \dfrac{V'}{U} + \dfrac{UV'}{U} + \dfrac{V'}{U} = (\mu - \lambda)V$

$$\left[V' - \frac{U'}{U}V\right] = (\mu - \lambda)V$$

or $\quad\quad$ $V' + \mu V^2 = (\mu - \lambda)\,V$

$$\left(\text{since } \xi' = \mu(1 - \xi)(1 - \eta) \Rightarrow -U' = \mu UV \Rightarrow -\frac{U'}{U} = \mu V\right)$$

or $\quad\quad$ $\dfrac{dV}{dt} - (\mu - \lambda)V = -\mu V^2$ $\quad\quad\quad\quad$ (11.135)

This is a linear differential equation of the Bernoulli's form $\dfrac{dy}{dx} + Py = Qy^m$ with Integrating factor (I.F)

$$= \exp\left(\int P\,dx\right)$$

Dividing eqn. (11.135) both sides by $-V^2$, we get

$$-\frac{1}{V^2}\frac{dV}{dt} + (\mu - \lambda)\frac{1}{V} = \mu$$

Let $\quad\quad$ $\dfrac{1}{V} = W \Rightarrow -\dfrac{1}{V^2}\dfrac{dV}{dt} = \dfrac{dW}{dt}$

$$\frac{dW}{dt} + (\mu - \lambda)W = \mu$$

Integrating Factor $=$ I.F. $= \exp\left(\int (\mu(t) - \lambda(t))\,dt\right) = e^{\rho(t)}$

where $\quad\quad$ $\rho(t) = \int (\mu(t) - \lambda(t))\,dt$

$\therefore$ $\quad\quad$ $e^{\rho(t)}\dfrac{dW(t)}{dt} + e^{\rho(t)}(\mu - \lambda)W = \mu e^{\rho(t)}$

$$\frac{d(We^{\rho(t)})}{dt} = \mu e^{\rho(t)}$$

Integrating from 0 to t, we get

$$We^{\rho(t)} = \int_0^t \mu e^{\rho(t)} dt + C \qquad (11.136)$$

Initially, when $t = 0$

$$\xi_l = P_0(t) \Rightarrow \xi_0 = P_0(0) = 0 \quad \text{(By assumption)}$$

Now

$$P_n(t) = (1 - \xi_t)(1 - \eta_t)(\eta_t)^{n-1}$$

$$\Rightarrow \qquad P_n(0) = (1 - \xi_0)(1 - \eta_0)(\eta_0)^{n-1}$$

$$= (1 - \eta_0)(\eta_0)^{n-1} \qquad (\because \xi_0 = 0)$$

For $n = 1$

$$P_1(0) = (1 - \eta_0)$$

$$\Rightarrow \qquad 1 = 1 - \eta_0$$

$$\Rightarrow \qquad \eta_0 = 0$$

$$\therefore \qquad \xi_0 = 0, \quad \eta_0 = 0$$

$$U_t|_{t=0} = (1 - \xi_t)|_{t=0} = 1 - \xi_0 = 1$$

$$V_t|_{t=0} = (1 - \eta_t)|_{t=0} = 1 - \eta_0 = 1$$

$$\therefore \qquad U = 1, \ V = 1 \text{ when } t = 0$$

From eqn. (11.136) we have

$$W = Ce^{-\rho(t)} + e^{-\rho(t)} \int_0^t \mu(\tau) e^{\rho(\tau)} d\tau \qquad (11.137)$$

Now consider

$$W_t|_{t=0} = \frac{1}{V_t}\bigg|_{t=0} = \frac{1}{1 - \eta_t}\bigg|_{t=0} = 1$$

$$\therefore \qquad W(0) = 1$$

$$\rho(t) = \int_0^t (\mu(\tau) - \lambda(\tau)) d\tau$$

$$\Rightarrow \qquad \rho(0) = 0$$

From eqn. (11.137), we have

$$W(0) = Ce^{-\rho(0)} + \int_0^t e^{\rho(\tau)} \mu(\tau) d\tau$$

$$\Rightarrow \qquad C = W(0) = 1$$

Hence

$$W = e^{-\rho(t)} \left[1 + \int_0^t \mu(\tau) e^{\rho(\tau)} \, d\tau \right] \qquad (11.138)$$

Next, we prove that

(A) $\qquad W(t) = 1 + e^{-\rho(t)} \int_0^t e^{\rho(\tau)} \lambda(\tau) d\tau$

(B) $\qquad W(t) = \dfrac{1}{2}(1 + e^{-\rho(t)}) + \dfrac{1}{2} e^{-\rho(t)} \int_0^t e^{\rho(\tau)} \{\lambda(\tau) + \mu(\tau)\} d\tau$

Proof: To prove (A), let us consider the integral on the Right Hand Side of (11.138) viz,

$$\int_0^t \mu(\tau) e^{\rho(\tau)} \, d\tau$$

Integrating by parts,

$$\Rightarrow \qquad \left[e^{\rho(\tau)} \left[\int \mu(\tau) \, d\tau \right] \right]_0^t - \int_0^t \left\{ \frac{d}{d\tau} [e^{\rho(\tau)}] \int \mu(\tau) \, d\tau \right\} d\tau$$

$$\left[\text{Since} \qquad \rho(\tau) = \int \{\mu(\tau) - \lambda(\tau)\} \, d\tau \right.$$

$$\int_0^t \{\mu(\tau) - \lambda(\tau)\} \, d\tau = \rho(\tau) \Big|_0^t = \rho(t) - \rho(0)$$

$$\Rightarrow \qquad \int_0^t \mu(\tau) \, d\tau = \rho(t) + \int_0^t \lambda(\tau) \, d\tau \qquad (\because \rho(0) = 0) \Big]$$

$$\therefore \int_0^t \mu(\tau) e^{\rho(\tau)} d\tau = \left[e^{\rho(t)} \int \mu(\tau) \, d\tau \right]_0^t - \int_0^t \left\{ \frac{d}{d\tau} (e^{\rho(\tau)}) \int \mu(\tau) \, d\tau \right\} d\tau$$

$$= e^{\rho(t)} \left[\rho(t) + \int_0^t \lambda(\tau) \, d\tau \right] - \int_0^t \left\{ \frac{de^{\rho(\tau)}}{d\tau} \int \mu(\tau) \, d\tau \right\} d\tau$$

$$= e^{\rho(t)} \left[\rho(t) + \int_0^t \lambda(\tau) \, d\tau \right] - \int_0^t \frac{d(e^{\rho(\tau)})}{d\tau} \left[\rho(t) + \int_0^t \lambda(\tau) \, d\tau \right] d\tau$$

$$= e^{\rho(t)} \rho(t) + e^{\rho(t)} \int_0^t \lambda(\tau) \, d\tau - \int_0^t \left\{ \frac{d}{d\tau} e^{\rho(\tau)} \right\} \rho(\tau) \, d\tau$$

$$- \int_0^t \left\{ \frac{d}{d\tau} (e^{\rho(\tau)}) \int_0^t \lambda(\tau) \, d\tau \right\} d\tau$$

since $\displaystyle \int_0^t e^{\rho(\tau)} \lambda(\tau) \, d\tau = e^{\rho(t)} \int \lambda(\tau) \, d\tau - \int_0^t \frac{d}{d\tau} (e^{\rho(\tau)}) \left(\int \lambda(\tau) \, d\tau \right) d\tau$

$\therefore \qquad \displaystyle \int_0^t \mu(\tau) e^{\rho(\tau)} \, d\tau = e^{\rho(t)} \rho(t) + \int_0^t e^{\rho(\tau)} \lambda(\tau) \, d\tau - \int_0^t \left\{ \frac{d}{d\tau} e^{\rho(\tau)} \right\} \rho(\tau) \, d\tau$

$$= I + II - III \text{ (say)}.$$

Now $\qquad III = \displaystyle \int_0^t \left[\frac{d}{d\tau} e^{\rho(\tau)} \right] \rho(\tau) \, d\tau$

$$= \int_0^t \left[e^{\rho(\tau)} \frac{d\rho(\tau)}{d\tau} \right] \rho(\tau) \, d\tau$$

Integrating above by parts, we get

$$III = \left[\rho(\tau) \int \left[e^{\rho(\tau)} \frac{d\rho(\tau)}{d\tau} \right] d\tau \right]_0^t - \int_0^t \frac{d\rho(\tau)}{d\tau} \int \left[e^{\rho(\tau)} \frac{d\rho(\tau)}{d\tau} \, d\tau \right] d\tau$$

$$= \left[\rho(\tau) \int \frac{d}{d\tau} e^{\rho(\tau)} \, d\tau \right]_0^t - \int_0^t \left\{ \frac{d\rho(\tau)}{d\tau} \int \frac{d}{d\tau} [e^{\rho(\tau)}] \, d\tau \right\} d\tau$$

$$= \rho(\tau) e^{\rho(\tau)} \,]_0^t - \int_0^t \frac{d\rho(\tau)}{d\tau} e^{\rho(\tau)} \, d\tau$$

$$= \rho(t) e^{\rho(t)} - \int_0^t \frac{d}{d\tau} (e^{\rho(\tau)}) \, d\tau \qquad\qquad (\because \rho(0) = 0)$$

$$= \rho(t) e^{\rho(t)} - e^{\rho(\tau)} \big|_0^t$$

$$= \rho(t) e^{\rho(t)} - [e^{\rho(t)} - 1]$$

$$= 1 + \rho(t) e^{\rho(t)} - e^{\rho(t)} \qquad\qquad (11.139)$$

$\therefore \qquad \displaystyle \int_0^t \mu(\tau) e^{\rho(\tau)} \, d\tau = \int_0^t e^{\rho(\tau)} \lambda(\tau) \, d\tau + e^{\rho(t)} - 1 \qquad\qquad (11.140)$

Now putting (11.140) in (11.138) we get

$$W = e^{-\rho(t)} \left[\int_0^t e^{\rho(\tau)} \lambda(\tau) \, d\tau + e^{\rho(t)} \right]$$

or
$$W = e^{-\rho(t)} \left[\int_0^t e^{\rho(\tau)} \lambda(\tau) \, d\tau \right] + 1 \tag{11.141}$$

which establishes result (A).

Similarly, from (11.138), we have

$$W(t) = e^{-\rho(t)} \left[1 + \int_0^t e^{\rho(\tau)} \mu(\tau) \, d\tau \right]$$

$$= e^{-\rho(t)} + e^{-\rho(t)} \int_0^t e^{\rho(\tau)} \mu(\tau) \, d\tau \tag{11.142}$$

From (11.141), we have

$$W(t) = 1 + e^{-\rho(t)} \int_0^t e^{\rho(\tau)} \lambda(\tau) \, d\tau \tag{11.143}$$

Adding (11.142) and (11.143) we get

$$W(t) = \frac{1}{2}(1 + e^{-\rho(t)}) + \frac{e^{-\rho(t)}}{2} \int_0^t e^{\rho(\tau)} (\lambda(\tau) + \mu(\tau)) \, d\tau$$

which proves result (B)

Next we recall

$$\frac{U'}{U} = -\mu V = -\frac{\mu}{W}$$

and also the differential equation

$$W' + (\mu - \lambda) W = \mu$$

with
$$\rho'(t) = (\mu(t) - \lambda(t))$$

$\Rightarrow$
$$W' + \rho'(t)W = \mu$$

$\Rightarrow$
$$-\frac{W'}{W} - \rho' = -\frac{\mu}{W}$$

$\Rightarrow$
$$\frac{U'}{U} = -\mu V = -\frac{\mu}{W} = -\frac{W'}{W} - \rho'$$

Integrating both sides $\Rightarrow$

$$\log U = -\log W - \rho + C$$

we have at $t = 0$, $U = V = W = 1$ and $\rho = 0$.

$$\Rightarrow \qquad C = 0$$

$$\log U = \log \frac{1}{W} - \rho$$

$$\Rightarrow \qquad U = \frac{e^{-\rho}}{W}$$

$$\therefore \qquad \xi = 1 - U = 1 - \frac{e^{-\rho}}{W}$$

and $$\qquad \eta = 1 - V = 1 - \frac{1}{W}$$

$$\Rightarrow \qquad \left. \begin{aligned} \xi(t) &= 1 - \frac{e^{-\rho(t)}}{W} \\[2mm] \eta(t) &= 1 - \frac{1}{W} \end{aligned} \right\} \qquad\qquad (11.144)$$

and $$\qquad W = e^{-\rho(t)} \left\{ 1 + \int_0^t e^{\rho(\tau)} \mu(\tau)\, d\tau \right\} \qquad\qquad (11.145)$$

It may be noted that the results (11.144) and (11.145) together with (11.141) may completely determine the population process $P_n(t)$.

Mean and variance of n_t can be obtained immediately. We have

$$\xi_t = 1 - \frac{e^{-\rho(t)}}{W}$$

and $$\qquad \eta_t = 1 - \frac{1}{W}$$

$$\Rightarrow \qquad \frac{1 - \xi_t}{1 - \eta_t} = e^{-\rho(t)}$$

$$\Rightarrow \qquad E(n_t) = e^{-\rho(t)} \qquad\qquad (11.146)$$

Also $$\qquad P_0(t) = \xi_t \quad \text{and} \quad P_n(t) = (1 - \xi_t)(1 - \eta_t)\eta_t^{n-1}$$

$$\therefore \qquad E(n_t) = 0 \times \xi_t + \sum_{n_t = 1}^{\infty} (1 - \xi_t)(1 - \eta_t)\eta_t^{n-1} n_t$$

$$= (1 - \xi_t)(1 - \eta_t) \sum_{n_t = 1}^{\infty} n_t\, \eta_t^{n-1}$$

$$= (1 - \xi_t)(1 - \eta_t)[1 + 2\eta_t + 3\eta_t^2 + \ldots]$$

$$= \frac{(1 - \xi_t)(1 - \eta_t)}{(1 - \eta_t)^2} = \frac{(1 - \xi_t)}{(1 - \eta_t)}$$

$$\text{Var}(n_t) = E(n_t^2) - [E(n_t)]^2$$

$$= \sum_{n_t = 1}^{\infty} (1 - \xi_t)(1 - \eta_t) \eta_t^{n-1} n_t^2 - \left[\frac{(1 - \xi_t)}{(1 - \eta_t)}\right]^2$$

$$= (1 - \xi_t)(1 - \eta_t)[1^2 + 2^2\eta_t + 3^2\eta_t^2 + \ldots] - \left[\frac{(1 - \xi_t)}{(1 - \eta_t)}\right]^2$$

$$= (1 - \xi_t)(1 - \eta_t)[1 + 2^2\eta_t + 3^2\eta_t^2 + \ldots] - \left[\frac{(1 - \xi_t)}{(1 - \eta_t)}\right]^2$$

$$= (1 - \xi_t)(1 - \eta_t)[(1 - \eta_t)^{-2} + 2\eta_t(1 - \eta_t)^{-3}] - \left[\frac{(1 - \xi_t)}{(1 - \eta_t)}\right]^2$$

$$= \frac{(1 - \xi_t)(1 - \eta_t)}{(1 - \eta_t)^2}\left[1 + \frac{2\eta_t}{(1 - \eta_t)}\right] - \left[\frac{(1 - \xi_t)}{(1 - \eta_t)}\right]^2$$

$$= \frac{(1 - \xi_t)(1 - \eta_t)}{(1 - \eta_t)^2}\frac{(1 + \eta_t)}{(1 - \eta_t)} - \frac{(1 - \xi_t)^2}{(1 - \eta_t)^2}$$

$$= \frac{(1 - \xi_t)(1 + \eta_t)}{(1 - \eta_t)^2} - \frac{(1 - \xi_t)^2}{(1 - \eta_t)^2}$$

$$= \frac{(1 - \xi_t)[1 + \eta_t - 1 + \xi_t]}{(1 - \eta_t)^2}$$

$$= \frac{(1 - \xi_t)(\xi_t + \eta_t)}{(1 - \eta_t)^2} \tag{11.147}$$

Alao

$$1 - \xi_t = \frac{e^{-\rho}}{W}, \quad \xi_t + \eta_t = 2 - \frac{(1 + e^{-\rho})}{W}$$

$$(1 - \eta_t)^2 = \frac{1}{W^2}$$

Hence

$$\left(\frac{e^{-\rho}}{W}\right)\left(\frac{2W - 1 - e^{-\rho}}{W}\right) \Big/ \frac{1}{W^2} = \text{Var}(\eta_t) \tag{11.148}$$

$$\Rightarrow e^{-\rho}(2W - 1 - e^{-\rho}) = \frac{(1 - \xi_t)(\xi_t + \eta_t)}{(1 - \eta_t)^2} = \text{Var}(n_t) \tag{11.149}$$

Also we have

$$2W = (1 + e^{-\rho}) + e^{-\rho} \int_0^t e^{\rho(\tau)} \{\lambda(\tau) + \mu(\tau)\} \, d\tau \tag{11.150}$$

11.8.1 Some Demographic Applications of Kendall's Birth and Death Process

An application of Kendall's birth and death process is to identify the process when the mean growth rate is given. Especially, sometimes the problem arises to achieve a certain stipulated mean growth rate and assuming that the population behaviour is described by a certain process (say Gaussian Process or Logistic process), there lies a problem of choosing birth and death parameters enabling to achieve the stipulated growth rate under the process. To be specific, let the mean growth rate follows a Logistic Law given by

$$\bar{n}_t = n_0 \frac{\alpha}{1 + (\alpha - 1) e^{-\beta t}}; \quad \alpha > 0, \beta > 0 \tag{11.151}$$

where $n_0 = n_t|_{t=0} \equiv$ the initial population size

$$\Rightarrow \qquad \frac{d \log \bar{n}_t}{dt} = [\lambda(t) - \mu(t)] = \frac{(\alpha - 1)\beta}{e^{\beta t} + (\alpha - 1)} = \rho(t) \tag{11.152}$$

is the mean growth rate at time t is independent of the initial population size.

$$\Rightarrow \qquad \lambda(t) = \frac{(\alpha - 1)\beta}{e^{\beta t} + (\alpha - 1)} + \mu(t) \tag{11.153}$$

Suppose $\mu(t) = \mu_0$ and $\lambda(t) = \lambda_0$ holds as $t \to \infty$ then

$$\lambda_0 = \left[\lim_{t \to \infty} \frac{(\alpha - 1)\beta}{e^{\beta t} + (\alpha - 1)} + \mu_0 \right]$$

$$\Rightarrow \qquad \lambda_0 = \mu_0 \text{ holds for very large } t \tag{11.154}$$

Therefore, if the population is stable and the mean growth rate takes place in the logistic form, then the population is stationary also, i.e., $\lambda_0 = \mu_0$. On the other hand, if the population is quasi stable in the form viz. $\lambda(t)$ depends on t but $\mu(t) = \mu_0$ is independent of t ; then we have

$$\lambda(t) = \left[(\alpha - 1)\beta e^{-\beta t} \frac{\bar{n}_t}{n_0 \alpha} + \mu_0 \right]$$

This gives an idea of the level of birth rate needed to maintain a mean or desirable size of the population given its death parameter which is stable.

For example, for population like in India where birth parameter is time dependent whereas death rate can approximately be assumed to be more or less stable, the problem lies in deciding the level of birth rate (by suitable family limitation programme) for the maintenance of optimal population size.

11.9 A Bisexual Population Growth Model of Goodman (1953)

So far, we have considered, the population growth of only one sex, usually female. L.A. Goodman (1953) presented a stochastic model of population growth based on both sexes. We begin with a population having two sub-population, the F-subpopulation and M-subpopulation with the sex-specific birth and death rates.

Let there be M males and F females at time t with B_m and D_m are the birth and death rates of males and B_f and D_f are the birth and death rates of females.

Again, let

$FB_m \, \delta t$ = Prob [that a female gives birth to a male child at time δt]

$FB_f \, \delta t$ = Prob [that a female gives birth to a female child at time δt]

then
$$B_m = KD_m \qquad (11.155)$$

where K is the ratio of males to females at birth.

Further,

$p_{00}(t)$ is the probability of extinction at time t

$p_{01}(t)$ = Prob[zero male and one female at time t]

and so on.

Now the problem is to find the probability distribution of number of males and females at the end of the period $(0, t + dt)$. Consider

$p_{m,f}(t + dt)$ = probability that there will be M males and F females at the end of the period $(0, t + dt)$]

= prob [M males and F females at time t and no births and no deaths in $(t, t + dt)$]

+ prob [$(M - 1)$ males and F females at time t and one male birth in $(t, t + dt)$]

+ prob[$(M$ males and $F - 1$ females at time t and one female birth in $(t, t + dt)$]

+ prob[$M + 1$ males and F females at time t and one male death in $(t, t + dt)$]

+ prob [M males and $F + 1$ females at time t and one female death in $(t, t + dt)$]

$$
\begin{aligned}
p_{m,f}(t + dt) = {} & [1 - (MD_m + FD_f + MB_m + FB_f) \, dt] \, P_{m,f}(t) \\
& + B_m F \, p_{m-1,f}(t) \, dt \\
& + B_f (F - 1) \, p_{m,f-1}(t) \, dt \\
& + D_m (M + 1) \, P_{m+1,f}(t) \, dt \\
& + D_f (F + 1) \, p_{m,f+1}(t) \, dt \, ; \, m = 0, 1, 2, \dots; \, f = 0, 1, 2 \dots \quad (11.156)
\end{aligned}
$$

Subtracting $p_{m,f}(t)$ from both sides of (11.156) and dividing throughout by dt, we obtain by taking limit

$$\lim_{dt \to 0} \frac{p_{m,f}(t + dt) - p_{m,f}(t)}{dt} = -(MD_m + FD_f + MB_m$$

$$+ FB_f) p_{m,f}(t) + B_m F p_{m-1,f}(t)$$

$$+ B_f(F - 1) p_{m,f-1}(t)$$

$$+ D_m(M + 1) p_{m+1,f}(t)$$

$$+ D_f(F + 1) p_{m,f+1}(t);$$

$$m = 0, 1, 2, \dots \text{ and } f = 0, 1, 2 \dots \tag{11.157}$$

The infinite set of differential-difference equation (11.157) can be converted into a single equation in the probability generating function. Defining as usual the probability generating function $\phi(s_1, s_2, t)$ as

$$\phi(s_1, s_2, t) = \sum_{m,f=0}^{\infty} p_{m,f}(t) s_1^m s_2^f, \ |s_i| \leq 1, i = 1, 2 \tag{11.158}$$

(11.157) becomes

$$\frac{dp_{m,f}(t)}{dt} = -(MD_m + FD_f + MB_m + FB_f) p_{m,f}(t) + B_m F p_{m-1,f}(t)$$

$$+ B_f(F - 1) p_{m,f-1}(t) + D_m(M + 1) p_{m+1,f}(t)$$

$$+ D_f(F + 1) p_{m,f+1}(t) \tag{11.159}$$

Multiplying (11.159) by $s_1^m s_2^f$ and summing over all m and f, we obtain by making use of (11.158),

$$\frac{\partial \phi}{\partial t} = -D_m s_1 \frac{\partial \phi}{\partial s_1} - (D_f + B_m + B_f) s_2 \frac{\partial \phi}{\partial s_2}$$

$$+ B_m s_1 s_2 \frac{\partial \phi}{\partial s_2} + B_f s_2^2 \frac{\partial \phi}{\partial s_2} + D_m \frac{\partial \phi}{\partial s_1} + D_f \frac{\partial \phi}{\partial s_2}$$

$$= D_m(1 - s_1) \frac{\partial \phi}{\partial s_1} - [B_m s_1 s_2 + B_f s_2^2 + D_f - (D_f + B_m + B_f) s_2] \frac{\partial \phi}{\partial s_2}$$

$$\tag{11.160}$$

Differentiating (11.160) with respect to s_1,

$$\frac{\partial^2 \phi}{\partial t \, \partial s_1} = -D_m \frac{\partial \phi}{\partial s_1} + D_m (1 - s_1) \frac{\partial^2 \phi}{\partial s_1^2} + B_m s_2 \frac{\partial \phi}{\partial s_2}$$

$$+ [B_m s_1 s_2 + B_f s_2^2 + D_f - (D_f + B_m + B_f) s_2] \frac{\partial^2 \phi}{\partial s_1 \, \partial s_2} \tag{11.161}$$

Putting $s_1 = s_2 = 1$, we get

$$\frac{\partial^2 \phi}{\partial t \, \partial s_1}\bigg|_{s_1 = s_2 = 1} = -D_m \frac{\partial \phi}{\partial s_1}\bigg|_{s_1 = s_2 = 1} + B_m \frac{\partial \phi}{\partial s_2}\bigg|_{s_1 = s_2 = 1}$$

$$\frac{\partial \psi_m}{\partial t} = -D_m \, \psi_m + B_m \, \psi_f \tag{11.162}$$

where $\quad \dfrac{\partial \phi}{\partial s_1}\bigg|_{s_1 = s_2 = 1} = \psi_m \;$ and $\; \dfrac{\partial \psi}{\partial s_2}\bigg|_{s_1 = s_2 = 1} = \psi_f$

ψ_m is the expected number or the first moment of males at time t and ψ_f that of females. Again differentiating (11.162) with respect to s_2

$$\frac{\partial^2 \phi}{\partial t \, \partial s_2} = D_m(1 - s_1)\frac{\partial^2 \phi}{\partial s_1 \, \partial s_2} + [B_m \, s_1 + 2B_f \, s_2 - (D_f + B_m + B_f)]\frac{\partial \phi}{\partial s_2}$$

$$+ [B_m \, s_1 \, s_2 + B_f \, s_2^2 + D_f - (D_f + B_m + B_f) \, s_2]\frac{\partial^2 \phi}{\partial s_2^2} \tag{11.163}$$

$$\frac{\partial^2 \phi}{\partial t \, \partial s_2}\bigg|_{s_1 = s_2 = 1} = [B_m + 2B_f - (D_f + B_m + B_f)]\frac{\partial \phi}{\partial s_2}\bigg|_{s_1 = s_2 = 1}$$

$$\Rightarrow \qquad \frac{\partial \psi_f}{\partial t} = (B_f - D_f)\psi_f \tag{11.164}$$

Similarly $\dfrac{\partial \psi_m}{\partial s_1}\bigg|_{s_1 = s_2 = 1} = E(m_t \, (m_t - 1)), \; \dfrac{\partial \psi_m}{\partial s_2}\bigg|_{s_1 = s_2 = 1} = E(f_t \, (f_t - 1))$

$$\frac{\partial \psi_m}{\partial s_1 \partial s_2}\bigg|_{s_1 = s_2 = 1} = E(m_t \, f_t)$$

where $E(m_t)$ and $E(f_t)$ represents the expected number of males and females at any time t.

$$\Rightarrow \qquad E(m_t^2) = \frac{\partial \psi_m}{\partial s_1}\bigg|_{s_1 = s_2 = 1} + \psi_m$$

$$E(f_t^2) = \frac{\partial \psi_f}{\partial s_2}\bigg|_{s_1 = s_2 = 1} + \psi_f$$

In view of this we have

$$\frac{\partial E(m_t^2)}{\partial t} = -2D_m \, \psi_m^2 + 2B_m \, \psi_{mf} + D_m \, \psi_m + B_m \, \psi_f \tag{11.165}$$

$$\frac{\partial E(m_t \, f_t)}{\partial t} = (B_f - D_f - D_m) \, \psi_{mf} + B_m \, \psi_f^2 \tag{11.166}$$

$$\frac{\partial E(f_t^2)}{\partial t} = 2(B_f - D_f)\psi_f^2 + (B_f + D_f)\psi_f \qquad (11.167)$$

After solving these five equations (11.162), (11.164), (11.165), (11.166) and (11.167) we get,

$$\psi_m = \frac{B_m}{D_m + B_f - D_f} \exp[(B_f - D_f)t] - \exp(-D_m t) \qquad (11.168)$$

$$\varphi_f = \exp[(B_f - D_f) t] \qquad (11.169)$$

with the initial condition of one female. If we start with F_0 females and M_0 males at the start we get

$$E(m_t) = \psi_m = M_0 \exp(-D_m t) + F_0 \frac{B_m}{D_m + B_f - D_f} [\exp[(B_f - D_f) t]$$

$$- \exp(-D_m t)] \qquad (11.170)$$

and
$$E(f_t) = \psi_f = F_0 \exp[(B_f - D_f)t] \qquad (11.171)$$

11.10 Models of Population Growth under Random Environment

(A Stochastic Version of the Logistic Population Model)
It is evident that the deterministic theory of population growth outlined in some of the preceeding sections relating to logistic and stable and quasi stable population models is not really adequate for the description of all the phenomena one might wish to study, for it fails to take into account the role of chance fluctuations in the development of the process. Extensive work in this respect has been done by Feller [5]. Feller's characterization of the 'built in boundaries' [5] as simplified by Keilson (1965) and reproduced by Goel and Dyn [6] is given in the following lines.

Let us take the stochastic version of Verhulst deterministic logistic model given in (11.12) viz,

$$\frac{dN(t)}{dt} = rN(t)\left[1 - \left(\frac{N(t)}{\pi}\right)^{\alpha}\right]\bigg/\alpha$$

For simplicity, let $\alpha = 1$ and $N(t) = X(t) = x$.
Then we have

$$\frac{dx}{dt} = rx\left[1 - \frac{x}{\pi}\right] \qquad (11.172)$$

Consider a stochastic form of r, the growth rate under a fluctuating environmental condition. Then $r = \bar{r} + \sigma F(t)$

where $\bar{r} = E(r)$, $\sigma = \sqrt{\text{Var}(r)}$ and $F(t)$ represents some noise.

$$\Rightarrow \qquad \frac{dx}{dt} = [\bar{r} + \sigma F(t)] \, x \left(1 - \frac{x}{\pi}\right)$$

$$= \bar{r}x \left(1 - \frac{x}{\pi}\right) + \sigma x \left(1 - \frac{x}{\pi}\right) F(t)$$

$$= \alpha(x) + \beta(x) \, F(t) \qquad (11.173)$$

where $\qquad \alpha(x) = \bar{r}x \left(1 - \frac{x}{\pi}\right) \qquad (11.174)$

and $\qquad \beta(x) = \sigma x \left(1 - \frac{x}{\pi}\right) \qquad (11.175)$

A process given by

$$\frac{dx}{dt} = \alpha(x) + \beta(x) \, F(t)$$

is called Gaussian Delta continuous process.

Let $\qquad a(x) = \alpha(x) + \frac{1}{4} \frac{\partial}{\partial x} \beta(x)^2$

and $\qquad b(x) = [\beta(x)]^2$

Then the Fokkar Planck equation (F.P. equation) is given by

$$\frac{\partial P}{\partial t} = -\frac{\partial}{\partial x} [a(x) \, P] + \frac{1}{2} \frac{\partial^2}{\partial x^2} [b(x) \, P] \qquad (11.176)$$

where $P = p\,[x\,|\,y,\,t]$ for any y and t i.e. the probability that the r. v. will take the value of x at time t given that it takes a value y at time zero.

Given the Fokkar Planck equation the stochastic differential equation of Gaussian delta continuous process is uniquely given by

$$\frac{\partial x(t)}{\partial t} = a(x) - \frac{1}{4} \frac{\partial b(x)}{\partial x} + [b(x)]^{-1/2} \, F(t) \qquad (11.177)$$

The process which are completely determined by the coefficients $a(x)$ and $b(x)$ of the F.P. equation with no additional boundary conditions are known as Unrestricted processes. Such processes have unlimited state spaces if $b(x) > 0$ and $a(x)$ is infinite. Again such processes may have some 'built in' boundaries in case (i) $b(x)$ vanishes at some point or (ii) $a(x)$ becomes infinite. These processes are continuous counterpart of the birth and death process. The type of boundary[+] is determined by the integrability of the following two functions.

$$h_1(x) = \pi(x) \int\limits_{x_0}^{x} [b(\xi) \, \pi(\xi)]^{-1} \, d\xi \qquad (11.178)$$

[+] Inaccessible, absorbing or reflecting.

and
$$h_2(x) = [b(x)\pi(x)]^{-1} \int_{x_0}^{x} \pi(\xi)\, d\xi \tag{11.179}$$

Over the interval $I = [x_0, r]$ where x_0 is some interior point of the state space of the process and r is the boundary.*

The function $\pi(x)$ is defined as

$$\pi(x) = \exp\left[-2\int_{x_0}^{x} \frac{a(\xi)}{b(\xi)}\, d\xi\right] \tag{11.180}$$

Further, for $y \le z \le x$

$$\mu_1(z\,|\,z_0) = 2\int_{x_0}^{z} h_1 dx = \text{average time to reach } z \text{ starting from } x_0 \text{ where}$$
$$z_0 \text{ is a reflecting boundary} \tag{11.181}$$

and

$$\mu_2(x_0\,|\,z) = 2\int_{x_0}^{z} h_2 dx = \text{average time to reach } x_0 \text{ starting from } z \text{ where}$$
$$z \text{ is the reflecting boundary.} \tag{11.182}$$

we have in the case of Logistic process

$$b(x) = [\beta(x)]^2 = \sigma^2 x^2 \left(1 - \frac{x}{\pi}\right)^2$$

$$b(0) = 0$$

and
$$b(\pi) = 0$$

Hence the process is an unrestricted singular process. The boundaries $x = 0$, $x = \pi$ are therefore singular.

Also near $x = 0$

$$a(x) = \alpha(x) + \frac{1}{4}\frac{\partial}{\partial x}\beta(x)^2$$

$$= \bar{r}x\left(1 - \frac{x}{\pi}\right) + \frac{1}{4}\frac{\partial}{\partial x}\sigma^2 x^2\left(1 - \frac{x}{\pi}\right)^2$$

$$= x\left(1 - \frac{x}{\pi}\right) + \left[\bar{r} + \frac{\sigma^2}{2}\left(1 - \frac{2x}{\pi}\right)\right].$$

For small x

$$a(x) \equiv \left(\bar{r} + \frac{\sigma^2}{2}\right)x \tag{11.183}$$

and
$$b(x) \equiv \sigma^2 x^2 \tag{11.184}$$

---

* r may be infinite in case of absence of "built-on" finite boundary.

Further, since $h_1(x)$ and $h_2(x)$ are both not integrable over I, it follows that the boundaries $x = 0$ and $x = \pi$ are inaccessible natural boundaries and cannot reach in finite time, a result in conformity with the deterministic process. The process may therefore describe a population which is far from extinction and which fluctuates about some mean value $(< \pi)$ due to fluctuations in the net growth rate.

The steady state probability density function of the unrestricted process $P(x \mid y, \infty)$ is given by

$$P(x \mid y, \infty) = \frac{c}{b(x)} \exp\left(2 \int\limits_{-\infty}^{x} [a(\xi_1)/b(\xi_1)] \, d\xi \right) \qquad (11.185)$$

where c is a normalizing constant; when no boundary is in exit boundary or a regular absorbing boundary. Again

$$P(x \mid y, \infty) = [b(x) \, \pi(x)]^{-1} / [\int b(x) \, \pi(x)]^{-1} \, dx$$

when boundaries are on exit or absorbing. Hence the steady state probability density function of the process is given by

$$P(x \mid y, \infty) = cx^{[(2\bar{r}/\sigma^2) - 1]} \left(1 - \frac{x}{\pi} \right)^{-\left(\frac{2x}{\sigma^2} + 1 \right)} \qquad (11.186)$$

Particular Cases

(i) If $2\bar{r}/\sigma^2 < 1$, then the density function is U shaped indicating that the density approaches 0 or π

(ii) If $2\bar{r}/\sigma^2 > 1$, then the density function is monotonically increasing and in J-shaped concentration of population around π.

To derive the time dependent probability density of the process, we introduce the variable,

$$z = \frac{1}{\sigma} \log\left[x / \left(1 - \frac{x}{\pi} \right) \right]$$

$$dz = \left[\sigma x \left(1 - \frac{x}{\pi} \right) \right]^{-1} dx$$

The equation (11.173) becomes

$$\frac{dz}{dt} = \frac{\bar{r}}{\sigma} + F(t)$$

which is nothing but the stochastic differential equation for the unrestricted Weinner Process.

The Fokkar Plank equation satisfied by the conditional probability density of Z given by, say $g(z \mid z_0 \, t)$ is

$$\frac{dy}{dt} = -\frac{\partial}{\partial z} [(\bar{r}/\sigma) \, g] + \frac{1}{2} \frac{\partial^2 g}{\partial z^2}$$

with the boundary condition

$$\lim_{z \to \pm \infty} g(z \,|\, z_0\, t) = 0$$

corresponding to the inaccessible boundaries

$$x = 0 \,(z = -\infty) \text{ and } x = \pi \,(z = +\infty)$$

Thus , the Gaussian distribution is given by

$$g\,(z\,|\,z,t) = \frac{1}{(2\pi t)^{1/2}} \exp{-\frac{1}{2t}\left(z - z_0 - \frac{\bar{r}t}{\sigma}\right)^2}$$

with peak at

$$z = z_0 + \frac{\bar{r}t}{\sigma}$$

Eventually this gives

$$P(x\,|\,y,t) = \frac{1}{\sigma\sqrt{2\pi\, tx}\,\left(1 - \frac{x}{\pi}\right)} \exp\left[\left(-\frac{1}{2t}\left(\frac{1}{\sigma}\log\frac{x}{y} - \frac{1}{\sigma}\right)\right.\right.$$

$$\left.\left.\log\left(1 - \frac{x}{\pi}\right)\Big/\left(1 - \frac{y}{\pi}\right) - \frac{\gamma t}{\sigma}\right)^2\right] \qquad (11.187)$$

which is the density function determining the behaviour of the population.

Note that when $\pi \to \infty$ either the population is far from saturation or the supply of resource is unlimited.

In this case

$$x = e^{\sigma z}$$

$$\Rightarrow \qquad E(X) = \int_0^\pi x\, P(x\,|\,y,t)\, dx$$

$$= \int_{-\infty}^\infty \exp(\sigma Z)\, g(z_0\,|\,z_0, t)\, dZ$$

$$= \exp\left(\frac{\sigma^2 t}{2}\right)\exp(\sigma(z_0 + \bar{r}t/\sigma)) = ye^{rt}\left[\exp\left(\frac{\sigma^2 t}{2}\right)\right]$$

as compared to $x = ye^{rt}$ for the deterministic case in absence of random fluctuations.

Similarly,

$$\text{Var}\,(X) = E\,(X^2) - [E(X)]^2$$

$$= E(Y^2)\, e^{2rt}\, e^{\sigma^2 t}\, [e^{\sigma^2 t} - 1]$$

$$= [E(X)]^2\, [e^{\sigma^2 t} - 1]$$

as compared to zero variance in the deterministic case. The coefficient of variation is given by

$$\frac{[\text{Var}\,(X)]^{1/2}}{E(X)} = (e^{\sigma^2 t} - 1)^{1/2} \tag{11.188}$$

As t increases the coefficient of variation also increases showing that the average rate cannot describe the growth of the population.

11.11 Conclusion

The discussion in respect of the deterministic and stochastic models unfortunately cannot have unlimited scope in view of the fact that further generalization of the model is naturally warranted with Mathematical complications and the consequent inflexibility of the models, so generalised, for making the same more realistic. For example, a population expert would like to generalize Leslie's original formulation (vide art 11.6) and might reasonably suppose that the number of individuals born during the interval $(t, t + 1)$ has the Poisson

distribution with mean $\displaystyle\sum_{X=0}^{m} F_x\, n_{xt}$ and the number in the age group $(x - x + 1)$

at the end of the interval has a binomial distribution with probability p_{x-t} and the index parameter $n_{x-1,\,t}$. Such generalisations are useful in the context of population problems in India. Similarly Lotka's stable population and Coale's quasi stable population analysis are also capable of being generalized which will definitely be useful in the context of population problem. Again such generalizations would considerably cut down the flexibility of the models because of Mathematical complexities. However, the complexities arising out of these kind of complications may partially be overcome by employing computer solutions and modern simulation techniques.

References

1. Bartlett, M. S. (1960): Stochastic Population models — *Mathuen Monograph*
2. Biswas, S. (1986) : Population Dynamics and its perspective — An invited paper in the 73rd Session (Statistics Section) of *All India Science Congress*, Delhi, 1986
3. Coale, Ansley (1976): *Growth and Structure of Human Population* — Princeton University Press.
4. Feller, William (1949) : *An Introduction to the Probability theory and its application*, Vol II, John Wiley & Sons.
5. Feller, William (1951) : Diffusion Processes in genetics — Proceedings of the symposium in Mathematical Statistics and Probability—Second Berkeley Symposium, page 56, Feller William (1952): The Parabolic differential equations and the Associated semi-group of transformations — *Ann of Math*. Vol 55, page 46, 227.
6. Goel. N. S. and Dyn, Nira Richter (1979): *Stochastic models in Biology*, Academic Press.
7. Goodman, L. A. (1953): Population growth of the sexes, *Biometrika*, Vol 9 page 212–225.

8. Keilson, J. (1965): A review of transient behaviour in regular diffusion and birth and death processes—*J. Appl Probability* Vol. 1, page 297.

9. Keilson, J. (1965): A review of transient behaviour in regular diffusion and birth and death processes, Part II, *J. Appl. Probability* Vol. 2, page 405.

10. Kendall. D. G. (1948): A generalised birth and death Processes —*Annals of Mathematical Statistics*, Vol 19, page 1–15.

11. Leslie, P. H. (1945): On the use of matrices in certain Population Mathematics *Biometrika*, Vol 33, page 183–212.

12. Leslie, P. H. (1948): Some further notes is Matrices in Population Mathematics *Biometrika*, Vol 35, page 213–245.

13. Levy, P. — Processes Semi-Markovians'—*Proc. Inter. Cong. Math.* (Amsterdam) Vol 3, page 46–426.

14. Lewis, E. G (1942): On the generation and the growth of Population. Sankhya Vol 36, page 93–95.

15. Lopez, Alvardo (1961): *Problems in Stable Population theory*, Princeton, N. J. Office of the research.

16. Lotka, A. J., Dullin L. I. and Spiegelman (1949): *Length of life*, New York. Ronald Press.

17. Moran, P. A. P. (1961): *Statistical processes in evolutionary theory*, Oxford, Clarendon Press.

18. Nicholson, A. J. (1957): The self adjustment of population to change, cold spring *Harbor Symposium on quantitative Biology* Vol 22, page 551–598.

19. Nishlet, R. M. and Gurney, W. SC (1982): *Modelling fluctuating Population* -John. Wiley & Sons.

20. Keyfitz, Nathan (1966): A life table that agrees with the data, *Journal of American Statistical Association*, Vol 61, page 305–312.

21. Keyfitz, Nathan (1966): A life table that agrees with the date II *Journal of American Statistical Association*, Vol 63, page 1253–68.

22. Keyfitz, N. (1977): *Introduction to the Mathematics of Population with revisions*, Addison Wisley, London.

23. Smith, W. L. (1954): Asymptotic renewal theorem: *Proc. Royal. Society* Edin A 64, page 9–48.

24. Taka'cs, L. (1954): On the secondary process generated by Poisson Process and their application in Physics, *Acta-Math-Acad-Science* Vol. 5, page 203–236.

25. Taka'cs, L. (1960): *Stochastic Process*, London, Mathuen.

Stochastic Processes on Survival and Competing Risk Theory

12.0   Survival Analysis: An Introduction

The statistical analysis of data relating to the duration of life of a component of a physical or a biological system is the main concern of 'Survival Analysis'. The basic feature of this branch of statistics (or Biostatistics) is the incomplete data or censoring in the data. Because of practical difficulties on infeasibilities of following up a patient several times till his death, or because of a patient withdrawing himself from clinical trials, the data remain incomplete or censored. Therefore survival analysis techniques attract the research workers providing appropriate tools on other fields too, for the analysis of incomplete data.

Survival data basically are of two types viz. (i) Parallel type (ii) Series or Competing type.

Parallel type of data occur in the analysis of multi-component parallel systems, particularly parallel paired components like two kidneys, two lungs, two eyes, etc. However, since the survival of the system depends on the longevity of the last surviving component in this case, paried data are available irrespective of whether they are censored or not. Although, the data being censored may provide further complications for the analysis.

On the other hand, series type of data generally arise in case of several competing risks say $R_1, R_2, ..., R_k$ which may be imagined to be competing with each other for taking the life of an individual or a system. But since death or failure may occur because of a particular risk (in presence of other risks) we do not get paried or multivariate data on the longevities due to several risks separately based on the same patients. Hence there exists no sample space and as such Bivariate or Multivariate models to measure the longevities under competing risk is not possible at least theoretically. This may happen more particularly in respect of the requirement paired data, for the longevity of a patient under the joint competing risk of Diabetes and Tuberculosis, or Cancer and Cardio respiratory disease. Since death may occur ultimately to one of the two risks, therefore it is not possible to get the paired data or bivariate sample space. The same difficulty remains when we extend to multivariate sample space too.

The traditional approach in Survival Analysis was based on Parametric techniques. Earlier statistical approaches for system survival, especially for Engineering application were concentrated on Parametric models. Investigations, by and large, reveal that parametric technique often become complicated and too indegant for providing estimational techniques for complex survival models for inevitable incompleteness in the data.

It is therefore, precisely to avoid the estimational difficulties arising from incomplete data [as it happens in several clinical trials] there has been of late, a shift of focus to Non-parametric methods of Research. This, no doubt, gave considerable impetus to the Research and Development of new methods in Non-parametric methods applied on Univariate and Bivariate survival data (with univariate or bivariate censoring), Leurgans, Tsai and Crowley (1984) have pointed out that there are self-consistent estimators [to be defined later in this chapter] which do not converge in probability to the correct limit. They have further maintained that self-consistency, which is otherwise considered to be a satisfactory criterion is, by itself, inadequate to provide a satisfactory criterion for a good estimation despite the established result that bivariate self-consistent estimators are asymptotically consistent. (Munoz (1980), Campbell(1981)). Moreover, asymptotic consistency, even if valid, does not produce much advantage in the analysis of survival data of clinical trials which are usually based on extremely small sample.

Besides these controversial issues in the applicability of Non-parametric techniques in the analysis of Bivariate or Multivariate survival data, there has been a common deficiency in their applications on survival data. It may be noted that under the parametric or non-parametric formulation, survival is merely expressed as a function of age and time only. However, survival is a function of several covariates or other external factors (such as blood pressure, heart condition, blood sugar level, diet and general health etc. and finally the type of treatment taken) some of which are measurable while most of others are qualitative in nature (called covariates).

As described in 4.10 this deficiency can be overcome by using a recently developed technique known as the technique of Partial likelihood by Cox (1975). The model used is semi-parametric in structure. The speciality of the method has in the fact that the parameters relating to the covariates are estimable independently of the base line parameters which describe the system longevity with age or time only. Again as discussed in 4.10 & 4.16 the method of partial likelihood has certain established properties of the classical maximum likelihood estimator.

The present chapter is intended to provide an outline of survival analysis and competing risk theory by parametric, non-parametric methods. We initially start with the indices of measuring the competing risk, following the line of approach of Chiang [2].

12.1 Measurement of Competing Risks

We assume a finite number of risks, say, $R_1, R_2, ..., R_k$ which may be competing with each other to take away the life of an individual or a failure of the system.

To measure the various types of competing risks we introduce the following indices, viz.,
(i) Crude probability
(ii) Net probability (Type A and Type B)
(iii) Partially crude probability

12.1.1 Definition of Hazard Rate $\mu(t, \delta)$

We define the hazard rate due to risks R_δ as the conditional probability of dying in $(t, t + dt]$ as $\Delta t \to 0$, in $x_i < t \leq x_{i+1}$ given that the individual is surviving at x_i. It is denoted by $\mu(t, \delta)$. Obviously, we have under the orderliness condition i.e. probability of more than one failure in $(x_t, x_t + \delta x_t)$, $\Delta t \to 0$ is negligible, $\sum_\delta \mu(t; \delta) = \mu(t)$, where $\mu(t)$ is the total hazard rate, (vide Elandt Johnson (1975))*.

12.1.2 Crude Probability

It is defined as the probability of dying in a risk say $R_\delta (\delta = 1, 2, ..., k)$ in an interval $(x_i, x_{i+1}]$ in presence of all other risks in the population given that the individual is surviving at $X = x_i$.

It is denoted by $Q_{i\delta}$.

12.1.3 Net Probability (Type A)

This is defined as the probability of dying in an interval $(x_i, x_{i+1}]$ by a risk R_δ when all other risks are off from the population given that the individual is surviving at $X = x_i$.

It is denoted by $q_{i\delta}$.

12.1.4 Net Probability (Type B)

This is defined as the probability of dying in $(x_i, x_{i+1}]$ given that the person is surviving at $X = x_i$, when a particular risk R_δ is off from the population. It is denoted as $q_{i\cdot\delta}$.

12.1.5 Partially Crude Probability

This is defined as the probability of dying in $(x_i, x_{i+1}]$ in R_δ in presence of all

---

* Elandt Johnson (1975) — Coditional failure time distributions under competing risk theory with dependent failure times and proportional Hazard rates. Institute of Statistics, Mimeo Series No. 1105, University of North Carolina.

other risks, when another risk say R_ε $(R_\delta \neq R_\varepsilon)$ is off from the population and given that the individual is surviving at $X = x_i$. It is denoted as $Q_{i\delta \cdot \varepsilon}(\delta \neq \varepsilon)$; δ, ε = 1, 2, 3, ... k.

12.2 Inter-relation of the Probabilities

$$Q_{i\delta}, q_{i\delta}, q_{i \cdot \delta}, Q_{i\delta \cdot \varepsilon}$$

We have two basic assumptions, viz.

$$q_i = Q_{i1} + Q_{i2} + ... + Q_{ik}$$

and

$$\mu(t) = \mu(t;1) + \mu(t; 2) + ... + \mu(t ; k)$$

where $\mu(t; \delta)$ represents the hazard rate due to risk R_δ.

12.2.1 Expression of $Q_{i\delta}$

Probability of surviving in $(x_i, t]$ where $x_i < t \leq X_{i+1}$ is given by

$$\exp\left(- \int_{x_i}^t \mu(\tau) d\tau \right) ; \tag{12.1}$$

where $\mu(t)$ is the hazard rate.

Again the probability of dying in $(t, t + \delta t)$ in risk R_δ is

$$\mu(t; \delta)\, \delta t \tag{12.2}$$

Therefore, the probability of surviving in $(x_i, t]$ and then dying in R_δ between $(t, t + dt]$ where $x_i < t \leq x_{i+1}$ is given by

$$\int_{x_i}^{x_{i+1}} \exp\left(- \int_{x_i}^t \mu(\tau)\, d\tau \right) \cdot \mu(t ; \delta)\, dt$$

Hence

$$Q_{i\delta} = \int_{x_i}^{x_{i+1}} \exp\left(- \int_{x_i}^t \mu(\tau)\, d\tau \right) \mu(t ; \delta)\, dt \tag{12.3}$$

$$= \int_{x_i}^{x_{i+1}} \exp\left(- \int_{x_i}^t \mu(\tau)\, d\tau \right) \left(\frac{\mu(t;\delta)}{\mu(t)} \right) \cdot \mu(t ; \delta)\, dt \tag{12.4}$$

Now, we have

$$\mu(t;1) + \mu(t;2) + ... + \mu(t;k) = \mu(t),$$

since the risks $R_1, R_2, ..., R_k$ are independent.

Chiang assumed that although $\mu(t;\delta)$ $(\delta = 1, 2, ... k)$ depends on t but $\dfrac{\mu(t;\delta)}{\mu(t)}$ $(\delta = 1, 2, ... k)$ is taken as independent of t, in some interval $(x_i, x_i + 1]$ (12.5) which is known as *Chiang's proportionality assumption.*

In view of Chiang's proportionality assumption, $\dfrac{\mu(t;\delta)}{\mu(t)}$ in (12.4) may be taken as independent of time and we can rewrite (12.4) as:

$$Q_{i\delta} = \frac{\mu(t;\delta)}{\mu(t)} \int\limits_{x_i}^{x_{i+1}} \exp\left(-\int\limits_{x_i}^{t} \mu(\tau)\, d\tau\right) \mu(t)\, dt$$

$$= -\frac{\mu(t;\delta)}{\mu(t)} \int\limits_{x_i}^{x_{i+1}} \frac{d}{dt}\left(\exp\left(-\int\limits_{x_i}^{t} \mu(\tau)\, d\tau\right)\right) dt$$

$$= -\frac{\mu(t;\delta)}{\mu(t)} \left[\exp\left(-\int\limits_{x_i}^{x_{i+1}} \mu(\tau)\, d\tau\right) - \exp\left(-\int\limits_{x_i}^{x_i} \mu(\tau)\, d\tau\right)\right]$$

$$= -\frac{\mu(t;\delta)}{\mu(t)} \left[\exp\left(-\int\limits_{x_i}^{x_{i+1}} \mu(\tau)\, d\tau\right) - 1\right]$$

$$\Rightarrow \quad Q_{i\delta} = \frac{\mu(t;\delta)}{\mu(t)} \left[1 - \exp\left(-\int\limits_{x_i}^{x_{i+1}} \mu(\tau)\, d\tau\right)\right]$$

Also $\exp\left(-\int\limits_{x_i}^{x_{i+1}} \mu(\tau)\, d\tau\right) = p_i = $ Probability of surviving upto x_{i+1} given that the person has survived upto x_i.

$$\Rightarrow \quad Q_{i\delta} = (1 - p_i) \cdot \frac{\mu(t;\delta)}{\mu(t)}$$

$$\Rightarrow \quad Q_{i\delta} = q_i \cdot \frac{\mu(t;\delta)}{\mu(t)}, \text{ where } q_i = 1 - p_i$$

$$\Rightarrow \quad \frac{Q_{i\delta}}{q_i} = \frac{\mu(t;\delta)}{\mu(t)} = c_{i\delta} \text{ (say)} \qquad (12.6)$$

$$\Rightarrow \quad Q_{i\delta} = q_i \cdot c_{i\delta}$$

$$\frac{Q_{i\varepsilon}}{q_i} = \frac{\mu(t;\varepsilon)}{\mu(t)}$$

$$\Rightarrow \quad \frac{Q_{i\delta}}{Q_{i\varepsilon}} = \frac{\mu(t;\delta)}{\mu(t;\varepsilon)}, \text{ the probabilities are proportional.}$$

12.2.2 Relationship between Crude Probability and Net Probability (Type A)

Probabilities of surviving in $(x_i, x_{i+1}]$ given that R_δ is the only risk is given

by

$$\exp\left(-\int_{x_i}^{x_{i+1}} \mu(t;\delta)\, dt\right) \tag{12.7}$$

Probability of dying in $(x_i, x_{i+1}]$ is given by

$$q_{i\delta} = 1 - \exp\left(-\int_{x_i}^{x_{i+1}} \mu(t;\delta)\, dt\right)$$

$$= 1 - \exp\left(-\int_{x_i}^{x_{i+1}} \left(\frac{\mu(t;\delta)}{\mu(t)}\right)\mu(t)\, dt\right)$$

$$= 1 - \exp\left(-\frac{\mu(t;\delta)}{\mu(t)}\int_{x_i}^{x_{i+1}}\mu(t)\, dt\right)$$

$$= 1 - \left[\exp\left(-\int_{x_i}^{x_{i+1}}\mu(t)\, dt\right)\right]^{\frac{\mu(t;\delta)}{\mu(t)}}$$

by Chiang's proportionality assumption (12.5)

$$= 1 - (p_i)^{\frac{\mu(t;\delta)}{\mu(t)}} \tag{12.8}$$

Also from (12.6)

$$\frac{\mu(t;\delta)}{\mu(t)} = \frac{Q_{i\delta}}{q_i}$$

Thus using (12.8)

$$\Rightarrow \qquad q_{i\delta} = 1 - (p_i)^{\frac{Q_{i\delta}}{q_i}} \tag{12.9}$$

which gives relationship between Net probability (type A) and Crude probabiity.

12.2.3 Relationship between Net Probability (Type B) and Crude Probability

We have,

$q_{i.\delta}$ = Probability of dying in $(x_i, x_{i+1}]$ where R_δ is off from the population given that the person is surviving at $X = x_i$ is given by

$$q_{i \cdot \delta} = 1 - \exp\left(-\int_{x_i}^{x_{i+1}} [\mu(t) - \mu(t; \delta)] \, dt\right)$$

$$= 1 - \exp\left[-\int_{x_i}^{x_{i+1}} \mu(t) \left(1 - \frac{\mu(t; \delta)}{\mu(t)}\right) dt\right]$$

$$= 1 - \left[\exp\left(-\int_{x_i}^{x_{i+1}} \mu(t) \, dt\right)\right]^{\left(1 - \frac{\mu(t; \delta)}{\mu(t)}\right)}$$

(by Chiang's proportionality assumption (12.5))

$\Rightarrow \qquad q_{i \cdot \delta} = 1 - (p_i)^{1 - \frac{Q_{i\delta}}{q_i}}$ (by using (12.6))

$\Rightarrow \qquad q_{i \cdot \delta} = 1 - (p_i)^{\frac{q_i - Q_{i\delta}}{q_i}}$ (12.10)

This gives a relationship between crude probability $Q_{i\delta}$ and Net probability (type *B*) $q_{i \cdot \delta}$.

Also $\qquad q_i = \sum_{\delta=1}^{k} Q_{i \cdot \delta}$

We can rewrite (12.10) as

$$q_{i \cdot \delta} = 1 - (p_i)^{\left(\sum_{\delta=1}^{k} Q_{i\delta} - Q_{i\delta}/q_i\right)}$$ (12.11)

12.2.4 Relationship between Partially Crude Probability $(Q_{i\delta} \cdot \varepsilon)$ $(\delta \neq \varepsilon)$ and Crude Probability $(Q_{i\delta})$

Let us consider $Q_{i\delta \cdot 1}$ $(\delta \neq 1)$. This is the probability of dying in $(x_i, x_{i+1}]$ in risk R_δ given that the risk '1' viz. R_1 is off from the population $(\delta \neq 1)$. We have, the probability of surviving from $(x_i, t]$ where $x_i < t \leq x_{i+1}$, when R_1 is off from the population is

$$\exp\left(-\int_{x_i}^{t} [\mu(t) - \mu(t; 1)] \, dt\right)$$ (12.12)

The probability of dying in between $(t, t + \delta t]$ given the materialisation of (12.12) is

$$\mu(t\;;\delta)\,dt \quad \text{where } x_i < t \le x_i + 1 \tag{12.13}$$

Therefore the simultaneous materialization of (12.12) and (12.13) when $x_i < t \le x_{i+1}$ is

$$Q_{i\delta\cdot 1} = \int_{x_i}^{x_{i+1}} \left\{ \exp\left(-\int_{x_i}^{t} [\mu(\tau) - \mu(\tau;1)]\,d\tau \right) \right\} \cdot \mu(t;\delta)\,dt$$

$$= \int_{x_i}^{x_{i+1}} \left\{ \exp\left(-\int_{x_i}^{t} [\mu(\tau) - \mu(\tau;1)]\,d\tau \right) \right\} \frac{\mu(t;\delta)}{\mu(t) - \mu(t;1)} (\mu(t) - \mu(t;1))\,dt$$

Therefore, by Chinag's proportionality assumption (12.5)

$$Q_{i\delta\cdot 1} = \frac{\mu(t;\delta)}{\mu(t) - \mu(t;1)} \int_{x_i}^{x_{i+1}} \exp\left(-\int_{x_i}^{t} [\mu(\tau) - \mu(\tau;1)]\,d\tau \right) \times (\mu(t) - \mu(t;1))\,dt$$

$$Q_{i\delta\cdot 1} = \frac{\left[\dfrac{\mu(t;\delta)}{\mu(t)} \right]}{\left[1 - \dfrac{\mu(t;1)}{\mu(t)} \right]} \int_{x_i}^{x_{i+1}} \exp\left(-\int_{x_i}^{t} [\mu(\tau) - \mu(\tau;1)]\,d\tau \right) \times (\mu(t) - \mu(t;1))\,dt$$

$$= \frac{\left[\dfrac{Q_{i\delta}}{q_i} \right]}{\left[1 - \dfrac{Q_{i1}}{q_i} \right]} \int_{x_i}^{x_{i+1}} \exp\left(-\int_{x_i}^{t} [\mu(\tau) - \mu(\tau;1)]\,d\tau \right) \times (\mu(t) - \mu(t;1))\,dt$$

$$= -\frac{Q_{i\delta}}{q_i - Q_{i1}} \int_{x_i}^{x_{i+1}} \frac{d}{dt}\left[\exp\left(-\int_{x_i}^{t} [\mu(\tau) - \mu(\tau;1)\,d\tau] \right) \right] dt$$

$$= \frac{Q_{i\delta}}{q_i - Q_{i1}} \left[-\exp\left(-\int_{x_i}^{t} [\mu(\tau) - \mu(\tau;1)] \right) d\tau \right]_{t=x_i}^{t=x_{i+1}}$$

$$= \frac{Q_{i\delta}}{(q_i - Q_{i1})} \left[-\exp\left(-\int_{x_i}^{x_{i+1}} [\mu(\tau) - \mu(\tau;1)]\,d\tau \right) + 1 \right]$$

$$Q_{i\delta\cdot 1} = \frac{Q_{i\delta}}{q_i - Q_{i1}} \left[1 - \exp\left(-\int_{x_i}^{x_{i+1}} [\mu(\tau) - \mu(\tau;1)]\,d\tau \right) \right]$$

Further we have $\exp\left(-\int_{x_i}^{x_{i+1}}[\mu(\tau) - \mu(\tau;1)]\,d\tau\right)$ = Probability of surviving in

$(x_i, x_{i+1}]$ when R_1 is off from the population

$\Rightarrow \qquad \exp\left(-\int_{x_i}^{x_{i+1}}[\mu(\tau;2) + \mu(\tau;3) + ... + \mu(\tau;k)]\,d\tau\right)$

$$= p_{i2} + p_{i3} + ... + p_{ik}$$

$\Rightarrow \qquad 1 - \exp\left(-\int_{x_i}^{x_{i+1}}[\mu(\tau;2) + \mu(\tau;3) + ... + \mu(\tau;k)]\,d\tau\right)$

$$= 1 - (p_{i2} + p_{i3} + ... + p_{ik})$$

$$= 1 - (p_i - p_{i1}) = 1 - p_{i\cdot1} = q_{i\cdot1} = 1 - (p_i)^{\left(\frac{q_i - Q_{i1}}{q_i}\right)}$$

Thus $\qquad Q_{i\delta\cdot1} = \dfrac{Q_{i\delta}}{q_i - Q_{i1}}[1 - p_i^{(q_i - Q_{i1})/q_i}] \qquad\qquad (12.14)$

(12.14) gives the relationship between Partially crude probability and crude probability as in general

$$Q_{i\delta\cdot j} = \left[1 - p_i^{\left(\sum_i Q_{i\delta} - Q_{ij}\right)/\sum_i Q_{i\delta}}\right]\frac{Q_{i\delta}}{q_i - Q_{ij}} \text{ for } j \ne \delta \qquad (12.15)$$

Example 12.1 Show that $q_{i\delta} > Q_{i\delta}$
We have

$$q_{i\delta} = 1 - (p_i)^{\frac{Q_{i\delta}}{q_i}}$$

$$= 1 - (p_i)^{\frac{\mu(t;\delta)}{\mu(t)}} \quad \text{(by 12.6)}$$

$$= 1 - (1 - q_i)^{\frac{\mu(t;\delta)}{\mu(t)}}$$

$$= 1 - \left(1 - \frac{\mu(t;\delta)}{\mu(t)}q_i + \frac{\mu(t;\delta)}{\mu(t)}\left(\frac{\mu(t;\delta)}{\mu(t)} - 1\right)\frac{q_i^2}{2!} + ...\right)$$

$$q_{i\delta} = \frac{\mu(t;\delta)}{\mu(t)}q_i - \frac{\mu(t;\delta)}{\mu(t)}\left[\frac{\mu(t;\delta)}{\mu(t)} - 1\right]\frac{q_i^2}{2!} + ...$$

$$\Rightarrow \quad q_{i\delta} = \frac{\mu(t;\delta)}{\mu(t)} q_i + \left(1 - \frac{\mu(t;\delta)}{\mu(t)}\right) \frac{\mu(t;\delta)}{\mu(t)} \frac{q_i^2}{2!} + \cdots$$

Since all the terms on the right hand side are positive, it follows that

$$q_{i\delta} > \frac{\mu(t;\delta)}{\mu(t)} q_i = Q_{i\delta} \quad \text{from (12.6)}$$

Example 12.2 $Q_{i\delta} > Q_{i\varepsilon} \Rightarrow q_{i\delta} > q_{i\varepsilon}$ and $q_{i.\delta} < q_{i.\varepsilon}$

We have $\qquad q_{i\delta} = 1 - (pi)^{\frac{Q_{i\delta}}{q_i}} \quad$ from (12.9)

Given $\qquad\qquad\qquad Q_{i\delta} > Q_{i\varepsilon}$

$$\Rightarrow \quad \frac{Q_{i\delta}}{q_i} > \frac{Q_{i\varepsilon}}{q_i} \quad \text{and} \quad 1 - \frac{Q_{i\delta}}{q_i} < 1 - \frac{Q_{i\varepsilon}}{q_i}$$

$$\Rightarrow \quad (p_i)^{\frac{Q_{i\delta}}{q_i}} < (p_i)^{\frac{Q_{i\varepsilon}}{q_i}}, \; p_i^{1 - \frac{Q_{i\delta}}{q_i}} > p_i^{1 - \frac{Q_{i\varepsilon}}{q_i}}$$

$$\Rightarrow \quad 1 - (p_i)^{\frac{Q_{i\delta}}{q_i}} > 1 - (p_i)^{\frac{Q_{i\varepsilon}}{q_i}}, 1 - p_i^{1 - \frac{Q_{i\delta}}{q_i}} < 1 - p_i^{1 - \frac{Q_{i\varepsilon}}{q_i}}$$

$$\Rightarrow \quad q_{i\delta} > q_{i\varepsilon} \quad \text{and} \quad q_{i.\delta} < q_{i.\varepsilon}$$

12.3 Estimation of Crude, Net and Partially Crude Probabilities

Consider the joint distribution of deaths due to various risks R_δ ($\delta = 1, 2,$ ...k) as well as the survivors in the age group (x_i, x_{i+1}) as follows:

	Deaths in various Risks					Survivors	Total
	1	2	...	δ	...	k		
Frequency	d_{i1}	d_{i2}	...	$d_{i\delta}$	...	d_{ik}	l_{i+1}	l_i
Probability	Q_{i1}	Q_{i2}	...	$Q_{i\delta}$	...	Q_{ik}	p_i	1

Since in the above set up

$$l_i = l_{i+1} + d_{i1} + d_{i2} + \ldots + d_{ik}$$
$$1 = p_i + Q_{i1} + Q_{i2} + \ldots + Q_{ik}$$

it follows that the joint distribution of $l_{i+1}, d_{i1}, d_{i2}, \ldots d_{ik}$ given l_i is multinomial given by

$$L = \frac{l_i!}{l_{i+1}! d_{i1}! d_{i2}! \ldots d_{ik}!} (p_i)^{l_{i+1}} (Q_{i1})^{d_{i1}} (Q_{i2})^{d_{i2}} \ldots (Q_{ik})^{d_{ik}} \quad (12.16)$$

$$\log L = \log C + \sum_{\delta=1}^{k} d_{i\delta} \log Q_{i\delta} + l_{i+1} \log p_i$$

where C is a constant independent of parameters

$$C = \frac{l_i!}{l_{i+1}!d_{i1}!\ldots d_{ik}!}$$

$$\frac{\partial \log L}{\partial Q_{i\delta}} = \frac{d_{i\delta}}{Q_{i\delta}} \cdot \frac{\partial Q_{i\delta}}{\partial Q_{i\delta}} + \frac{l_{i+1}}{p_i} \frac{\partial p_i}{\partial Q_{i\delta}} = 0$$

$$\Rightarrow \left[\frac{d_{i\delta}}{Q_{i\delta}} \cdot (+1) - \frac{l_{i+1}}{p_i} \right] = 0 \left(\text{since} \sum Q_{i\delta} = q_i = (1 - p_i), \frac{\partial Q_{i\delta}}{\partial p_i} = -1 \right)$$

$$\Rightarrow \qquad \frac{l_{i+1}}{p_i} = \frac{d_{i\delta}}{Q_{i\delta}} = \frac{\sum d_{i\delta}}{\sum Q_{i\delta}} = \frac{d_i}{q_i}$$

$$\Rightarrow \qquad \hat{p}_i = \left(\frac{\hat{q}_i}{d_i} \right) \cdot l_{i+1}$$

$$\Rightarrow \qquad \frac{\hat{p}_i}{\hat{q}_i} = \frac{l_{i+1}}{d_i}$$

$$\Rightarrow \qquad \frac{\hat{p}_i}{1 - \hat{p}_i} = \frac{l_{i+1}}{d_i}$$

$$\Rightarrow \qquad \hat{p}_i = \frac{l_{i+1}}{d_i + l_{i+1}}$$

$$\Rightarrow \qquad \hat{p}_i = \frac{l_{i+1}}{l_i} \qquad\qquad (12.17)$$

$$\therefore \quad \frac{\partial \log L}{\partial Q_{i\delta}} = 0 \Rightarrow \frac{d_{i\delta}}{Q_{i\delta}} = \frac{l_{i+1}}{p_i}$$

$$\Rightarrow \frac{Q_{i\delta}}{d_{i\delta}} = \frac{p_i}{l_{i+1}}$$

$$\Rightarrow \hat{Q}_{i\delta} = \hat{p}_i \cdot \frac{d_{i\delta}}{l_{i+1}} = \frac{l_{i+1}}{l_i} \cdot \frac{d_{i\delta}}{l_{i+1}} = \frac{d_{i\delta}}{l_i} \qquad (12.18)$$

$$(\text{By using } (12.17))$$

Since $d_{i\delta}$, l_{i+1} are multinomial variates, therefore

$$E(d_{i\delta}) = l_i Q_{i\delta}, \ \text{Var}(d_{i\delta}) = l_i Q_{i\delta}(1 - Q_{i\delta})$$

$$\text{Cov}(d_{i\delta}, d_{i\epsilon}) = -l_i Q_{i\delta} Q_{i\epsilon}$$

$$E(l_{i+1}) = l_i p_i, \ \text{Var}(l_{i+1}) = l_i p_i(1 - p_i)$$

We can check that the Maximum Likelihood estimators are unbiased.

$$E(\hat{Q}_{i\delta}) = E\left(\frac{d_{i\delta}}{l_i}\right) = E\left(\frac{1}{l_i} E(d_{i\delta}|l_i)\right)$$

$$= E\left(\frac{1}{l_i} l_i Q_{i\delta}\right) = Q_{i\delta} \qquad (12.19)$$

$$E(\hat{p}_i) = E\left(\frac{l_{i+1}}{l_i}\right) = E\left(\frac{1}{l_i} E(l_{i+1}|l_i)\right)$$

$$= E\left(\frac{1}{l_i} \cdot l_i \, p_i\right) = p_i \qquad (12.20)$$

Thus both $\hat{p}_i$ and $\hat{Q}_{i\delta}$ are unbiased for p_i and $Q_{i\delta}$.

$$\mathrm{Var}(\hat{Q}_{i\delta}) = E(\hat{Q}_{i\delta}^2) - \left[E(\hat{Q}_{i\delta})\right]^2$$

$$E(\hat{Q}_{i\delta}^2) = E\left(\frac{d_{i\delta}^2}{l_i^2}\right) = E\left(\frac{1}{l_i} E\left(\frac{d_{i\delta}^2}{l_{i\infty}}\bigg| l_i\right)\right)$$

$$= E\left(\frac{1}{l_i} \frac{(l_i Q_{i\delta}(1 - Q_{i\delta}) + l_i^2 Q_{i\delta}^2)}{l_i}\right)$$

$$= E\left(\frac{1}{l_i}(Q_{i\delta}(1 - Q_{i\delta}) + l_i Q_{i\delta}^2)\right)$$

$$= E\left(\frac{1}{l_i}\right) Q_{i\delta}(1 - Q_{i\delta}) + Q_{i\delta}^2$$

$$\therefore \quad \mathrm{Var}(\hat{Q}_{i\delta}) = E\left(\frac{1}{l_i}\right) Q_{i\delta}(1 - Q_{i\delta}) + Q_{i\delta}^2 - Q_{i\delta}^2$$

$$= E\left(\frac{1}{l_i}\right) Q_{i\delta}(1 - Q_{i\delta}) \qquad (12.21)$$

$$\mathrm{Cov}(\hat{Q}_{i\delta}, \hat{Q}_{i\varepsilon}) = E(\hat{Q}_{i\delta} \hat{Q}_{i\varepsilon}) - E(\hat{Q}_{i\delta})(\hat{Q}_{i\varepsilon}), (\delta \neq \varepsilon)$$

$$= E\left(\frac{d_{i\delta}}{l_i} \cdot \frac{d_{i\varepsilon}}{l_i}\right) - Q_{i\delta} Q_{i\varepsilon}$$

$$= E\left(\frac{1}{l_i} E\left(\frac{d_{i\delta} d_{i\varepsilon}}{l_i}\bigg| l_i\right)\right) - Q_{i\delta} Q_{i\varepsilon}$$

$$= E\left[\frac{1}{l_i} \mathrm{Cov}\left(\frac{d_{i\delta} d_{i\varepsilon}}{l_i}\bigg| l_i\right)\right] + E\left(\frac{d_{i\delta}}{l_i}\bigg| l_i\right) E\left(\frac{d_{i\varepsilon}}{l_i}\right) - Q_{i\delta} Q_{i\varepsilon}$$

$$= E\left(\frac{1}{l_i}\left(\frac{l_i Q_{i\delta} Q_{i\varepsilon}}{l_i}\right)\right) + \left(\frac{l_i Q_{i\delta}}{l_i}\right)\left(\frac{l_i Q_{i\varepsilon}}{l_i}\right) - Q_{i\delta} Q_{i\varepsilon}$$

$$= -E\left(\frac{1}{l_i}\right) Q_{i\delta} Q_{i\varepsilon} + Q_{i\delta} Q_{i\varepsilon} - Q_{i\delta} Q_{i\varepsilon}$$

$$= -E\left(\frac{1}{l_i}\right) Q_{i\delta} Q_{i\varepsilon}, (\delta \neq \varepsilon) \tag{12.22}$$

Finally, we can show

$$\text{Cov}(\hat{Q}_{i\delta}, \hat{Q}_{j\varepsilon}) = 0 \text{ for } i \neq j, \delta \neq \varepsilon$$

Since $\text{Cov}(\hat{Q}_{i\delta}, \hat{Q}_{j\varepsilon}) = E(\hat{Q}_{i\delta} \hat{Q}_{j\varepsilon}) - E(\hat{Q}_{i\delta}) E(\hat{Q}_{j\varepsilon})$

$$= E\left(\frac{d_{i\delta}}{l_i} \cdot \frac{d_{j\varepsilon}}{l_j}\right) - Q_{i\delta} Q_{j\varepsilon}$$

$$= E\left(\frac{d_{i\delta}}{l_i}\right) E\left(\frac{d_{j\varepsilon}}{l_j}\right) - Q_{i\delta} Q_{j\varepsilon}$$

$\Bigg($Since $\dfrac{d_{i\delta}}{l_i}, \dfrac{d_{j\varepsilon}}{l_j}$ are independent random variables corresponding to two

multinomial populations given by

$$\sum_\delta d_{i\delta} + l_{i+1} = l_i \text{ and } \sum_\delta Q_{i\delta} = q_i \Rightarrow \sum_\delta Q_{i\delta} + p_i = 1$$

and $$\sum_\varepsilon d_{j\varepsilon} + l_{j+1} = l_j$$

and $$\sum_\varepsilon Q_{j\varepsilon} = q_j \Rightarrow \sum_\varepsilon Q_{j\varepsilon} + p_j = 1\Bigg)$$

$\therefore \quad \text{Cov}(\hat{Q}_{i\delta}, \hat{Q}_{j\varepsilon}) = Q_{i\delta} Q_{j\varepsilon} - Q_{i\delta} Q_{j\varepsilon} = 0 \text{ for } i \neq j, \delta \neq \varepsilon \tag{12.23}$

12.4 (Aliter) Neyman's Modified χ^2 Method

Neyman's modified χ^2 is given by

$$\chi_0^2 = \sum_i \frac{(O_i - E_i)^2}{O_i}$$

where O_i and E_i represent the observed and the expected frequencies at the *i*th cell of the multinomial table given as:
given as

Death \ Risk	Deaths due to risk					Survivors	Total
	R_1	R_2	...	R_δ	... R_K		
	d_{i1}	d_{i2}	...	d_{id}	... d_{ik}	l_{i+1}	l_i
Probabilities	Q_{i1}	Q_{i2}	...	Q_{id}	... Q_{ik}	p_i	1

$$\chi_0^2 = \sum_{\delta=1}^{k} \frac{(d_{i\delta} - l_i Q_{i\delta})^2}{d_{i\delta}} + \frac{(l_{i+1} - l_i p_i)^2}{l_{i+1}} \tag{12.24}$$

$$\frac{\partial \chi_0^2}{\partial Q_{i\delta}} = 0 \Rightarrow \frac{2 (d_{i\delta} - l_i Q_{i\delta})(-l_i)}{d_{i\delta}} + \frac{2 (l_{i+1} - l_i p_i)}{l_{i+1}}(-l_i) \frac{\partial p_i}{\partial Q_{i\delta}} = 0$$

$$\Rightarrow \quad \frac{d_{i\delta} - l_i Q_{i\delta}}{d_{i\delta}} = \frac{l_{i+1} - l_i p_i}{l_{i+1}} \left(\text{since } \frac{\partial Q_{i\delta}}{\partial p_i} = -1 \right)$$

$$\Rightarrow \quad 1 - l_i \cdot \frac{Q_{i\delta}}{d_{i\delta}} = 1 - \frac{l_i}{l_{i+1}} \cdot p_i$$

$$\Rightarrow \quad l_i \frac{Q_{i\delta}}{d_{i\delta}} = \frac{l_i}{l_{i+1}} \cdot p_i$$

$$\Rightarrow \quad \frac{Q_{i\delta}}{d_{i\delta}} = \frac{p_i}{l_{i+1}}$$

or $$\tilde{Q}_{i\delta} = d_{i\delta} \frac{\tilde{p}_i}{l_{i+1}} \tag{12.25}$$

where $\tilde{Q}_{i\delta}$ and $\tilde{p}_i$ are the modified minimum χ^2 estimators of $Q_{i\delta}$ and p_i. Again,

$$\frac{\partial \chi_0^2}{\partial p_i} = \frac{2 (d_{i\delta} - l_i Q_{i\delta})(-l_i)}{d_{i\delta}} \cdot \frac{\partial Q_{i\delta}}{\partial p_i} + \frac{2 (l_{i+1} - l_i p_i)}{l_{i+1}}(-l_i) = 0$$

$$\Rightarrow \quad \frac{d_{i\delta} - l_i Q_{i\delta}}{d_{i\delta}} = \frac{l_{i+1} - l_i p_i}{l_{i+1}} \left(\text{since } \frac{\partial Q_{i\delta}}{\partial p_i} = -1 \right)$$

$$\Rightarrow \quad 1 - \frac{l_i Q_{i\delta}}{d_{i\delta}} = 1 - \frac{l_i}{l_{i+1}} p_i$$

$$\Rightarrow \quad \frac{Q_{i\delta}}{d_{i\delta}} = \frac{p_i}{l_{i+1}}$$

$$\therefore \quad \frac{\sum_{\delta} Q_{i\delta}}{\sum_{\delta} d_{i\delta}} = \frac{p_i}{l_{i+1}} \Rightarrow \frac{p_i}{1 - p_i} = \frac{l_{i+2}}{d_i}$$

$$\left(\because \sum_{\delta} Q_{i\delta} = q_i \text{ and } \sum_{\delta} d_{i\delta} = d_i \right)$$

$$\Rightarrow \qquad \tilde{p}_i = \frac{l_{i+1}}{l_{i+1} + d_i} = \frac{l_{i+1}}{l_i} \qquad (12.26)$$

is the modified χ^2 estimator of $\tilde{p}_i$ which is same as $\hat{p}_i$, the maximum likelihood estimator.

Putting (12.26) in (12.25) $\Rightarrow$

$$\tilde{Q}_{i\delta} = \frac{l_{i+1}}{l_i} \cdot \frac{d_{i\delta}}{l_{i+1}} = \frac{d_{i\delta}}{l_i} \qquad (12.27)$$

which again makes $\tilde{Q}_{i\delta}$, the same as $\hat{Q}_{i\delta}$ the maximum likelihood estimator in (12.18).

Thus the maximum likelihood and Modified Minimum χ^2 provide one and the same result.

12.5 Estimation of the Partially Crude Probability $(Q_{i\delta\cdot\varepsilon})$

We have,

$$Q_{i\delta\cdot\varepsilon} = \frac{Q_{i\delta}}{(q_i - Q_{i\varepsilon})}\left[1 - (p_i)^{\frac{(q_i - Q_{i\varepsilon})}{q_i}}\right] \qquad \forall\ \delta \neq \varepsilon, \delta = 1, 2, \ldots k$$

(from (12.15))

By the invariance property of Maximum Likelihood estimator, the m.l.e. of $Q_{i\delta\cdot\varepsilon}$ is given by

$$\hat{Q}_{i\delta\cdot\varepsilon} = \frac{\hat{Q}_{i\delta}}{(\hat{q}_i - \hat{Q}_{i\varepsilon})}\left[1 - (\hat{p}_i)^{\frac{(\hat{q}_i - \hat{Q}_{i\varepsilon})}{\hat{q}_i}}\right]$$

$$= \frac{\left(\frac{d_{i\delta}}{l_i}\right)}{\left(\frac{d_i}{l_i} - \frac{d_{i\varepsilon}}{l_i}\right)}\left[1 - \left(\frac{l_{i+1}}{l_i}\right)^{\left(\frac{d_i}{l_i} - \frac{d_{i\varepsilon}}{l_i}\right)\frac{l_i}{d_i}}\right]$$

$$= \frac{d_{i\delta}}{d_i - d_{i\varepsilon}}\left[1 - \left(\frac{l_{i+1}}{l_i}\right)^{\frac{d_i - d_{i\varepsilon}}{d_i}}\right] \qquad (12.28)$$

Similarly the maximum likelihood estimators of $q_{i\delta}$ and $q_{i\cdot\delta}$ may be constructed.

12.6 Independent and Dependent Risks

Let $R_\delta (\delta = 1, 2, \ldots k)$ be k competing risks and the r.v. $Y_\delta (\delta = 1, 2, \ldots k)$ be an individual's length of life under the risk R_δ (assuming that R_δ is the only operative risk function), $Y_\delta \geq 0$.

Let

$$P_\delta(x) = P\{Y_\delta \le x\} \text{ is the c.d.f. of } Y_\delta$$

and
$$p_\delta(x) \, dx = P\{x \le Y_\delta \le x + dx\}$$

where $p_\delta(x)$ represents the density function of Y_δ.

Now one may note that Y_δ can be observed when all the risks excepting R_δ be eliminated.

If we denote

$$Z = \min(Y_1, Y_2 ..., Y_K) = \min Y_\delta \tag{12.29}$$

then we may note that Z is an observable, whereas Y_δ's ($\delta = 1, 2, ...$) are not observables.

Also $\qquad Z > x \implies Y_\delta > x$ and we denote

$$1 - F_Z(x) = \overline{F}_Z(x) = P(Z > x)$$

$$= P\{Y_1 > x, Y_2 > x, ..., Y_k > x\} \tag{12.30}$$

$\overline{F}$ (rather than F) being used to emphasize that only Z is observable (unlike Y_δ). $\overline{F}_Z(x)$ is called the survival function of the r.v. Z.

Also if we define

$$\begin{aligned} h_Z(x) dx &= P\{\text{minimum } Z \text{ of } k \text{ theoretical life times} \\ &\qquad Y_\delta(\delta = 1, 2, ...k) \text{ well lie between } (x, x + dx) \,|\, \text{ no} \\ &\qquad \text{individual dies before } x\} \\ &= \frac{f_Z(x) \, dx}{\overline{F}_Z(x)}, \text{ where } f_Z(x) = \frac{d}{dx}(\overline{F}_Z(x)) \end{aligned}$$

$$\implies \qquad h_Z(x) = \frac{f_Z(x)}{\overline{F}_Z(x)} \tag{12.31}$$

then $h_Z(x)$ is known as force of decrement, instantaneous force of mortality, or hazard rate.

Let $g_\delta(x) \, \delta x$ ($\delta = 1, 2, ...k$) denote the probability of failure from R_δ in an interval $(x, x + \delta x)$ in the presence of all other $(k - 1)$ risks, given that the individual is alive at x then $g_\delta(x)$ is a measure of the crude probability of dying.

Assuming the probability of not more than one failure being possible in $(x, x + dx)$ then

$$h_Z(x) = \sum_{i=1}^{k} g_i(x) \tag{12.32}$$

since $h_Z(x) \, dx$ is the probability of failing in either of the causes in

$$(x, x + dx]$$

If we assume R_δ to act independently then

$$P\{Z > x\} = P\{Y_1 > x\} \, P\{Y_2 > x\} ... P\{Y_k > x\}$$

$$\Rightarrow \qquad \overline{F}_z(x) = \prod_{i=1}^{K} \overline{P}_i(x) \qquad (12.33)$$

$$g_i(x)\, dx = P\{x \le Y_i \le x + dx\} \cdot \frac{P\{Y_1 > x, Y_2 > x \dots Y_{i-1} > x, Y_{i+1} > x \dots, Y_K > x\}}{P\{Y_i > x, Y_2 > x \dots Y_i > x, Y_{i+1} > x \dots, Y_K > x\}}$$

$$(12.34)$$

$$\Rightarrow \qquad g_i(x) = \frac{p_{i}(x)}{\overline{F}_z(x)} \prod_{\substack{j=1 \\ j \ne 1}}^{K} \overline{P}_j(x)$$

$$= p_i(x) \frac{\overline{P}_1(x)\,\overline{P}_2(x) \dots \overline{P}_{i-1}(x)\,\overline{P}_{i+1}(x) \dots \overline{P}_K(x)}{\overline{P}_1(x) \dots \overline{P}_{i-1}(x)\,\overline{P}_i(x)\,\overline{P}_{i+1}(x) \dots \overline{P}_K(x)}$$

$$= \frac{p_i(x)}{\overline{P}_i(x)} = h_i(x), \text{say} \qquad (12.35)$$

where $h_i(x)$ is called the *i*th case specific failure rate or marginal intensity function.
Thus

$$g_i(x) = h_i(x),\ i = 1, 2, \dots K$$

$$h_z(x) = \sum_{i=1}^{K} g_i(x) = \sum_{i=1}^{K} h_i(x) \qquad (12.36)$$

We can now denote crude, net, partially crude probability as follows:–
Crude probability of death in an interval $(a, b]$ is given by

$$Q_\delta(a, b) = \int_a^b \exp\left(-\int_a^x h_z(t)\, dt \right) g_\delta(x)\, dx \qquad (12.37)$$

Also $\qquad \displaystyle \int_a^x h_z(t)\, dt = -\int_a^x \frac{d}{dt} \log \overline{F}_z(t)\, dt$

$$= -\log \frac{\overline{F}_z(x)}{\overline{F}_z(a)} \qquad (12.38)$$

Hence $\qquad \displaystyle Q_\delta(a, b] = \int_a^b \frac{\overline{F}_z(x)}{\overline{F}_z(a)} g_\delta(x)\, dx$ by putting (12.38) in (12.37)

$$Q_\delta(a, b] = \frac{1}{\overline{F}_z(a)} \int_a^b \overline{F}_z(x)\, g_\delta(x)\, dx \qquad (12.39)$$

Also $\quad \displaystyle\int_a^b h_\delta(x)\, dx = -\int_a^b \frac{d}{dx} \log \overline{P}_\delta(x)\, dx = -\log \frac{\overline{P}_\delta(b)}{\overline{P}_\delta(a)}$

$$\Rightarrow \qquad \exp\left(-\int_a^b h_\delta(x)\, dx\right) = \frac{\overline{P}_\delta(b)}{\overline{P}_\delta(a)}; \delta = 1, 2, ..., K \qquad (12.40)$$

$= 1 - P$ [failure in $(a, b]$ when R_δ is the only risk]
$= 1 - $ Net probability (Type A) of failure in $(a, b]$ due to R_δ.
$= 1 - q_\delta(a, b]$

$$\Rightarrow \qquad q_\delta(a, b] = 1 - \exp\left(-\int_a^b h_\delta(x)\, dx\right) \qquad (12.41)$$

Finally, denote $\exp\left(-\displaystyle\int_a^b h_z^{(-j)}(t)\, dt\right)$ where $h_z^{(-j)}(t)$ is the hazard rate on elimination of R_j

$= P$ [surviving all the risks but R_j in $(a, x]$ having survived upto a]

$$\Rightarrow \int_a^b \exp\left(-\int_a^x h_z^{(-j)}(t)\, dt\right) g_\delta^{(-j)}(x)\, dx \text{ where } g_\delta^{(-j)} \text{ represents the}$$

probability of failure in $(x, x + dx)$ in R_δ when R_j is off.

$= P$ [of failure from R_δ $(R_\delta \neq R_j)$ in $(a, b]$ in the presence of all the risks excepting R_j]
$= $ partially crude probability of death in R_δ in $(a, b]$ when R_j is off from the population.

Hence partially crude probability is

$$Q_{\delta \cdot j}_{\delta = j}(a, b] = \int_a^b \exp\left(-\int_a^x h_z^{(-j)}(t)\, dt\right) g_\delta^{(-j)}(x)\, dx \qquad (12.42)$$

12.6.1 Theory of Independent Risks

We assume that the risks to which each individual is exposed act independently. Further, let

$$X_i = \min_i Y_i = \text{observed life time of the individual given}$$
$$\text{that the death is due to cause } R_i (i = 1, 2, ...k)$$
$$\pi_i = P\{\text{the failure is due to cause } R_i\}$$

$$= P\{Y_i = \min_i Y_i\} \tag{12.43}$$

$$\pi_i > 0, \ \sum_{i=1}^{k} \pi_i = 1$$

where, we can without loss of generality, omit the causes for which $\pi_i = 0$, i.e., for every risk there is a non zero probability of failure.

Then the p.d.f. $f_i(x)$ of X_i assuming the independence of Y_i is given by

$$f_i(x)dx = P\{\text{of death in } (x, x + dx) \mid R_i\}$$

$$\times \ P\{\text{surviving after time } x \text{ due to risks}$$

$$R_1, R_2, \ldots R_{i-1}, R_{i+1} \ldots R_K\}$$

$$= \frac{p_i(x)\,dx}{\pi_i} \ P\{Y_1 > x, Y_2 > x \ldots, Y_{i-1} > x, Y_{i+1} > \ldots Y_k > x\}$$

$$= \frac{p_i(x)\,dx}{\pi_i} \prod_{\substack{l=1 \\ l \neq i}}^{k} P_l\{Y_1 > x\} \text{ since } Y_i\text{'s are independent r.v.'s.}$$

$$f_i(x) = \frac{p_i(x)}{\pi_i} \prod_{\substack{l=1 \\ l \neq i}}^{k} \overline{P}_l(x) \tag{12.44}$$

where $f_i(x)$ is called the death density function, i.e., $f_i(x)\,dx$ represents the conditional probability of dying in $(x, x + dx]$ due to cause R_i in the presence of all other risks in the population, given that death will occur due to risk R_i.

Then

$$f_i(x) = \frac{p_i(x)}{\pi_i} \cdot \frac{\prod_{l=1}^{k} \overline{P}_l(x)}{\overline{P}_l(x)}$$

$$= \frac{1}{\pi_i} \frac{p_i(x)}{\overline{P}_i(x)} \prod_{l=1}^{k} \overline{P}_l(x)$$

$$f_i(x) = \frac{1}{\pi_i} h_i(x)\,\overline{F}_z(x) \tag{12.45}$$

where

$$\overline{F}_z(x) = P\{z > x\}$$

$$= P\{\min Y_l > x\}$$

$$= P\{Y_1 > x \cap Y_2 < x \ldots \cap Y_K > x\} \tag{12.46}$$

As defined in (12.44), death density function due to risk R_i is

$$f(t \mid R_i)\,\delta t = f_i(t)$$

$$= P\,[\text{dying in } (t, t + \delta t) \mid \text{cause of death } R_i]$$

$$= \frac{S(t)\,\lambda_i(t)\,dt}{\int\limits_0^\infty S(t)\,\lambda_i(t)\,dt}$$

where $\qquad S(t)$ = probability of surviving at least upto t

$$= P[T \geq t]$$

Now under Chiang's probability assumption, we have

$$\lambda_i(t) = C_i\lambda(t) \text{ where } \lambda(t) = \sum_{i=1}^k \lambda_i(t)$$

where C_i is independent of t.
Therefore

$$f_i(t) = \frac{S(t)\,C_i\,\lambda(t)\,dt}{C_i\int\limits_0^\infty S(t)\,\lambda(t)\,dt} = \frac{S(t)\,\lambda(t)\,dt}{\int\limits_0^\infty S(t)\,\lambda(t)\,dt}$$

$$\int\limits_0^\infty f_i(t)\,dt = 1 \Rightarrow \frac{\int\limits_0^\infty S(t)\,\lambda(t)\,dt}{\int\limits_0^\infty S(t)\,\lambda(t)\,dt} = 1$$

Note that $\qquad \int\limits_0^\infty S(t)\,\lambda_i(t)\,dt \neq 1$

but $\qquad \int\limits_0^\infty S(t)\,\lambda(t)\,dt = 1$

Example 12.3
Let $\lambda_1, \lambda_2, ..., \lambda_k$ are the death intensities corresponding to $R_1, R_2 ... R_k$ respectively then the probability of dying in $R_i (i = 1, 2, ... k)$ is $\dfrac{\lambda_i}{\lambda}$ where

$$\lambda = \sum_{i=1}^k \lambda_i$$

Solution: Let the hazard rate corresponding to the risk R_i be λ_i

$\Rightarrow \qquad$ survival function $= \exp(-\lambda_i t) = \overline{P_i}(t)$

$\qquad$ density function $= p_i(t) = \lambda_i \exp(-\lambda_i(t))$

$$f_i(t) = \frac{p_i(t)}{\pi_i} \prod_{\substack{j=1 \\ j \neq i}}^{k} \bar{P}_j(t)$$

$$= \frac{\lambda_i e^{-\lambda_i t}}{\pi_i} \prod_{\substack{j=1 \\ j \neq i}}^{k} e^{-\lambda_j t}$$

$$= \frac{\lambda_i e^{-\lambda_i t}}{\pi_i} \exp\left(-t \sum_{j=1}^{k} \lambda_j\right) \Big/ e^{-\lambda_i t}$$

$$= \frac{\lambda_i e^{-\lambda_i t}}{e^{-\lambda_i t} \pi_i} e^{-\lambda t} \text{ where } \lambda = \sum_i \lambda_i$$

$$\Rightarrow \qquad f_i(t) = \frac{\lambda_i}{\pi_i} e^{-\lambda t}$$

$$\Rightarrow \qquad \int_0^\infty f_i(t)\, dt = \frac{\lambda_i}{\pi_i} \int_0^\infty e^{-\lambda t}\, dt = \frac{\lambda_i}{\pi_i} \frac{e^{-\lambda t}}{-\lambda}\Big|_0^\infty$$

$$= \frac{\lambda_i}{\lambda} \frac{1}{\pi_i}$$

Since $\qquad \displaystyle\int_0^\infty f_i(t)\, dt = 1 \Rightarrow \frac{\lambda_i}{\lambda} \frac{1}{\pi_i} = 1$

$$\Rightarrow \frac{\lambda i}{\lambda} = \pi_i = P[\text{dying in } R_i]$$

$$(i = 1, 2, \ldots k)$$

12.6.2 Dependent Risks

Defining $f_i(x)$ the death density function due to risk R_i $(i = 1, 2, \ldots k)$ as the conditional probability of dying in $(x, x + dx)$ given that the person dies in R_i, we have

$$f_i(x)\, dx = \frac{p_i(x)}{\pi_i} \{P(Y_1 > x)\, P(Y_2 > x) \ldots P(Y_{i-1} > x)$$

$$P(Y_{i+1} > x) \ldots P(Y_k > x)\}\, dx$$

$$= \frac{p_i(x)}{\pi_i} \prod_{\substack{l=1 \\ l \neq i}}^{k} P(Y_l > x) \text{ holds where the risks } R_i\text{'s are independent}$$

If, however, the risks are dependent, then we can write

$$f_i(x) = \frac{\int\limits_{x}^{\infty\infty}\int\limits_{x}...\int\limits_{x}^{\infty} p(y_1, y_2 ... y_{i-1}, x, y_{i+1}, ... y_k)\, dy_1\, dy_2 ... dy_{i-1} dx\, dy_{i+1} ... dy_k}{\pi_i}$$

$$\Rightarrow \quad f_i(x) = \frac{p_i(x)}{\pi_i} \int\limits_{x}^{\infty\infty}\int\limits_{x} ... \int\limits_{x}^{\infty} p(y_1\, y_2 ... y_{i-1}\, y_{i+1} ... y_k | y_i = x) \times \prod_{\substack{j=1 \\ j \neq i}}^{k} dy_i$$

(12.47)

12.6.3 Bivariate Dependent Risks

A special case : $k = 2$

$$f_1(x) = \frac{p_1(x)}{\pi_1} \int\limits_{x}^{\infty} p(y_2 | Y_1 = x)\, dy_2 \qquad (12.48)$$

$$f_2(x) = \frac{p_2(x)}{\pi_2} \int\limits_{x}^{\infty} p(y_1 | Y_2 = x)\, dy_1 \qquad (12.49)$$

Further a case of interest when

$Y_1, Y_2 \sim BVN(\mu_1, \mu_2, \sigma_1^2, \sigma_2^2, \rho)$ $(BVN \Rightarrow$ Bivariate normal distribution)

$$\pi_1 = P[\text{dying in } R_1]$$
$$= P[Y_2 - Y_1 > 0]$$

$$Y_2 - Y_1 \sim N(\mu_2 - \mu_1, \sigma_2^2 + \sigma_1^2 - 2\rho\sigma_1\sigma_2)$$

$$\pi_1 = \frac{1}{\sqrt{2\pi}\sqrt{\sigma_1^2 + \sigma_2^2 - 2\rho\,\sigma_1\,\sigma_2}} \int\limits_{0}^{\infty} \exp\left[-\frac{1}{2}\left(\frac{\tau - (\mu_2 - \mu_1)}{\sqrt{\sigma_1^2 + \sigma_2^2 - 2\rho\,\sigma_1\,\sigma_2}}\right)\right] d\tau$$

Put $$\frac{\tau - (\mu_2 - \mu_1)}{\sqrt{\sigma_1^2 + \sigma_2^2 - 2\rho\,\sigma_1\,\sigma_2}} = \xi$$

Then $\pi_1 = \dfrac{1}{\sqrt{2\pi}} \displaystyle\int\limits_{\frac{\mu_1 - \mu_2}{\sqrt{\sigma_1^2 + \sigma_2^2 - 2\rho\,\sigma_1\,\sigma_2}}}^{\infty} \exp(-\xi^2/2)\, d\xi = 1 - \Phi(\xi) = \overline{\Phi}(\xi)$ (12.50)

where $$\xi = \frac{(\mu_1 - \mu_2)}{\sqrt{\sigma_1^2 + \sigma_2^2 - 2\rho\,\sigma_1\,\sigma_2}} \qquad (12.51)$$

Next consider

$$f_1(x) = \frac{p_1(x)}{\pi_1} \int\limits_x^\infty p(y_2|Y_1 = x)\, dy_2$$

$$= \frac{p_1(x)}{\pi_1} I \qquad (12.52)$$

where

$$I = \int\limits_x^\infty p(y_2|Y_1 = x)\, dy_2$$

Consider

$$I = \int\limits_x^\infty p(y_2|Y_1 = x)\, dy_2$$

Since the r.v.

$$[Y_2|Y_1 = x] \sim N\left(\mu_2 + \rho\frac{\sigma_2}{\sigma_1}(x - \mu_1),\, \sigma_2\sqrt{1-\rho^2}\right)$$

$$\therefore \quad I = \frac{1}{\sqrt{2\pi}\,\sigma_2\sqrt{1-\rho^2}} \int\limits_x^\infty \exp{-\frac{1}{2}\left\{\frac{y_2 - \left(\mu_2 + \rho\frac{\sigma_2}{\sigma_1}(x-\mu_1)\right)}{\sigma_2\sqrt{1-\rho^2}}\right\}^2}\, dy_2$$

Put

$$\frac{y_2 - \left(\mu_2 + \rho\frac{\sigma_2}{\sigma_1}(x-\mu_1)\right)}{\sigma_2\sqrt{1-\rho^2}} = z$$

Again when $y_2 = x \Rightarrow z = \dfrac{x - \left(\mu_2 + \rho\frac{\sigma_2}{\sigma_1}(x-\mu_1)\right)}{\sigma_2\sqrt{1-\rho^2}}$ \qquad (12.53)

$$= \frac{x\left(1 - \rho\frac{\sigma_2}{\sigma_1}\right) - \left(\mu_2 - \rho\frac{\sigma_2}{\sigma_1}\mu_1\right)}{\sigma_2\sqrt{1-\rho^2}}$$

$$= \frac{x - \left(\dfrac{\mu_2}{1 - \rho\frac{\sigma_2}{\sigma_1}} - \dfrac{\rho\frac{\sigma_2}{\sigma_1}\mu_1}{1 - \rho\frac{\sigma_2}{\sigma_1}}\right)}{\dfrac{\sigma_2\sqrt{1-\rho^2}}{1 - \rho\frac{\sigma_2}{\sigma_1}}}$$

$$= \left[x - \frac{\sigma_1 \mu_2}{\sigma_1 - \rho\sigma_2} - \frac{\rho\sigma_2 \mu_1}{\sigma_1 - \rho\sigma_2} \right] \Big/ \frac{\sigma_2 \sqrt{1-\rho^2}}{1 - \rho\dfrac{\sigma_2}{\sigma_1}}$$

$$= \left[x - \left(\left(\frac{\sigma_1}{\sigma_1 - \rho\sigma_2} \right)\mu_2 - \left(\frac{\rho\sigma_2}{\sigma_1 - \rho\sigma_2} \right)\mu_1 \right) \frac{\sigma_1 - \rho\sigma_2}{\sigma_1 \sigma_2 \sqrt{1-\rho^2}} \right]$$

$$= [x - \mu_1'] / \sigma_1' \tag{12.54}$$

$$\mu_1' = \left(\frac{\sigma_1}{\sigma_1 - \rho\sigma_2} \right)\mu_2 - \left(\frac{\rho\sigma_2}{\sigma_1 - \rho\sigma_2} \right)\mu_1$$

$$= \alpha_1 \mu_2 + (1 - \alpha_1)\mu_1$$

where $\alpha_1 = \dfrac{\sigma_1}{\sigma_1 - \rho\sigma_2}$ and $1 - \alpha_1 = \dfrac{\rho\sigma_2}{\sigma_1 - \rho\sigma_2}$

and $\quad \sigma_1' = \dfrac{\sigma_1 \sigma_2 \sqrt{1-\rho^2}}{\sigma_1 - \rho\sigma_2} \tag{12.55}$

$$= \frac{\sigma_1}{\sigma_1 - \rho\sigma_2} \sigma_2 \sqrt{1-\rho^2}$$

$$= \alpha_1 \sigma_2 \sqrt{1-\rho^2}$$

where $\quad \alpha_1 = \left(\dfrac{\sigma_1 - \rho\sigma_2}{\sigma_1} \right)^{-1} = \left(1 - \rho\dfrac{\sigma_2}{\sigma_1} \right)^{-1}$

$$\therefore \quad \sigma_1' = \left(1 - \rho\frac{\sigma_2}{\sigma_1} \right)^{-1} \sigma_2 \sqrt{1-\rho^2} \tag{12.56}$$

Case I   Suppose $\rho\sigma_2 \neq \sigma_1$
From (12.52), we have

$$f_1(x) = \frac{1}{\pi_1} \frac{1}{\sigma_1 \sqrt{2\pi}} \exp\left(-\frac{1}{2}\left(\frac{x - \mu_1}{\sigma_1} \right)^2 \right) \frac{1}{\sqrt{2\pi}\, \sigma_2 \sqrt{1-\rho^2}}$$

$$\int\limits_x^\infty \exp\left(-\frac{\left(y_2 - \left(\mu_2 + \rho\dfrac{\sigma_2}{\sigma_1}(x - \mu_1) \right) \right)^2}{2\sigma_2^2(1-\rho^2)} \right) dy_2$$

$$= \frac{\phi\left(\dfrac{x - \mu_1}{\sigma_1} \right)}{\pi_1 \sigma_1} \frac{1}{\sqrt{2\pi}} \int\limits_{\frac{x - \mu_1'}{\sigma_1'}}^\infty \exp\left(-\frac{z^2}{2} \right) dZ$$

$$
= \frac{\phi\!\left(\dfrac{x-\mu_1}{\sigma_1}\right)}{\pi_1 \sigma_1}\,\overline{\Phi}\!\left(\frac{x-\mu_1'}{\sigma_1'}\right)
\tag{12.57}
$$

where $\rho\sigma_2 \neq \sigma_1$.

Case II Next suppose $\rho\sigma_2 = \sigma_1$

From (12.53) we have by putting $\rho\sigma_2 = \sigma_1$

$$
z = \frac{(x-\mu_2 - x + \mu_1)}{\sigma_2\sqrt{1-\rho^2}} = \frac{\mu_1 - \mu_2}{(\sigma_2^2 - \sigma_1^2)^{1/2}}
$$

$$
\therefore \quad f_1(x) = \frac{p_1(x)}{\pi_1}\int_x^\infty p(y_2 | Y_1 = x)\,dy_2
$$

$$
= \frac{\dfrac{1}{\sqrt{2\pi}\sigma_1}\exp\left[-\dfrac{1}{2}\left(\dfrac{x-\mu_1}{\sigma_1}\right)^2\right]}{\pi_1}
$$

$$
\int_x^\infty \exp\left(-\frac{\left(y-\left(\mu_2 + \rho\dfrac{\sigma_2}{\sigma_1}(x-\mu_1)\right)^2\right)}{2\sigma_2^2(1-\rho^2)}\right)dy\cdot\frac{1}{\sqrt{2\pi}\,\sigma_2\sqrt{1-\rho^2}}
$$

then

$$
f_1(x) = \frac{\dfrac{1}{\sqrt{2\pi}\sigma_1}\exp\left[-\dfrac{1}{2\sigma_1^2}(x-\mu_1)^2\right]}{\pi_1}\frac{1}{\sqrt{2\pi}\,\sigma_2\sqrt{1-\rho^2}}
$$

$$
\int_x^\infty \exp\left(-\frac{((y-\mu_2)-(x-\mu_1))^2}{2\sigma_2^2\left(1-\dfrac{\sigma_1^2}{\sigma_2^2}\right)}\right)dy
$$

$$
= \frac{\phi\!\left(\dfrac{x-\mu_1}{\sigma_1}\right)}{\sigma_1 \pi_1}\cdot\frac{1}{\sqrt{2\pi}\,\sigma_2\sqrt{1-\rho^2}}
$$

$$
\int_x^\infty \exp\left(-\frac{((y-\mu_2)-(x-\mu_1))^2}{2\left(\sqrt{\sigma_1^2 - \sigma_2^2}\right)^2}\right)dy
$$

Putting $(y-\mu_2)-(x-\mu_1)/(\sigma_1^2 - \sigma_2^2)^{1/2} = \xi$

we have

$$f_1(x) = \frac{\phi\left(\dfrac{x-\mu_1}{\sigma_1}\right)}{\pi_1 \sigma_1} \left\{ \frac{1}{\sqrt{2\pi}} \int\limits_{\frac{\mu_1-\mu_2}{(\sigma_2^2-\sigma_1^2)^{1/2}}}^{\infty} \exp(-\xi^2/2)\,d\xi \right\} \qquad (12.58)$$

Again $\pi_1 = \dfrac{1}{\sqrt{2\pi}} \int\limits_{\frac{\mu_1-\mu_2}{(\sigma_2^2-\sigma_1^2)^{1/2}}}^{\infty} \exp(-\xi^2/2)\,d\xi$ by putting $\rho\sigma_2 = \sigma_1$ in (12.50)

$$\qquad (12.59)$$

Hence from (12.58) and (12.59), we have

$$f_1(x) = \frac{\phi\left(\dfrac{x-\mu_1}{\sigma_1}\right)}{\pi_1 \sigma_1} \cdot \pi_1 = \frac{\phi\left(\dfrac{x-\mu_1}{\sigma_1}\right)}{\sigma_1} \qquad (12.60)$$

12.7   Applications

Comparison of expected longevity under Competing and Non-competing risk set up:

We illustrate the concept by taking Freund's bivariate exponential distribution given by

$$\begin{aligned} f(x,y) &= \alpha\beta' e^{-\beta'y-(\alpha+\beta-\beta')x} ; 0 < x < y \\ &= \beta\alpha' e^{-\alpha'x-(\alpha+\beta-\beta')y} ; 0 < y < x \end{aligned} \Bigg\} \qquad (12.61)$$

$$0 \le x, y < \infty$$
$$\alpha, \beta, \alpha', \beta' \ge 0$$

The parameters $(\alpha, \beta, \alpha', \beta')$ of Freund's Bivariate exponential distribution cannot be interpreted in the same way as in the case of Bilateral paired components in case of Bivariate risks R_1 and R_2 when R_is are competing amongst each other to cause death of an individual or failure of the system. However, one may consider Freund's Bivariate exponential distribution as the joint survival distribution under competing risks R_1 and R_2; where X(a r.v.) stands for the longevity of R_2 in presence of R_1. Our object is to obtain the expected longevity under R_1 in presence of R_2 and that of R_2 in presence of R_1. Also compare the same with the expected longevities under R_1 and R_2 respectively. Following the definition of the death density function $f_i(x)$ (i = 1, 2) due to risk R_1 as given in (12.52), we formulate

$$f_1(x) = \frac{p_1(x)}{\pi_1} \int\limits_x^{\infty} p(y|x)\,dy$$

where π_1 = unconditional probability of dying due to risk R_1

$p_1(x) = P$ [dying due to R_1 in $(x, x + \delta x)$ as $\delta x \to 0$].

Similarly, one can define the death density function f_2 due to risk R_2.

Obviously, $\pi_1 = P[Y - X > 0]$

Using Freund's model, we have

$$\pi_1 = \int_0^{\infty} \int_x^{\infty} \alpha \beta' e^{-\beta' y - (\alpha + \beta - \beta')x} dy \, dx \tag{12.62}$$

$$\pi_1 = \frac{\alpha}{\alpha + \beta} \tag{12.63}$$

Also, $p_1(x) = \dfrac{\alpha' \beta}{\alpha + \beta - \alpha'} e^{-\alpha' x} + \dfrac{(\alpha - \alpha')(\alpha + \beta)}{\alpha + \beta - \alpha'} e^{-(\alpha + \beta)x}$ \hfill (12.64)

$$\int_x^{\infty} p(y \mid x) \, dy = \int_x^{\infty} \frac{f(x, y)}{p_1(x)} dy \tag{12.65}$$

$$\int_x^{\infty} p(y \mid x) \, dy = \frac{\alpha \beta' (\alpha + \beta - \alpha') e^{-(\alpha + \beta)x}}{\beta' [(\alpha - \alpha')(\alpha + \beta) e^{-(\alpha + \beta)x} + \alpha' \beta e^{-\alpha' x}]}$$

The death density function is given be

$$f_1(x) = \frac{\alpha' \beta e^{-(\alpha + \beta - \alpha')x} + (\alpha - \alpha')(\alpha + \beta) e^{-2(\alpha + \beta)x}}{(\alpha - \alpha') e^{-(\alpha + \beta)x} + \dfrac{\alpha \beta'}{\alpha + \beta} e^{-\alpha' x}} \tag{12.66}$$

Therefore, the expected longevity as a result of failure due to risk R_1 in presence of the risk R_2 is given by

$$E(\widetilde{X}) = \int_0^{\infty} x \, f_1(x) \, dx \tag{12.67}$$

$$\Rightarrow E(\widetilde{X}) = \int_0^{\infty} \frac{\alpha' \beta \, x \, e^{-(\alpha + \beta - \alpha')x} + (\alpha - \alpha')(\alpha + \beta) \, x \, e^{-2(\alpha + \beta)x}}{(\alpha - \alpha') e^{-(\alpha + \beta)x} + \dfrac{\alpha' \beta}{\alpha + \beta} e^{-\alpha' x}} \tag{12.68}$$

$$\Rightarrow E(\widetilde{X}) = \left[\frac{1}{\alpha + \beta} - \frac{(\alpha' \beta)^2}{(\alpha + \beta)(\alpha - \alpha')(2\alpha' - \alpha - \beta)^2} \right] \tag{12.69}$$

Similarly, one may construct the death density function due to risk R_2, in presence of the risk R_1. Defining the same as: Given that a person will die in R_2, the conditional probability of dying in $(y, y + \delta y)$ as $\delta y \to 0$ in presence of first risk (R_1). In other words, the conditional probability of dying is due to risk

R_2 in presence of risk of R_1

$$f_2(y) = \frac{p_2(y)}{\pi_2} \int_y^\infty p(x \mid y) dx \tag{12.70}$$

π_2 = unconditional probability of dying due to risk R_2

$\Rightarrow \qquad \pi_2 = P$ [that individual's length of life due to risk R_1 is greater than the individual's length of life due to risk R_2]

$$= P[X - Y > 0] \tag{12.71}$$

$\therefore \qquad p_2(y) = P$ [dying due to risk R_2 in $(y, y + \delta y)$ as $\delta y \to 0$]

$$\pi_2 = \int_0^\infty \int_y^\infty \alpha' \beta \, e^{-\alpha' x - (\alpha + \beta - \alpha')y} dx dy \tag{12.72}$$

$$\pi_2 = \frac{\beta}{\alpha + \beta} \tag{12.73}$$

$$p_2(y) = \frac{\alpha \beta'}{\alpha + \beta - \beta'} e^{-\beta' y} + \frac{(\beta - \beta')(\alpha + \beta)}{\alpha + \beta - \beta'} e^{-(\alpha + \beta)y} \tag{12.74}$$

$$\int_y^\infty p(x \mid y) dx = \int_y^\infty \frac{f(x, y)}{p_2(y)} dx \tag{12.75}$$

$$\int_y^\infty p(x \mid y) dx = \frac{\alpha \beta' (\alpha + \beta - \beta') e^{-(\alpha + \beta)y}}{\alpha' [\alpha \beta' e^{-\beta' y} + (\beta - \beta')(\alpha + \beta) e^{-(\alpha + \beta)y}]} \tag{12.76}$$

This determines the conditional failure death density function due to risk R_2 in presence of the other risk R_1 as

$$f_2(y) = \frac{\alpha \beta' e^{-(\alpha + \beta - \beta')y} + (\beta - \beta')(\alpha + \beta) e^{-2(\alpha + \beta)y}}{(\beta - \beta') e^{-(\alpha + \beta)y} + \frac{\alpha \beta'}{\alpha + \beta} e^{-\beta' y}} \tag{12.77}$$

The mean of longevity under the failure due to R_2 under the presence of R_1 is given by

$$E(\widetilde{Y}) = \int_0^\infty y \, f_2(y) \, dy \tag{12.78}$$

$$= \int_0^\infty \frac{\alpha \beta' y \, e^{-(\alpha + \beta - \beta')y} + (\beta - \beta')(\alpha + \beta) y \, e^{-2(\alpha + \beta)y}}{(\beta - \beta') e^{-(\alpha + \beta)y} + \frac{\alpha \beta'}{\alpha + \beta} e^{-\beta' y}} \tag{12.79}$$

$$= \left[\frac{1}{\alpha + \beta} - \frac{(\alpha \beta')^2}{(\alpha + \beta)(\beta - \beta')^2 (2\beta' - \beta - \alpha)^2} \right] \qquad (12.80)$$

Thus $E(\widetilde{X})$ represents the mean longevity on account of failure due to R_1 in presence of R_2 and $E(\widetilde{Y})$ the mean longevity on account of failure due to R_2 in presence of R_1, both under competing risks conditions.

Now we shall compare $E(\widetilde{X})$ with $E(X)$, where $E(X)$ is simply the longevity on account of failure due to R_1 without considering any associated secondary competing risk. Similarly $E(\widetilde{Y})$ with $E(Y)$, where $E(Y)$ is the expected longevity on account of failure due to R_2 without considering any associated competing risks. For this we again consider Freund's distribution which has four parameters viz, α, β, α' and β' in the model.

We consider the moment generating functions of r.v.'s X and Y given by

$$m(t_1\, t_2) = E[e^{t_1 x + t_2 y}] = \int\!\int e^{(t_1 x + t_2 y)} f(x, y)\, dy\, dx$$

$$= \int_0^\infty \int_x^\infty \alpha \beta'\, e^{-\beta' y - (\alpha + \beta - \beta')}\, e^{t_1 x + t_2 y}\, dy\, dx$$

$$+ \int_0^\infty \int_y^\infty \alpha' \beta\, e^{-\alpha' x - (\alpha + \beta - \alpha')y}\, e^{t_1 x + t_2 y}\, dx\, dy$$

$E(X)$ = coefficient of t_1 in the expansion of $m(t_1, t_2)$. It can be seen that,

$$E(X) = \mu_1(X) = \frac{\alpha' + \beta}{\alpha'(\alpha + \beta)} = \frac{1}{\alpha + \beta}\left[1 + \frac{\beta}{\alpha'}\right] \geq \frac{1}{\alpha + \beta} \qquad (12.81)$$

and $E(Y)$ = co-efficient of t_2 in the expansion of $m(t_1, t_2)$

$$\therefore \qquad E(Y) = \mu_1(Y) = \frac{\alpha + \beta'}{\beta'(\alpha + \beta)} = \frac{1}{\alpha + \beta}\left[1 + \frac{\alpha}{\beta'}\right] \geq \frac{1}{\alpha + \beta} \qquad (12.82)$$

Comparing (12.69) with (12.81) it follows that the second term on the **R.H.S.** of (12.69) being non-negative it follows that

$$E(\widetilde{X}) < E(X)$$

$$(\because \alpha, \beta, \alpha', \beta' > 0)$$

Similarly comparing (12.80) with (12.82), we get

$$E(\widetilde{Y}) < E(Y)$$

This is a result which one may expect viz. the expected longevity under competing risk is less than that of the expected longevity under the same risk. The result is due to Biswas and Noor (1986). However, Monthir (1990) points out the result is not true for all Bivariate models. For example, for a distribution considered by Morgenstern (1956) and later on by Gumbel (1960) given by the p.d.f.

$$f(x, y) = [(1 + \alpha x)(1 + \alpha y) - \alpha] e^{-x - y - \alpha xy}; x, y \geq 0, 0 \leq \alpha \leq 1 \quad (12.83)$$

with
$$E(X) = E(Y) = 1$$

$$E(\widetilde{X}) > E(X) \text{ and } E(\widetilde{Y}) > E(X)$$

whereas a reverse behaviour is noted viz. $E(\widetilde{X}) < E(X)$ and $E(\widetilde{Y}) < E(Y)$. Condition for $E(\widetilde{X}) > E(X)$ and $E(\widetilde{Y}) > E(X)$ is stated by Monther as

$$P(X \mid X > Y) < P(X \mid X < Y)$$

Further investigation is necessary to prove into the analytical condition admitting the desirability of a survival function viz.

$$E(\widetilde{X}) < E(X) \text{ and } E(\widetilde{Y}) < E(Y).$$

12.8 Analysis of Censored Data

Concept and Definition of Censoring:

In many of the experimental situations or in sample surveys collection of data remains incomplete because of physical conditions not permitting the collection of data beyond a certain period of time or beyond a fixed proportion of observation in the sample drawn.

To illustrate the idea, let us take a life testing problem. Suppose we are required to test the longevity of a certain variety of electric bulbs. Testing procedure of electric bulb is known as 'Destructive testing' since each testing of the bulb is continued till the expiry of its life. Hence each testing is not only time consuming to unlimited extent, but at the same time involves the high cost of spoilt bulbs. Further, if time saving is the concern of the experimenter he would naturally be inclined to stop the experiment of testing the longevity of the bulb after a fixed period of time. Naturally, the percentage of bulbs which will remain unexhausted (for which recording of longevitites will remain unknown) will be a random variable and we say a censored sample of type I has been drawn. On the other hand, if cost, is a primary concern to the experimenter then he would like to stop taking observations relating to the longevity of the bulbs after the data relating to the longevity of a fixed number (or a fixed proportion) of bulbs in the sample have been taken. In this case, the time to complete the experiment is a random variable. This is called a censored sample of type II. However, in any case with the censored sample the data remains incomplete. Nevertheless, we are interested to get useful information as the 'Mean longevity of the bulbs' or 'Variance of the longevity of the bulbs' based on this incomplete data. Such a procedure leads to the methods of censored distribution and the present discussion is devoted to the same problems.

The basic difference between 'Truncated' and 'Censored' sample lies in the fact that if the population from which sample has been drawn is itself truncated then the sample is called 'Truncated Sample'. On the other hand, if the sample itself is truncated then the sample is called censored sample (Hald 1949).

12.8.1 Type I Censoring

Let t_0 be some fixed number which is usually pre-assigned time to stop the experiment of observing values deliberately. Proportion of observations which exceeds t_0 are not capable of being recorded under this set up and these are censored. Thus in type I censoring we have the proportion of observations censored is a random variable while the time of censoring is fixed. In fact, here instead of observing $T_1, T_2, \dots T_n$ (n random variables) which may be the life times of n independent units, virtually, we observe $Y_1, Y_2, \dots, Y_n$ where

$$
\left.
\begin{aligned}
Y_i &= T_i \text{ if } T_i \leq t_0 \\
&= t_0 \text{ if } t_0 < T_i
\end{aligned}
\right\} \tag{12.84}
$$

We may note that the distribution of Y_i has non-zero probability at $Y_i = t_0$ given by

$$
P[T > t_0] > 0
$$

12.8.2 Type II Censoring

A type II censoring is obtained if we stop the observation as soon as a fixed proportion of the sample is covered. In this case one may note that the point (or time) of censoring is a r.v. whereas the proportion censored is fixed.

Let $r < n$ and the order statistics of n r. v. $T_1, T_2, \dots, T_n$ be

$$
T_{(1)} < T_{(2)} < \dots < T_{(n)}
$$

Here observation ceases at the rth observation ($r < n$). So we observe only

$$
T_{(1)} < T_{(2)} < \dots < T_{(r)}
$$

Suppose we go on testing the longevities of n electric bulbs. Really the ordered observations corresponding to the n bulbs can be denoted as

$$
X_{(1)} < X_{(2)} \dots X_{(n)}
$$

Whereas if we stop the experiment at the exhaustion of the rth bulb in order ($r < n_0$) i.e. we are able to observe $X_{(1)} < X_{(2)} \dots < X_{(r)}$ then it is a type II censoring. We may require to obtain the best estimate of the mean longevity of the bulb using censored data (Feller, [9]).

12.9 Random Censoring (Progressive Censoring)

Random censoring occurs in Biological or Medical applications in clinical trials. In a clinical trial, patients usually enter in the trial at different times and observed upto some time.

Then each patient is given a therapy. The problem lies in estimating the mean life times of patients based on progressively censored data which usually occur in the following forms viz.

(1) Loss to follow up, (2) Drop out, (3) Termination of the study.

The loss to follow up is due to non-availability of the patient to respond. The drop out of the patients may be due to the adverse effect of the experiment and the patient may prefer to withdraw himself from the experiment; and termination of the study may occur due to (a) death of the patient (which gives an uncensored observation) (b) the patient may remain alive even at the end of the study (which makes the observation censored) (c) patient may be lost in follow up (which makes the observation censored).

Example   12.4

(An example of   Progressively censored type I data)

We assume that each patient has the same death density function

$$f(t) = \lambda e^{-\lambda t}; \lambda > 0, t \geq 0 \tag{12.85}$$

Let the progressively censoring scheme for the ith patient is $(0, T_i]$.

Therefore, probability of the record of the ith patient (death time) remaining uncensored is

$$\int_0^{T_i} \lambda e^{-\lambda t} dt = 1 - e^{-\lambda T_i} = \theta_i, \text{say}$$

The problem is to estimate $\mu = \dfrac{1}{\lambda}$, the mean period of longevity from the progressively censored data $(0, T_i]$ for the ith patient ($i = 1, 2...$).

The contribution to the likelihood of the sample due to ith individual is

$$\left. \begin{array}{ll} f(t_i) = \mu^{-1} e^{-\mu^{-1} t_i} ; & 0 \leq t_i < T_i \\ \\ = e^{-\mu_i^{-1} T_i} ; & t_i > T_i \end{array} \right\} \tag{12.86}$$

The above equations hold because of an individual either dying at $t_i \leq T_i$ with density $\mu^{-1} e^{-\mu^{-1} t_i}$ or surviving beyond T_i with probability $e^{-\mu_i^{-1} T_i}$. In either case we can note only the probability of his survival beyond T_i which is $e^{-\mu_i^{-1} T_i}$.

The likelihood function of the sample is

$$L(\mu) = \left(\prod_{i=1}^{n} f(t_i) \right)^{\delta_i} (R(T_i))^{1-\delta_i} \text{ where } R(T_i) = e^{-\mu_i^{-1} T_i}$$

$$= \left(\prod_{i=1}^{n} \mu^{-1} e^{-\mu_i^{-1} t_i} \right)^{\delta_i} (e^{-\mu_i^{-1} T_i})^{1-\delta_i} \tag{12.87}$$

where δ_i = 1, if the ith individual (or patient) dies in the interval $0 \leq t_i \leq T_i$

= 0, if the ith individual (or patient) does not die in the interval $0 \leq t_i \leq T_i$

Taking logarithms on both sides of (12.87), we have

$$\log_e L = -\left[\sum_{i=1}^{n} \delta_i (\mu^{-1} t_i + \log_e \mu) + (1 - \delta_i) \mu^{-1} T_i \right] \tag{12.88}$$

$$\Rightarrow \frac{\partial \log L}{\partial \mu}\Bigg|_{\mu = \hat{\mu}} = 0$$

$$\Rightarrow \sum_{i=1}^{n} [\delta_i (\hat{\mu}^{-2} t_i + (-\hat{\mu}^{-1})) + (1 - \delta_i) \hat{\mu}^{-2} T_i] = 0$$

$$\Rightarrow \hat{\mu}^{-2} \sum_{i=1}^{n} [\delta_i t_i - \hat{\mu}^{-1} \sum_{i=1}^{n} \delta_i + \hat{\mu}^{-2} \sum_{i=1}^{n} T_i - \hat{\mu}^{-2} \sum_{i=1}^{n} \delta_i T_i = 0$$

$$\Rightarrow \hat{\mu}^{-2} \left[\sum_{i=1}^{n} \delta_i t_i + \sum_{i=1}^{n} (1 - \delta_i) T_i \right] = \hat{\mu}^{-1} \sum_{i=1}^{n} \delta_i$$

Putting $\sum_{i=1}^{n} \delta_i = d$ in the above,

$$\Rightarrow \hat{\mu} = \frac{1}{\hat{\lambda}} = d^{-1} \left[\sum_{i=1}^{n} \{\delta_i t_i + (1 - \delta_i) T_i\} \right] \tag{12.89}$$

$$\Rightarrow \hat{\lambda} = d \left[\sum_{i=1}^{n} \{\delta_i t_i + (1 - \delta_i) T_i\} \right]^{-1} \tag{12.90}$$

$\hat{\lambda}$ is the maximum likelihood estimator of λ by the invariance property of the maximum likelihood method.

Sampling variance for $\hat{\mu}$ is obtainable as

$$\text{Var}(\hat{\lambda}) = -\frac{1}{E\left(\frac{\partial^2 \log L}{\partial \lambda^2}\right)} = -\frac{1}{E_d \left\{ E_{\hat{\lambda}} \left(\frac{\partial^2 \log L}{\partial \lambda^2} \bigg| d \right) \right\}}$$

Here $\log_e L = -\sum_{i=1}^{n} \left[\delta_i \left(\lambda t_i + \log \frac{1}{\lambda} \right) + (1 - \delta_i) \lambda T_i \right]$

$$\frac{\partial \log L}{d \lambda} = -\sum_{i=1}^{n} \left\{ \delta_i \left[t_i + \lambda \left(-\frac{1}{\lambda^2} \right) \right] + (1 - \delta_i) T_i \right\}$$

$$\frac{\partial^2 \log L}{\partial \lambda^2} = -\frac{1}{\lambda^2} \sum_{i=1}^{n} \delta_i \Rightarrow \frac{\partial^2 \log L}{\partial \lambda^2} = \frac{1}{\lambda^2} \cdot d$$

Again
$$\frac{1}{E\left(-\dfrac{\partial^2 \log L}{\partial \lambda^2}\right)} = \frac{1}{E_d\left\{\left(\dfrac{1}{\lambda^2} d \,\middle|\, d\right)\right\}}$$

$$= \frac{1}{E_d\left(\dfrac{1}{\lambda^2} d\right)} = \frac{\lambda^2}{E(d)} \tag{12.91}$$

Now consider the r.v.

$\delta_i \mid T_i = 1$, if the person does not survive upto T_t with probability $(1 - e^{-\lambda T_i})$

$\quad\quad = 0$, if the person survives upto T_i with probability $e^{-\lambda T_i}$

$$\tag{12.92}$$

$$E(\delta_i \mid T_i) = 1 - \exp(-\lambda T_i) = Q_i$$

where
$$d = \sum_{i=1}^{n} \delta_i$$

$$\Rightarrow \quad E(d) = E\left(\sum_{i=1}^{n} \delta_i\right) = E_{T_i}\left(\sum_{i=1}^{n} E(\delta_i \mid T_i)\right)$$

$$= E_{T_i}\left(\sum_{i=1}^{n} Q_i\right) = \sum_{i=1}^{n} Q_i, \text{ since } T_i's \text{ are fixed.} \tag{12.93}$$

Thus by putting (12.93) in (12.91), we have

$$\mathrm{Var}\,(\hat{\lambda}) = \frac{\lambda^2}{\displaystyle\sum_{i=1}^{n} Q_i}$$

$$\Rightarrow \quad \mathrm{Var}\,(\hat{\mu}) = \frac{\mu^2}{\displaystyle\sum_{i=1}^{n} Q_i} \tag{12.94}$$

Example 12.5

(An example of type II censored data)

We assume that each patient has the same exponential density function and has the same point of entry into the study and the study is terminated after the survival time of the dth patient ($n \geq d$). Here n and d are both fixed.

Thus $t_{(d)}$, the survival time of the dth patient is assumed to be a r.v. The likelihood function $L(\theta')$ for the kth parameter case

$$\theta' = (\theta_1, \theta_2, \dots \theta_k)$$

is

$$L(\theta') = d! \frac{n!}{d!(n-d)!} \prod_{i=1}^{d} f(t_{(i)}; \theta') [R(t_{(d)}; \theta')]^{n-d}$$

where

$$R(t_{(d)}; \theta') = \int_{t_{(d)}}^{\infty} f(t; \theta') dt \qquad (12.95)$$

For the exponential case where $\lambda = \mu^{-1}$ is the parameter of interest, Halperin (1952) shows

$$L(\lambda) = \frac{n!}{(n-d)!} \lambda^d \exp\left[-\lambda \sum_{i=1}^{d-1} t_{(i)} \right] [\exp - \lambda t_{(d)}]^{n-d+1}$$

$$= \frac{n!}{(n-d)!} \lambda^d \exp\left[-\lambda \left(\sum_{i=1}^{d-1} t_{(i)} + (n-d+1) t_{(d)} \right) \right] \qquad (12.96)$$

The maximum likelihood estimator by standard procedure is

$$\hat{\lambda} = \frac{d}{\sum_{i=1}^{d-1} t_{(i)} + (n-d+1) t_{(d)}} \qquad (12.97)$$

and

$$\hat{\lambda} = \frac{d}{y} \Rightarrow d = \hat{\lambda} y \Rightarrow 2d = (2\hat{\lambda}) y \qquad (12.98)$$

has a chi-square distribution with $2d$ degrees of freedom, where

$$y = \sum_{i=1}^{d-1} t_{(i)} + (n-d+1) t_{(d)}$$

Halperin (1952) also obtained the mean and variance of $\hat{\lambda}$ and the mean is as follows:

$$E(\hat{\lambda}) = \frac{d}{d-1} \lambda$$

$\Rightarrow \hat{\lambda}$ is biased for λ

We shall show in the next section that

$$\hat{\mu} = \frac{1}{\hat{\lambda}} = \frac{\sum_{i=1}^{d-1} t_{(i)} + (n-d+1) t_{(d)}}{d} \qquad (12.99)$$

is the Best Linear Unbiased Estimator (BLUE) of μ and under certain assumption, it is the best estimator of μ.

12.9.1 Derivation of a Best Linear Unbiased Estimator based on Type II Censored Data from an Exponential Population

Assumptions:

(i) A sample n is drawn from a negative exponential population

$$f(t\,;\mu) = \frac{1}{\mu}e^{-t/\mu}, \mu > 0$$

$$= \lambda e^{-\lambda t}, \text{ if } \lambda = \frac{1}{\mu}$$

(ii) The observations become available in order so that

$$t_{(1)} < t_{(2)} < ...< t_{(d)} ...< t_{(n)}$$

where $t_{(i)}$ is the ith ordered survival time of the n observations.

(iii) The life testing is discontinued as soon as $t_{(d)}$ is recorded, i.e., observations $t_{(1)} < t_{(2)} < ...< t_{(d)}$ are available

We define a linear function of the form

$$U = \mu'_1\, t_{(1),} + \mu'_2\,(t_{(2)} - t_{(1)}) + ... + \mu'_d\,(t_{(d)} - t_{(d-1)}) \qquad (12.100)$$

where $$t_{(1)}, t_{(2)} - t_{(1)}, ..., t_{(d)} - t_{(d-1)}$$

are mutually independent r.v.s from the property of the exponential distribution (Feller) (1968).

We note that,

$$P\{t_{(1)} > t\} = P\{t_{(1)} > t, t_{(2)} > t, ..., t_{(n)} > t\}$$
$$= (e^{-\lambda t})^n = e^{-n\lambda t} \qquad (12.101)$$

Similarly

$$P\{t_{(2)} - t_{(1)} > t\} = (e^{-\lambda t})^{n-1} = e^{-(n-1)\lambda t} \qquad (12.102)$$

In general

$$P\{t_{(k)} - t_{(k-1)} > t\} = e^{-(n-k+1)\lambda t}, k < n \qquad (12.103)$$

$$\Rightarrow \qquad P\{t_{(1)} < t\} = 1 - e^{-n\lambda t}$$

$$\Rightarrow \quad P\{t < t_{(1)} \le t + \delta t\} = \frac{d}{dt}(1 - e^{-n\lambda t})\delta t$$

$$\Rightarrow \qquad f(t_{(1)}) = n\lambda\, e^{-n\lambda t} \qquad (12.104)$$

$$E\,(t_{(1)}) = \frac{1}{n\lambda}, \text{Var}\,(t_{(1)}) = \frac{1}{(n\lambda)^2}$$

Similarly

$$E\left(t_{(2)} - t_{(1)}\right) = \frac{1}{(n-1)\lambda}, \text{Var}\left(t_{(2)} - t_{(1)}\right) = \frac{1}{[(n-1)\lambda]^2}$$

In general,

$$E\left(t_{(k)} - t_{(k-1)}\right) = \frac{1}{(n-k+1)\lambda} \tag{12.105}$$

$$\text{Var}\left(t_{(k)} - t_{(k-1)}\right) = \frac{1}{[(n-k+1)\lambda]^2} \tag{12.106}$$

Next let us unconditionally minimize

$$\phi = \frac{\mu_1'^2}{(n\lambda)^2} + \frac{\mu_2'^2}{[(n-1)\lambda]^2} + ... + \frac{\mu_k'^2}{[(n-k+1)\lambda]^2} + ... + \frac{\mu_d'^2}{[(n-d+1)\lambda]^2}$$

$$- \frac{\eta}{\lambda}\left[\frac{\mu_1'}{n} + \frac{\mu_2'}{(n-1)} + ... + \frac{\mu_d'}{(n-d+1)} - 1\right], k = 1, 2,... \tag{12.107}$$

where η is the Lagrangian multiplier.

$$\frac{\partial \phi}{\partial \mu_k'} = 0 \Rightarrow \frac{2\mu_k'}{[(n-k+1)\lambda]^2} - \frac{\eta}{(n-k+1)\lambda} = 0 \tag{12.108}$$

$$\Rightarrow \quad \mu_k' = \frac{\eta\lambda(n-k+1)}{2}, k = 1, 2, ...d \tag{12.109}$$

$$\frac{\partial \phi}{\partial \eta} = 0 \Rightarrow \frac{\mu_1'}{n} + ... + \frac{\mu_k'}{n-k+1} + ... + \frac{\mu_d'}{(n-d+1)} = 1$$

$$\Rightarrow \quad \frac{\eta\lambda}{2} + ... + \frac{\eta\lambda}{2} + ... + \frac{\eta\lambda}{2} = 1$$

$$\Rightarrow \quad \frac{d\eta\lambda}{2} = 1$$

$$\Rightarrow \quad \eta = \frac{2}{d\lambda} \tag{12.110}$$

(12.109) and (12.110) can be put together as

$$\mu_k' = \frac{n-k+1}{d}, k = 1, 2, ..., d \tag{12.111}$$

Thus introducing (12.111) in (12.100), we have

$$U = \frac{n}{d}t_{(1)} + \frac{n-1}{d}(t_{(2)} - t_{(1)}) + ... +$$

$$\frac{n-d+2}{d}(t_{(d-1)} - t_{(d-2)}) + \frac{n-d+1}{d}(t_{(d)} - t_{(d-1)})$$

$$= \left(\frac{n}{d} - \frac{n-1}{d}\right)t_{(1)} + \left(\frac{n-1}{d} - \frac{n-2}{d}\right)t_{(2)} + \dots$$

$$+ \left(\frac{n-d+2}{d} - \frac{n-d+1}{d}\right)t_{(d-1)} + \left(\frac{n-d+1}{d}\right)t_{(d)}$$

$$= \frac{1}{d}t_{(1)} + \frac{1}{d}t_{(2)} + \dots + \frac{1}{d}t_{(d-1)} + \frac{n-d+1}{d}t_{(d)}$$

$$\Rightarrow \qquad \hat{\mu} = \frac{1}{d}\sum_{i=1}^{d-1} t_{(i)} + \frac{n-d+1}{d}t_{(d)} \tag{12.112}$$

which is same as $\hat{\mu}$ by Halperin (vide 12.99).

12.9.2   Best Estimator of μ (Exponential Distribution Parameter) based on a Censored Sample of Ordered Observations

Let $t_{(1)} < t_{(2)} \dots < t_{(k)}$ be k ordered observations in a sample of size n where $(n - k)$ observations are censored from an exponential distribution with parameter $\frac{1}{\mu}$ given by

$$f(t) = \frac{1}{\mu} e^{-t/\mu} ; \ 0 < \frac{1}{\mu} < \infty, t \geq 0. \tag{12.113}$$

From (12.112), it has been shown that Best Linear Unbiased Estimator (BLUE) of μ is given by

$$\hat{\mu} = \sum_{i=1}^{K} \frac{n-i+1}{k} \{t_{(i)} - t_{(i-1)}\}$$

Also $(n - i + 1)(t_{(i)} - t_{(i-1)})$ are i.i.d. r.v.s exponentially distributed with parameters $\frac{1}{\mu}$.

Let

$$P\{T > t\} = e^{-t/\mu}$$

$$P\{(n-i+1)(t_{(i)} - t_{(i-1)}) > t\} = P\{T > t\} = e^{-t/\mu} \tag{12.114}$$

Characteristic function (C.F.) of $T = E(e^{iST}) = \frac{1}{\mu}\int_0^\infty e^{-iSt} e^{-t/\mu} dt$

$$= (1 - i\mu S)^{-1}$$

Therefore, the characteristic function of $(n - i + 1)(t_{(i)} - t_{(i-1)})$ is also $(1 - i\mu S)^{-1}$ by (12.115)

$$\therefore \text{ C.F. of } \sum_{i=1}^{k}(n-i+1)(t_{(i)} - t_{(i-1)}) = (1 - i\mu S)^{-k}$$

Since $(n-i+1)(t_{(i)} - t_{(i-1)})$, $i = 1,2,..k$ being **i.i.d. r.v.s.** Therefore,

the C.F. of $\hat{\mu} = \sum_{i=1}^{k} \left(\frac{n-i+1}{k}\right) t_{(i)} - t_{(i-1)}$ is given by

$$\left(1 - \frac{i\mu S}{k}\right)^{-k}$$

which shows that the distribution of $\hat{\mu}$ is k fold convolution of the distribution

which is exponential with parameter $\frac{k}{\mu}$ i.e. $\left(\frac{k}{\mu} \exp (k/\mu)\, \hat{\mu}\right)$.

The Laplace transform of the k-fold convolution of the exponential distribution with parameter $\frac{k}{\mu}$ is given by

$$\left(\frac{\dfrac{k}{\mu}}{S + \dfrac{k}{\mu}}\right)^{k}$$

$[\because$ Laplace transform of an exponential variable with parameter $\dfrac{1}{\mu}$ is $\left(\dfrac{1/\mu}{S+1/\mu}\right)$;

therefore the Laplace transform of the k fold convolution of the exponential

distribution with parameter $\dfrac{k}{\mu}$ is $\left(\dfrac{\dfrac{k}{\mu}}{S + \dfrac{k}{\mu}}\right)^{k}$] whose inverse transform gives the

distribution of $\hat{\mu}$ as

$$f(\hat{\mu}) = \frac{\exp\left(-\dfrac{k}{\mu}\hat{\mu}\right)\left(\dfrac{k}{\mu}\right)^{k} \hat{\mu}^{k-1}}{\Gamma(k)} \qquad (12.115)$$

$$E(\hat{\mu}) = \frac{1}{\Gamma(k)} \int_0^\infty \exp\left(-\frac{k}{\mu}\hat{\mu}\right)\left(\frac{k}{\mu}\right)^{k} \hat{\mu}^{k-1}\, \hat{\mu}\, d\hat{\mu} \qquad (12.116)$$

$$= \mu$$

$\Rightarrow \hat{\mu}$ is unbiased for μ.

$$\text{Var}(\hat{\mu}) = E(\hat{\mu}^2) - [E(\hat{\mu})]^2$$

$$E(\hat{\mu}^2) = \int_0^\infty \frac{\exp\left(-\dfrac{k}{\mu}\hat{\mu}\right)\left(\dfrac{k}{\mu}\right)^{k} \hat{\mu}^{k-1}\hat{\mu}^2\, d\hat{\mu}}{\Gamma(k)}$$

$$= \left(\frac{\mu}{k}\right)^2 k\,(k+1)$$

$$\text{Var}\,(\hat{\mu}) = \frac{\mu^2}{k^2}\,k\,(k+1) - \mu^2 = \frac{\mu^2}{k} \tag{12.117}$$

Also Cramer Rao lower bound for a regular unbiased estimator is given by

$$\frac{1}{E\left(\dfrac{\partial \log f}{\partial \mu}\right)^2} \text{ where } f \text{ is given by (12.115).}$$

This gives

$$\left(\frac{\partial \log f}{\partial \mu}\right)^2 = E\left(\frac{k}{\mu} + \frac{k}{\mu^2}\,\hat{\mu}\right)^2$$

$$= k^2\,E\left(\frac{1}{\mu^2} + \frac{\hat{\mu}^2}{\mu^4} - \frac{2\hat{\mu}}{\mu^3}\right)$$

$$= \frac{k^2}{\mu^2} + \frac{E\,(\hat{\mu}^2)}{\mu^4}\,k^2 - \frac{2\mu}{\mu^3}\,k^2$$

$$= \frac{k^2}{\mu^2} - \frac{2k^2}{\mu^2} + \frac{E\,(\hat{\mu}^2)}{\mu^4}\,k^2$$

$$= -\frac{k^2}{\mu^2} + \frac{\left(\mu^2 + \dfrac{\mu^2}{k}\right)}{\mu^4}\,k^2$$

$$\left[\because E\,(\hat{\mu}^2) = \frac{\mu^2}{k^2}\,k\,(k+1)\right]$$

$$= \frac{k}{\mu^2}$$

$$\Rightarrow \qquad E\left(\frac{\partial \log f}{\partial \mu}\right)^2 = \frac{k}{\mu^2} \tag{12.118}$$

Thus the Cramer Rao lower bound is given by $\dfrac{\mu^2}{k}$ which is the variance of $\hat{\mu}$.
This shows that $\hat{\mu}$ which is the BLUE is also the Minimum Variance Unbiased Estimator (MVUE) of μ which is the best estimator.

12.10 Parametric and Non Parametric Survival Models

(i) *Exponential Model*
Let $\quad f(t)$ = density function, $F(t)$ = c.d.f. function,
$\qquad R(t)$ = Survival or Reliability function.
Here $\quad f(t) = \lambda e^{-\lambda t}$, $\quad F(t) = 1 - e^{-\lambda t}$
$\qquad R(t) = e^{-\lambda t}$

Hazard rate $\qquad h(t) = \dfrac{f(t)}{1 - F(t)} = \lambda$

$$E(T) = \frac{1}{\lambda}, \text{Var}(T) = \frac{1}{\lambda^2} \qquad (12.119)$$

(ii) *Gamma Model*

$$f(t) = \frac{\lambda^v}{\Gamma(v)} e^{-\lambda t} t^{v-1}, \lambda > 0, v > 0 \qquad (12.120)$$

$$E(T) = \frac{v}{\lambda}, \text{Var}(T) = \frac{v}{\lambda^2}$$

$$h(t) = \frac{\dfrac{\lambda^v}{\Gamma(v)} e^{-\lambda t} t^{v-1}}{\left[1 - \displaystyle\int_0^t \frac{\lambda^v}{\Gamma(v)} e^{-\lambda v} \tau^{v-1} d\tau \right]} \qquad (12.121)$$

$$R(t) = \exp\left(-\int_0^t h(\tau)\, d\tau \right)$$

$\Rightarrow \qquad \log R(t) = -\displaystyle\int_0^t h(\tau)\, d\tau$

$\Rightarrow \qquad -\log(1 - F(t)) = \displaystyle\int_0^t h(\tau)\, d\tau$

$\Rightarrow \qquad \dfrac{f(t)}{1 - F(t)} = h(t)$

$\Rightarrow \qquad \dfrac{f(t)}{R(t)} = h(t)$

$\Rightarrow \qquad R(t) = \dfrac{f(t)}{h(t)}$ where $h(t)$ is obtainable from (12.121)

(iii) *Weibull Model*

We write $h(t) = \alpha\lambda(\lambda t)^{\alpha - 1}, \lambda > 0$

If $\alpha > 1$, we have increasing failure rate (I.F.R.)
If $\alpha = 1$, $h(t) = \lambda$, we have constant failure rate (Poisson)(C.F.R.)
If $\alpha < 1$, we have decreasing failure rate (D.F.R.)

$$\Rightarrow \qquad R(t) = \exp\left(-\int_0^t h(\tau)\, d\tau\right)$$

$$= \exp\left(-\alpha\lambda \int_0^t (\lambda\tau)^{\alpha-1}\, d\tau\right)$$

$$= \exp(-(\lambda t)^\alpha),\ \alpha > 0,\ \lambda > 0 \qquad (12.122)$$

$$f(t) = h(t)\, R(t) = \alpha\lambda\, (\lambda t)^{\alpha-1} \exp\left(-\lambda t\right)^\alpha \qquad (12.123)$$

$$E(t) = \int_0^\infty \exp(-(\lambda\tau)^\alpha)\, d\tau$$

is not expressible in closed form so also Var (T). For Weibull model $R(t)$ and $\lambda(t)$ are useful in survival analysis.

(iv) *Rayleigh Model*

Hazard rate $\qquad h(t) = \lambda_0 + \lambda_1 t$

$$R(t) = \exp\left(-\left(\lambda_0 t + \frac{\lambda_1}{2} t^2\right)\right)$$

$$f(t) = (\lambda_0 + \lambda_1 t) \exp\left(-\left(\lambda_0 t + \frac{\lambda_1}{2} t^2\right)\right) \qquad (12.124)$$

A generalised Rayleigh Model is given by the hazard function

$$h(t) = \sum_{i=0}^{k} \lambda_i t^j$$

(v) *Log Normal*

$$T_i \sim \Lambda(\mu, \sigma^2),\ \Lambda \equiv \text{log normally distributed}$$

$$\Rightarrow \qquad \log T_i \sim N(\mu, \sigma^2)$$

$$R(t) = 1 - P[T \le t] = 1 - P[\log T \le \log t]$$

$$= 1 - P\left[\frac{\log T}{\sigma} \le \frac{\log - \mu}{\sigma}\right]$$

$$= 1 - \Phi\left(\frac{\log t - \mu}{\sigma}\right)$$

where $\quad \Phi(t) = \dfrac{1}{2\pi} \displaystyle\int_{-\infty}^{t} \exp\left(-\frac{z^2}{2}\right) dz \qquad (12.125)$

(vi) *Bath tub type of Survival Model*

A bath type of survival model is given by the hazard function

$$h(t) = \lambda_1, \qquad \text{for } t_0 = 0 \leq t \leq t_1$$
$$= \lambda_2, \qquad \text{for } t_1 \leq t < t_2$$
$$= \lambda_3, \qquad \text{for } t_2 \leq t < t_3$$
$$= \lambda_{k-1}, \qquad \text{for } t_{k-2} \leq t < t_{k-1} \qquad (12.126)$$
$$= \lambda_k, \qquad \text{for } t \geq t_{k-1}.$$

As may be seen $h(t)$ has discontinuities at $t = t_1, t_2, t_3, \dots t_{k-1}$. By choosing λ_i's suitably, $h(t)$ may correspond to IFR, DFR or IFR followed by DFR or DFR followed by IFR (while middle range hazard rate being more or less constant) is leading to a bath tub type of model. Here

$$
\left.
\begin{aligned}
R(t) &= e^{-\lambda_1 t}; & 0 \leq t < t_1 \\
&= e^{-\lambda_1 t_1} e^{-\lambda_2 (t_2 - t_1)}; & t_1 \leq t < t_2 \\
&= e^{-\lambda_1 t_1} e^{-\lambda_2 (t_2 - t_1)} e^{-\lambda_3 (t - t_2)}; & t_2 \leq t < t_3 \\
&\vdots \\
&= e^{-\lambda_1 t_1} e^{-\lambda_2 (t_2 - t_1)} e^{-\lambda_3 (t - t_2)} \cdots e^{-\lambda_{k-2} (t_{k-2} - t_{k-1})} \\
&\quad e^{-\lambda_{k-1}(t - t_{k-1})}; & t_{k-2} \leq t < t_{k-1}
\end{aligned}
\right\} \qquad (12.127)
$$

and finally

$$R(t) = e^{-\lambda_1 t_1} e^{-\lambda_2 (t_2 - t_1)} \cdots e^{-\lambda_{k-1}(t_{k-2} - t_{k-2})} e^{-\lambda_k (t - t_{k-1})}; \text{for } t \geq t_{k-1}.$$

Accordingly the density function is given by

$$
\left.
\begin{aligned}
f(t) &= \lambda_1 e^{-\lambda_1 t}; & 0 \leq t < t_1 \\
&= e^{\lambda_1 t_1} e^{-\lambda_2 t_1} \lambda_2 e^{-\lambda_2 t} = \lambda_2 e^{-\lambda_2 (t - t_1)} e^{-\lambda_1 t_1}, \\
&\qquad \text{for } t_1 \leq t < t_2 \\
&\cdots \qquad\qquad \cdots \qquad\qquad \cdots \\
&= \lambda_k e^{-\lambda_k (t - t_{k-1})} e^{-\lambda_1 t_1} e^{-\lambda_2 (t_{k-1} - t_{k-2})} e^{-\lambda_k (t - t_{k-1})} \\
&\qquad\qquad \text{for } t \geq t_{k-1}
\end{aligned}
\right\} \qquad (12.128)
$$

12.11 Other Bivariate Exponential Models (Freund, Marshall, Olkin, Downton)

The present section attempts to make a broad survey of different parametric and non-parametric approaches on the analysis of the paired organs representing the paired failure times of the bilateral parallel paired systems e.g. two kidneys, two lungs and other bilateral organs.

Failure distribution of parallel paired organs constituting bilateral systems like kidneys, lungs, eyes, ears etc. constitute a very important information in survival analysis.

The basic problem that arises in the survival of the paired organs is that if one organ in the bilateral system fails then the other surviving organ is subject to increase failure rate. For example, it is often seen that persons who have a kidney damaged often due to a disease say "Renal Calculii" commonly exhibit increased failure rate for the remaining kidney. On the other hand, if a kidney is removed because of an accident then the failure rate of the remaining kidney may not increase or flare up so abruptly as in the former case (Gross and Clark (1973)). It appears that human bilateral systems behave like two-component parallel systems in reliability such as " two aircraft engine".

The first modelling of this kind of bivariate survival distribution was given by Morgenstern (1956) and subsequently studied by Gumbell (1960). In both the cases, loss of memory property (LMP) of univariate exponentials is not realised. Freund (1961), while assuming that each of the two components has an exponential distribution with means $\dfrac{1}{\alpha}$ and $\dfrac{1}{\beta}$, until the first failure occurs, at which point, the surviving component still has an exponential life time distribution of the Freund (1961) of a bi-component parallel system is given by

$$f(x, y) = \begin{cases} \alpha\beta' \, e^{-\beta' y - (\alpha + \beta - \beta')x}; \, 0 < x < y \\[2mm] \beta\alpha' \, e^{-\alpha' x - (\alpha + \beta - \alpha')y}; \, 0 < y < x \end{cases} \qquad (12.129)$$

$$0 \le x, y < \infty; \alpha, \beta, \alpha', \beta' \ge 0$$

Biswas and Nair (1984) have extended Freund's model for a parallel bi-component system by allowing the repairement of one component for a fixed dead period π during which surviving component carries out the operation with increased hazard rate $\lambda'(\lambda' > \lambda)$. The probability of the system failure during the time of the first repairement of a component is given by

$$p^{(1)} = 2\left(\frac{\lambda}{\lambda + \lambda'}\right)(1 - e^{-\lambda'\pi}) + \frac{\lambda'}{\lambda}(1 - e^{-\lambda'\pi}) + \lambda'\pi \qquad (12.130)$$

Gross and Clark (1973) model is given by the failure density function

$$f(t) = \frac{2\lambda_0\lambda_1}{\lambda_1 - 2\lambda_0}(e^{-2\lambda_0 t} - e^{-\lambda_1 t}), \, t \ge 0, \lambda_1 > 2\lambda_0. \qquad (12.131)$$

following the differential equation

$$p'_0(t) = -2\lambda_0 p_0(t) \text{ and } p'_1(t) = -2\lambda_0 p_0(t) - \lambda_1 p_1(t) \qquad (12.132)$$

being the Kolmogorov's equations.

An innovative approach of constructing bivariate exponential models retaining the loss of memory property of the univariate exponential as well as the condition that the marginals of the bivariate exponential are exponentially distributed random variables was evolved by Marshall and Olkin (1966) which is given as follows:

Given that there are three kinds of shocks of which the first category of shocks can affect both components with hazard rate λ_{12}, we have

$$F(y_1, y_2) = 1 - P(Y_1 > y_1) - P(Y_2 > y_2) + P((Y_1 > y_1) \cap (Y_2 > y_2))$$

$$= 1 - e^{-(\lambda_1 + \lambda_{12})} - e^{-(\lambda_2 + \lambda_{12})y_2} + e^{-\lambda y_1 - \lambda_2 y_2 - \lambda_{12} \max(y_1, y_2)} \qquad (12.133)$$

where Y_1 and Y_2 are the life times of the first and second components.

It can be seen that $\qquad \lambda = \lambda_1 + \lambda_2 + \lambda_{12}$

and $\qquad \overline{F}(y_1, y_2) = P(Y_1 > y_1 \cap Y_2 > y_2)$

$$= \left(\frac{\lambda_1 + \lambda_2}{\lambda}\right) = \overline{F}_a(y_1, y_2) + \frac{\lambda_{12}}{\lambda} \overline{F}_s(y_1, y_2) \qquad (12.134)$$

where $\overline{F}_a$ and $\overline{F}_s$ represent the absolute continuous and singular part.

The presence of the singular part is due to the fact that if X, Y are jointly bivariate exponentials then $P[X = Y]$ is realised to be non-zero, under Marshall and Olkin set up whereas the line $X = Y$ has the two dimensional Lebesgue measure zero. Marshall-Olkin have further generalised their work by introducing a fatal shock model (1967) in which shocks are further categorised as "fatal" or "non-fatal".

A similar bivariate model based on the concept of successive damage system leading to bivariate exponential was evolved by Downton (1970). The basic assumptions for the development of the bivariate exponential model by downton (1970) are comprised of;

(i) The inter-arrival distribution of shocks (which can affect either or both the organs in a paired system) is exponentially distributed i.e. corresponds to a Poisson process with parameters λ_1 and λ_2' say.

(ii) Shocks are non-fatal in the sense, that the number of shocks needed to cause failure to either of the components corresponds to a Geometric distribution (bivariate geometric distribution with parameters p_1 and p_2).

The joint distribution of the components is bivariate exponential with marginals as exponentials. The bivariate exponential has no loss of memory property. The regression of the life time of one component on the life time of the other components is linear but non-homoscedastic, which acquired a much simpler form than that of a similar regression obtained by Marshall and Olkin (1967) based on their model.

According to Downton (1970), the joint density function $f(t_1, t_2)$ of the component life times is given by

$$f(t_1, t_2) = \left(\frac{\mu_1 \mu_2}{1 - \rho}\right) \exp\left(\frac{\mu_1 t_1 + \mu_2 t_2}{1 - \rho}\right) I_0\left(\frac{2\sqrt{\rho \mu_1 \mu_2 t_1 t_2}}{1 - \rho}\right) \qquad (12.135)$$

$$; \mu_1, \mu_2 > 0, 0 \le \rho \le 1$$

where $\qquad I_n(t) = \sum_{k=0}^{\infty} \frac{t^{n+2k}}{k!(n+k)!}, n \ge 0, \qquad (12.136)$

being the Bessel function of order n.

$$\mu_1 = \frac{\lambda_1}{1 + \alpha + r}, \quad \mu_2 = \frac{\lambda_2}{1 + \beta + r} \tag{12.137}$$

$$\alpha + r = \frac{p_1}{1 - p_1}, \quad \beta + r = \frac{p_2}{1 - p_2} \tag{12.138}$$

$$\rho = \frac{\alpha\beta + \beta r + r\alpha + r + r^2}{(1 + \alpha + r)(1 + \beta + r)} \tag{12.139}$$

(λ_1, λ_2) and (p_1, p_2) are the parameters of the marginal exponential distribution and geometric distribution respectively.

The inherent difference in the result of Marshall and Olkin (1966, 1967) and Downton (1970) is explained clearly by Block and Basu (1974), the study of which has resulted to several characterisations of different bivariate exponential models.

However different variants of bivariate exponential models (Freunds, Marshall Olkin, Downton, Gross and Clark etc.) providing joint survival distribution while being quite, prolific, it may be seen at the same time that the extent of literatures for non-exponential bivariate survival models is relatively meagre. Time dependent non-exponential models (Weibull Models etc.) would have been more realistic to describe the process of parallel paired component.

12.12   Non Parametric Survival Methods

Classical methods of estimating the reliability $R(t)$ at time t are based on the Actuarial approaches or the Life table approaches. Below we present an outline of the same. Let us divide $(0, t]$ into sub intervals $I_k = (\tau_{K-1}, \tau_K]$.

12.12.1   Reduced Sampling Method

Denoting by $\quad n_k$ = # persons surviving at the beginning of I_k $(k = 1, 2, 3, ...)$
$\qquad\qquad l_k$ = # persons lost to follow up during I_k
$\qquad\qquad w_k$ = # persons withdrawn through I_k given that the persons are surviving at the beginning of the I_k $(k = 1, 2, 3, ...)$
we have with the usual life table notations, let

$$n = n_1 - \sum_{i=1}^{k} l_i - \sum_{i=1}^{k} w_i \tag{12.140}$$

where n represents the # persons who were exposed to the risk of dying upto the beginning of I_k and $d = \sum_{i=1}^{k} d_i$ = # persons died upto I_k.

$$\overline{P}(T > \tau_k] = R(\hat{\tau}) = \left(1 - \frac{d}{n}\right), \tag{12.141}$$

the estimated survival probability upto the beginning of $(I_k]$.

12.12.2 Actuarial method

With the same set up as in the sample method, we have

$$R(\tau_k) = P[T > \tau_k]$$

$$= \prod_{i=1}^{k} P(T > \tau_i \mid T > \tau_{i-1})$$

$$= \prod_{i=1}^{k} \left(1 - \frac{d_i}{n_i}\right) \tag{12.142}$$

subject to $\qquad P(T > \tau_i \mid T > \tau_{i-1}) = P(T > \tau_i)$

However, if there is loss given by l_i and withdrawal of w_i persons in $(I_i]$ then (12.142) may be rewritten as

$$R(\tau_k) = \prod_{i=1}^{k} \left(1 - \frac{d_i}{n_i'}\right) \tag{12.143}$$

where $\qquad n_i' = \left\{n_i - \frac{1}{2}(l_i + w_i)\right\}, i = 1, 2, ..., k$

Variance of the Actuarial Estimator
Denoting by

$$P(T > \tau_i \mid T > \tau_{i-1}) = P_{\tau_{i-1}, \tau_i}$$

we have

$$\log \overline{R}(\tau_n) = \sum_{i=1}^{n} \log P_{\tau_{i-1}, \tau_i} \tag{12.144}$$

Using Delta method (Vide Appendix A–1) for obtaining the large sample variance of an estimator

$$\text{Var}(\log \hat{P}_{\tau_{i-1}, \tau_i}) \cong \text{Var}(\overline{P}_{\tau_{i-1}, \tau_i}) \left(\frac{d}{d P_{\tau_{i-1}, \tau_i}} \log P_{\tau_{i-1}, \tau_i}\right)^2$$

$$= \frac{(P_{\tau_{i-1}, \tau_i})(1 - P_{\tau_{i-1}, \tau_i})}{n_i} \cdot \frac{1}{(P_{\tau_{i-1}, \tau_i})^2}$$

$$= \frac{q_{\tau_{i-1}, \tau_i}}{n_i \, (p_{\tau_{i-1}, \tau_i})} = \frac{d_i}{n_i \, (n_i - d)} \tag{12.145}$$

Assuming $\log \hat{p}_{\tau_{i-1}, \tau_i}$'s are all independent

$$\text{Var} \, (\log \hat{R}_n(t)) = \sum_{i=1}^{n} \frac{q_{\tau_{i-1}, \tau_i}}{n_i \, (p_{\tau_{i-1}, \tau_i})}$$

where $R_n(t) = P \, [\tau_n > t]$

$$\text{Var} \, (R_n \, (\tau_n)) \equiv \text{Var} \, (\log \, \hat{R}_n(\tau_n)) \left(\frac{\partial \log \hat{R}_n \, (t)}{\partial R_n \, (t)} \right)^2$$

$$= R_n \, (\tau_n)^2 \sum_{i=1}^{n} \frac{q_{\tau_{i-1}, \tau_i}}{n_i \, (p_{\tau_{i-1}, \tau_i})}$$

Thus $\qquad \text{Var} \, (\hat{R}_n \, (\tau_n)) = (R_n \, (\tau_n))^2 \sum_{i=1}^{n} \frac{d_i}{n_i} \frac{n_i}{n_i \, (n_i - d_i)}$

$$= (R_n \, (\tau_n))^2 \sum_{i=1}^{n} \frac{d_i}{n_i \, (n_i - d_i)} \tag{12.146}$$

12.12.3 Kaplan Meier Estimator of the Survival Probability (Product Limit Estimator)

Kaplan and Meier (1958) evolved a generalization of the Actuarial estimator in the sense that the intervals (I_i) unlike the Actuarial estimator need not be equal. The estimator is known as the product limit (PL) estimator.

We denote by $O \equiv$ Censored observation and by $X \equiv$ Uncensored observation.

Let $C_1, C_2, ..., C_n$ be the random censoring times of n patients having a common c.d.f. $G(.)$; C_i's are thus i.i.d. r.v.s. We denote by $T_1, T_2, ...T_n$ as the actual life times. Define a r.v. $Y_i (i = 1, 2)$ such that

$$Y_i = T_i \text{ if } C_i \geq T_i$$

$$= C_i \text{ if } C_i < T_i$$

and let $Y_{(1)} < Y_{(2)} ... < Y_{(n)}$ be n ordered observations.

Note that $\qquad Y_i = \min \, (T_i, C_i) = T_i \wedge C_i$

and
$$\delta_i = I\,(T_i \le C_i) = 1 \text{ if } T_i \le C_i$$
$$= 0 \text{ if } T_i > C_i$$

Next we consider the pairs

$$(Y_{(1)}, \delta_{(1)})\,(Y_{(2)}, \delta_{(2)})..\,(Y_{(n)}, \delta_{(n)})$$

where $\delta_{(i)}$ is the value of δ_i corresponding to $Y_{(i)}$.

Note that $\delta_{(i)}$'s are not ordered but placed in the above form by the abuse of notation.

Assuming no ties, we have the Kaplan Meier estimator of the Survival probability is given by

$$\hat{R}(t) = \hat{P}(T > t) = \prod_{Y_{(i)} < t} \hat{p}_i$$

where $p_i = P[T > Y_{(i)} \mid T > Y_{(i-1)}]$ where T is the longevity of an individual

$$= \prod_{u\,:\,Y_{(i)} \le t} \left(1 - \frac{1}{n_i}\right) \qquad (12.147)$$

where the last product runs over all uncensored observations:

$$\Rightarrow \qquad \hat{R}(t) = \prod_{Y_{(i)} \le t} \left(1 - \frac{1}{n_i}\right)^{\delta_{(i)}} \qquad (12.148)$$

where n_i = # persons in the risk set denoted by

$$R(y_{(i)}) = \text{\# persons alive upto time } y_{(i)-0}$$

Also
$$n_1 = n,\, n_2 = n - 1,\, ...\, n_i = n - i + 1$$

Therefore
$$\hat{R}(t) = \prod_{Y_{(i)} \le t} \left(\frac{n-i}{n-i+1}\right)^{\delta_{(i)}} \qquad (12.149)$$

is the Kaplan Meier estimator of the survival probability upto $T = t$ based on progressively censored sample.

12.12.4 Variance of the Kaplan Meier Estimator

$$\hat{R}(t) = P(T > t) = \prod_{Y_{(i)} < t} \hat{p}_i = \prod_{u\,:\,Y_{(i)} \le t} \left(1 - \frac{1}{n_i}\right)$$

$$\Rightarrow \qquad \log \hat{R}(t) = \sum_{u\,:\,Y_{(i)} \le t} \log \bar{p}_i$$

$$\therefore \qquad \frac{1}{\hat{R}(t)} \frac{\partial \hat{R}(t)}{\partial t} = \frac{1}{\hat{p}_i}$$

$$\Rightarrow \qquad \frac{\partial \hat{R}(t)}{\partial t} = \hat{R}(t) \cdot \frac{1}{\hat{p}_i}$$

$$\text{Var}\,(\hat{R}(t)) = \sum_{Y_{(i)} \le t} \left(\frac{\partial \hat{R}(t)}{\partial t} \right)^2_E \text{Var}\,(\hat{p}_i) \quad \text{(vide appendix A.I)}$$

$$= \sum_{Y_{(i)} \le t} \left(\frac{\hat{R}(t)}{p_i} \right)^2_E \frac{p_i\, q_i}{n_i}$$

$$= \sum_{Y_{(i)} \le t} \frac{(R(t))^2}{p_i} \frac{q_i}{n_i}$$

$$\therefore \qquad \text{Var}\,(\hat{R}(t)) = (\hat{R}(t))^2 \sum_{u:Y_{(i)} \le t} \frac{\left(\dfrac{1}{n-i+1} \right)}{\left(1 - \dfrac{1}{n-i+1} \right)} \frac{1}{(n-i+1)}$$

$$= (\hat{R}(t))^2 \sum_{u:Y_{(i)} \le t} \frac{1}{(n-i)} \frac{1}{(n-i+1)}$$

$$\text{Var}\,(\hat{R}(t)) = (\hat{R}(t))^2 \sum_{Y_{(i)} \le t} \frac{\delta_{(i)}}{(n-i)\,(n-i+1)} \qquad (12.150)$$

For Kaplan Meier Estimator with the data and its standard error, the reader is referred to Miller (1981).

12.13　Self Consistency and Kaplan-Meier Estimator

An estimator of Survival function is said to be self consistent if the proportion estimated to survive beyond t is equal to the proportion of subjects observed to survive past t; plus the sum for all individuals censored before t of the estimated conditional probability of surviving past 't' given survival to be the censoring time (Leurgans et al. (1982))

That is, an estimator $R(t)$ is said to be self consistent (denoted by $\hat{S}_c(t)$) if

$$\hat{S}_c(t) = \frac{1}{n} \left[\sum_{i=1}^{n} 1 \cdot I\,(T_i > t) + \sum_{i=1}^{n} 0 \cdot I\,(Y_i \le t, \delta_{(i)} = 1) \right.$$

$$\left. + \sum_{i=1}^{n} \frac{\hat{S}_c(t)}{\hat{S}_c\, Y_{(i)}} I\,(Y_i \le t, \delta_{(i)} = 0) \right] \qquad (12.151)$$

Efron (1967) has shown a self consistent estimator as defined above gives rise to Kaplan Meier estimator. Hence method of self consistency gives an easy way to construct Kaplan Meier estimator.

12.14 'Self consistency'in Bivariate Survival Estimator

Munoz (1984) has extended the definition of 'self-consistency' for a bivariate survival function estimator for a randomly censored sample as follows:

$$R(t_1, t_2) = P(T_1 > t_1, T_2 > t_2)$$

$$= \frac{1}{n} \left[\sum_{i=1}^{n} I\,(T_{1i} > t_1, T_{2i} > t_2) \right.$$

$$+ \sum_{\delta_i = (1,0)} \frac{\hat{R}(t_1 - 0, t_2) - \hat{R}\,(t_1, t_2)}{\hat{R}\,(t_1 - 0, y_{2i}) - \hat{R}\,(t_1, y_{2i})}$$

$$t_1 \le y_{1i} = y_{2i} < t_2$$

$$+ \sum_{\delta_i = (0,1)} \frac{\hat{R}(t_1\, t_2 - 0) - \hat{R}\,(t_1, t_2)}{\hat{R}\,(y_{1i}, t_2 - 0) - \hat{R}\,(y_{1i}, t_2)}$$

$$t_2 < y_{2i} = y_{1i} < t_1$$

$$\left. + \sum_{\delta_i = (0,0)} \frac{\hat{R}\,(t_1, t_2)}{\hat{R}\,(y_{1i}, y_{2i})} \right]$$

$$y_{1i} < t_1, y_{2i} > t_2 \tag{12.152}$$

$$y_{1i} = y_{2i}$$

where $y_{1i} = \min\,(T_{1i}, C_i)$, $y_{2i} = \min\,(T_{2i}, C_i)$, C_i being the random censoring times; $(i = 1, 2. \ldots n\,)$ T_{1i} and T_{2i} being the life times of first and second components respectively and $\delta_i = (\delta_{1i}, \delta_{2i})$ where

$$\delta_{1i} = 1 \text{ if } T_{1i} \wedge C_i = T_{1i}$$

$$= 0 \text{ if } T_{1i} \wedge C_i = C_i$$

Similarly for δ_{2i} also.

It may be noted that the second summand on the right hand side of (12.152) estimates

$$P(T_{1i} > t_1, T_{2i} > t_2 \,|\, T_{1i} = t_1, T_{2i} > y_{2i}),$$

the third summand estimates

$$P(T_{1i} > t_1, T_{2i} > t_2 \,|\, T_{1i} = y_{1i}, T_{2i} = t_2),$$

and the last one estimates

$$P(T_{1i} > t_1, T_{2i} > t_2 \,|\, T_{1i} = y_{1i}, T_{2i} > y_{2i}).$$

It follows that, an estimator in order to be self consistent has to assign positive probability to lines associated with one component censored observations and to regions associated with two component censored observations.

There s also another way of looking at the self consistency which may be given as follows:

Given an arbitrary survival function estimator say $\hat{R}^{(0)}$. One can use (12.152) to obtain a new estimator $\hat{R}^{(1)}$ from the left hand side of (12.152) as a function of $\hat{R}^{(0)}$. One can continue with this iteration scheme and stop at the mth stage when

$$\hat{R}^{(m-1)} = \hat{R}^{(m)}$$

and $\hat{R}^{(m)}$ becomes a self consistent estimator. Again a bivariate self consistent survival function estimator is a Bivariate Kaplan Meier estimator. (Efron (1967)) and this gives us a simple way of constructing Bivariate Kaplan Meier Estimator.

12.15 Generalised Maximum Likelihood Estimator (G.M.L.E.)

Kaplan Meier estimator apart from having the property of 'self consistency' is a Generalised Maximum Likelihood estimator (G.M.L.E.) which reduces to 'Ordinary Maximum likelihood estimator' if the family of the probability measures corresponding to the joint distribution of failure times is dominated. However, in general the family has no 'dominated' character, therefore, we define a G.M.L.E. due to Keifer and Wolfwitz (1956) as follows:

Let $\underset{\sim}{X}$ be the observation vector and P be the class of probability measures. Let $\hat{P}$ be the estimate of P suppose we define

$$f(x, \hat{P}, P) = \frac{d\,\hat{P}(x)}{d\,(\hat{P}+P)} \tag{12.153}$$

as Radon Nikodym derivative.

$$\Rightarrow \qquad f(x, P, \hat{P}) = \frac{d\,P(x)}{d\,(\hat{P}+P)}, \text{ if } \hat{P}(x) > P(x)$$

$$\Rightarrow \qquad f(x, \hat{P}, P) > f(x, P, \hat{P}). \tag{12.154}$$

$\hat{P}$ is called G.M.L.E. of P.

Johansen (1978) noted that if the family is dominated, the definition of G.M.L.E. reduces to a classical one and what is more relevant in the present situation is that if $\hat{P}$ gives positive probability to the data then

$$\hat{P}(x) > P(x) \,\forall\, P \in \mathcal{P} \text{ such that } P(x) > 0$$

In the next place we prove a specific result viz. G.M.L.E. is the Kaplan Meier Estimator.

12.16   A Result on the G.M.L.E.

Theorem: A G.M.L.E. is a Kaplan Meier Estimator. We prove the result for univariate survival function only.

Let Y_i's and δ_i's are defined as earlier and $Y_{(1)} < Y_{(2)} \dots < Y_{(n)}$ are ordered observations and $\delta_{(i)}$'s corresponding to $Y_{(i)}$'s.

Then define

$$L = P[(y_{(1)}, \delta_{(1)}), (y_{(2)}, \delta_{(2)}), \dots (y_{(n)}, \delta_{(n)})]$$

$$= \prod_{i=1}^{n} P(T_i = y_{(i)})^{\delta_{(i)}} P(T_i > y_{(i)})^{1-\delta_{(i)}}$$

Let

$$P(T_i = y_{(i)}) = p_i \text{ if } \delta_{(i)} = 1$$

and

$$P(y_{(j)} < T_i < y_{(j+1)}) = p_j \text{ if } \delta_{(i)} = 0, j = i, i+1, \dots, n.$$

Then

$$L = \prod_{i=1}^{n} \left\{ p_i^{\delta_{(i)}} \left(\sum_j p_j \right)^{1-\delta_{(i)}} \right\}$$

Putting $\lambda_i = \dfrac{p_i}{\displaystyle\sum_{j=i}^{n} p_j} \Rightarrow 1 - \lambda_i = \left\{ \displaystyle\sum_{j=i+1}^{n} p_j \bigg/ \displaystyle\sum_{j=i}^{n} p_j \right\}$ and $\lambda_n = 1$

$$\left(\because \lambda_n = \frac{p_n}{\displaystyle\sum_j p_j} = \frac{p_n}{p_n} = 1 \right)$$

We have

$$L = \prod_{i=1}^{n} \lambda_i^{\delta_{(i)}} (1 - \lambda_i)^{n-i}$$

$$\because \quad \prod_{j=1}^{i-1}(1 - \lambda_i) = \left[\frac{\displaystyle\sum_{j=2}^{n} p_j \ \sum_{j=3}^{n} p_j \ \cdots \ \sum_{j=i}^{n} p_j}{\displaystyle\sum_{j=1}^{n} p_j \ \sum_{j=2}^{n} p_j \ \cdots \ \sum_{j=i-1}^{n} p_j} \right] = \sum_{j=i}^{n} p_j \bigg/ \sum_{j=1}^{n} p_j$$

$$= \sum_{j=i}^{n} p_j \quad \left(\because \sum_{j=1}^{n} p_j = 1 \right)$$

and therfore
$$L = \left(\prod_{i=1}^{n} \lambda_i \sum_{j=i}^{n} p_j \right)^{\delta_{(i)}} \left(\sum_{j=i}^{n} p_j \right)^{1-\delta_{(i)}}$$

$$= \prod_{i=1}^{n} \lambda_i^{\delta_{(i)}} \left(\sum_{j=i}^{n} p_j \right)^{\delta_{(i)}} \left(\prod_{j=1}^{i-1} 1 - \lambda_j \right)^{-\delta_{(i)}} \prod_{j=1}^{i-1} (1-\lambda_j)$$

$$\left(\because \sum_{j=i}^{n} p_j = \prod_{j=1}^{i-1} (1-\lambda_j) \right)$$

$$= \prod_{i=1}^{n} \lambda_i^{\delta_{(i)}} \left\{ \prod_{j=1}^{i-1} (1-\lambda_i)^{-\delta_{(i)}} \prod_{j=1}^{i-1} (1-\lambda_j)^{-\delta_i} \right\} \prod_{j=1}^{i-1} (1-\lambda_j)$$

$$= \prod_{i=1}^{n} \lambda_i^{\delta_{(i)}} \prod_{j=1}^{i-1} (1-\lambda_j)$$

$$= (\lambda_1^{\delta_{(1)}} \cdot 1)(\lambda_2^{\delta_{(2)}} (1-\lambda_1)) .. (\lambda_n^{\delta_n} (1-\lambda_1) ... (1-\lambda_{n-1}))$$

$$= \lambda_1^{\delta_{(i)}} \lambda_2^{\delta_{(n)}} ... \lambda_n^{\delta_{(n)}} (1-\lambda_1)^{n-1} (1-\lambda_2)^{n-2} ... (1-\lambda_{n-2})^2 (1-\lambda_{n-1})$$

$$= \prod_{i=1}^{n} \lambda_i^{\delta_{(i)}} (1-\lambda_i)^{n-i} \tag{12.155}$$

Now setting $\dfrac{\partial \log L}{\partial \lambda_i} = 0 \, (i = 1, 2, ... n)$

$$\Rightarrow \qquad \hat{\lambda}_i = \frac{\delta_{(i)}}{\delta_{(i)} + (n-i)} = \frac{\delta_{(i)}}{n-i+1}$$

which corresponds to the maximization of each product in (12.155).

It gives $\hat{p}_i = \dfrac{\delta_{(i)}}{\delta_{(i)} + (n-i)} \left\{ \prod_{j=1}^{i-1} (1-\lambda_j) \right\} = \dfrac{\delta_{(i)}}{n-i+1} \prod_{j=1}^{i-1} \left(1 - \dfrac{\delta_{(j)}}{n-j+1} \right)$

which corresponds to the self consistent estimator in the univariate case.

References

1. Breslow, N. and Crowley, J. (1974): A large sample study of the life table and product limit estimates under random censorship, *Annals of Statistics*, Vol. 2 page 437– 453

2. Biswas,S. and Nair,G (1984): A generalisation of Freund's model a repairable paired component based on bivariate Geiger-Muller (G.M.) counter, Microelectronics and reliability, vol.24, No. 4, pp 671-75.

3. Biswas, S and Nair,G (1986): A palm probabilistic technique on the estimation of expected time to failure at specified shock based on the data of earlier shocks, J. Ind. Soc. Ag. Statist. , vol, XXXVIII, No. 2, pp 240-248.

4. Block and Basu (1974): A continuous Bivariate exponential extension – JASA, Vol 60, page 1031-37.

5. Campbell. George (1981): Non parametric bivariate estimation with randomly censored data, Biometrika, vol. 68, No.2, pp.417-422.

6. Chiang. C.L. (1968): *Introduction to Stochastic Processes in Biostatistics*, John Wiley, New York.

7. Cox, D.R. (1972): Regression models and life tabels, *Journal of the Royal Statistics Society*, series B, Vol, 34, page 187–220.

8. Cox, D.R. and Isham, V. (1980): Point Processes London, Chapman and Hall, *Monograph on statistics and applied probability*.

9. Cox, D.R. and Oakes, D. (1982): Analysis of survival data. Chapman and Hall, London. New York, *Monograph on statistics and applied probability*.

10. David, H.A. And Moeschberger, M.L. (1978): *The Theory of competing risks, Monograph* No. 39, Charles Griffin & Co., Ltd.

11. Downton, F. (1970) Bivariate exponeutial distribution in reliability theory. J.R.S.S. Series B. vol. 32. pp 408-17.

12 Efron. B. (1967): The two sample problem with censored data. *Proceedings of the fifth Berkeley Symposium on Mathematical Statistics and Probability*, vol. 2V, 831-853, University of California Press, Berkeley, California.

13. Elandt. R.C, and Johnson, N.L. (1980): *Survival models and data analysis*, john Wiley & Sons., New York.

14. Feller, W. (1968): *An introduction to Probability Theory and its applications*, Vol II, John Wiley, New York.

15. Freund J.E. (1961): A bivariate extension of the exponeutial distribution, J.A.S.A. ,vol. 56, pp 971-77.

16. Gross, A.J and Clark, V.A.(1975): *Survival distributions: Reliability application in biomedical Sciences*, John Wiley, New York.

17. Gumbel, E.J. (1960): Bivariate exponential distribution, J.A.S.A., vol, 55 pp 698–707

18. Hald (1949): Maximum likelihood estimation of the parameters of a normal distribution which is truncated at a known point, SKand Aktar, Tidskar, pp 119-34.

19. Johansen, S. (1978): The product limit estimator of a Maximum likelihood estimator, *Scandinavian Journal of Statistics*, vol. 5, Page 195–199.

20. Kalbfleisch, J. and Prentice. R.L. (1980): *The statistical analysis of failure time data.*. John Wiley& sons, New York.

21. Kalbfleisch, J. and Prentice R.L. (1979): Hazard rate models with Covariates, *Biometrics*, Vol. 35, page 25–39.

22. Kaplan, E.L. and Meier, P. (1958) Non-Parametric estimation from incomplete observations *journal of the american Statistical association*, Vol. 53, Page 457– 481.

23. Keifer, J, and Wolfwitz (1956): Consistency of the maximum likelihood estimator in the presence of infinitely many incident parameters, *Annals of Mathematical Statistics*, Vol 27. Page 887-906.

24. Leurgans, S. Tsai, Eei-Yann and Crowley, J. (1984): Freund's bivariate exponential distribution and censoring, IMS lecture Notes, Monograph Series, Vol 2, pp 230 –242.

25. Marshall and Olkin (1966): A multivariate exponential distribution, J. Appl. Probability, Vol 4.

26. Marshall and Olkin (1967): A multivariate exponential distribution, JASA, Vol 62, page 30-44.

27. Miller, Rupert Jr. (1981): Survival Analysis—Wiley series in Probability and Mathematical Statistics, *Applied Probability and Statistics,* John Wiley & sons.

28. Miller (1981): Biostatistics Case hand book – John Wiley & sons.

29. Montheir (1990): Competing Risk Theory on Reliability, Ph.D Thesis, University of Delhi.

30. Morgenstern (1956): Ein fache Beispiels zwlidimen sioneler Vestieilungen-Mitt Math statist. Vol 8, page 234-5.

31. Munoz. A. (1980): Non-Parametric estimation from censored Bivariate observations. Technical report No. 60. August 1980, *Division of Biostatistics*, Stanford University, Stanford, California.

32. Munoz, A. (1980): Consistency of the self consistent estimator of the distribution function from censored operations. Technical report No. 61. August 1980. *Division of Biostatistics*, Stanford University, Stanford. California.

Stochastic Processes in Genetics

13.0 Introduction

Statistical Genetics provides another fertile area for the application of Stochastic Processes. The distribution of gametes (and Genotypes) over generations under random mating or different forms of inbreeding are obtained by using techniques of Simple Markov processes. The diffusion models in Population Genetics conform to special type of Processes given by Fokker-Plank differential equations. While we plan to show a few illustrative applications it is necessary to have elementary background of Physical basis of heredity for understanding the physical basis of the Stochastic models.

13.1 Physical Basis of Heredity

It is a familiar fact that a child resembles his parents. The only physical connection between the father and his child is a sperm cell while that between a mother and the child is a single egg cell.

Our body contains a multitude of cells. Assexual cells are called 'somatic cells'; whereas sexual cells are called 'germ cells'. The nucleus of a cell contains a series of rod like microscopic structures which are called chromosomes. Any species has its own fixed number of chromosomes. For example, the characteristic number of chromosomes in 'Drosophila' (a kind of fruit fly) is four. That of wheat is 7 and house mouse 20 and man 23. One set of such different chromosomes is called 'Genome'. Organisms containing two sets of such genomes are called 'Diploids'. Thus we have Monoploids, Diploids and Polyploids. Human being is an example of diploid organism. Out of the 23 pairs of chromosomes 22 pairs are homologous (i.e. they are identical and indistinguishable physically) while the 23rd pair is non-homologous if the individual is a male. Two dissimilar chromosomes in this case are denoted by X and Y respectively. For males, X has been inherited from mother while Y from father. In case of females, the homologous sexual chromosomes are denoted by X and X both having been inherited from each of the parent (one X for each). The 22 homologous pairs of chromosomes are called autosomes. In the sexual reproduction processes one of the two chromosomes from a single reproductive cell is passed on from either parents and the combination is either XX (female zygote) or *XY* (male zygote). Genes are located in chromosomes which are ultimate carriers of heredity. During Mitosis (i.e. formation of Germ Cells) when two homologous chromosomes are seggregated genes attached to them may pass on alongwith the chromosomes or they may join another chromosomes. The former phenomenon is called *linkage* and the latter is known as *recombination* or *crossing over*. For example, let the two genes located at the locus A^L (giving rise to a particular character say color of the eye) are *A* and *a* (Given that the two varieties are available in the population). Varieties are called alleles. Similarly, let at another locus of a chromosomes B^L (giving rise to another character, say color of the hair) possess two varieties (or alleles say *B* and *b*). Then we may consider to types of individuals whose two homologous chromosomes contain are either *AB* and *ab* i.e.

Suppose of the two individuals whose two locii at two homologous chromosomes are located as above are female and male respectively. On the union of them the gametic constitution of the offspring or genotype) is *ABab*. When the same offspring becomes adult he/she will develop either of the following four types of gametes viz.

$$AB, ab, Ab \text{ and } aB$$

The first and second gametes are like parental gametes whereas the third and the fourth kind of gametes are not parental gametes and these cases correspond to recombination or crossing over. In the above case, we have four types of gametes *AB*, *Ab*, *aB* and *ab* in the population with respect two characters A^L and B^L.

Conceptually, we can think of V different types of gametes with respect to a single or multiple characters. Let us denote them by $(\gamma_1, \gamma_2 ... \gamma_\nu)$. This is known as gamete vector or gametic output of a population. Assuming any γ_i ($i = 1, 2, ... \nu$) can combine with any γ_j ($j = 1, 2... V$) randomly i.e. if $P(\gamma_i) = g_i$ and $P(\gamma_j) = g_j$, if the probability of the genotype $A_i A_j$ denoted by g_{ij} is $g_i g_j$, then the mating is said to be *random*. If however $P(A_i A_j) = g_{ij} \neq g_i g_j$ then the mating is said to be nonrandom.

13.2 Distribution of Genotypes under Random Mating

Suppose, $P(AB) = P(\gamma_1) = g_1, P(Ab) = P(\gamma_2) = g_2,$

$$P(aB) = P(\gamma_3) = g_3 \text{ and } P(ab) = P(\gamma_4) = g_4$$

at the starting of parental generation; then the problem is to obtain the probability distribution of γ_is ($i = 1, 2, 3, 4$) in the *n*th generation under random mating. Obviously, this is a problem on Simple Markov chain; each generation is a step of the Markov chain. The Markovity is decided by the fact that the probability of a particular genotype in the rth generation ($r = 1, 2.....n$) depends only on the joint probability generation of the gametes in the $(r - 1)$th generation.

Now it can easily be seen that γ_1 when combines with γ_1 gives rise to only γ_1 types of gemetes; whereas γ_1 when combining with γ_2 or γ_3 gives rise to only parental type of gemetes i.e. either (γ_1, γ_2) or (γ_1, γ_3) on the other hand γ_1 combining with γ_4 can give rise to two different gametes other than the parental types viz Ab and aB. Assuming gametes of parental types can be formed with probability λ and non parental type by $1 - \lambda$ then the probability of four types of gametes under random mating for the next generation of $\gamma_1, \gamma_2, \gamma_3$ and γ_4 are $1 - \lambda/2, 1 - \lambda/2, \lambda/2, \lambda/2$ respectively.

Again the probability of forming γ_1 type of gametes due to combination of all types of gametes are given as follows

Table 13.1

Paternal Gamete	Maternal Gamete			
	AB	Ab	aB	ab
AB	1	½	½	½ $(1 - \lambda)$
Ab	½	0	$\lambda/2$	0
aB	½	$\lambda/2$	0	0
ab	½ $(1 - \lambda)$	0	0	0

The above is known as the segregation matrix for type 1 (γ_1) gamete and is denoted by $C_{1(ij)}$ i.e. the probability of forming type 1 gamete given that the same is formed out of combination of γ_i and γ_j types of parental gametes ($i, j = 1, 2, 3, 4$).　　　　　　　　　　　　　　　　　　(13.1)
Similarly one can verify

$$C_{2(ij)} = \begin{bmatrix} 0 & \frac{1}{2} & 0 & \lambda/2 \\ \frac{1}{2} & 1 & \frac{1}{2}(1-\lambda) & \frac{1}{2} \\ 0 & \frac{1}{2}(1-\lambda) & 0 & 0 \\ \lambda/2 & \frac{1}{2} & 0 & 0 \end{bmatrix} \qquad (13.2)$$

$$C_{3(ij)} = \begin{bmatrix} 0 & 0 & \frac{1}{2} & \lambda/2 \\ 0 & 0 & \frac{1}{2}(1-\lambda) & 0 \\ \frac{1}{2} & \frac{1}{2}(1-\lambda) & 1 & \frac{1}{2} \\ \lambda/2 & 0 & \frac{1}{2} & 0 \end{bmatrix} \qquad (13.3)$$

and $$C_{4(ij)} = \begin{bmatrix} 0 & 0 & 0 & \frac{1}{2}(1-\lambda) \\ 0 & 0 & \lambda/2 & \frac{1}{2} \\ 0 & \lambda/2 & 0 & \frac{1}{2} \\ \frac{1}{2}(1-\lambda) & \frac{1}{2} & \frac{1}{2} & 1 \end{bmatrix} \qquad (13.4)$$

Thus if we denote by $g_i(j)$ the probability of i type of gamete in the j-th generation ($i = 1, 2, 3, 4$) and $j = 1, 2, 3...$, then, under random mating

$$g_1^{(1)} = \sum_{i,j} C_i(ij)\, g_{ij} = \sum C_i(ij)\, g_i\, g_j \tag{13.5}$$

$g_1^{(1)} = g'\, C_{1(ij)} \cdot g$ where g represents the vector (g_1, g_2, g_3, g_4) for the starting generation.

i.e.
$$g_1^{(1)} = (g_1, g_2, g_3, g_4)\, C_{1(ij)} \begin{bmatrix} g_1 \\ g_2 \\ g_3 \\ g_4 \end{bmatrix}$$

$$= g_1(g_1 + g_2 + g_3 + g_4) - \lambda(g_1 g_4 - g_2 g_3)$$

$$= g_1 - \lambda(g_1 g_4 - g_2 g_3)$$

$$= g_1 - \lambda(g_1 g_4 - g_2 g_3), \text{ on simplification}$$

$$= g_1 - \lambda\, \Delta_0 \quad \text{where } \Delta_0 = \begin{vmatrix} g_1 & g_2 \\ g_3 & g_4 \end{vmatrix} \tag{13.6}$$

Similarly,
$$g_2^{(1)} = g'\, C_{2(ij)}\, g = g_2 + X\, \Delta_0 \tag{13.7}$$

$$g_3^{(1)} = g'\, C_{3(ij)}\, g = g_3 - X\, \Delta_0 \tag{13.8}$$

Similarly proceeding in the same way, we have,

$$\left. \begin{aligned} g_1^{(2)} &= g_1(1) + \lambda\, \Delta_1 \\ g_2^{(2)} &= g_2(1) + \lambda\, \Delta_1 \\ g_3^{(2)} &= g_3(1) + \lambda\, \Delta_1 \end{aligned} \right\} \tag{13.9}$$

where
$$\Delta_1 = \begin{vmatrix} g_1(1) & g_2(1) \\ g_3(1) & g_4(1) \end{vmatrix}$$

$$= g_1(1)\, g_4(1) - g_2(1)\, g_3(1)$$

$$= (g_1 - \lambda\, \Delta_0)(g_4 - \lambda\, \Delta_0) - (g_2 + \lambda\, \Delta_0)(g_3 + \lambda\, \Delta_0)$$

$$= (g_1 g_4 - g_2 g_3) - \lambda\, \Delta_0 (g_1 + g_2 + g_3 + g_4)$$

$$= \Delta_0 - \lambda\, \Delta_0 = (1 - \lambda)\, \Delta_0$$

Thus
$$\Delta_1 = (1 - \lambda)\, \Delta_0 \tag{13.10}$$

Following the argument in the same way, we notice

$$\left. \begin{aligned} \Delta_2 &= (1-\lambda)\, \Delta_1 = (1-\lambda)^2\, \Delta_0 \\ \Delta_3 &= (1-\lambda)^3\, \Delta_0 \\ &\dots\dots\dots\dots\dots\dots \\ \Delta_n &= (1-\lambda)^n\, \Delta_0 \end{aligned} \right\} \tag{13.11}$$

Now
$$g_1^{(n)} = g_1^{(n-1)} - \lambda \Delta_{n-1}$$

$$= g_1^{(n-2)} - \lambda \Delta_{n-2} - \lambda \Delta_{n-1}$$

$$= g_1^{(n-3)} - \lambda \Delta_{n-3} - \lambda \Delta_{n-2} - \lambda \Delta_{n-1}$$

$$\vdots$$

$$= g_1 - \lambda \Delta_0 - \lambda \Delta_1 - \ldots - \lambda \Delta_{n-1}$$

$$= g_1 - \lambda (\Delta_0 + \Delta_1 + \ldots + \Delta_{n-1})$$

$$= g_1 - \lambda \Delta_0 (1 + (1-\lambda) + \ldots + (1-\lambda)^{n-1})$$

$$= g_1 - \lambda \Delta_0 \frac{(1-(1-\lambda)^n)}{(1-(1-\lambda))}$$

$$= g_1 - \Delta_0 (1-(1-\lambda)^n)$$

In the same way,

$$g_2^{(n)} = g_2 + \Delta_0 (1-(1-\lambda)^n)$$
$$g_3^{(n)} = g_3 + \Delta_0 (1-(1-\lambda)^n)$$
$$\text{and} \quad g_4^{(n)} = g_4 + \Delta_0 (1-(1-\lambda)^n) \qquad (13.12)$$

A particular case when $\lambda = \frac{1}{2}$ i.e. when linkage is not distinguished is given by

$$\Delta_n = (\tfrac{1}{2})^n \Delta_0$$

and
$$g_1^{(n)} = g_1 - \Delta_0 (1-(\tfrac{1}{2})^n)$$

$$g_2^{(n)} = g_2 + \Delta_0 (1-(\tfrac{1}{2})^n)$$

$$g_3^{(n)} = g_3 + \Delta_0 (1-(\tfrac{1}{2})^n)$$

and
$$g_4^{(n)} = g_4 - \Delta_0 (1-(\tfrac{1}{2})^n)$$

It may be noted that as $n \to \infty$,

$$g_1^{(\infty)} = g_1 - \Delta_0, \ g_2^{(\infty)} = g_2 + \Delta_0, \ g_3^{(\infty)} = g_3 + \Delta_0$$

and
$$g_4^{(\infty)} = g_4 - \Delta_0$$

i.e.,
$$g_1^{(\infty)} + g_2^{(\infty)} = g_1 + g_2$$

$$g_3^{(\infty)} + g_4^{(\infty)} = g_3 + g_4 \text{ etc.} \qquad (13.13)$$

13.3 Hardy Wienberg Law

An infinite population with random mating and no selection for two alleles (say

A and *a*) will reach a stable equilibrium after one generation; such that the relative gene frequencies for *A* and *a* are *p* and *q* respectively and the relative genotypic frequencies for *AA*, *Aa* (or *aA*) and *aa* are p^2, $2pq$ and q^2.

13.3.1 Validity of Hardy Weinberg Law Under Overlapping Generation

We consider a model with continuous time parameter in which the genotypic probabilities for *AA*, *Aa* (for *aA*) and *aa* at time *t* are $P(t)$, $R(t)$, and $Q(t)$ respectively.

Suppose that in any time interval $(t, t + \delta t)$ a fraction δt of the population chosen at random dies and is replaced by a new fraction of size δt which is produced by the random mating of the existing population. We then have the following Kolmogorov equations,

$$\left.\begin{aligned}
P(t + \delta t) &= P(t)(1 - \delta t) + (P(t) + \tfrac{1}{2}R(t))^2\,\delta t \\[2mm]
R(t + \delta t) &= R(t)(1 - \delta t) + 2(P(t) + \tfrac{1}{2}R(t))(Q(t) + \tfrac{1}{2}R(t))\,\delta t \\[2mm]
\text{and}\quad Q(t + \delta t) &= Q(t)(1 - \delta t) + (Q(t) + \tfrac{1}{2}R(t))^2\,\delta t
\end{aligned}\right\}$$

$$(13.14)$$

This gives, in the limit when $\delta t \to 0$

$$\frac{d\,P(t)}{dt} = -P(t) + (P(t) + \tfrac{1}{2}R(t))^2$$

$$\frac{d\,R(t)}{dt} = -R(t) + 2(P(t) + \tfrac{1}{2}R(t))(Q(t) + \tfrac{1}{2}R(t))$$

and

$$\frac{d\,Q(t)}{dt} = -Q(t) + (Q(t) + \tfrac{1}{2}R(t))^2$$

It may immediately be seem that

$$\frac{d\,P(t)}{dt} + \frac{1}{2}\frac{d\,R(t)}{dt} = 0 \qquad\qquad (13.15)$$

$$(\because\ P(t) + Q(t) + R(t) = 1)$$

$\Rightarrow \qquad P(t) + \tfrac{1}{2}R(t) = $ a time independent constant $= p$, say

Similarly,

$$\frac{d}{dt}(Q(t) + \tfrac{1}{2}R(t)) = 0 \qquad\qquad (13.16)$$

$\Rightarrow \qquad Q(t) + \tfrac{1}{2}R(t) = $ a time independent constant $= q$, say

so that $\qquad\qquad p + q = P(t) + Q(t) + R(t) = 1$

Thus we have,

$$\left. \begin{array}{l} \dfrac{d\,P(t)}{dt} = -P(t) + p^2 \\[2ex] \dfrac{d\,R(t)}{dt} = -R(t) + 2pq \\[2ex] \dfrac{d\,Q(t)}{dt} = -Q(t) + q^2 \end{array} \right\} \qquad (13.17)$$

Solution of the above systems of linear differential equations give

$$\left. \begin{array}{l} P(t) = (P(0) - p^2)\,e^{-t} + p^2 \\[2ex] Q(t) = (Q(0) - q^2)\,e^{-t} + q^2 \\[2ex] \text{and} \qquad R(t) = (R(0) - 2pq)\,e^{-t} + 2pq \end{array} \right\} \qquad (13.18)$$

As $t \to \infty$ we have $P(t) = p^2$, $R(t) = 2pq$ and $Q(t) = q^2$.

Thus Hardy Weinberg Law for overlapping generations only holds asymptotically.

13.4 Mating under Various Types of Selection

Sometimes instead of random mating special type of selective matings are made for better autosomal as well as X-linked inheritance. This kind of choices are made for Agricultural and Animal breeding experiments. For human being also genetic counselling is made by advice of proper selecting mating among individuals or groups to avoid the inheritance of X-linked genes and associated complications. Below we present several cases of interesting applications of simple Markov processes.

Let us consider a single locus say A^L and two alleles A and a, giving rise to three different genotypes AA, Aa (or aA) and aa. Let $g_1^{(n)}$, $g_2^{(n)}$ and $g_3^{(n)}$ be the probabilities of these genotypes in the nth generation respectively. Suppose the genes (and the genotypes) refer to types of plants in an agricultural experiment. It is also desired to cross each plant of the above three types viz AA, Aa or aa with AA only for better yield. Suppose this kind of experimentation is carried on for n generations then the problem lies to find out the proportion of plants of each type in the nth generation.

We have $\qquad g_1^{(n)} + g_2^{(n)} + g_3^{(n)} = 1 \, (n = 0, 1, 2, \dots)$

where $n = 0$ refer to the original distribution of plants called the genotypic frequency in the parental generation.

Looking into the following table concerning the genotypic distribution of the offsprings conditional on the genotypic distribution of parents we have the following recurrence relations between $g_i^{(n)}$ and $g_i^{(n-1)}$ involving $g_j^{(n-1)}$ where $j \neq i$.

Table 13.2 Genotypes of offsprings and Parents

Genotype of Offsprings	Genotypes of Parents					
	$AA \times AA$	$AA \times Aa$	$AA \times aa$	$Aa \times Aa$	$Aa \times aa$	$aa \times aa$
AA	1	$\frac{1}{2}$	0	$\frac{1}{4}$	0	0
Aa	0	$\frac{1}{2}$	1	$\frac{1}{2}$	$\frac{1}{2}$	0
aa	0	0	0	$\frac{1}{4}$	$\frac{1}{2}$	1

$$g_1^{(n)} = g_1^{(n-1)} + \frac{1}{2} g_2^{(n-1)}$$

$$g_2^{(n)} = g_3^{(n-1)} + \frac{1}{2} g_2^{(n-1)}$$

and $$g_3^{(n)} = 0.$$

These equations can be combined by a single matrix equation viz.,

$$G^{(n)} = M G^{(n-1)} \tag{13.19}$$

where $$G^{(n)} = (g_1^{(n)}, g_2^{(n)}, g_3^{(n)})'$$

and $G^{(n-1)}$ is defined accordingly.

$$M = \begin{bmatrix} 1 & \frac{1}{2} & 0 \\ 0 & \frac{1}{2} & 1 \\ 0 & 0 & 0 \end{bmatrix}$$

Now we can write

$$G^{(n)} = MG^{(n-1)} = M^2 G^{(n-2)} = M^3 G^{(n-3)} = \ldots = M^n G^{(0)}$$

Therefore, on getting M^n we can get $G^{(n)}$ explicitly in terms of $G^{(0)}$. For getting nth power of M we require to diagonalize M i.e. we have to obtain a nonsingular matrix such that

$$PMP^{-1} = D, \text{ a diagonal matrix}$$

$$\Rightarrow \qquad M = PDP^{-1}$$

This gives $\qquad M^n = PD^n P^{-1}$ for $n = 1, 2, 3\ldots.$ $\tag{13.20}$

Since the characteristic roots of M are $X_1 = 1$, $X_2 = \frac{1}{2}$ and $X_3 = 0$ and the corresponding eigen vectors are $(1, 0, 0)'$, $(1, -1, 0)'$ and $(1, -2, 1)'$ therefore

$$PD^n P^{-1} = \begin{bmatrix} 1 & 1 & 1 \\ 0 & -1 & -2 \\ 0 & 0 & 1 \end{bmatrix} \begin{bmatrix} 1 & 0 & 0 \\ 0 & \frac{1}{2} & 0 \\ 0 & 0 & 0 \end{bmatrix}^n \begin{bmatrix} 1 & 1 & 1 \\ 0 & -1 & -2 \\ 0 & 0 & 1 \end{bmatrix}^{-1}$$

$$= \begin{bmatrix} 1 & 1-(\frac{1}{2})^n & 1-(\frac{1}{2})^{n-1} \\ 0 & (\frac{1}{2})^n & (\frac{1}{2})^{n-1} \\ 0 & 0 & 0 \end{bmatrix} \tag{13.21}$$

Noting $g_1^{(0)} + g_2^{(0)} + g_3^{(0)} = 1$, we have

$$
\left.
\begin{aligned}
g_1^{(n)} &= 1 - (\tfrac{1}{2})^n \, g_2^{(0)} - (\tfrac{1}{2})^{n-1} \, g_3^{(0)} \\
g_2^{(n)} &= (\tfrac{1}{2})^n \, g_2^{(0)} - (\tfrac{1}{2})^{n-1} \, g_3^{(0)} \\
g_3^{(n)} &= 0
\end{aligned}
\right\}
\tag{13.22}
$$

As $n \to \infty$ we note that $g_1^{(\infty)} = 1, g_2^{(\infty)} = 0, g_3^{(\infty)} = 0$ showing that if we make plant breeding according to that program we shall be ultimately realising only one type of plant viz. *AA*.

Example 13.1

Obtain the distribution of genotypes in the nth generation starting with initial genotypes AA, Aa and aa given that each genotype is crossed with a genotype of its own type (selfing) and observe the distribution as n tends to infinity.

Hints : In this case M is given by,

$$
M = \begin{bmatrix} 1 & \tfrac{1}{4} & 0 \\ 0 & \tfrac{1}{2} & 0 \\ 0 & \tfrac{1}{4} & 1 \end{bmatrix}
$$

Its characteristic roots are $\lambda_1 = 1, \lambda_2 = \lambda_3 = \tfrac{1}{2}$

The corresponding characteristic vectors are

$$
(1, 0, -1)', \ (0, 0, -1)' \text{ and } (1, -2, 1)'
$$

This gives

$$
\begin{aligned}
g_1^{(n)} &= g_1^{(0)} + (\tfrac{1}{2} - (\tfrac{1}{2})^{n+1}) \, g_2^{(0)} \\
g_2^{(n)} &= (\tfrac{1}{2})^n \\
g_3^{(n)} &= g_3^{(0)} + (\tfrac{1}{2} - (\tfrac{1}{2})^{n+1}) \, g_2^{(0)}
\end{aligned}
$$

13.5 Autosomal Inheritance

Often in human genetics it is seen that genes belonging to the same locus say A^L viz A and a one may be dominant over the other. If A is dominant over a (in which case we will call a to be recessive gene) then the physical expression of Aa will be the same as A. In this case we say the genotype AA has the phenotype A. This situation is in the case of human blood group (ABO system). While A, B are codominants both of them are dominant over O. Thus among the genotypes AA, AO, BO, BB and OO both AA and AO will have phenotype A; similarly BO and BB will be B phenotypically. While the genotype AB will be AB phenotypically because A and B being both codominants to each other. Finally the genotype OO will be of phenotype O (O being a recessive gene)

Sometimes of the two alleles at the locus A^L say A and a one gene(usually recessive) may be defective in the sense that it may produce genetic diseases (like Albinism, Sicklecell Anaemia, Cystic fibrosis, Cooley's anaemia, Tay-Sachs disease etc.). It may happen that if a is the defective gene then the genotype AA as well as Aa will be both normal but Aa will be a carrier of the disease because of a. On the other hand aa is directly afflicted with the disease. Under the case, genetic counselling only permits genotypes Aa to marry only AA (while AA can marry AA as well as Aa) and aa's are prohibited to marry them. With the above technique of Markov chain we can find out the proportion of genotypes in the nth generation.

We consider a simple example below.

Example 13.2   Let the proportion of AA be $g_1^{(0)}$ and Aa be $g_2^{(0)}$ at the start subject to $g_1^{(0)} + g_2^{(0)} = 1$ then obtain $g_1^{(n)}$ and $g_2^{(n)}$ and observe the distribution as n tends to infinity.
We find $G^{(n)} = M\, G^{(n+1)}$ where $G^{(n)} = (g_1^{(n)}, g_2^{(n)})$ and $G^{(n-1)}$ defined accordingly.

In this case $M = \begin{bmatrix} 1 & \frac{1}{2} \\ 0 & \frac{1}{2} \end{bmatrix}$ The characteristic roots of M are 1 and ½ respectively.

The corresponding eigen vectors are $(1, 0)'$ and $(0, 1)'$ respectively. This gives $g_1^{(n)} = 1 - (\frac{1}{2})^n g_2^{(0)}$ and $g_2^{(n)} = (\frac{1}{2})^n g_2^{(0)}$. Therefore, as n tends to infinity $g_1^{(n)}$ tends to unity showing that this genetic counselling will ultimately remove the recessive gene from the population. Also since $g_2^{(n)} = (\frac{1}{2}) g_2^{(n-1)}$ it follows that the carriers are reduced by half in every generation by this programme.

13.6   Sex linked (*X*-linked) Inheritance

Let there be V types of genes in the X chromosome of an individual. In otherwords, there will be V types of maternal and paternal gametes in a population with respect to X-linked genes. Let $f_i^{(n)}$ and $m_i^{(n)}$ be the paternal and maternal gene frequency (relative) of the type i ($i = 1, 2... V$) in the nth generation. Then,

$$f_i^{(n)} = m_i^{(n-1)} \tag{13.23}$$

$$m_i^{(n)} = \frac{1}{2}\left(\sum_{j=1}^{V} m_i^{(n-1)} f_j^{(n-1)} + \sum_{j=1}^{V} m_j^{(n-1)} f_i^{(n-1)} \right) \tag{13.24}$$

Using (13.23) and (13.24) we have,

$$m_i^{(n)} = \frac{1}{2}(m_i^{(n-1)} + f_i^{(n-1)}) \tag{13.25}$$

Again using (13.23) and (13.25) we have

$$f_i^{(n)} = \frac{1}{2}(m_i^{(n-2)} + f_i^{(n-2)})$$

$$= \frac{1}{2}(f_i^{(n-1)} + f_i^{(n-2)}) \quad \text{by (13.23)}$$

$$\Rightarrow \qquad (f_i^{(n)} - f_i^{(n-1)}) = -\frac{1}{2}(f_i^{(n-1)} - f_i^{(n)}) \qquad (13.26)$$

Putting $f_i^{(1)} - f_i^{(0)} = \Delta_0(i)$
which gives

$$f_i^{(2)} - f_i^{(1)} = \frac{1}{2}(f_i^{(1)} + f_i^{(0)}) - f_i^{(1)}$$

$$= \frac{1}{2}(f_i^{(0)} - f_i^{(1)}) = -\frac{1}{2}\Delta_0(i)$$

Similarly, $\qquad f_i^{(3)} - f_i^{(2)} = \frac{1}{2}(f_i^{(2)} + f_i^{(1)}) - f_i^{(2)}$

$$= \frac{1}{2}(f_i^{(1)} - f_i^{(2)}) = \frac{1}{2^2}\Delta_0(i)$$

Proceeding in this way

$$f_i^{(n)} - f_i^{(n-1)} = \left(-\frac{1}{2}\right)^{n-1}\Delta_0(i)$$

Now $\qquad f_i^{(n)} = f_i + (f_i^{(1)} - f_i) + (f_i^{(2)} - f_i^{(1)}) + \ldots + (f_i^{(n)} - f_i^{(n-1)})$

$$= f_i + \Delta_0(i) - \frac{1}{2}\Delta_0(i) + \left(\frac{1}{2}\right)^2\Delta_0(i) + \ldots + \left(-\frac{1}{2}\right)^{n-1}\Delta_0(i)$$

$$= f_i + \Delta_0(i)\frac{\left(-\frac{1}{2}\right)^n - 1}{\left(-\frac{1}{2}\right) - 1}$$

$$f_i^{(n)} = f_i + \Delta_0(i)(-2/3)\left(\left(-\frac{1}{2}\right)^n - 1\right)$$

Or $\qquad f_i^{(n)} = f_i + \frac{2}{3}\Delta_0(i)\left(1 - \frac{1}{(-2)^n}\right) \qquad (13.27)$

Precisely, in a similar way

$$m_i^{(n)} = f_i + \frac{2}{3}\Delta_0(i)\left(1 - \frac{1}{(-2)^{n+1}}\right) \qquad (13.28)$$

$$(\text{since } m_i^{(n)} = f_i^{(n+1)})$$

As $n \to \infty$, we get

$$f_i^{(\infty)} = f_i + \frac{2}{3}\Delta_0(i)$$

$$= f_i + 2/3(f_i^{(1)} - f_i^{(0)})$$

$$= \frac{1}{3}f_i + \frac{2}{3}m_i \qquad (13.29)$$

Similarly, $\qquad m_i^{(\infty)} = \frac{1}{3}f_i + \frac{2}{3}m_i \qquad (13.30)$

Example 13.3

Given the genotypes of the parents with respect to X-linked gene with two alleles A and a as in the following check the distribution of offspring as given in the table.

Table 13.3 Genotype of the Parents (Father, Mother) w.r.t. X linked genes

Genotpe of the offspring	(A, AA)	(A, Aa)	(A, aa)	(a, AA)	(a, Aa)	(a, aa)
A	1	½	0	1	½	0
a	0	½	1	0	½	1
AA	1	½	0	0	0	0
Aa	0	½	1	1	½	0
aa	0	0	0	0	½	1

Example 13.4

Using the table given in example 13.3 if the probability of the following matings of genotypes w.r.to X linked genes (A, AA) (A, Aa) (A, aa) (a, AA) (a, Aa) and (a, AA) at the start be $g_1^{(0)}, g_2^{(0)}, g_3^{(0)}, g_4^{(0)}, g_5^{(0)}$ and $g_6^{(0)}$ respectively, obtain $g_i^{(n)}$ $(i = 1, 2.....6)$ given that matings are restricted between offsprings of the same parents.

Denoting $\qquad G^{(n)} = (g_1^{(n)}, g_2^{(n)} .... g_6^{(n)})$

and $G^{(n-1)}$ accordingly we have

$$G^{(n)} = M G^{(n-1)}$$

where

$$\qquad (A, AA) \; (A, Aa) \; (A, aa) \; (a, AA) \; (a, Aa) \; (a, aa)$$

$$
M = \begin{bmatrix}
1 & ¼ & 0 & 0 & 0 & 0 \\
0 & ¼ & 0 & 1 & ¼ & 0 \\
0 & 0 & 0 & 0 & ¼ & 0 \\
0 & ¼ & 0 & 0 & 0 & 0 \\
0 & ¼ & 1 & 0 & ¼ & 0 \\
0 & 0 & 0 & 0 & ¼ & 1
\end{bmatrix}
\begin{matrix}
(A, AA) \\
(A, Aa) \\
(A, aa) \\
(a, AA) \\
(a, Aa) \\
(a, aa)
\end{matrix}
$$

Check that M' is a transition matrix of the Markov chain.

This gives
$$G^{(n)} = M^n G^{(0)}$$

The characteristic root of M are $1, 1, \frac{1}{2}, -\frac{1}{2}, \frac{1}{4}(1+\sqrt{5}), \frac{1}{4}(1-\sqrt{5})$ and the corresponding characteristic vectors are

$(1, 0, 0, 0, 0, 0,)', (0, 0, 0, 0, 0, 1)', (-1, 2, -1, 1, -2, 1)', (1, -6, -3, 3, 6, -1)'$

$$(\tfrac{1}{4}(-3-\sqrt{5}), 1, \tfrac{1}{4}(-1+\sqrt{5}), \tfrac{1}{4}(-1+\sqrt{5}), 1, \tfrac{1}{4}(-3-\sqrt{5})'$$

$$(\tfrac{1}{4}(-3+\sqrt{5}), 1, \tfrac{1}{4}(-1-\sqrt{5}), \tfrac{1}{4}(-1-\sqrt{5}), 1, \tfrac{1}{4}(-3-\sqrt{5})'$$

respectively.

Following the procedure in 13.3 we get as n tends to infinity

$$g_1^{(\infty)} = \frac{2}{3}, \ g_2^{(\infty)} = 0, \ g_3^{(\infty)} = 0, \ g_4^{(\infty)} = 0, g_5^{(\infty)} = 0$$

and
$$g_6^{(\infty)} = \frac{1}{3}$$

Thus if the inbreeding (Brother $\times$ Sister mating) continues then ultimately we will have matings of two types of homogeneous genotypes viz $A \times AA$ and $a \times aa$. Thus inbreeding produces homozygosity.

13.7 Stochastic Processes in the Change of Gene Frequencies

Let us denote $\phi(x, t)$ the probability distribution of the gene frequency (relative) at time t. We consider the class interval which correspond to the gene frequency x. Assuming that x lies in the middle of the infinitesimal interval $(x - h/2, x + h/2)$ for small h, we have $\phi(x, t)$ which represents the probability that the population has gene frequency in the interval $(x - h/2, x + h/2)$. We can imagine that after a time interval dt the population with gene frequency x (on the average) will move to another class of gene frequencies on account of systematic as well as random factors. Since the gene frequency is essentially discrete and its change over time is effected to visualize by a continuous Stochastic Process, therefore we shall impose the restriction of sufficiently small dt so that the movement of the gene frequency during dt in an adjacent class which is sufficiently close to the earlier class.

Let $s(x, t) \, dt$ represents the probability that the population moves to a higher class of gene frequency i.e. $(x + h)$ say by systematic effect during an infinitesimal interval $(t, t + dt)$. Also let $r(x, t) \, dt$ is the probability that the gene frequency outside the class by random fluctuation. Since the effect is assumed to be random we assume that the gene frequency will go to class $(x - h)$ in half the cases and to $(x + h)$ in another half of the cases. Also the probability of higher displacements on either side is neglected. Thus the population will have gene frequency in the interval $(x - h/2, x + h/2)$ during a period of dt following t is obtained by considering the gene frequencies between the adjacent classes.

In what follows, we have

$$\phi\,(x,\,t+\delta t)\,h \;=\; \phi\,(x,\,t)\,h\,(1-(s\,(x,\,t)+r\,(x,\,t))\,\delta t$$

$$+\,h\,\phi(x-h,\,t)\frac{r}{2}\,(x-h,\,t)\,\delta t + h\phi(x+h,\,t)\frac{r}{2}\,(x+h,\,t)\,\delta t$$

$$+\,h\,\phi\,(x-h,\,t)\,s\,(x-h,\,t)\,\delta t \qquad\qquad (13.31)$$

Denoting by $V(x,t)$, the variance of the change in gene frequency due to random effect i.e. the change in x per unit time which is dt, it follows that

$$V(x,\,t)\,\delta t \;=\; h^2\cdot\frac{1}{2}\,r(x,t)\,\delta t + (-h)^2\cdot\frac{1}{2}\cdot r(x,t)\,\delta t$$

$$-(\frac{h}{2}r\,(x,\,t)\,\delta t + \frac{h}{2}\,r\,(x,\,t)\,\delta t)^2$$

$$\Rightarrow\qquad V(x,\,t) \;=\; h^2\,r\,(x,\,t) \qquad\qquad (13.32)$$

Similarly denoting by $M\,(x,\,t)\,dt$, the mean change of x per unit of time i.e. dt we have

$$M(x,\,t)\,\delta t \;=\; h\,s\,(x,\,t)\,\delta t$$

i.e.,$\qquad\qquad M(x,\,t) \;=\; h\,s\,(x,\,t) \qquad\qquad (13.33)$

On substitution of (13.32) and (13.33) on the right hand side of (13.31) we have,

$$\frac{\phi(x,\,t+\delta t)-\phi(x,\,t)}{\delta t}$$

$$=\;\frac{1}{2}\frac{\dfrac{V\,\phi(x+h,\,t)-V\,\phi(x,\,t)}{h}-\dfrac{V\,\phi(x,\,t)-V\,\phi(x-h,\,t)}{h}}{h}$$

$$-\left[\frac{M\,V\,\phi(x,\,t)-M\,\phi(x-h,\,t)}{h}\right]$$

where $\qquad V\,\phi\,(x,\,t) \;=\; V(x,\,t)\,\phi\,(x,\,t)$

$$M\,\phi\,(x,\,t) \;=\; N(x,\,t)\,\phi\,(x,\,t)$$

Taking limit as $\delta t \to 0$ and $h \to 0$ we have,

$$\frac{\partial\,\phi(x,t)}{\partial\,t} \;=\; \frac{1}{2}\frac{\partial^2}{\partial x^2}\,(V\,(x,t)\,\phi(x,t)) - \frac{\partial}{\partial x}\,(M\,(x,t)\,\phi(x,t)) \qquad (13.34)$$

which is known as the Fokker-Plank diffusion equation which was first introduced in population genetics by Sewell Wright(1945) and later on in connection with the problem of gene fixation by Kimura (1957).

13.8 Tendency for a Population Towards Homozygosity under Random Mating

We consider a population of M haploid individuals and write $2N$ for M without much loss of generalisation. We further consider that generations are non-overlapping so that the parent population j 'a' genes and $M - j$ 'A' genes. The next generation is then produced by an independent choice of new individuals of size M which are either a or A with probabilities

$$p_j = jM^{-1} \text{ and } q_j = (M - j) M^{-1} \text{ respectively.}$$

The probability distribution viz. out of M the chance of getting k number of a genes and $(n - k)$ number of A genes in the next generation (from j number of a and $(M - j)$ number of A) is clearly Binomial and given by

$$p_{kj} = \binom{M}{k} p_j^k q_j^{M-k} \tag{13.35}$$

and the matrix $((p_{kj}))$ is the transition probability matrix of the Markov of the Markov chain of the corresponding process

Now the chain has two absorbing states viz. $k = 0$ and $k = M$ and Feller (vide Moran P.A.P. (1961)) has obtained the probabilities of absorption in these two states given the initial value j and the asymptotic rate at which the absorption occurs.

Feller has shown the characteristic roots of (p_{kj}) $k, j = 0, 1, 2, ...M$ are of the form

$$\lambda_r = \binom{M}{r} r! \, M^{-r} \tag{13.36}$$

Subject to $\quad \lambda_0 = \lambda_1 = 1 \text{ and } \lambda_2 > \lambda_3 > ..... > \lambda_M$

For a rigorous proof, the reader is referred to Moran (1961).

Feller has also obtained the left characteristic vectors of (p_{kj}) but the right characteristic vectors have so far not been obtained.

Let $P = (p_{kj})$ be the transition matrix representing the transition probability from j to k and X'_r and Y_r ($r = 0, 1, ... M$) be the left and right characteristic vectors corresponding to the eigen value λ_r such that

$$X'_r P = \lambda_r X'_r \text{ holds}$$

and $\qquad\qquad P Y_r = \lambda_r Y_r$

where X'_r being a row vector of order $1 \times (M + 1)$ and Y_r being a column vector of order $1 \times (M + 1)$.

Suppose that p_0 is a column vector describing the probability distribution of the initial state of the system. Thus if initially there are j number of a genes and $(M - j)$ number of A genes, p_0 will have zero elements everywhere excepting at the jth place where the element is unity. Thus the probability distribution of the states at the nth generation will be given by a vector.

$$p_n = P^n p_0$$

By using the spectral decomposition of P we can express p_n in terms of λ_r and the characteristic vectors. Since $Y_0, Y_1 ... Y_M$ are linearly independent, we have

$$p_n = \sum_{r=0}^{M} \lambda_r Y_r$$

Therefore, $\quad p_n = P^n p_0 =$ probability distribution of the state of the system at time n

$$= \sum_{r=0}^{M} \lambda_r^n Y_r X_r'/\delta_r \cdot p_0$$

$$= \sum_{r=0}^{M} \lambda_r^n \left(\frac{Y_r X_r'}{\delta_r}\right) \alpha_r Y_r$$

$$= \sum_{r=0}^{M} \lambda_r^n \alpha_r Y_r \left(\frac{X_r' Y_r}{\delta_r}\right)$$

$$= \sum \lambda_r^n \alpha_r Y_r \quad (\because X_r' Y_r = \delta_r) \tag{13.37}$$

Thus using (13.36) and (13.37) we have,

$$p_n = \alpha_0 Y_0 + \alpha_1 Y_1 + \dots + \sum_{r=2}^{M} \alpha_r \lambda_r^n Y_r$$

$$\cong \alpha_0 Y_0 + \alpha_1 Y_1 \quad \text{as } n \to \infty$$

Also it is easily verified that

$$Y_0 = \begin{bmatrix} 1 \\ 0 \\ 0 \\ \vdots \\ 0 \end{bmatrix} \quad \text{and} \quad Y_1 = \begin{bmatrix} 0 \\ 0 \\ 0 \\ \vdots \\ 1 \end{bmatrix}$$

where α_0 and α_1 are the ultimate absorption probabilities in the state $k = 0$ and $k = M$ respectively. Thus the population will be tending to homozygosity with probability α_0 in favour of $k = 0$ and α_1 in favour of $k = M$ respectively. Feller has obtained the value of α_0 and α_1 directly. For details the reader is referred to Moran (1961).

13.9 An Alternative Proof of Population tending to Homozygosity under Random Mating

Let K_t be the number of 'A' gene (or the gamete carrying A type of Gene) and $M - K_t$ be the number of 'a' type of gene at any time t in the population.

We consider the conditional random variable $k_{t+1} \mid k_t$

We have $\quad \text{Var}(K_{t+1} \mid K_t) = E(K_{t+1} \mid K_t)^2 - E^2(K_{t+1} \mid K_t)$

$$= E(K_{t+1} \mid K_t)^2 - \left(M \cdot \frac{K_t}{M}\right)^2$$

Then the expectation of the above conditional variance is given by

$$E\left(\text{Var}\left(K_{t+1}\mid K_t\right)\right) = E\left(K_{t+1}\right)^2 - E\left(K_t\right)^2 \tag{13.38}$$

$$\left(\because E(K_{t+1}\mid K_t) = M\frac{K_t}{M} = K_t\right)$$

Again $$\text{Var}\left(K_{t+1}\mid K_t\right) = M\frac{K_t}{M}\left(1-\frac{K_t}{M}\right) = K_t\left(M-K_t\right)M^{-1}$$

$$E\left(\text{Var}\left(K_{t+1}\mid K_t\right)\right) = M^{-1}E\left(K_t(M-K_t)\right)$$

$$= E\left(K_t\right) - M^{-1}E\left(K_t^2\right) \tag{13.39}$$

Comparing (13.38) and (13.39) we have

$$E\left(K_{t+1}\right)^2 - E\left(K_t^2\right) = E\left(K_t\right) - M^{-1}E\left(K_t^2\right)$$

$$\Rightarrow \qquad E\left(K_{t+1}^2\right) = \left(1 - M^{-1}\right)E\left(K_t^2\right) + E\left(K_t\right) \tag{13.40}$$

Now consider

$$E\left(K_{t+1}\left(M - K_{t+1}\right)\right)$$

$$= M E\left(K_{t+1}\right) - E\left(K_{t+1}^2\right) \tag{13.41}$$

Again since $E\left(K_{t+1}\mid K_t\right) = K_t$, therefore

$$E\left(E\left(K_{t+1}\mid K_t\right)\right) = E\left(K_{t+1}\right) = E\left(K_t\right) \tag{13.42}$$

Putting (13.42) in (13.41) we have

$$E\left(K_{t+1}\left(M - K_{t+1}\right)\right) = M E\left(K_t\right) - E\left(K_{t+1}^2\right) \tag{13.43}$$

Putting (13.40) in (13.43) we have

$$E\left(K_{t+1}\right)\left(M - K_{t+1}\right)) = M E\left(K_t\right) - \left(1 - M^{-1}\right)E\left(K_t^2\right) - E\left(K_t\right)$$

$$= (M-1)E\left(K_t\right) - \left(1 - M^{-1}\right)E\left(K_t^2\right)$$

$$= M E\left(K_t\right) - E\left(K_t^2\right) - E\left(K_t\right) + M^{-1}E\left(K_t^2\right)$$

$$= \left(M E\left(K_t\right) - E\left(K_t^2\right)\right) - M^{-1}\left(M E\left(K_t\right) - E\left(K_t^2\right)\right)$$

$$= \left(1 - M^{-1}\right)\left(M E\left(K_t\right) - E\left(K_t^2\right)\right) \tag{13.44}$$

Also $$E\left(K_t\left(M - K_t\right)\right) = M E\left(K_t\right) - E\left(K_t^2\right) \tag{13.45}$$

Comparing (13.44) and (13.45) we have,

$$E\left(K_{t+1}\left(M - K_{t+1}\right)\right) = \left(1 - M^{-1}\right)E\left(K_t\left(M - K_t\right)\right)$$

$$\Rightarrow \quad E\left(\frac{K_{t+1}}{M}\frac{\left(M - K_{t+1}\right)}{M}\right) = \left(1 - M^{-1}\right)E\left(K_t/M\right)\left(M - K_t\right)/M) \tag{13.46}$$

Left hand side of (13.46) represent the proportion of heterozygotes i.e., *Aa*, at time $(t+1)$ and the right hand side represents the same at time t. This shows the proportion of heterozygotes decreases by $(1-M^{-1})$ per unit time even under random mating.

13.10   A Diffusion Model

The random variable in the Markov chain denoted by K in Art. 13.8 is dicrete. Assuming M to be large the r.v. $K/M = X$ lies in $(0, 1)$. We denote by suffix t the time at which X is observed. One can measure t in units of M generations (i.e. the size of the haploid population). Thus for large M, t (like X_t) behaves like a continuously variable quantity and we have in the limit

$$\text{Var}(X_{t+h} - X_t) = h\,\text{Var}(X_t) = h\,X_t\frac{(1-X_t)}{M} = h\,X_t\,(1-X_t)\ (\because M = 1)$$

(13.47)

Let us denote

$$\lim_{h\to 0} h^{-1}\,E\,(X_{t+h} - X_t) = a\,(X_t)$$

(13.48)

and

$$\lim_{h\to 0} h^{-1}\,\text{Var}\,(X_{t+h} - X_t) = B\,(X_t)$$

From (13.48) it follows that $a\,(X_t)$ is a measure of the rate of drift which acquires a non zero value when a selection occurs. $B\,(X_t)$ is a measure of the rate of spread by diffusion.

Taking h to be small let us consider the distribution $\phi\,(x, t + h)$ in terms of $(t + h)$. Denoting the probability distribution of a jump of size 'u' starting from the point x as $\psi\,(u, x)$, we then have

$$\phi(x, t + h) = \int \phi(x - u\,; t)\,\psi\,(u, x - u)\,du$$

Assuming h to be small compared to I but large compared to M^{-1} so that the jump has a mean and variance of the order of h (i.e., $E\,(u) = \text{Var}\,(u) \cong h$, u being a Poisson variable) the range of integration thus can be taken as infinite as $h \to 0$ Therefore,

$$\phi\,(x, t + h) = \int \phi(x - u, t)\,\psi(u, x - u)\,du$$

$$= \int\left[\phi(x, t) - u\,\phi_x(x, t) + \frac{1}{2}u^2\phi_{ux}(x, t) + 0(u^3)\right]$$

$$\cdot\left[\phi(u, x) - u\,\psi(u, x) + \frac{u^2}{2}\phi_{xx}(u, x) + 0(u^3)\right]du \quad (13.49)$$

$$\phi\,(x, t + h) = \int \phi(x, t)\,\psi(u, x)\,du - \int u\psi(u, x)\,\phi(x, t)\,du$$

$$+ \int u\phi_x(x, t)\,\psi(u, x)\,du + \int u^2\phi_x(x, t)\,\psi(u, x)\,du$$

$$+ \int \frac{u^2}{2}\phi_{xx}(x, t)\,\psi(u, x)\,du$$

Putting
$$\int \psi(u, x)\, du = 1$$

we have
$$\int u\psi(u, x)\, du = h\, a(x) + 0(h)$$

$$\int u^2 \psi(u, x)\, du = h\, B(x) + 0(h)$$

Therefore
$$\left. \begin{aligned} \int u\psi_x(u, x)\, du &= ha_x(x) + 0(h) \\ \text{and} \quad \int u^2 \psi_x(u, x)\, du &= hB_x(x) + 0(h) \end{aligned} \right\} \tag{13.50}$$

$$\Rightarrow \quad \phi(x, t+h) - \phi(x, t) = -h\, a_x(x)\, \phi(x, t) + \tfrac{1}{2} B_{xx}\, \phi(x, t)\, h$$
$$- h\, \phi_x(x, t)\, a(x) + h\, \phi_x(x, t)\, B_x(x) + h/2\, \phi_{xx}(x, t)\, B(x) + 0(h^2)$$

$$\lim_{h \to 0} \frac{\phi(x, t+h) - \phi(x, t)}{h} = -a_x(x)\, \phi(x, t) + \tfrac{1}{2} B_{xx}\, \phi(x, t)$$

$$- \phi_x(x, t)\, a(x) + \phi_x(x, t)\, B_x(x) + \tfrac{1}{2} \phi_{xx}(x, t)\, B(x)$$

$$\phi_t(x, t) = -\left(\phi(x, t)\, a_x(x) + \phi_x(x, t)\, a(x) \right)$$

$$+ \left(\tfrac{1}{2} \phi(x, t)\, B_{xx} + \tfrac{1}{2} \phi_{xx}(x, t)\, B(x) + \phi_x(x, t)\, B_x(x) \right)$$

or
$$\frac{d\phi}{dt} = -\frac{d}{dx}\left(\phi(x, t)\, a(x) \right) + \frac{d^2}{dx^2}\left(\tfrac{1}{2} \phi(x, t)\, B(x) \right) \tag{13.51}$$

$$\left(\text{since } \frac{d^2}{dx^2}\left(\tfrac{1}{2} \phi(x, t)\, B(x) \right) = \tfrac{1}{2} \left(B_{xx}\, \phi(x, t) + B x(x)\, \phi_x(x, t) \right. \right.$$

$$+ \phi_{xx}\, B(x) + \phi_x\, B_x(x) \Big) + \left(\tfrac{1}{2} B_{xx}\, \phi(x, t) + B_x(x)\, \phi_x(x, t) + \tfrac{1}{2} \phi_{xx}\, B(x) \right)$$

Now putting $a(x) = 0$ and $B(x) = x(1 - x)$

$$\left(\because E(X_{t+h} - X_t) = 0, \ \text{Var}(X_{t+h} - X_t) = hX_t(1 - X_t) \right)$$

We get

$$\frac{d\phi(x, t)}{dt} = \frac{d^2}{dx^2}\left(\tfrac{1}{2} x(1 - x)\, \phi(x, t) \right) \tag{13.52}$$

This is the Fokker Plank diffusion equation.

References

1. Crow J. F. and Kimura M (1970): *An Introduction to Population Genetics Theory*, Harper and Row, New York.
2. Elandt Johnson R.C. (1971): *Probability models and Statistical methods in Genetics*, John Wiley, New York.
3. Fisher R.A. (1958): *The genetical theory of natural selection*, 2nd edition, Dover Press, New York.

4. Jain J.P. and Prabhakar V.T. (1992): *Genetics of Population*, South Asia Publishers Pvt. Ltd, Delhi.

5. Kempthrone O. (1957): *An introduction to Genetic Statistics*: John Wiley & Sons New York.

6. Mather K and Jinks S.L (1970): *Biometrical Genetics* , Chapman and Hall Ltd.

7. Mather K (1963): *The measurement of Linkage in Heridity*, Mathuen Monograph. London, New York. John Wiley & Sons.

8. Moran P.A.P. (1961): *Statistical Processes in Evolutionary Theory*, Oxford, Clarendon Press.

9. Nishbet R.M and Gurney W. S.C. (1982): *Modeling of Fluctuating and Population* John Wiley & Sons.

10. Narain Prem (1990): *Statistical Genetics* , Wiley Eastern Limited, New Delhi.

Appendix

A.1 Delta Method for Large Sample Standard Error

Let the r.v. $T \sim N(\mu, \sigma^2)$ we require the large sample standard error of some functions of T say $\psi(T)$.

Assuming $\psi(T)$ is capable of being expressed as a Taylor series about $T = \mu$, we have

$$\psi(T) = \psi(\mu) + (T - \mu)\psi(\mu) + \dots$$

Ignoring terms involving higher powers of $(T - \mu)$ we have,

$$E(\psi(T) \cong \psi(\mu) \tag{A.1.1}$$

$$E(\psi(T) - \psi(\mu))^2 \cong E(T - \mu)^2 [\psi'(u)]^2$$

$$\text{Var}(\psi(T)) \cong \text{Var}(T)\left(\frac{\partial\psi}{\partial u}\right)^2 \tag{A.1.2}$$

Note that the results (A.1.1) and (A.1.2) hold approximately irrespective of whether T is normally distributed or not.

Delta method has a multivariate analogue. Suppose T_1 and T_2 are two random variables which are jointly distributed with parameters $(\mu_1, \mu_2, \sigma_1^2, \sigma_2^2, \rho_{12})$ and we want the large sample standard error of $\psi(T_1, T_2)$ some functions of T_1 and T_2.

Then by Taylor's expansion

$$\psi(T_1, T_2) = \psi(\mu_1, \mu_2) + (T_1 - \mu_1)^2 \frac{\partial}{\partial T_1}\psi(T_1, T_2) + (T_2 - \mu_2)^2 \frac{\partial}{\partial T_2}\psi(T_1, T_2) + \dots$$

Following the earlier argument for the large sample

$$\psi(T_1, T_2) \sim N\left(\psi(\mu_1, \mu_2), \text{Var}(T_1)\left(\frac{\partial}{\partial T_1}\psi\right)^2\right.$$

$$\left. + 2\text{Cov}(T_1, T_2)\left(\frac{\partial}{\partial T_1}\psi \frac{\partial}{\partial T_2}\psi\right) + \text{Var}(T_2)\left(\frac{\partial}{\partial T_2}\psi\right)^2\right.$$

The delta method is highly advantageous. For example, we could use it to get an approximate value for $\text{Var}\left(\dfrac{T_1}{\overline{T}_2}\right)$ or $\text{Var}\,(\overline{T}_1, \overline{T}_2)$ where $\overline{T}_1$ and $\overline{T}_2$ are the sample means of T_1 and T_2 respectively.

A.2 Crámer Rao Inequality

Suppose we want to estimate a parameter θ occurring in the distribution $f(x, \theta, \theta', \theta'' \ldots)$ by means of a statistic (a function of the sample observations) constructed from a sample of size n.

Let $x_1, x_2 \ldots, x_n$ be n independent observations on the r.v. X.

T is said to be *unbiased estimator* of θ if $E\,(T) = \theta$ whatever may be θ^1, θ^{11} ... An estimator is called the *best* or the *minimum variance unbiased estimator* (MVUE) if $E_T\,[(T - \theta)^2]$ is minimum.

Crámer Rao inequality states, that under certain regularity conditions (vide Crámer Harold – Mathematical Statistics; Princeton University Press)

$$\text{Var}\,(T) \ge \frac{1}{nE\left(\dfrac{\partial \log f}{\partial \theta}\right)^2}$$

where $f = f(x, \theta, \theta', \theta'' \ldots)$.

It follows that the variance of the unbiased estimator satisfying the regularity condition cannot be less than that of Cramer Rao lower bound viz.

$$\frac{1}{nE\left(\dfrac{\partial \log f}{\partial \theta}\right)^2}$$

A.3 Rao Blackwell Theorem

The theorem is useful in obtaining the minimum variance unbiased estimator (MVUE) of a parameter (which may be a vector) by a process known as 'Blackwellization'. Before we state the theorem, two fundamental characteristics of an estimator viz. (1) Sufficiency (2) Completeness should be defined.

An estimator is said to be *sufficient* if it provides all the information contained in a sample in respect of the parameter θ.

Let $\hat{\theta}$ and $\tilde{\theta}$ be two estimators of θ subject to the condition that one is not a function of the other. $\hat{\theta}$ is said to be sufficient if the conditional density function of $\tilde{\theta}$ given $\hat{\theta}$ is independetn of θ. A simple criterion of checking the sufficiency of an estimator is known as *Fisher-Neyman factorization theorem* which runs as follows:

Let $x_1 \ldots x_n$ be a random sample of size n from a population with density function $g\,(x\,;\theta)$, $a < x < b$, when a and b does not depend upon θ. Then $\hat{\theta}$ is a sufficient statistic for θ if the joint density function of the sample is capable of being expressed in the form

$$g(x_1, x_3 \dots x_n ; \theta) = g_1(\hat{\theta}, \theta) \, g_2(x_1 \, x_2 \dots x_n) \tag{A 3.1}$$

where $g_1(\hat{\theta}, \theta)$ is the density function of $\hat{\theta}$ (the sufficient statistics) involving the parameter θ and $g_2(x_1 \, x_2 \dots x_n)$ is the conditional density function of $g(x_1 \, x_2 \dots x_n ; \theta)$ which does not involve θ, if ($\hat{\theta}$ is sufficient. Conversely if the representation (A. 4. 1) holds then $\hat{\theta}$ is sufficient for θ.

Completeness of a Statistic
A sufficient statistic T is said to be complete if $E_\theta(T) = 0 \Rightarrow T = 0$, almost everywhere, except for a set of points with probability measure zero. A statistic which is unbiased and complete is necessarily unique and is the minimum variance unbiased estimator (MVUE) of its expectation provided MVUE exists.

Rao Blackwell theorem
Let T be any sufficient statistic for θ (where T and θ may be vector valued) and T_1 be any other statistic

Let
$$h(t) = E(T_1 \mid T = t)$$

Then
$$E(T_1 - g(\theta))^2 \geq E(h(T) - g(\theta))^2$$

where $g(\theta)$ is any parametric function of θ.

A particular case
If T_1 is unbiased for $g(\theta)$ implying $h(t)$ is also unbiased for $g(\theta)$ then we have
$\text{Var}(T_1) \geq \text{Var}(h(T))$

Blackwellization
The implication of the result is that if we are given an unbiased estimator, say, T_1 of $g(\theta)$ we can improve upon T_1 by taking $E(T_1 \mid T_2 = t) = h(t)$ based on the sufficient statistic T_2. This process of finding an improved estimator using the concept of unbiasedness and then reducing the variance by taking the conditional mean of the unbiased statistic given a sufficient statistic is known as 'Blackwellization' $h(T)$ is better than T_1 but the best if $h(T)$ is complete.

A.4 Central Limit Theorems

(i) *Lindeberg Levy Central Limit Theorems*
Consider a sequence of identically independently distributed random variables $X_1, X_2 \dots$ each having a finite mean m and finite standard deviation $\sigma \neq 0$. Let $\{X_n\}$ be a sequence of distribution functions where

$$Z_n = \frac{Y_n - nm}{\sigma \sqrt{n}}$$

and
$$Y_n = X_1 + X_2 + \dots + X_n$$

and $F_n(Z)$ is the cumulative distribution function of Z_n, i.e.,

$$P[Z_n \leq Z] = F_n(Z)$$

then $\lim\limits_{n \to \infty} F_n(Z) = \dfrac{1}{\sqrt{2\pi}} \int\limits_{-\infty}^{Z} \exp(-t^2/2)\,dt = \Phi(z)$ for $\forall\ Z$.

(ii) *Liapounv's form of the Central limit theorem*

Let $X_1\ X_2\ \dots\ X_n$ be n independent random variables and $E(X_r) = \mu_r$ and $\mathrm{Var}(X_r) = \sigma_r^2; r = 1, 2\dots n$. Assume that the third absolute moments v_r of X_r about its mean μ_r given by

$$v_r^3 = E(|X_r - \mu_r|)^3$$

exists and is finite $\forall\ r$. If $v^3 = v_1^3 + v_2^3 + \dots + v_r^3$ and if $\lim\limits_{n \to \infty} \dfrac{v}{\Sigma} = 0$ then

$\sum\limits_{r=1}^{n} X_r$ is asymptotically normally distributed with mean Λ and standard deviation Σ where

$$\Lambda = \mu_1 + \mu_2 + \dots + \mu_n$$

and $$\Sigma^2 = \sigma_1^2 + \sigma_2^2 + \dots + \sigma_n^2$$

A.5   A Technique of Converting Deterministic Population Model to Stochastic Model

Considerable effort has been directed to construct Stochastic Population models by Kendall (1949), Bartlett (1955). However, whenever an otherwise deterministic model is made Stochastic, Mathematical complications become often too great to run into the analytic solutions of the models.

However, if both fertility and mortality components are assumed to be independent of age as well as the size of the population then, of course, reasonably adequate solution exists but again the same is far from realistic. Even when the logistic model is made Stochastic considerable complications arise. However, a simple exercise relating to the conversion of deterministic into stochastic model due to Bartlett is useful to throw light in this respect. Let us consider the following transitions with the probability rates in infinitesimal time dt given below.

Transition	Probability
$N \to N+1$	$a_1 N - b_1 N^2$
$N \to N-1$	$a_2 N + b_2 N^2$

The above transitions are based on the Differential equation of the Logistic Curve (vide art 11.2)

$$\frac{dN(t)}{dt} = N(B(N) - D(N))$$

$$= N(\alpha - \beta N) \text{ where } B(N) - D(N) = \alpha - \beta N$$

$$\Rightarrow \qquad dN(t) = (\alpha - \beta N)\, dt \qquad\qquad (A.5.1)$$

We make this Deterministic differential equation by introducing an error term, converting the above equation in the form.

$$dN = [(\alpha N - \beta N^2)\, dt] + dz_1 - dz_2 \qquad\qquad (A.5.2)$$

where the 1st term under [ ] represents the systematic part of the random or stochastic changes dN (dN can take really the value 0 or 1) and dz_1 and dz_2 have consequently zero means. But since dz_1 and dz_2 may reasonably be regarded as Poisson variables as means $(a_1 N - b_1 N^2)\, dt$ and $(a_2 N + b_2 N^2)\, dt$ respectively. We may regard the variance of dz_1 and dz_2 to be same as their means as given above.

We have

$$\begin{aligned}
\text{Var}\,(dz) &= \text{Var}\,(dz_1 - dz_2) \\
&= \text{Var}\,(dz_1) + \text{Var}\,(dz_2) \\
&= (a_1 N - b_1 N^2)\, dt + (a_2 N + b_2 N^2)\, dt \\
&= [(a_1 + a_2) N - (b_1 - b_2) N^2]\, dt \qquad\qquad (A.5.3)
\end{aligned}$$

In next place, we introduce a change of variable from N to u given by

$$N = \frac{\alpha (1+u)}{\beta} \qquad\qquad (A.5.4)$$

$$\Rightarrow \qquad \alpha - \beta N = \alpha - \alpha (1+u) = -\alpha u$$

Also from (A. 5.4) and (A. 5.3)

$$\frac{dN}{dt} = \frac{\alpha}{\beta}\frac{du}{dt} = N(\alpha - \beta N) + \frac{dz}{dt}\ \text{where}\ dz = dz_1 - dz_2 \qquad (A.5.5)$$

Putting (A.5.4) in (A.5.5)

$$\frac{dN}{dt} = -N\alpha u + \frac{dz}{dt}$$

$$\Rightarrow \qquad \frac{\alpha}{\beta}\frac{du}{dt} = -N\alpha u + \frac{dz}{dt}$$

$$\Rightarrow \qquad du = -N\alpha u \frac{\beta}{\alpha}\, dt + \left(\frac{\beta}{\alpha}\frac{dz}{dt}\right) dt$$

$$= \frac{\beta}{\alpha}\, dt \left[-N\alpha u + \frac{dz}{dt}\right]$$

$$= \frac{\beta}{\alpha}\, dt \left[-\frac{\alpha (1+u)\,\alpha u}{\beta} + \frac{dz}{dt}\right]$$

$$= -\alpha (1+u)\, u\, dt + \left(\frac{\beta}{\alpha}\cdot dz\right)$$

$$du = -\alpha(1+u)u\,dt + \frac{\beta}{\alpha}\,dz \tag{A.5.6}$$

The representation (A.5.6) is quite useful. This may show that how the solution of stochastic equation differs from that of the deterministic equation (in which case we put $dz = 0$). Suppose the process has started down its ultimate value $\frac{\alpha}{\beta}$ when $dz = 0$

$$N \to \frac{\alpha}{\beta} \Rightarrow u \to 0 \left(\because N = \frac{\alpha}{\beta}(1+u) \right)$$

we can write (A.5.6) as

$$du = -\alpha u\,dt + \frac{\beta}{\alpha}\,dz \tag{A.5.7}$$

Further Var (error component) $= \mathrm{Var}\left(\frac{\beta}{\alpha}\,dz\right) = \frac{\beta^2}{\alpha^2}\,\mathrm{Var}\,(dz)$

$\Rightarrow$ Var (error component) $= \dfrac{\beta^2}{\alpha^2}\left[(a_1+a_2)N - (b_1-b_2)N^2\right]dt$

$$= \frac{\beta^2}{\alpha^2}\left[(a_1+a_2)\frac{\alpha(1+u)}{\beta} - (b_1-b_2)\frac{\alpha^2}{\beta^2}(1+u^2)\right]dt$$

As $1+u \to 1$, we have the same

$$= \frac{\beta^2}{\alpha^2}\left[(a_1+a_2)\frac{\alpha}{\beta} - (b_1-b_2)\frac{\alpha^2}{\beta^2}\right]dt$$

$$= \gamma\,dt \text{ say}$$

where $\gamma = \left\{ \frac{\beta}{\alpha}(a_1+a_2) + (b_2-b_1) \right\}$

Strictly speaking for $a_2 > 0$ as the death rate $a_2 N + b_2 N^2 > 0$ even for large N the population under these conditions become extinct. But at the same time it may be observed that the time of extinction becomes enormously large under this condition and the complication of Population getting extinct can be avoided in all practical sense.
Also

$$u + du = u - \alpha u\,dt + \frac{\beta}{\alpha}\,dz \text{ (from A.5.7)}$$

Now under statistical equilibrium

$$u(t+dt) = u + du \text{ is same as } u(t) = u$$

which gives

$$\text{Var}\,(U) \;=\; E\,(U + dU)^2 \;=\; E\,(U - \alpha U\,dt)^2 + E\left(\frac{\beta}{\alpha}\,dz\right)^2$$

$$= E\,(U)^2 + \alpha^2\,E\,(U^2)\,(dt)^2 - 2\alpha E\,(U^2)\,dt + \text{Var}\left(\frac{\beta}{\alpha}\,dz\right)$$

$$\because \quad \text{Var}\left(\frac{\beta}{\alpha}\,dz\right) = E\left(\frac{\beta}{\alpha}\,dz\right)^2 - \underbrace{\left[E\left(\frac{\beta}{\alpha}\,dz\right)\right]^2}_{=\,0}$$

Neglecting the term $\alpha^2\,E\,(U^2)\,(dt)^2$ because of the orderliness of the Poisson Process, we have

$$\sigma_u^2 \;=\; \sigma_u^2 - 2\sigma_u^2\,\alpha\,dt + \gamma\,dt$$

$$\Rightarrow \qquad \sigma_u^2 \;=\; \sigma_u^2(1 - 2\alpha\,dt) + \gamma\,dt$$

$$\Rightarrow \qquad 1 \;=\; 1 - 2\alpha\,dt + \frac{\gamma}{\sigma_u^2}\,dt$$

$$\Rightarrow \qquad \frac{\gamma}{\sigma_u^2} \;=\; 2\alpha$$

$$\Rightarrow \qquad \sigma_u^2 \;=\; \frac{\gamma}{2\alpha}$$

Let us consider the asymptotic stochastic mean of the population given by $\bar{n}_a$. The first approximation of the mean is $\dfrac{\alpha}{\beta}$.

We next proceed to calculate approximately the mean. If μ is the true mean and $X(t) = N(t) - \mu$.

We have approximately

$$X\,(t + \Delta t) - X(t) \;=\; (\alpha N_i - \beta N_i^2)\,dt + dz_1 - dz_2$$

$$E\,[X(t) + \Delta(t) - X(t)] \;=\; [\alpha E\,(N_i) - \beta E\,(N_i^2)]\,dt + E\,(dz)$$

$$0 \;=\; \alpha\mu - \beta\,(\sigma_N^2 + \mu^2) + 0$$

$$\Rightarrow \qquad \alpha\mu \;=\; \beta\,(\sigma_N^2 + \mu^2)$$

$$\Rightarrow \qquad \alpha \;=\; \beta \cdot \frac{\sigma_N^2}{\mu} + \beta\mu$$

$$\Rightarrow \qquad \left[\alpha - \beta\,\frac{\sigma_N^2}{\mu}\right]\frac{1}{\beta} \;=\; \mu$$

$\bar{n}_\infty \sim \dfrac{\alpha}{\beta} - \dfrac{\sigma_N^2}{\mu}$ is the second approximation of the asymptotic mean of the population.

Hence, the population lies between

$$\left[\frac{\alpha}{\beta} - \frac{\sigma_N^2}{\mu} \pm 3 \sqrt{\frac{\gamma}{2\alpha} \frac{d}{\beta}} \right]$$

with probability more than 99%.

Now if the stipulated population has to lie in the above range, then what should be the choice of α and β i.e. birth and death rates?

In this case, we have

$$a_1 - a_2 = \alpha$$

$$-b_1 + b_2 = \beta$$

$$\frac{\alpha}{\beta} = \frac{a_1 - a_2}{-b_1 + b_2}$$

$$\sigma_N^2 = \frac{\gamma}{2\alpha} \left(\frac{\alpha}{\beta} \right)^2$$

where

$$\gamma = (a_1 + a_2) \frac{\beta}{\alpha} + (b_2 - b_1)$$

$$\alpha = \beta\mu + \frac{\beta}{\mu} \sigma_N^2$$

and $\overline{n_\infty} = \dfrac{\alpha}{\beta} - \dfrac{\sigma_N^2}{\mu}$ where $\bar{n}_\infty$ is the steady state population size.

Index